DISCRETE AND CONTINUOUS
FOURIER TRANSFORMS

ANALYSIS, APPLICATIONS
AND FAST ALGORITHMS

DISCRETE AND CONTINUOUS FOURIER TRANSFORMS

ANALYSIS, APPLICATIONS AND FAST ALGORITHMS

Eleanor Chu

University of Guelph
Guelph, Ontario, Canada

CRC Press
Taylor & Francis Group
Boca Raton London New York

CRC Press is an imprint of the
Taylor & Francis Group, an **informa** business

A CHAPMAN & HALL BOOK

Chapman & Hall/CRC
Taylor & Francis Group
6000 Broken Sound Parkway NW, Suite 300
Boca Raton, FL 33487-2742

© 2008 by Taylor & Francis Group, LLC
Chapman & Hall/CRC is an imprint of Taylor & Francis Group, an Informa business

First issued in paperback 2019

No claim to original U.S. Government works

ISBN 13: 978-0-367-45269-8 (pbk)
ISBN 13: 978-1-4200-6363-9 (hbk)

**Visit the Taylor & Francis Web site at
http://www.taylorandfrancis.com**

**and the CRC Press Web site at
http://www.crcpress.com**

Contents

List of Figures

List of Tables

Preface

The topics in this book were selected to build a solid foundation for the application of Fourier analysis in the many diverging and continuously evolving areas in the digital signal processing enterprise. While Fourier transforms have long been used systematically in electrical engineering, the wide variety of modern-day applications of the discrete Fourier transform (DFT) on digital computers (made feasible by the fast Fourier transform (FFT) algorithms) motivates people in all branches of the physical sciences, computational sciences and engineering to learn the DFT, the FFT algorithms, as well as the many applications that directly impact our life today. To understand how the DFT can be deployed in any application area, one needs to have the core knowledge of Fourier analysis, which connects the DFT to the continuous Fourier transform, the Fourier series, and the all important sampling theorem. The tools offered by Fourier analysis enable us to correctly deploy and interpret the DFT results.

This book presents the fundamentals of Fourier analysis and their deployment in signal processing by way of the DFT and the FFT algorithms in a logically careful manner so that the text is self-contained and accessible to senior undergraduate students, graduate students, and researchers and professionals in mathematical science, numerical analysis, computer science, physics, and the various disciplines in engineering and applied science. The contents of this book are divided into two parts and fourteen chapters with the following features, and the cited topics can be selected and combined in a number of suggested ways to suit one s interest or the need of a related course:

• From the very beginning of the text a large number of graphical illustrations and worked examples are provided to help explain the many concepts and relationships; a detailed table of contents makes explicit the logical arrangement of topics in each chapter, each section, and each subsection.

• Readers of this book are not required to have prior knowledge of Fourier analysis or signal processing. To provide background, the basic concepts of signals and signal sampling together with a practical introduction to the DFT are presented in Chapters 1 and 2, while the mathematical derivation of the DFT is deferred to Chapter 4.

• The coverage of the Fourier series in Chapter 3 (Sections 3.1 3.8) is self-contained, and its relationship to the DFT is explained in Section 3.11. Section 3.9 on orthogonal projections and Section 3.10 on the convergence of Fourier series (including a detailed study of the Gibbs phenomenon) are more mathematical, and they can be skipped in the rst reading.

• The DFT is formally derived in Chapter 4, and a thorough discussion of the relationships between the DFT spectra and sampled signals under various circumstances is presented with supporting numerical results and graphical illustrations. In Section 4.7 I provide instructional MATLAB®[1] codes for computing the DFT formulas per se, while the fast algorithms for

---

[1] MATLAB is a registered trademark of The MathWorks, Inc.

computing the DFT are deferred to Part II of the book.

• The continuous Fourier transform is introduced in Chapter 5. The concepts and results from Chapters 1 through 3 are used here to derive the sampling theorem and the Fourier transform pair. Worked examples of the Fourier transform pair are then given and the properties of Fourier transform are derived. The computing of Fourier transform from discrete-time samples is investigated, and the relationship between sampled Fourier transform and Fourier series coef cients is also established in this chapter.

• Chapter 6 is built on the material previously developed in Chapters 3 and 5. The topics covered in Chapter 6 include the Dirac delta function, the convolution theorems concerning the Fourier transform, and the periodic and discrete convolution theorems concerning the Fourier series. I then show how these mathematical tools interplay to model the sampling process and develop the sampling theorem directly.

• With the foundations laid in Chapters 1 through 6, the Fourier transform of an ideally sampled signal is now formally de n ed (in mathematical terms) in Chapter 7, which provides the theoretical basis for appropriately constructing and deploying digital signal processing tools and correctly interpreting the processed results in Chapters 8 through 10.

• In Chapter 8 the data-weighting window functions are introduced, the analysis of the possibly distorted DFT spectra of windowed sequences is pursued, and the various scenarios and consequences related to frequency detection are demonstrated graphically using numerical examples.

• Chapter 9 covers discrete convolution algorithms, including the linear convolution algorithm, the periodic (and the equivalent circular or cyclic) convolution algorithm, and their implementation via the DFT (computed by the FFT). The relationship between the chirp Fourier transform and the cyclic convolution is also established in this chapter.

• The application of the DFT in digital ltering and lters is the topic of Chapter 10. The Gibbs phenomenon is also revisited in this chapter from a ltering viewpoint.

• Since the FFTs are the fast algorithms for computing the DFT and the associated convolution, the Fourier analysis and digital ltering of sampled signals in Part I of the book are based solely on the DFTs, and Part II of the book is devoted to covering the FFTs exclusively. While Part II of this book is self-contained, the material in Chapters 11 through 13 is more advanced than the previous book:

> Eleanor Chu and Alan George, *Inside the FFT Black Box: Serial and Parallel Fast Fourier Transform Algorithms*, CRC Press, 2000.

• In Chapter 11 the many ways to organize the mixed-radix DFT computation through index mapping are explored. This approach allows one to study the large family of mixed-radix FFT algorithms in a systematic manner, including the radix-2 special case. While this chapter can be read on its own, it also paves the way for the more specialized prime factor FFT algorithms covered in Chapter 13.

• In Chapter 12 a connection is established between the multi-factor mixed-radix FFT algorithms and the Kronecker product factorization of the DFT matrix. This process results in a sparse matrix formulation of the mixed-radix FFT algorithm.

• In Chapter 13 the family of prime factor FFT algorithms is presented. To cover the mathematical theory behind the prime factor algorithm, the relevant concepts from elementary number theory concerning the properties of integers are introduced, and the Chinese Remainder Theorem (CRT) is proved, because CRT and CRT-related index maps are responsible for the number-theoretic splitting of the DFT matrix, which gives rise to the prime factor algorithm.

• Chapter 14 provides full details of the mathematics behind Bluestein s FFT, which is a (deceptively simple) fast algorithm for computing the DFT of arbitrary length and is particularly useful when the length is a large prime number. The MATLAB® implementation of Bluestein s FFT is given, and numerical and timing results are reported.

Acknowledgments

My interest in the subject area of this book has arisen out of my research activities conducted at the University of Guelph, and I thank the Natural Sciences and Engineering Research Council of Canada for continued research grant support. Writing a book of this scope demands one s dedication to research and commitment of time and effort over multiple years, and I thank my husband, Robert Hiscott, for his understanding, consistent encouragement, and unwavering support at all fronts.

I thank the reviewers of my book proposal and draft manuscript for their helpful suggestions and insightful comments, which led to many improvements.

I extend my sincere thanks and appreciation to Robert Stern (Executive Editor) and his staff at Chapman & Hall/CRC Press for their ongoing enthusiastic support of my writing projects.

Eleanor Chu
Guelph, Ontario

About the Author

Eleanor Chu, Ph.D., received her B.Sc. from National Taiwan University in 1973, her B.Sc. and M.Sc. from Acadia University, Canada, in 1980 and 1981, respectively, and her M.Math and Ph.D. in Computer Science from the University of Waterloo, Canada, in 1984 and 1988, respectively.

From 1988 to 1991 Dr. Chu was a research assistant professor of computer science at the University of Waterloo. In 1991 she joined the faculty at the University of Guelph, where she has been Professor of Mathematics since 2001. Dr. Chu is the principal author of the book *Inside the FFT Black Box: Serial and Parallel Fast Fourier Transform Algorithms* (CRC Press, 2000). She has published journal articles in the broad area of computational mathematics, including scienti c computing, matrix analysis and applications, parallel computing, linear algebra and its applications, supercomputing, and high-performance computing applications.

Part I

Fundamentals, Analysis and Applications

Chapter 1

Analytical and Graphical Representation of Function Contents

Our objective in this chapter is to introduce the fundamental concepts and graphical tools for analyzing time-domain and frequency-domain function contents. Our initial discussion will be restricted to linear combinations of explicitly given sine and cosine functions, and we will show how the various representations of their frequency contents are connected to the Fourier series representation of periodic functions in general.

1.1 Time and Frequency Contents of a Function

Let us consider a familiar trigonometric function $x(t) = 5\cos(2\pi t)$. By plotting $x(t)$ versus t over the interval $0 \le t \le 4$, one obtains the following diagram.

Figure 1.1 A time-domain plot of $x(t) = 5\cos(2\pi t)$ versus t.

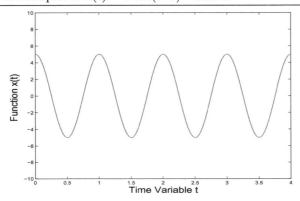

The graph is the time-domain representation of $x(t)$. We observe that when t varies from 0 to 1, the angle $\theta = 2\pi t$ goes from 0 radians to 2π radians, and the cosine function completes

3

one cycle. The same cycle repeats for each following time intervals: $t \in [1, 2]$, $t \in [2, 3]$, and so on. The time it takes for a periodic function $x(t)$ to complete one cycle is called the *period*, and it is denoted by T. In this case, we have $T = 1$ unit of time (appropriate units may be used to suit the application in hand), and $x(t + T) = x(t)$ for $t \geq 0$.

While the function $x(t)$ is fully speci ed in its analytical form, the graph of $x(t)$ reveals how the numerical function values change with time. Since a graph is plotted from a table of pre-computed function values, the cont ents of the graph are the numbers in the table. However, compared to reading a large table of data, reading the graph is a much more convenient and effective way to s ee the trend or pattern represented by the data, the approximate locations of minimum, maximum, or zero function values. With this understanding, the time-domain (or time) content of $x(t)$ (in this simple case) is the graph which plots $x(t)$ versus t.

For a single sinusoidal function like $x(t) = 5\cos(2\pi t)$, one can easily tell from its time-domain graph that it goes through one cycle (or 2π radians) per unit time, so its frequency is $f = 1$. It is also apparent from the same graph that the amplitude of $x(t) = 5\cos(2\pi t)$ is $A = 5$. However, *strictly for our future needs*, let us formally represent the frequency-domain (or frequency) content of $x(t)$ in Figure 1.1 by a two-tuple $(f, A) = (1, 5)$ in the amplitude-versus-frequency stem plot given below. The usefulness of the frequency-domain plot will be apparent in the next section.

Figure 1.2 A frequency-domain plot of $x(t) = 5\cos(2\pi t)$.

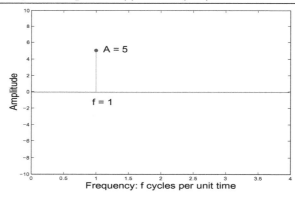

1.2   The Frequency-Domain Plots as Graphical Tools

We next consider a function synthesized from a linear combination of several cosine functions each with a different amplitude as well as a different frequency. For example, let

$$x(t) = x_1(t) + x_2(t) + x_3(t)$$
$$= A_1 \cos(2\pi f_1 t) - A_2 \cos(2\pi f_2 t) + A_3 \cos(2\pi f_3 t)$$
$$= 5\cos(2\pi t) - 7\cos(4\pi t) + 11.5\cos(6\pi t).$$

We see that the rst component function $x_1(t) = 5\cos(2\pi t)$ can be written as $x_1(t) = A_1 \cos(2\pi f_1 t)$ with amplitude $A_1 = 5$, and frequency $f_1 = 1$. Similarly, the second component function $x_2(t) = -7\cos(4\pi t)$ can be written as $x_2(t) = A_2 \cos(2\pi f_2 t)$ with amplitude

$A_2 = -7$ and frequency $f_2 = 2$. For $x_3(t) = 11.5 \cos(6\pi t)$, we have $A_3 = 11.5$ and $f_3 = 3$. The function $x_1(t)$ was fully explained in the last section. In the case of $x_2(t)$, the cosine function completes one cycle when its angle $\theta = 4\pi t$ goes from 0 radians to 2π radians, which implies that t changes from 0 to 0.5 units. So the period of $x_2(t)$ is $T_2 = 0.5$ units, and its frequency is $f_2 = \frac{1}{T_2} = 2$ cycles per unit time. The expression in the form

$$x_k(t) = A_k \cos(2\pi f_k t)$$

thus explicitly indicates that $x_k(t)$ repeats f_k cycles per unit time. Now, we can see that the time unit used to express f_k will be canceled out when f_k is multiplied by t units of time. Therefore, $\theta = 2\pi f_k t$ remains dimension-less, and the same holds regardless of whether the time is measured in seconds, minutes, hours, days, months, or years. Note that the equivalent expression $x_k(t) = A_k \cos(\omega_k t)$ is also commonly used, where $\omega_k \equiv 2\pi f_k$ radians per unit time is called the *angular frequency*.

In the time domain, a graph of the composite $x(t)$ can be obtained by adding the three graphs representing $x_1(t)$, $x_2(t)$, and $x_3(t)$ as shown below. The time-domain plot of $x(t)$ reveals a periodic composite function with a common period $T = 1$: the graph of $x(t)$ for $t \in [0, 1]$ is seen to repeat four times in Figure 1.3.

Figure 1.3 Time-domain plots of $x(t)$ and its components.

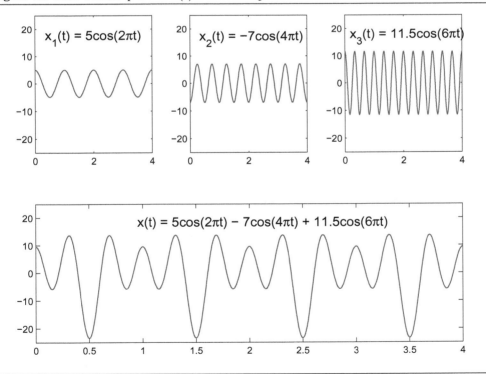

In the frequency domain, suppose that the two-tuple (f_k, A_k) represents the frequency content of $x_k(t)$, the collection $\{(f_1, A_1), (f_2, A_2), (f_3, A_3)\}$ de nes the frequency content of $x(t) = x_1(t) + x_2(t) + x_3(t)$. *Note that when $x(t)$ is composite, we speak of the individual frequencies and amplitudes of its components and they collectively represent the frequency*

content of $x(t)$. The frequency plot of $x(t)$ is obtained by superimposing the three component stem plots as shown in Figure 1.4.

Figure 1.4 The time and frequency-domain plots of composite $x(t)$.

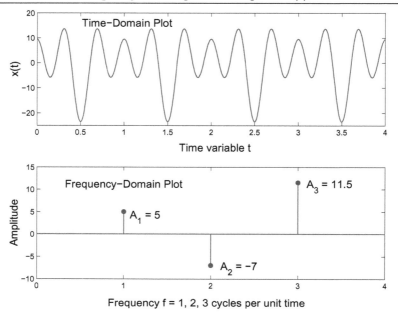

Now, with the time-domain plot and the frequency-domain plot of $x(t)$ both available, we see that when $x(t)$ is composite, the frequency content of $x(t)$ can no longer be deciphered from the time-domain plot of $x(t)$ versus t — one cannot visually decompose the graph of $x(t)$ into its component graphs. The reverse is also true: the time-domain plot shows the behavior of $x(t)$, which cannot be inferred from the frequency plot alone. Therefore, the time-domain and the frequency-domain plots are both needed, and they carry different but complementary information about the function $x(t)$.

1.3   Identifying the Cosine and Sine Modes

In general, a function may have both sine and cosine components, and the two modes must be explicitly identi ed in expressing the frequency content. For the previous example, the function $x(t) = \sum_{k=1}^{n=3} A_k \cos 2\pi f_k t$ has three cosine components, so each two-tuple in its frequency content $\{(f_1, A_1), (f_2, A_2), (f_3, A_3)\}$ implicitly represents the amplitude and the frequency of a pure cosine mode, and they are shown together in a single frequency plot. However, the function $y(t)$ below consists of two cosine and three sine components,

$$y(t) = 5.3 \cos(4\pi t) - 3.2 \sin(6\pi t) - 2.5 \cos(14\pi t) - 2.1 \sin(4\pi t) + 9.5 \sin(8\pi t),$$

so the subset of two-tuples $\{(2, 5.3), (7, -2.5)\}$ and its stem plot represent its pure cosine mode,   whereas the other subset of two-tuples $\{(2, -2.1), (3, -3.2), (4, 9.5)\}$ and a separate stem plot represent its pure sine mode.   When we allow zero amplitude and use the same range of frequencies in both modes, we obtain the following expression:

$$(1.1) \qquad y(t) = \sum_{k=1}^{n} A_k \cos(2\pi f_k t) + B_k \sin(2\pi f_k t).$$

The frequency content of $y(t)$ can now be conveniently represented by a set of three-tuples $\{ (f_1, A_1, B_1), (f_2, A_2, B_2), \ldots, (f_n, A_n, B_n) \}$, with the understanding that A_k is the amplitude of a pure cosine mode at frequency f_k, and B_k is the amplitude of a pure sine mode at f_k. We still need two separate stem plots: one plots A_k versus f_k, and the other one plots B_k versus f_k. The time-domain and frequency-domain plots of the sum of eleven cosine and eleven sine component functions are shown in Figure 1.5, where for $1 \le k \le 11$, $f_k = k$, with amplitudes $0 < A_k \le 2$ and $0 < B_k \le 3$ randomly generated. The time-domain plot of $x(t)$ again reveals a periodic composite function with a common period $T = 1$; the graph of $x(t)$ for $t \in [0, 1]$ is seen to repeat four times in Figure 1.5.

Figure 1.5 An example: the sum of 11 cosine and 11 sine components.

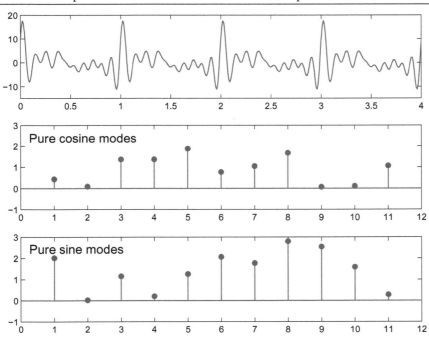

1.4 Using Complex Exponential Modes

By using complex arithmetics, Euler s formula $e^{j\theta} = \cos\theta + j\sin\theta$, where $j \equiv \sqrt{-1}$, and the resulting identities

$$\cos\theta = \frac{e^{j\theta} + e^{-j\theta}}{2}, \quad \sin\theta = \frac{e^{j\theta} - e^{-j\theta}}{2j},$$

we can express $y(t)$ in terms of complex exponential modes as shown below.

(1.2)

$$
\begin{aligned}
y(t) &= \sum_{k=1}^{n} A_k \cos(2\pi f_k t) + B_k \sin(2\pi f_k t) \\
&= \sum_{k=1}^{n} A_k \left(\frac{e^{j2\pi f_k t} + e^{-j2\pi f_k t}}{2} \right) + B_k \left(\frac{e^{j2\pi f_k t} - e^{-j2\pi f_k t}}{2j} \right) \\
&= \sum_{k=1}^{n} \left(\frac{A_k - jB_k}{2} \right) e^{j2\pi f_k t} + \left(\frac{A_k + jB_k}{2} \right) e^{-j2\pi f_k t} \\
&= \sum_{k=1}^{n} X_k e^{j2\pi f_k t} + X_{-k} e^{j2\pi f_{-k} t}, \qquad \left(\text{Note: } X_{\pm k} \equiv \frac{A_k \mp jB_k}{2}, \quad f_{-k} \equiv -f_k \right) \\
&= X_0 + \sum_{k=1}^{n} X_k e^{j2\pi f_k t} + X_{-k} e^{j2\pi f_{-k} t}, \qquad \text{(Note: the term } X_0 \equiv 0 \text{ is added)} \\
&= \sum_{k=-n}^{n} X_k e^{j2\pi f_k t}.
\end{aligned}
$$

When the complex number $X_{\pm k}$ is expressed in rectangular coordinates as $(\operatorname{Re}(X_{\pm k}), \operatorname{Im}(X_{\pm k}))$, the frequency contents of $y(t)$ are commonly expressed by two sets of two-tuples: $(f_{\pm k}, \operatorname{Re}(X_{\pm k}))$ and $(f_{\pm k}, \operatorname{Im}(X_{\pm k}))$. The example in Figure 1.5 is shown again in Figure 1.6 using the exponential mode. When comparing the two gures, note that $\operatorname{Re}(X_{\pm k}) = A_k/2$ and $\operatorname{Im}(X_{\pm k}) = \mp B_k/2$.

Figure 1.6 Time plot and complex exponential-mode frequency plots.

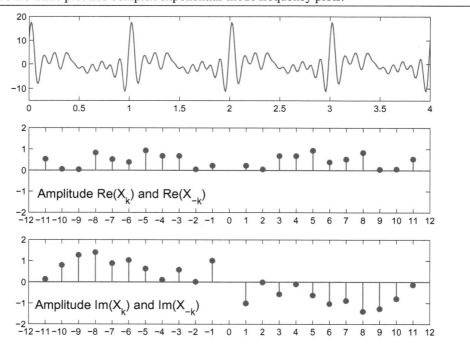

Note that in order to simplify the terms in the summation, we have added the term $X_0 \equiv 0$, and for $1 \le k \le n$, we have de ned

(1.3) $\quad X_k = \dfrac{A_k - jB_k}{2}, \quad X_{-k} = \dfrac{A_k + jB_k}{2}, \quad f_{-k} = -f_k, \text{ and } \omega_{-k} = -\omega_k = -2\pi f_k.$

In the present context, since the negative frequencies are simply the consequence of applying trigonometric identities in our derivation of an alternative mathematical formula, they do not change the original problem. For example, if one uses the identity $\cos(\theta_k) = \cos(-\theta_k)$, when $\theta_k = 2\pi f_k t$, $-\theta_k = 2\pi(-f_k)t$ occurs, and it causes the presence of negative frequency $-f_k$. (Note that a nonzero $X_0 = X_0 e^{j2\pi f_0 t}$ term at $f_0 = 0$ models a DC (direct current) term in electrical circuit applications.)

Alternatively we may express the complex amplitude $X_{\pm k}$ using polar coordinates, namely,

$$X_k = |X_k| e^{j\phi_k} = |X_k| \left(\cos \phi_k + j \sin \phi_k \right),$$

(1.4)

$$X_{-k} = |X_{-k}| e^{j\phi_{-k}} = |X_{-k}| \left(\cos \phi_{-k} + j \sin \phi_{-k} \right),$$

where

$$|X_{\pm k}| = \frac{\sqrt{A_k^2 + B_k^2}}{2}, \quad \text{with each } \phi_{\pm k} \text{ chosen to satisfy both}$$

$$\cos \phi_{\pm k} = \frac{A_k}{\sqrt{A_k^2 + B_k^2}}, \quad \sin \phi_{\pm k} = \frac{\mp B_k}{\sqrt{A_k^2 + B_k^2}}.$$

Note that each angle $\phi_{\pm k}$ is unique in the quadrant determined by the rectangular coordinates $(A_k, \mp B_k)$ of the complex number $2X_k$. In Figure 1.7, the frequency plots show $|X_{\pm k}|$ and $\phi_{\pm k}$ versus $f_{\pm k}$. In the next section we show that $\phi_{\pm k}$ may also be interpreted as the phase shift angle.

1.5 Using Cosine Modes with Phase or Time Shifts

Instead of separating the pure cosine and pure sine modes, we may use a pure cosine mode combined with phase shift angles, which is represented by a single set of three-tuples $(f_k, \hat{\phi}_k, D_k)$ as de ned below.

$$
\begin{aligned}
y(t) &= \sum_{k=1}^{n} A_k \cos(2\pi f_k t) + B_k \sin(2\pi f_k t) \\
&= \sum_{k=1}^{n} \sqrt{A_k^2 + B_k^2} \left(\frac{A_k}{\sqrt{A_k^2 + B_k^2}} \cos(2\pi f_k t) + \frac{B_k}{\sqrt{A_k^2 + B_k^2}} \sin(2\pi f_k t) \right) \\
&= \sum_{k=1}^{n} D_k \left(\cos \hat{\phi}_k \cos(2\pi f_k t) + \sin \hat{\phi}_k \sin(2\pi f_k t) \right) \\
&= \sum_{k=1}^{n} D_k \cos(2\pi f_k t - \hat{\phi}_k),
\end{aligned}
$$

(1.5)

where

$$D_k \equiv \sqrt{A_k^2 + B_k^2}, \quad \text{with } \hat{\phi}_k \text{ satisfying both } \cos \hat{\phi}_k = \frac{A_k}{D_k} \text{ and } \sin \hat{\phi}_k = \frac{B_k}{D_k}.$$

Figure 1.7 Time plot and complex exponential-mode frequency plots.

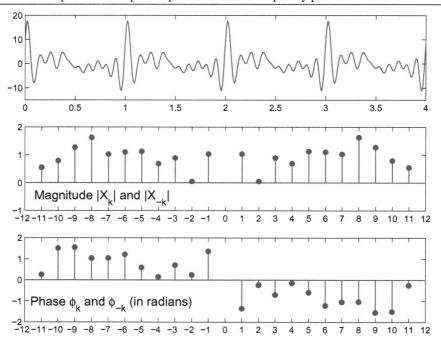

Therefore, each component function $y_k(t)$ may always be interpreted as a pure cosine mode *shifted* by a phase angle of $\hat{\phi}_k$ radians.

The phase shifts may be interpreted as t ime shifts   by rewriting Equation (1.5) as

$$
\begin{aligned}
y(t) &= \sum_{k=1}^{n} D_k \cos(2\pi f_k t - \hat{\phi}_k) \\
&= \sum_{k=1}^{n} D_k \cos\left(2\pi f_k \left(t - \frac{\hat{\phi}_k}{2\pi f_k}\right)\right) \\
&= \sum_{k=1}^{n} D_k \cos\left(2\pi \frac{1}{T_k} \left(t - \frac{\hat{\phi}_k}{2\pi \left(\frac{1}{T_k}\right)}\right)\right). \quad \left(\because f_k \equiv \frac{1}{T_k}\right)
\end{aligned}
$$

(1.6)

When it is known that the fundamental frequency $f_1 = \frac{1}{T}$ and that $f_k = kf_1 = \frac{k}{T}$ for $1 \le k \le n$, Equation (1.6) is commonly presented with time shifts t_k de ned below.

(1.7)
$$
y(t) = \sum_{k=1}^{n} D_k \cos\left(2\pi \frac{k}{T}(t - t_k)\right), \quad \text{where } t_k \equiv \frac{\hat{\phi}_k}{2\pi\left(\frac{k}{T}\right)}.
$$

Since $2|X_k| = \sqrt{A_k^2 + B_k^2}$, which is equal to $|D_k|$ in Equations (1.6) and (1.7), we immediately obtain the following relationship.

$$y(t) = \sum_{k=1}^{n} X_k e^{j2\pi f_k t} + X_{-k}^{-j2\pi f_k t}, \text{ where } X_{\pm k} = \frac{A_k \mp jB_k}{2}, \quad j \equiv \sqrt{-1},$$

(1.8)
$$= \sum_{k=1}^{n} 2|X_k| \cos(2\pi f_k t - \hat{\phi}_k)$$

$$= \sum_{k=1}^{n} 2|X_k| \cos\left(2\pi \frac{k}{T}(t - t_k)\right), \text{ if } f_k = \frac{k}{T}, \text{ and } t_k \equiv \frac{\hat{\phi}_k}{2\pi\left(\frac{k}{T}\right)}.$$

Remark 1 In the literature any function of the form

(1.9) $$f(t) = D_k \sin(2\pi f_k t + \phi_k),$$

where D_k, f_k and ϕ_k are real constants, is said to be *sinusoidal*. Using the trigonometric identity

$$\cos\left(\theta - \tfrac{1}{2}\pi\right) = \cos\theta \cos\tfrac{1}{2}\pi + \sin\theta \sin\tfrac{1}{2}\pi = \sin\theta$$

with $\theta = 2\pi f_k t + \phi_k$, we can also express (1.9) as a cosine function:

$$f(t) = D_k \sin(2\pi f_k t + \phi_k) = D_k \cos\left(2\pi f_k t + \phi_k - \tfrac{1}{2}\pi\right).$$

Hence, a sinusoidal function can be written in two forms which differ by $\tfrac{1}{2}\pi$ in the phase angle:

(1.10) $$D_k \sin(2\pi f_k t + \phi_k) = D_k \cos(2\pi f_k t + \hat{\phi}_k), \text{ where } \hat{\phi}_k = \phi_k - \frac{1}{2}\pi.$$

In particular, both $\sin(2\pi f_k t)$ and $\cos(2\pi f_k t)$ are sinusoidal functions by this de nition.

Remark 2 Any component function of the form

(1.11) $$g_k(t) = A_k \sin(2\pi f_k t) + B_k \cos(2\pi f_k t)$$

is said to be a sinusoidal component, because we have shown at the beginning of this section that it can be expressed as $g_k(t) = D_k \cos\left(2\pi f_k t - \hat{\phi}_k\right)$, with D_k and $\hat{\phi}_k$ determined by A_k and B_k.

Remark 3 The easiest way to add two or more sinusoidal functions of the same frequency is provided by form (1.11). For example, given $f(t) = 5\sin(1.2t) + 2\cos(1.2t)$ and $g(t) = \sin(1.2t) + \cos(1.2t)$, we obtain the sum by adding the corresponding coef cients:

$$h(t) = f(t) + g(t) = 6\sin(1.2t) + 3\cos(1.2t).$$

Therefore, the sum of two or more sinusoidal functions of frequency f_k is again a sinusoidal function of frequency f_k.

Remark 4 Be aware that sinusoidal functions may be given in disguised forms: e.g., $f(t) = \sin(1.1t)\cos(1.1t)$ is the disguised form of the sinusoidal $f(t) = \tfrac{1}{2}\sin(2.2t)$; $g(t) = 1 - 2\sin^2 t$ is the disguised form of the sinusoidal $g(t) = \cos 2t$.

1.6   Periodicity and Commensurate Frequencies

Recall that when we present the frequency-domain plots for specic examples of

$$y(t) = \sum_{k=1}^{n} A_k \cos(2\pi f_k t) + B_k \sin(2\pi f_k t),$$

we have let $f_k = k$ cycles per unit time, and we plot the amplitudes A_k and B_k versus k. In such examples we automatically have uniform spacing with $\triangle f = f_{k+1} - f_k = 1$, and we have $f_k = kf_1$ with $f_1 = 1$ being the fundamental frequency. Since the time period T of composite $y(t)$ is the shortest duration over which each sine or cosine component completes an integer number of cycles, we determine T by the LCM (least common multiple) of the individual periods. From $f_k = kf_1$ and $T_k = 1/f_k$, we obtain $T_1 = kT_k$, so T_1 is the LCM of the individual periods. Accordingly, the time period T of the composite $y(t)$ is the reciprocal of the fundamental frequency f_1. Note that f_1 is the GCD (greatest common divisor) of the individual frequencies.

In general, $f_k \neq k$, and we need to distinguish periodic $y(t)$ from non-periodic $y(t)$ by examining its frequency contents. The conditions and results are given below.

1. The function $y(t)$ is said to be a *commensurate* sum if the ratio of *any* two individual periods (or frequencies) is a rational fraction   ratio of integers with common factors canceled out.

   Example 1.1 The function

   $$y(t) = 4.5 \cos\left(2\pi f_\alpha t\right) + 7.2 \cos\left(2\pi f_\beta t\right) = 4.5 \cos\left(1.2\pi t\right) + 7.2 \cos\left(1.8\pi t\right)$$

   is a commensurate sum, because $f_\alpha = 0.6$ Hz, $f_\beta = 0.9$ Hz, and the ratio $f_\alpha/f_\beta = 2/3$ is a rational fraction.

2. A commensurate $y(t)$ is periodic with its fundamental frequency being the GCD of the individual frequencies and its common period being the LCM of the individual periods.

   Example 1.2 We continue with Example 1.1: the fundamental frequency of the function $y(t) = 4.5 \cos\left(1.2\pi t\right) + 7.2 \cos\left(1.8\pi t\right)$ is $f_o = \text{GCD}(0.6, 0.9) = 0.3$ Hz; and the fundamental period is $T_o = 1/f_o = 3\frac{1}{3}$ seconds. We get the same result from $T_o = \text{LCM}\left(\frac{1}{0.6}, \frac{1}{0.9}\right) = \text{LCM}\left(\frac{5}{3}, \frac{10}{9}\right) = 3\frac{1}{3}$. It can be easily veried that $y(t + T_o) = y(t)$.

   Example 1.3 When $f_k = k/T$, the fundamental frequency is $f_1 = 1/T$, and the composite function

   $$y(t) = \sum_{k=1}^{n} A_k \cos\frac{2\pi kt}{T} + B_k \sin\frac{2\pi kt}{T}$$

   is commensurate and periodic with common period T, i.e., $y(t + T) = y(t)$. Since we have uniform spacing $\triangle f = f_{k+1} - f_k = 1/T$, we may still plot A_k and B_k versus k with the understanding that k is the index of equispaced f_k; of course, one may plot A_k and B_k versus the values of f_k if that is desired. (Note that $f_k = k/T = k$ if $T = 1$.)

3. A non-commensurate $y(t)$ is not periodic, although all its components are periodic. For example, the function

$$y(t) = \sin(2\pi t) + 5\sin(2\sqrt{3}\pi t)$$

is not periodic because $f_1 = 1$ and $f_2 = \sqrt{3}$ are not commensurate.

1.7   Review of Results and Techniques

In the preceding sections we show that a sum of sinusoidal modes can be expressed in a number of ways. While the various formulas are mathematically equivalent, one form could be more convenient than another depending on the manipulations required for a particular application. Also, it is not uncommon that while one form is more suitable for describing a physical problem, another form is more desirable for a computational purpose. These formulas are summarized below.

Form 1 Using pure cosine and sine modes

$$(1.12) \qquad y(t) = \sum_{k=1}^{n} A_k \cos(2\pi f_k t) + B_k \sin(2\pi f_k t).$$

If the angular frequency $\omega_k = 2\pi f_k$ is used, we obtain

$$(1.13) \qquad y(t) = \sum_{k=1}^{n} A_k \cos(\omega_k t) + B_k \sin(\omega_k t).$$

A common case: when $y(t) = y(t+T)$ with $f_k = k/T$, this fact is explicitly recognized by expressing

$$(1.14) \qquad y(t) = \sum_{k=1}^{n} A_k \cos \frac{2\pi k t}{T} + B_k \sin \frac{2\pi k t}{T}.$$

Form 2 Using complex exponential modes

$$(1.15) \qquad y(t) = \sum_{k=-n}^{n} X_k e^{j2\pi f_k t}.$$

Form 3 Using cosine modes with phase shifts

$$(1.16) \qquad y(t) = \sum_{k=1}^{n} D_k \cos(2\pi f_k t - \hat{\phi}_k).$$

Form 4 Using cosine modes with time shifts

$$(1.17) \qquad y(t) = \sum_{k=1}^{n} D_k \cos\big(2\pi f_k (t - t_k)\big).$$

Form 5 Using complex exponential modes with phases

$$(1.18) \qquad y(t) = \sum_{k=-n}^{n} \big(|X_k| e^{j\phi_k}\big) e^{j2\pi f_k t}.$$

A reminder: The definitions $f_k = \frac{1}{T_k}$ and $\omega_k = 2\pi f_k$ may be used to express $y(t)$ in terms of T_k (individual period) or ω_k (individual angular frequency) in all forms. Also, when $f_k = k/T$, this fact is commonly recognized wherever f_k is used.

To convert one form to another, one may use the relationship between the coefficients as summarized below.

Relation 1 Define $X_0 \equiv 0$ when A_0 and B_0 are missing. For $1 \leq k \leq n$,

$$X_{\pm k} = \frac{A_k \mp jB_k}{2}, \quad \text{and } f_{-k} = -f_k.$$

Relation 2

$$|X_{\pm k}| = \frac{\sqrt{A_k^2 + B_k^2}}{2}, \quad \text{and the phase angle } \phi_{\pm k} \text{ satisfies both}$$

$$\cos \phi_{\pm k} = \frac{A_k}{\sqrt{A_k^2 + B_k^2}} \text{ and } \sin \phi_{\pm k} = \frac{\mp B_k}{\sqrt{A_k^2 + B_k^2}}.$$

A reminder: ϕ_k is unique in the quadrant determined by the rectangular coordinates $(A_k, -B_k)$ of the complex number $2X_k$; ϕ_{-k} is unique in the quadrant determined by the rectangular coordinates (A_k, B_k) of the complex number $2X_{-k}$.

Relation 3 For $1 \leq k \leq n$,

$$D_k = \sqrt{A_k^2 + B_k^2} = 2|X_{\pm k}|, \quad t_k = \frac{\hat{\phi}_k}{2\pi f_k}, \quad \text{where } \hat{\phi}_k = \phi_{-k}.$$

Relation 4 For $1 \leq k \leq n$,

$$A_k = X_k + X_{-k} = 2\operatorname{Re}(X_k); \quad B_k = j(X_k - X_{-k}) = -2\operatorname{Im}(X_k).$$

We also identify the mathematical techniques used in deriving the various results in this section:

Technique 1 Euler's identity in three forms:

$$e^{j\theta} = \cos\theta + j\sin\theta, \quad \cos\theta = \frac{e^{j\theta} + e^{-j\theta}}{2}, \quad \text{and } \sin\theta = \frac{e^{j\theta} - e^{-j\theta}}{2j}.$$

Examples of future use:

- Prove $\displaystyle\sum_{k=-n}^{n} e^{jk\theta} = \frac{\sin\left(n + \frac{1}{2}\right)\theta}{\sin\frac{\theta}{2}}.$ (Chapter 3, Section 3.10.2, page 84)

- Prove $\displaystyle\int_{-\pi}^{\pi} \sum_{k=-n}^{n} e^{jk\theta}\, d\theta = 2\pi.$ (Chapter 3, Section 3.10.2, page 85)

- Prove $\displaystyle\int_{-\pi}^{\pi} \frac{\sin\left(n + \frac{1}{2}\right)\theta}{\sin\frac{\theta}{2}}\, d\theta = 2\pi.$ (Chapter 3, Section 3.10.2, page 85)

- Prove $\displaystyle\frac{1}{2f_c}\int_{-f_c}^{f_c} e^{j2\pi ft}\, df = \frac{\sin 2\pi f_c t}{2\pi f_c t}.$ (Chapter 5, Example 5.4, page 171)

Technique 2 Trigonometric identities and their alternate forms:

$$\cos(\alpha \pm \beta) = \cos\alpha\cos\beta \mp \sin\alpha\sin\beta, \qquad \sin(\alpha \pm \beta) = \sin\alpha\cos\beta \pm \cos\alpha\sin\beta,$$

$$\cos\alpha\cos\beta = \frac{\cos(\alpha+\beta)+\cos(\alpha-\beta)}{2}, \qquad \sin\alpha\cos\beta = \frac{\sin(\alpha+\beta)+\sin(\alpha-\beta)}{2},$$

$$\sin\alpha\sin\beta = \frac{\cos(\alpha-\beta)-\cos(\alpha+\beta)}{2}, \qquad \cos\alpha\sin\beta = \frac{\sin(\alpha+\beta)-\sin(\alpha-\beta)}{2}.$$

Examples of future use:

- Letting $\alpha = \beta$, we immediately have the useful identities

$$\cos 2\alpha = \cos^2\alpha - \sin^2\alpha, \quad \sin 2\alpha = 2\sin\alpha\cos\alpha;$$

$$\cos^2\alpha = \frac{1+\cos 2\alpha}{2}, \quad \sin^2\alpha = \frac{1-\cos 2\alpha}{2}.$$

- Letting $\alpha = m\theta$ and $\beta = n\theta$, it is straightforward to apply the identities given above to prove the following results for future use.

$$(1.19) \qquad \int_{-\pi}^{\pi} \cos m\theta \cos n\theta \, d\theta = \begin{cases} 0, & \text{if } m \neq n; \\ \pi, & \text{if } m = n \neq 0; \\ 2\pi, & \text{if } m = n = 0. \end{cases}$$

$$(1.20) \qquad \int_{-\pi}^{\pi} \sin m\theta \sin n\theta \, d\theta = \begin{cases} 0, & \text{if } m \neq n; \\ \pi, & \text{if } m = n \neq 0; \\ 0, & \text{if } m = n = 0. \end{cases}$$

$$(1.21) \qquad \int_{-\pi}^{\pi} \cos m\theta \sin n\theta \, d\theta = 0.$$

1.7.1   Practicing the techniques

To practice the techniques in nontrivial settings, we show how to manipulate some trigonometric series encountered in Fourier analysis in the examples that follow.

Example 1.4 Derive the following identity:

$$(1.22) \qquad \sum_{\ell=1}^{n} \sin(2\ell-1)\theta = \frac{\sin^2 n\theta}{\sin\theta},$$

and show that this identity is valid at $\theta = 0$ by the *limit convention*. (When this convention is used, the value of a function at a point where a denominator vanishes is understood to be the limit, provided this limit is finite.)

$$\because \sin\theta \sum_{\ell=1}^{n} \sin(2\ell - 1)\theta$$

$$= \sin^2\theta + \sin\theta\sin 3\theta + \sin\theta\sin 5\theta + \cdots + \sin\theta\sin(2n-1)\theta$$

$$= \frac{1 - \cos 2\theta}{2} + \frac{\cos 2\theta - \cos 4\theta}{2} + \frac{\cos 4\theta - \cos 6\theta}{2} + \cdots + \frac{\cos(2n-2)\theta - \cos(2n)\theta}{2}$$

$$= \frac{1}{2} - \frac{\cos 2\theta}{2} + \frac{\cos 2\theta}{2} - \frac{\cos 4\theta}{2} + \cdots - \frac{\cos(2n-2)\theta}{2} + \frac{\cos(2n-2)\theta}{2} - \frac{\cos(2n)\theta}{2}$$

$$= \frac{1 - \cos 2n\theta}{2} \qquad \text{(only the first term and the last term remain)}$$

$$= \sin^2 n\theta. \qquad \left(\text{recall } \sin^2\alpha = \tfrac{1}{2}(1 - \cos 2\alpha)\right)$$

$$\therefore \sum_{\ell=1}^{n} \sin(2\ell - 1)\theta = \frac{\sin^2 n\theta}{\sin\theta}.$$

When $\theta = 0$, since the right side is in the indeterminate form $0/0$, we apply L'Hospital's rule to determine the limit:

$$\lim_{\theta\to 0} \frac{\sin^2 n\theta}{\sin\theta} = \lim_{\theta\to 0} \frac{2n\sin n\theta\cos n\theta}{\cos\theta} = 2n\sin 0 = 0.$$

Hence the two sides are equal at $\theta = 0$ by the limit convention.

Example 1.5 Using Euler's identity $e^{j\theta} = \cos\theta + j\sin\theta$, the finite sum of a geometric series in $z = e^{j\theta} \neq 1$, i.e.,

$$(1.23) \qquad \sum_{\ell=0}^{n} z^\ell = \frac{1 - z^{n+1}}{1 - z},$$

and the complex arithmetic identity

$$(1.24) \qquad \frac{c + jd}{a + jb} = \frac{(c + jd)(a - jb)}{(a + jb)(a - jb)} = \frac{ac + bd}{a^2 + b^2} + j\frac{ad - bc}{a^2 + b^2},$$

determine the closed-form sums of the following cosine and sine series:

$$(1.25a) \qquad \sum_{\ell=0}^{n} \cos \ell\theta = 1 + \cos\theta + \cdots + \cos n\theta = ?$$

$$(1.25b) \qquad \sum_{\ell=1}^{n} \sin \ell\theta = \sin\theta + \sin 2\theta + \cdots + \sin n\theta = ?$$

By letting $z = e^{j\theta}$ in the left side of (1.23), we identify the cosine series (1.25a) and the sine series (1.25b) as the real and imaginary parts:

$$\sum_{\ell=0}^{n} z^\ell = \sum_{\ell=0}^{n} e^{j\ell\theta} = \sum_{\ell=0}^{n} \cos \ell\theta + j\sin \ell\theta = \sum_{\ell=0}^{n} \cos \ell\theta + j\sum_{\ell=1}^{n} \sin \ell\theta. \quad (\because \sin 0 = 0)$$

By letting $z = e^{j\theta}$ in the right side of (1.23), we express

$$(1.26) \qquad \frac{1 - z^{n+1}}{1 - z} = \frac{1 - e^{j(n+1)\theta}}{1 - e^{j\theta}} = \frac{\{1 - \cos(n+1)\theta\} - j\sin(n+1)\theta}{(1 - \cos\theta) - j\sin\theta} = U + jV.$$

Accordingly, the real part U represents the cosine series, and the imaginary part V represents the sine series. To express U and V in (1.26), we use identity (1.24) with $c = 1 - \cos(n+1)\theta$, $d = -\sin(n+1)\theta$, $a = 1 - \cos\theta$, and $b = -\sin\theta$:

(1.27)
$$
\begin{aligned}
U &= \frac{\{1 - \cos(n+1)\theta\}(1 - \cos\theta) + \sin(n+1)\theta\sin\theta}{(1 - \cos\theta)^2 + \sin^2\theta} \\
&= \frac{1 - \cos(n+1)\theta - \cos\theta + \{\cos(n+1)\theta\cos\theta + \sin(n+1)\theta\sin\theta\}}{1 - 2\cos\theta + \{\cos^2\theta + \sin^2\theta\}} \\
&= \frac{1 - \cos(n+1)\theta - \cos\theta + \cos\big((n+1)\theta - \theta\big)}{1 - 2\cos\theta + 1} \\
&= \frac{1 - \cos\theta + \cos n\theta - \cos(n+1)\theta}{2 - 2\cos\theta};
\end{aligned}
$$

(1.28)
$$
\begin{aligned}
V &= \frac{-(1 - \cos\theta)\sin(n+1)\theta + \{1 - \cos(n+1)\theta\}\sin\theta}{(1 - \cos\theta)^2 + \sin^2\theta} \\
&= \frac{\{\sin(n+1)\theta\cos\theta - \cos(n+1)\theta\sin\theta\} - \sin(n+1)\theta + \sin\theta}{1 - 2\cos\theta + \{\cos^2\theta + \sin^2\theta\}} \\
&= \frac{\sin\big((n+1)\theta - \theta\big) - \sin(n+1)\theta + \sin\theta}{1 - 2\cos\theta + 1} \\
&= \frac{\sin\theta + \sin n\theta - \sin(n+1)\theta}{2 - 2\cos\theta}.
\end{aligned}
$$

We have thus obtained

(1.29)
$$
\sum_{\ell=0}^{n} \cos\ell\theta = \frac{1 - \cos\theta + \cos n\theta - \cos(n+1)\theta}{2 - 2\cos\theta};
$$

(1.30)
$$
\sum_{\ell=1}^{n} \sin\ell\theta = \frac{\sin\theta + \sin n\theta - \sin(n+1)\theta}{2 - 2\cos\theta}.
$$

Example 1.6 Derive the trigonometric identity

(1.31)
$$
\frac{1}{2} + \sum_{\ell=1}^{n} \cos\ell\theta = \frac{\sin\left(n + \frac{1}{2}\right)\theta}{2\sin\frac{1}{2}\theta},
$$

and show that it is valid at $\theta = 0$ by the limit convention.

Beginning with the identity (1.29), we obtain

$$
\begin{aligned}
\frac{1}{2} + \sum_{\ell=1}^{n} \cos\ell\theta &= \frac{1 - \cos\theta + \cos n\theta - \cos(n+1)\theta}{2 - 2\cos\theta} - \frac{1}{2} \\
&= \frac{2\sin^2\frac{1}{2}\theta + \cos\left(\left(n+\frac{1}{2}\right)\theta - \frac{1}{2}\theta\right) - \cos\left(\left(n+\frac{1}{2}\right)\theta + \frac{1}{2}\theta\right)}{4\sin^2\frac{1}{2}\theta} - \frac{1}{2} \\
&= \frac{2\sin^2\frac{1}{2}\theta + 2\sin\left(n+\frac{1}{2}\right)\theta\sin\frac{1}{2}\theta}{4\sin^2\frac{1}{2}\theta} - \frac{1}{2} \\
&= \frac{\sin\frac{1}{2}\theta + \sin\left(n+\frac{1}{2}\right)\theta}{2\sin\frac{1}{2}\theta} - \frac{1}{2} \\
&= \frac{\sin\left(n+\frac{1}{2}\right)\theta}{2\sin\frac{1}{2}\theta}.
\end{aligned}
$$

At $\theta = 0$, because $\cos \ell\theta = \cos 0 = 1$ for $1 \leq \ell \leq n$ in the left side, the sum is $n + \frac{1}{2}$. Here again the right side is in the indeterminate form $0/0$, we apply L Hospital s rule to determine the limit:

$$\lim_{\theta \to 0} \frac{\sin\left(n + \frac{1}{2}\right)\theta}{2 \sin \frac{1}{2}\theta} = \lim_{\theta \to 0} \frac{\left(n + \frac{1}{2}\right)\cos\left(n + \frac{1}{2}\right)\theta}{\cos \frac{1}{2}\theta} = n + \frac{1}{2}.$$

Hence the two sides are equal at $\theta = 0$ by the limit convention.

Example 1.7 Show that

$$(1.32) \qquad \sum_{\ell=0}^{n} \cos \frac{(2m + 1)\ell\pi}{n + 1} = 1.$$

If we let $\theta = \left(\frac{2m+1}{n+1}\right)\pi$ in the geometric series (1.23), the numerator in the right side can be further simpli ed :

$$\sum_{\ell=0}^{n} e^{j\ell\theta} = \frac{1 - e^{j(n+1)\theta}}{1 - e^{j\theta}}$$

$$(1.33) \qquad = \frac{2}{1 - \cos\theta - j\sin\theta} \qquad \left(\because \theta = \left(\tfrac{2m+1}{n+1}\right)\pi \therefore e^{j(n+1)\theta} = -1\right)$$

$$= \frac{2(1 - \cos\theta + j\sin\theta)}{(1 - \cos\theta)^2 + \sin^2\theta}$$

$$= \left[\frac{2 - 2\cos\theta}{1 - 2\cos\theta + 1}\right] + j\left[\frac{2\sin\theta}{1 - 2\cos\theta + 1}\right]$$

Recall from Example 1.5 that the real part of the series (1.33) represents the cosine series, we have thus proved the desired result:

$$\text{If } \theta = \frac{(2m + 1)\pi}{n + 1}, \text{ then } \sum_{\ell=0}^{n} \cos \ell\theta = \frac{2 - 2\cos\theta}{2 - 2\cos\theta} = 1.$$

Example 1.8 Show that, if the nonzero integer m is not a multiple of $n + 1$, we have

$$(1.34) \qquad \sum_{\ell=0}^{n} \cos \frac{(2m)\ell\pi}{n + 1} = 0.$$

We again let $\theta = \left(\frac{2m}{n+1}\right)\pi$ in the geometric series (1.23), we have

$$\sum_{\ell=0}^{n} e^{j\ell\theta} = \frac{1 - e^{j(n+1)\theta}}{1 - e^{j\theta}}$$

$$(1.35) \qquad = \frac{0}{1 - e^{j\theta}} \qquad \left(\because \theta = \left(\tfrac{2m}{n+1}\right)\pi \therefore e^{j(n+1)\theta} = 1\right)$$

$$= 0.$$

Example 1.9 Show that the following alternative expressions for the nite sum of the sine series can be obtained from identity (1.30) in Example 1.5.

$$(1.36) \qquad \sum_{\ell=1}^{n} \sin \ell\theta = \frac{\cos \frac{1}{2}\theta - \cos\left(n + \frac{1}{2}\right)\theta}{2 \sin \frac{1}{2}\theta};$$

(1.37)
$$\sum_{\ell=1}^{n} \sin \ell\theta = \frac{\sin\left(\frac{n+1}{2}\right)\theta \sin \frac{n}{2}\theta}{\sin \frac{1}{2}\theta}.$$

To derive the two mathematically equivalent results, we continue from (1.30):

$$\sum_{\ell=1}^{n} \sin \ell\theta = \frac{\sin \theta + \sin n\theta - \sin(n+1)\theta}{2 - 2\cos\theta}$$

$$= \frac{2\sin\frac{1}{2}\theta\cos\frac{1}{2}\theta + \sin\left(\left(n+\frac{1}{2}\right)\theta - \frac{1}{2}\theta\right) - \sin\left(\left(n+\frac{1}{2}\right)\theta + \frac{1}{2}\theta\right)}{4\sin^2\frac{1}{2}\theta}$$

$$= \frac{2\sin\frac{1}{2}\theta\cos\frac{1}{2}\theta - 2\cos\left(n+\frac{1}{2}\right)\theta\sin\frac{1}{2}\theta}{4\sin^2\frac{1}{2}\theta}$$

$$= \frac{\cos\frac{1}{2}\theta - \cos\left(n+\frac{1}{2}\right)\theta}{2\sin\frac{1}{2}\theta} \qquad \text{(this is the desired result (1.36))}$$

$$= \frac{\cos\left(\frac{n+1}{2} - \frac{n}{2}\right)\theta - \cos\left(\frac{n+1}{2} + \frac{n}{2}\right)\theta}{2\sin\frac{1}{2}\theta}$$

$$= \frac{\sin\left(\frac{n+1}{2}\right)\theta \sin\frac{n}{2}\theta}{\sin\frac{1}{2}\theta}. \qquad \text{(this is the desired result (1.37))}$$

1.8 Expressing Single Component Signals

Since many puzzling phenomena we encounter in analyzing or processing composite signals can be easily investigated through single-mode signals, they are indispensable tools in our continued study of signal sampling and transformations, and it pays to be very familiar (and comfortable) with expressing a single-mode signal in its various forms. Although we can formally put such a signal in one of the standard forms (with a single nonzero coef cient) and apply the full-force conversion formulas, it is much easier to forgo the formalities and work with the given signal directly, as demonstrated by the following examples.

Example 1.10 $f(t) = \cos(2\pi f_a t) = \cos(80\pi t)$ is a 40-Hertz sinusoidal signal, its amplitude is A = 1.0, its period is $T = 1/f_a = 1/40 = 0.025$ seconds, and it has zero phase. We express $f(t)$ in the complex exponential modes by applying Euler s formula directly:

$$f(t) = \cos(80\pi t) = \frac{1}{2}\left(e^{j80\pi t} + e^{-j80\pi t}\right) = 0.5e^{-j80\pi t} + 0.5e^{j80\pi t}.$$

The difference between $f(t)$ given above and $g(t) = \sin(80\pi t)$ lies in the phase angle, because the latter can be rewritten as a shifted cosine wave, namely, $g(t) = \cos(80\pi t - \pi/2)$. The phase can also be recognized directly from expressing $g(t)$ in the complex exponential modes:

$$g(t) = \sin(80\pi t) = \frac{1}{2j}\left(e^{j80\pi t} - e^{-j80\pi t}\right) = (0.5j)e^{-j80\pi t} + (-0.5j)e^{j80\pi t}.$$

$$= \left(0.5e^{j\pi/2}\right)e^{-j80\pi t} + \left(0.5e^{-j\pi/2}\right)e^{j80\pi t}.$$

The coef cients $\pm 0.5j$ each has nonzero imaginary part, which re ects a nonzero phase in the signal. The polar expression $\pm j = e^{\pm j\pi/2}$ reveals the phase explicitly.

Example 1.11 For $h(t) = 4\cos(7\pi t + \alpha)$, we have

$$h(t) = 4\cos(7\pi t + \alpha) = \frac{4}{2}\left(e^{j(7\pi t+\alpha)} + e^{-j(7\pi t+\alpha)}\right)$$
$$= \left(2e^{-j\alpha}\right)e^{-j7\pi t} + \left(2e^{j\alpha}\right)e^{j7\pi t}.$$

Observe that when the phase $\alpha \neq 0, \pi$, the coefficients $2e^{\pm j\alpha} = 2(\cos\alpha \pm j\sin\alpha)$ have nonzero imaginary part.

For $u(t) = 4\sin(7\pi t+\beta)$, we may apply Euler's formula directly to the given sine function to obtain

$$u(t) = 4\sin(7\pi t + \beta) = \frac{4}{2j}\left(e^{j(7\pi t+\beta)} - e^{-j(7\pi t+\beta)}\right)$$
$$= \left(2je^{-j\beta}\right)e^{-j7\pi t} + \left(-2je^{j\beta}\right)e^{j7\pi t}$$
$$= \left(2e^{-j(\beta-\pi/2)}\right)e^{-j7\pi t} + \left(2e^{j(\beta-\pi/2)}\right)e^{j7\pi t}.$$

The same expression can also be obtained if we use the result already available for $u(t) = 4\cos(7\pi + \alpha)$ with $\alpha = \beta - \pi/2$.

Example 1.12 For $v(t) = 3\cos(15\pi t)\cos(35\pi t)$, be aware that it hides two cosine modes. To bring them out, we use the trigonometric identity for $\cos\alpha\cos\beta$ (given under Technique 2 in the previous section) to obtain

$$v(t) = 3\cos(15\pi t)\cos(35\pi t) = 1.5\left(\cos(15+35)\pi t + \cos(15-35)\pi t\right)$$
$$= 1.5\left(\cos 50\pi t + \cos 20\pi t\right)$$
$$= 1.5e^{-j50\pi t} + 1.5e^{-j20\pi t} + 1.5e^{j20\pi t} + 1.5e^{j50\pi t}.$$

The two cosine modes may also be disguised as $s(t) = 3\sin(15\pi t)\sin(35\pi t)$, and they can again be obtained using the trigonometric identity for $\sin\alpha\sin\beta$ (given under Technique 2 in the previous section):

$$s(t) = 3\sin(15\pi t)\sin(35\pi t) = 1.5\left(\cos(15-35)\pi t - \cos(15+35)\pi t\right)$$
$$= 1.5(\cos 20\pi t - \cos 50\pi t)$$
$$= -1.5e^{-j50\pi t} + 1.5e^{-j20\pi t} + 1.5e^{j20\pi t} - 1.5e^{j50\pi t}.$$

1.9 General Form of a Sinusoid in Signal Application

When a cyclic physical phenomenon is described by a cosine curve, the general form used in many applications is the cosine mode with phase shift angle (or *phase* in short)

(1.38) $$x(t) = D_\alpha \cos(2\pi f_\alpha t - \phi_\alpha),$$

where the amplitude D_α, frequency f_α, and phase ϕ_α (in radians) provide useful information about the physical problem at hand. For example, suppose that it is justi able to model the variation of monthly precipitation in each appropriately identi ed geographic region by a co-sine curve with period $T_\alpha = 1/f_\alpha = 12$ months, then the amplitude of each tted cosine curve predicts the maximum precipitation for each region, and the phase (converted to time shift) predicts the date of maximum precipitation for each region. Graphically, the time shift t_α (computed from the phase ϕ_α) is the actual distance between the origin and the crest of the cosine curve when the horizontal axis is time, because $x(t) = D_\alpha$ when $2\pi f_\alpha t - \phi_\alpha = 0$ is satisfied by $t = t_\alpha = \phi_\alpha / 2\pi f_\alpha$.

Note that when a negative frequency $f_\alpha < 0$ appears in the general form, it is interpreted as the result of phase reversal as shown below.

$$
\begin{aligned}
x(t) &= D_\alpha \cos(2\pi f_\alpha t - \phi_\alpha) \\
&= D_\alpha \cos(-2\pi \hat{f}_\alpha t - \phi_\alpha) && (\because \hat{f}_\alpha = -f_\alpha > 0) \\
&= D_\alpha \cos(-(2\pi \hat{f}_\alpha t + \phi_\alpha)) \\
&= D_\alpha \cos(2\pi \hat{f}_\alpha t + \phi_\alpha) && (\because \cos(-\theta) = \cos\theta) \\
&= D_\alpha \cos(2\pi \hat{f}_\alpha t - (-\phi_\alpha)).
\end{aligned}
$$

For example, to obtain the time-domain plot of $x(t) = 2.5 \cos(-40\pi t - \pi/6)$, we simply plot $x(t) = 2.5 \cos(40\pi t - \phi)$ with $\phi = -\pi/6$ (reversed from $\pi/6$) in the usual manner.

1.9.1 Expressing sequences of discrete-time samples

When the sinusoid $x(t) = D_\alpha \cos(2\pi f_\alpha t - \phi_\alpha)$ is sampled at intervals of $\triangle t$ (measured in chosen time units), we obtain the discrete-time sinusoid

(1.39) $$ x_\ell \equiv x(\ell \triangle t) = D_\alpha \cos(2\pi f_\alpha \ell \triangle t - \phi_\alpha), \quad \ell = 0, 1, 2, \ldots $$

Observe that the sequence of discrete-time samples $\{x_0, x_1, x_2, \ldots\}$ can also be represented by the three-tuple $\{f_\alpha \triangle t, \phi_\alpha, D_\alpha\}$, where the product of the analog frequency f_α (cycles per unit time) and the sampling interval $\triangle t$ (elapsed time between consecutive samples) defines the *digital* (or *discrete*) frequency

$$ \mathbb{F}_\alpha \equiv f_\alpha \triangle t \ \text{(cycles per sample)}. $$

Therefore, a discrete-time sinusoid has the general form

(1.40) $$ x_\ell = D_\alpha \cos(2\pi \mathbb{F}_\alpha \ell - \phi_\alpha), \quad \ell = 0, 1, 2, \ldots $$

Since $f_\alpha = \mathbb{F}_\alpha / \triangle t$, the digital frequency can always be converted back to the analog frequency as desired. Furthermore, because

$$ \mathbb{F}_\alpha \equiv f_\alpha \triangle t = \frac{1}{m} f_\alpha (m \triangle t) = m f_\alpha \left(\frac{1}{m} \triangle t\right), $$

an m-fold increase (or decrease) in $\triangle t$ amounts to an m-fold decrease (or increase) in the analog frequency, i.e.,

$$ f_\beta = \frac{\mathbb{F}_\alpha}{m \triangle t} = \frac{1}{m} \left(\frac{\mathbb{F}_\alpha}{\triangle t}\right); \quad f_\gamma = \frac{\mathbb{F}_\alpha}{\frac{1}{m} \triangle t} = m \left(\frac{\mathbb{F}_\alpha}{\triangle t}\right). $$

Consequently, by simply adjusting $\triangle t$ at the time of output, the same set of digital samples may be converted to analog signals with different frequencies. This will provide further flexibility in the sampling and processing of signals.

Corresponding to the (analog) angular frequency $\omega_\alpha = 2\pi f_\alpha$ (radians per second), we have the *digital* (or *discrete*) angular frequency $\mathbb{W}_\alpha = 2\pi \mathbb{F}_\alpha$ (radians per sample); hence, we may also express the two general forms as

(1.41) $$ x(t) = D_\alpha \cos(\omega_\alpha t - \phi_\alpha), $$

and

(1.42) $x_\ell = D_\alpha \cos(\mathbb{W}_\alpha \ell - \phi_\alpha), \quad \ell = 0, 1, 2, \ldots$

1.9.2   Periodicity of sinusoidal sequences

While the period of the sinusoid $x(t) = D_\alpha \cos(2\pi f_\alpha t - \phi_\alpha)$ is always $T = 1/f_\alpha$, we cannot say the same for its sampled sequence for two reasons:

1. The discrete-time sample sequence may or may not be periodic depending on the sampling interval $\triangle t$;

2. If the discrete-time sample sequence is periodic, its period varies with the sampling interval $\triangle t$.

To nd out whether a discrete-time sinusoid is periodic and to determine the period (measured by the number of samples), we make use of the mathematical expression for the ℓth sample, namely,

$$x_\ell = D_\alpha \cos(2\pi \mathbb{F}_\alpha \ell - \phi_\alpha), \quad \ell = 0, 1, 2, \ldots,$$

and we recall that $\mathbb{F}_\alpha = f_\alpha \triangle t$. We now relate the discrete-time samples represented by the sequence $\{x_\ell\}$ to the period of its *envelope* function

$$x(t) = D_\alpha \cos(2\pi f_\alpha t - \phi_\alpha)$$

through the digital frequency $\mathbb{F}_\alpha$:

1. If we can express

$$\mathbb{F}_\alpha = f_\alpha \triangle t = \frac{\mathbb{K}}{\mathbb{N}},$$

where $\mathbb{K}$ and $\mathbb{N}$ are integers (with no common factor), then we have

$$x_\mathbb{N} = D_\alpha \cos(2\pi \mathbb{K} - \phi_\alpha),$$

and $x_\mathbb{N}$ is positioned exactly at the point where its envelope function $x(t)$ completes $\mathbb{K}$ cycles, and we may conclude that the discrete-time sample sequence $\{x_\ell\}$ is periodic with period $\mathbb{T} = \mathbb{N}$ samples m eaning that $x_{\ell+\mathbb{N}} = x_\ell$ for $0 \le \ell \le \mathbb{N} - 1$, and $x_\mathbb{N} = x_0$ is the rst sample of the next period.

2. The sequence $\{x_\ell\}$ is *not* periodic if we cannot express its digital frequency $\mathbb{F}$ as a rational fraction.

We demonstrate the different cases by several examples below.

Example 1.13 The discrete-time sinusoid $x_\ell = \cos(0.025\pi\ell - \pi/6)$ can be written as

$$x_\ell = \cos(2\pi \mathbb{F}_\alpha \ell - \pi/6)$$

with $\mathbb{F}_\alpha = 0.025/2 = 0.0125 = 1/80$, so the given sequence is periodic with period $\mathbb{N} = 80$ (samples). In this case, we have $\mathbb{K} = 1$, so the $\mathbb{N}$ samples are equally spaced over a single period of its envelope function.

Example 1.14 The discrete-time sinusoid $g_\ell = \cos(0.7\pi\ell + \pi/8)$ can be written as

$$g_\ell = \cos(2\pi\mathbb{F}_\alpha\ell + \pi/8)$$

with $\mathbb{F}_\alpha = 0.7/2 = 0.35 = 7/20$, so the given sequence is periodic with period $\mathbb{N} = 20$ (samples). In this case, we have $\mathbb{K} = 7$, so the $\mathbb{N}$ equispaced samples span *seven* periods of its envelope function.

Example 1.15 The discrete-time sinusoid $y_\ell = \cos(\sqrt{3}\pi\ell)$ is not periodic, because when we express

$$y_\ell = \cos(2\pi\mathbb{F}_\beta\ell),$$

we have $\mathbb{F}_\beta = \sqrt{3}/2$, which is not a rational fraction.

Example 1.16 The discrete-time sinusoid $z_\ell = \cos(2\ell + \pi/6)$ is not periodic, because when we express

$$z_\ell = \cos(2\pi\mathbb{F}_\gamma\ell + \pi/6),$$

we have $\mathbb{F}_\gamma = 1/\pi$, which is not a rational fraction.

Sampling and reconstruction of signals will be formally treated in Chapters 2, 5 and 6.

1.10 Fourier Series: A Topic to Come

In this chapter we limit our discussion to functions consisting of explicitly given sines and cosines, because their frequency contents are precisely de ned and easy to understand. To extend the de n itions and results to an arbitrary function $f(t)$, we must seek to represent $f(t)$ as a sum of sinusoidal modes   this process is called *Spectral Decomposition* or *Spectral Analysis*. The Fourier series refers to such a representation with frequencies speci ed at $f_k = k/T$ cycles per unit time for $k = 0, 1, 2, \ldots, \infty$. The *unknowns* to be determined are the amplitudes (or coef cients) A_k and B_k so that

(1.43)
$$f(t) = \sum_{k=0}^{\infty} A_k \cos\frac{2\pi kt}{T} + B_k \sin\frac{2\pi kt}{T}.$$

If we are successful, the Fourier series of $f(t)$ is given by the commensurate sum in the right-hand side, and we have $f(t+T) = f(t)$. That is, T is the common period of $f(t)$ and $f_1 = 1/T$ is the fundamental frequency of $f(t)$. Note that $f(t)$ completes one cycle over any interval of length T, including the commonly used $[-T/2, T/2]$.

Depending on the application context, the Fourier series of function $f(t)$ may appear in variants of the following forms:

1. Using pure cosine and sine modes with variable t,

(1.44)
$$f(t) = \frac{A_0}{2} + \sum_{k=1}^{\infty} A_k \cos\frac{2\pi kt}{T} + B_k \sin\frac{2\pi kt}{T}.$$

Note that $f(t)$ has a nonzero DC term, namely, $A_0/2$, for which we have the following remarks:

Remark 1. For $k = 0$, we have $\cos 0 = 1$ and $\sin 0 = 0$; hence, the constant term in (1.43) is given by $(A_0 \cos 0 + B_0 \sin 0) = A_0$.

Remark 2. By convention the constant (DC) term in the Fourier series (1.44) is denoted by $\frac{1}{2} A_0$ instead of A_0 so that one mathematical formula de nes A_k for all k, including $k = 0$. The analytical formulas which de ne A_k and B_k will be presented when we study the theory of Fourier series in Chapter 3.

A common variant uses $T = 2L$ with spatial variable x,

$$(1.45) \qquad f(x) = \frac{A_0}{2} + \sum_{k=1}^{\infty} A_k \cos \frac{\pi k x}{L} + B_k \sin \frac{\pi k x}{L}.$$

Note that $f(x + 2L) = f(x)$, and a commonly chosen interval of length $2L$ is $[-L, \, L]$.

2. Using cosine modes with phase shifts,

$$(1.46) \qquad f(t) = D_0 + \sum_{k=1}^{\infty} D_k \cos\left(\frac{2\pi k t}{T} - \hat{\phi}_k\right).$$

The individual terms $D_k \cos(\frac{2\pi k t}{T} - \hat{\phi}_k)$ a re called the harmonics of $f(t)$. Note that the spacing between the harmonic frequencies is $\triangle f = f_{k+1} - f_k = \frac{1}{T}$. Hence, periodic analog signals are said to have discrete spectra, and the spacing in the frequency domain is the reciprocal of the period in the time domain.

3. Using complex exponential modes with variable t,

$$(1.47) \qquad f(t) = \sum_{k=-\infty}^{\infty} X_k e^{j 2\pi k t / T}.$$

Note that $X_0 = A_0/2$ (see above).

4. Using pure cosine and sine modes with dimension-less variable $\theta = 2\pi t/T$ radians,

$$(1.48) \qquad g(\theta) = \frac{A_0}{2} + \sum_{k=1}^{\infty} A_k \cos k\theta + B_k \sin k\theta.$$

Since t varies from 0 to T, $\theta = 2\pi t/T$ varies from 0 to 2π, we have $g(\theta + 2\pi) = g(\theta)$. Note that $g(\theta)$ completes one cycle over any interval of length 2π, including the commonly used $[-\pi, \pi]$.

5. Using complex exponential modes with dimension-less variable $\theta = 2\pi t/T$ radians,

$$(1.49) \qquad g(\theta) = \sum_{k=-\infty}^{\infty} X_k e^{j k \theta}.$$

6. In Chapter 5, we will learn that the frequency contents of a nonperiodic function $x(t)$ are de ned by a continuous-frequency function $X(f)$, and we will also encounter the Fourier series representation of the periodically extended $X(f)$, which appears in the two forms given below. A full derivation of the continuous-frequency function $X(f)$ and its Fourier series (when it exists) will be given in Chapter 5.

Using pure cosine and sine modes with variable f (which represents the continuously varying frequency) and bandwidth F, that is to say, $f \in [-F/2, F/2]$,

$$(1.50) \qquad X(f) = \sum_{k=0}^{\infty} a_k \cos \frac{2\pi k f}{F} + b_k \sin \frac{2\pi k f}{F}.$$

Using complex exponential modes with variable f and bandwidth F,

$$(1.51) \qquad X(f) = \sum_{k=-\infty}^{\infty} c_k e^{j2\pi k f/F}.$$

Instead of using the variable $f \in [-F/2, F/2]$, a dimension-less variable $\theta = 2\pi f/F \in [-\pi, \pi]$ may also be used in the frequency domain. Corresponding to the two forms of $X(f)$ given above, we have

$$(1.52) \qquad G(\theta) = \sum_{k=0}^{\infty} a_k \cos k\theta + b_k \sin k\theta,$$

and

$$(1.53) \qquad G(\theta) = \sum_{k=-\infty}^{\infty} c_k e^{jk\theta}, \quad \text{where } \theta \in [-\pi, \pi].$$

Observe that because the Fourier series expression in θ may be used for both time-domain function $x(t)$ and frequency-domain function $X(f)$, the dimension-less variable θ is also known as a *neutral* variable. Since the Fourier series expression is signi can tly simpli ed by using the neutral variable θ, it is often the variable of choice in mathematical study of Fourier series.

The theory and techniques for deriving the Fourier series representation of a given function will be covered in Chapter 3.

1.11   Terminology

Analog signals Signals continuous in time and amplitude are called analog signals.

Temporal and spatial variables The *temporal* variable t measures time in chosen units; the *spatial* variable x measures distance in chosen units.

Period and wavelength The period T satis es $f(t + T) = f(t)$; the wavelength $2L$ satis es $g(x + 2L) = g(x)$.

Frequency and wave number The (rotational) frequency is de ned by $\dfrac{1}{T}$ (cycles per unit time); the wave number is de ned by $\dfrac{1}{2L}$ (wave numbers per unit length).

Sine and cosine modes A pure sine wave with a xed frequency f_k is called a sin e mode and it is denoted by $\sin(2\pi f_k t)$; similarly, a cosine mode is denoted by $\cos(2\pi f_k t)$.

Phase or phase shift It refers to the phase angle $\hat{\phi}_k$ (expressed in radians) in the *shifted* cosine mode $\cos(2\pi f_k t - \hat{\phi}_k)$ or $\cos(2\pi f_k x - \hat{\phi}_k)$.

Time and space shifts The time shift refers to t_k in $\cos\big(2\pi f_k(t - t_k)\big)$; the space shift refers to x_k in $\cos\big(2\pi f_k(x - x_k)\big)$.

Angular frequency It is defined by $\omega = 2\pi f$ (radians per unit time or unit length), where f refers to the rotational frequency $\dfrac{1}{T}$ or the wave number $\dfrac{1}{2L}$ defined above.

Neutral variables In the time domain, a neutral variable $\theta = 2\pi t/T$ (radians) varies from 0 to 2π when the time variable t goes from 0 to T (units of time); in the frequency domain, a neutral variable $\theta = 2\pi f/F$ (radians) varies from $-\pi$ to π when the frequency variable f goes from $-F/2$ to $F/2$ (cycles per unit time). The neutral variable θ is dimensionless, and it is always expressed in radians.

Digital or discrete frequency The digital (rotational) frequency $\mathbb{F}$ measures cycles per sample, and the digital angular frequency $\mathbb{W}$ measures radians per sample.

References

1. A. Ambardar. *Analog and Digital Signal Processing*. Brooks/Cole Publishing Company, Pacific Grove, CA, second edition, 1999.

2. R. W. Hamming. *Digital Filters*. Prentice-Hall, Inc., Englewood Cliffs, NJ, third edition, 1989.

3. J. N. Rayner. *An Introduction to Spectral Analysis*. Pion Limited, London, Great Britain, 1971.

4. H. J. Weaver. *Applications of Discrete and Continuous Fourier Analysis*. John Wiley & Sons, Inc., New York, 1983.

5. C. R. Wylie. *Advanced Engineering Mathematics*. McGraw-Hall Book Company, New York, fourth edition, 1975.

Chapter 2

Sampling and Reconstruction of Functions—Part I

In Chapter 1 we study the time and frequency contents of functions formed by combining explicitly given sines and cosines. In the real world we need to process signals which are available only as a sequence of samples collected at equally spaced intervals, and we will begin the discussion on recovering frequency contents from discrete-time samples in this chapter.

2.1 DFT and Band-Limited Periodic Signal

Suppose that the unknown signal $x(t)$ is periodic and it has Fourier series representation

$$(2.1) \qquad x(t) = \frac{A_0}{2} + \sum_{k=1}^{\infty} A_k \cos \frac{2\pi kt}{T} + B_k \sin \frac{2\pi kt}{T}.$$

Since $x(t + T) = x(t)$, we only need to sample the function over a single period. While there is no limit on the variable t (in the sense that $x(t)$ is de n ed everywhere in the in nite time domain), the range (or bandwidth) of frequencies may or may not be limited depending on whether there are *finite* or *infinite* number of terms in its Fourier series. Let us begin with the case when the Fourier series coef cients $A_k = B_k = 0$ for $k > n$. That is, $x(t)$ is *band-limited* (up to the maximum frequency $f_n = n/T$), and it is represented by a nite Fourier series of $N = 2n+1$ terms, namely,

$$(2.2) \qquad x(t) = \frac{A_0}{2} + \sum_{r=1}^{n} A_r \cos \frac{2\pi rt}{T} + B_r \sin \frac{2\pi rt}{T}.$$

To determine the frequency contents of $x(t)$, we may solve for the $N = 2n+1$ unknown coef cients A_r and B_r by setting up a system of N linear equations in N unknowns, provided that we are given N values of $x(t)$. When the samples of $x(t)$ are equally spaced over the period $[0, T]$, we have $x_\ell = x(t_\ell)$ with $t_\ell = \ell \triangle t = \ell \left(\frac{T}{N} \right)$ for $0 \le \ell \le N-1$, and the resulting system is given by

$$(2.3) \qquad x_\ell = \frac{A_0}{2} + \sum_{r=1}^{n} A_r \cos \frac{2\pi r\ell}{N} + B_r \sin \frac{2\pi r\ell}{N}, \quad \ell = 0, 1, \ldots, N-1.$$

If we change the variable from t to $\theta = \frac{2\pi t}{T}$, then $\theta_\ell = \frac{2\pi t_\ell}{T} = \ell\left(\frac{2\pi}{N}\right)$, and we obtain the alternate form

$$(2.4) \qquad x_\ell = \frac{A_0}{2} + \sum_{r=1}^{n} A_r \cos(r\theta_\ell) + B_r \sin(r\theta_\ell), \ \ \ell = 0, 1, \ldots, N-1.$$

For $N = 2n+1 = 7$, the mapping of $\{t_0, t_1, \ldots, t_{N-1}\}$ to $\{\theta_0, \theta_1, \ldots, \theta_{N-1}\}$ is shown in Figure 2.1.

Figure 2.1 Changing variable from $t \in [0, T]$ to $\theta = 2\pi t/T \in [0, 2\pi]$.

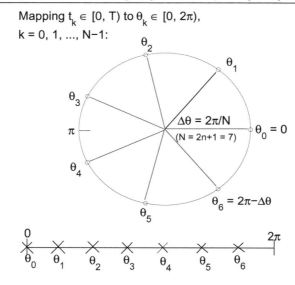

By using Euler s formula

$$\cos(r\theta_\ell) = \frac{e^{jr\theta_\ell} + e^{-jr\theta_\ell}}{2}, \quad \sin(r\theta_\ell) = \frac{e^{jr\theta_\ell} - e^{-jr\theta_\ell}}{2j}, \quad \text{where } j \equiv \sqrt{-1},$$

we obtain the system in complex exponential modes

$$(2.5) \qquad x_\ell = \sum_{r=-n}^{n} X_r \omega_N^{r\ell}, \text{ where } \omega_N \equiv e^{j2\pi/N}, \ \ell = 0, 1, \ldots, N-1.$$

Noting that $\omega_N^N = 1$ and $\omega_N^{N \pm r} = \omega_N^{\pm r}$, if we relabel $X_{-r}\omega_N^{-r\ell}$ by $X_{N-r}\omega_N^{(N-r)\ell}$ for $-n \le -r \le -1$, we obtain

$$(2.6) \qquad \boxed{x_\ell = \sum_{r=0}^{N-1} X_r \omega_N^{r\ell}, \ \ \omega_N \equiv e^{j2\pi/N}, \ \ell = 0, 1, \ldots, N-1,}$$

which leads to the DFT (discrete Fourier transform) formula (2.7) given below, by which we can transform the sequence of discrete samples $\{x_0, x_1, \ldots, x_{N-1}\}$ to the sequence of coef - cients $\{X_0, X_1, \ldots, X_{N-1}\}$ without solving a system of equations. (The DFT formulas and their derivation will be covered in full detail in Chapter 4.)

(2.7)
$$X_r = \frac{1}{N} \sum_{\ell=0}^{N-1} x_\ell \omega_N^{-r\ell}, \quad \text{for } r = 0, 1, \ldots, N-1.$$

In Figure 2.2, we give two examples of equally spaced $N = 2n+1$ samples and the computed DFT coefficients the computed X_r s are relabeled for $-n \leq r \leq n$ as given originally by Formula (2.5). Since $X_{\pm k}$ are the coefficients of the complex exponential modes $e^{\pm j2\pi kt/T}$, the corresponding frequencies $\pm f_k = \pm k/T$ are marked on the frequency grid. (Note that the $X_{\pm k}$ s in Figure 2.2 are all real-valued, because we have constructed signals $x_1(t)$ and $x_2(t)$ to have only cosine modes.)

Figure 2.2 Equally-spaced samples and computed DFT coefficients.

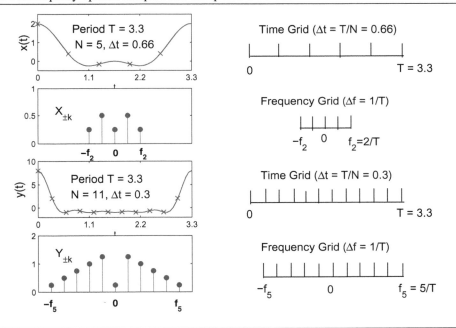

We defer the matrix formulation of the DFT until Chapter 4. It turns out that because of the special properties of the DFT matrix, the DFT coefficient X_r can be computed more efficiently using various fast Fourier transform algorithms (commonly known as the FFT). Interested readers are referred to our earlier book [13] and/or Part II of this book for the design, analysis, and implementation of a large collection of the FFT algorithms.

With the DFT coefficient X_r computed from (2.7), we can reconstruct the signal $x(t)$ using the complex exponential modes:

$$x(t) = \sum_{r=-n}^{n} X_r e^{j2\pi rt/T},$$

where $X_{-r} = X_{N-r}$ for $-n \leq -r \leq -1$ by reversing the relabeling operation. By applying the relations we developed in Chapter 1, the DC term and the amplitudes of the sine and cosine modes are immediately available from the computed X_r values as shown below.

1. The DC term $\dfrac{A_0}{2} = X_0$;

2. Noting that $N = 2n + 1$ and $X_{N-r} = X_{-r}$ for $1 \leq r \leq n$, we express

$$A_r = X_r + X_{N-r}, \quad B_r = j(X_r - X_{N-r}),$$

for $r = 1, 2, \ldots, n$.

With the values of A_r and B_r available, we can reconstruct the signal $x(t)$ using the pure cosine and sine modes:

$$x(t) = \frac{A_0}{2} + \sum_{k=1}^{n} A_k \cos \frac{2\pi kt}{T} + B_k \sin \frac{2\pi kt}{T}.$$

The process described above reveals a relationship between the number of samples and the number of complex exponential or real sinusoidal modes   the $N = 2n+1$ samples allow us to determine the coef cien t X_r for exactly $N = 2n+1$ complex exponential modes, from which we can recover the n cosine modes, n sine modes, and the DC term. This relationship is precise for the band-limited periodic function if we know the maximum frequency present in the signal. In our example, the frequencies $f_k = k/T$ range from $f_1 = 1/T$, which is the fundamental frequency of the signal, to the maximum $f_n = n/T$, with $n = 2$ and $n = 5$ in the two examples illustrated in Figure 2.2, in which we also show uniform spacing $\triangle f = 1/T$ on the frequency grid, together with uniform spacing $\triangle t = T/N$ on the time grid. The sampling rate $\mathbb{R}$ is de n ed to be $1/\triangle t$, which measures the number of samples per unit time.

The following relations can now be easily established from the de n itions:

Relation 1 (Reciprocity relation) The grid spacing $\triangle f$ in the frequency domain and the grid spacing $\triangle t$ are related inversely by the equation:

$$\triangle f \triangle t = \frac{1}{T} \frac{T}{N} = \frac{1}{N}.$$

Relation 2 (Maximum frequency and sampling rate/interval)

$$f_n = n\triangle f = \frac{n}{T} = \frac{n}{N\triangle t} \leq \frac{1}{2}\left(\frac{1}{\triangle t}\right)$$

or

$$\mathbb{R} = \frac{1}{\triangle t} \geq 2f_n.$$

This relation reveals that the maximum frequency we can possibly discover from the samples is one half of the sampling rate $\mathbb{R} = 1/\triangle t$. In the context of sampling theorem (to be presented in Chapter 5), the maximum frequency f_n so determined is formally referred to as the Nyquist frequency, and we have $\mathbb{R} = 1/\triangle t = \lceil 2f_n \rceil$. As illustrated in Figure 2.2, the range of frequencies $[-f_n, f_n]$ (corresponding to those shown in complex exponential mode) is called the fundamental interval or the Nyquist interval (with bandwidth $F = 2f_n$).

Relation 3 (Sample spacing and shortest period)

$$\triangle t \leq \frac{1}{2}\left(\frac{1}{f_n}\right) = \frac{T_n}{2}.$$

Since a mode at the maximum frequency f_n has the shortest period $T_n = 1/f_n$, when samples are spaced by $\triangle t \leq \frac{1}{2}T_n$, at least *two samples per cycle* are available for detecting the mode at this (known) frequency.

Digital Frequency and Relations 1–3 Recall that the product of analog frequency f_k and sampling interval $\triangle t$ defines the digital frequency

$$\mathbb{F}_k \equiv f_k \triangle t = \frac{k}{T}\triangle t = \frac{k}{N\triangle t}\triangle t = \frac{k}{N} \text{ (cycles per sample).}$$

Thus we have the uniform spacing $\triangle \mathbb{F} = 1/N$ on the digital frequency grid. The maximum digital frequency is

$$\mathbb{F}_n = f_n \triangle t = \frac{n}{N} \leq \frac{1}{2},$$

and we obtain the Nyquist interval $[-\mathbb{F}_n, \mathbb{F}_n] \subseteq [-\frac{1}{2}, \frac{1}{2}]$. The relationship between analog and digital frequency grids is illustrated by examples in Figure 2.3.

Figure 2.3 Analog frequency grids and corresponding digital frequency grids.

Analog Frequency Grid: Δf = 1/T (T = 3.3 = NΔt, N = 5)

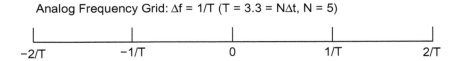

Digital Frequency Grid: ΔF = ΔfΔt = 1/N (N = 2n+1 = 5)

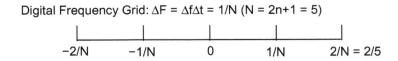

Analog Frequency Grid: Δf = 1/T (T = 7.26 = NΔt, N = 11)

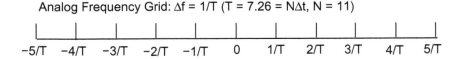

Digital Frequency Grid: ΔF = ΔfΔt = 1/N (N = 2n+1 = 11)

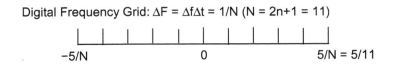

Note that *after* $\triangle t$ is absorbed into $\mathbb{F}_k$, we can only refer to the ℓth sample x_ℓ in the discrete-time domain, so the spacing is $\triangle \ell = 1$, and the reciprocity relation

$$\triangle \mathbb{F} \triangle \ell = \frac{1}{N}$$

is satisfied. To drive home this last point, we only need to evaluate

$$x(t) = \frac{A_0}{2} + \sum_{k=1}^{n} A_k \cos 2\pi f_k t + B_k \sin 2\pi f_k t$$

at $t = \ell \triangle t$ and express the value of the ℓth sample using $\mathbb{F}_k$ instead of f_k as shown below.

$$x_\ell = \frac{A_0}{2} + \sum_{k=1}^{n} A_k \cos 2\pi (\mathbb{F}_k)\ell + B_k \sin 2\pi (\mathbb{F}_k)\ell.$$

Accordingly, the data spacing in the discrete-time domain is $\triangle \ell = 1$ (with period $\mathbb{T} = N$) when the spacing in the frequency domain is measured by digital frequency instead of analog frequency.

Finally, since $\triangle \ell = 1$ and $2 \leq \dfrac{1}{\mathbb{F}_n}$, the relation $\triangle \ell \leq \dfrac{1}{2\mathbb{F}_n}$ is also satis ed .

We will consider the implications of these relations on signals which are either non-periodic or not band-limited (or both) in Chapter 5.

2.2   Frequencies Aliased by Sampling

In this section we study the sampling process in a less precise setting. We begin with the simplest case: suppose we are given two samples x_i and x_j which are spaced $\widetilde{T}/2$ units apart within an interval of $\widetilde{T}$ units, and we are required to determine a single-frequency component wave $\tilde{x}(t)$ which interpolates the two discrete samples. We learn from Chapter 1 that we may express $\tilde{x}(t)$ in the following forms:

$$\tilde{x}(t) = \alpha \cos(2\pi \tilde{f} t) + \beta \sin(2\pi \tilde{f} t) = \gamma \cos(2\pi \tilde{f} t - \hat{\phi}).$$

Note that we cannot apply the solution process developed in the last section to this (seemingly simple) special case without knowing the frequency (or period) of $\tilde{x}(t)$, because, as shown in Figure 2.4, multiple functions of different periods pass through the same two points spaced $\widetilde{T}/2$ units apart within an interval of $\widetilde{T} = 2$ units. To emphasize this potential problem in the current setting, we repeat its source three times (in three ways):

1. We do not know the period of $\tilde{x}(t)$;

2. We do not know the frequency $\tilde{f}$ of $\tilde{x}(t)$;

3. We do not know how many cycles $\tilde{x}(t)$ has completed over the interval $\widetilde{T}$.

Mathematically, the function $\tilde{x}(t)$ interpolating the two samples is no longer unique if the frequency $\tilde{f}$ is not speci ed. However, if we are required to have (at least) two samples per cycle, we will accept the $\tilde{x}(t)$ which completes one cycle over the interval $\widetilde{T}$, i.e., the frequency we can resolve for $\tilde{x}(t)$ is $\tilde{f} = \tilde{r}/\widetilde{T} = 1/\widetilde{T}$.

When we deal with discrete samples taken from a composite signal, the so-called aliased frequencies are equivalent in the sense that they contribute the same numerical values at the sample points. For example, as illustrated in Figure 2.5, the signal

$$y(\theta) = \cos(\theta) + 2\cos(3\theta) + 3\cos(5\theta)$$

cannot be distinguished from

$$x(\theta) = 6\cos(\theta)$$

based on the two values sampled at $\theta_1 = 0$ and $\theta_2 = \pi$, because $y(0) = x(0) = 6$ and $y(\pi) = x(\pi) = -6$. Consequently, if the samples actually come from $y(\theta)$, we would never

Figure 2.4 The function interpolating two samples is not unique.

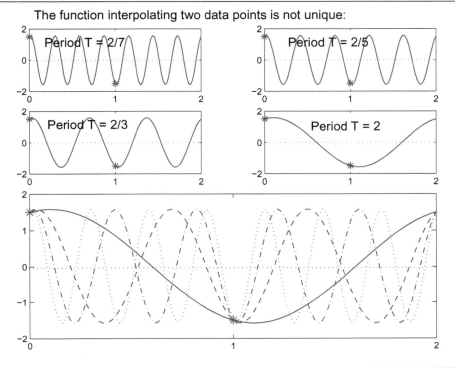

Figure 2.5 Functions $x(\theta)$ and $y(\theta)$ have same values at 0 and π.

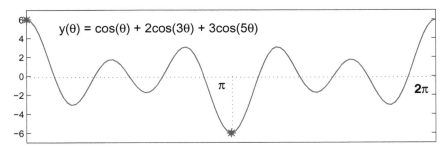

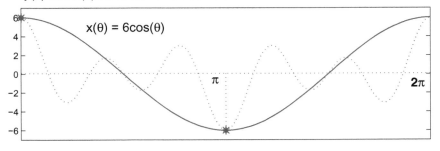

Figure 2.6 The aliasing of frequencies outside the Nyquist interval.

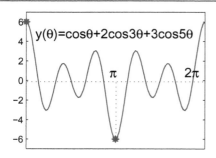

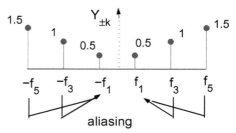

y(θ) in frequency domain

aliasing

due to sampling only y(0) and y(π)

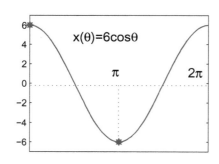

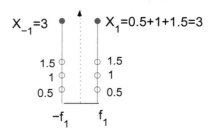

Nyquist interval: $[-f_1, f_1]$

n d out by taking only two samples over one period of $y(\theta)$. This consequence is shown in Figure 2.6.

In general, if a signal contains (higher) frequencies outside the Nyquist interval, they would be aliased to (lower) frequencies inside the Nyquist interval. Recall from the last section that when we are given $N = 2n+1$ equally spaced samples over one period (T) of a composite signal, the maximum frequency we can possibly resolve is the Nyquist frequency $f_n \leq 1/(2\triangle t) = N/(2T)$, and there are two samples available over the (shortest) period $T_n = 1/f_n$, which is shown in Figure 2.7. Our solution was precise because there were no

Figure 2.7 Sampling rate and Nyquist frequency.

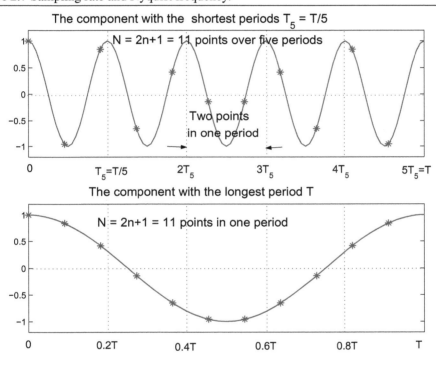

other (higher) frequencies present in the signal. When this is not the case, suppose that the Fourier series of $x(t)$ has more than $N = 2n + 1$ terms, then the Nyquist frequency (which is determined by the current sampling rate) is not the highest frequency present in the signal. Instead, the Nyquist frequency now represents the cutoff frequency. When this happens, because the (higher) frequencies outside the Nyquist interval cannot be resolved at the chosen sampling rate, their numerical values at the sample points would appear as contributions from the equivalent (lower) frequencies inside the Nyquist interval: a high frequency appears as (is aliased into) a low frequency, and the affected DFT coef cient is said to contain an aliasing error.

Relevant later sections: A precise accounting of the aliased frequencies in the DFT co-ef cients will be given in Section 3.11, and we shall verify the aliasing effect by concrete examples in Section 4.4.

2.3   Connection: Anti-Aliasing Filter

When the original signal contains frequencies outside the Nyquist interval, their contribution to the sample values would appear as a contribution from lower frequencies in the signal reproduced from the samples. Therefore, while the reproduced signal agrees with the original signal at discrete sample points in the time domain, they do not agree with each other in the frequency domain: the reproduced signal has potentially *fewer* modes and their amplitudes are potentially *different*. For example, the kth coefficient computed by the DFT includes the contribution not only from the original kth mode, but from all of the aliased modes as well. These differences cause distortion in the reproduced signal — the fewer aliased frequencies, the better the original signal is preserved.

Of course, aliasing of signals will not occur if the highest frequency present in the sampled signal can be limited to the Nyquist frequency, which is determined by a suitable choice of the sampling rate. To ensure this, a low-pass pre-filter or anti-aliasing filter may be used to band-limit the original signal *before* the samples are collected. The cutoff frequency of the anti-aliasing filter can be set according to the sampling rate, so in theory no component with frequencies higher than the Nyquist frequency remains in the filtered signal, and no aliasing will occur in the reconstructed signal.

Signal filtering is a topic covered in Chapter 10.

2.4   Alternate Notations and Formulas

Since Fourier analysis and its wide-ranging applications span across numerous areas over a long time in history, there exists a very large body of technical terms and formulas, and they are expressed using widely varying notations in the literature. In this section we revisit some familiar terms and formulas in this context.

In Table 2.1 we list the symbols we chose to adopt in this book in the second column, and we give examples of alternate symbols in the third column. Since the definitions of these terms are inter-related, we need to be consistent in our choice of notations to make their roles clear, transparent, and easily applicable in deriving future results. As revealed in Table 2.1, there is a certain degree of inconsistency in the terminology and symbols used in the literature, which should not cause too much difficulty once they are explicitly recognized and dealt with. At times we do have the need for an alternate notation — for example, we may use a single letter F to replace the (more meaningful) expression $2f_{max}$ when the same term is repeated in multiple places during a long mathematical proof or derivation — in this example, confusion can be avoided if $F = 2f_{max}$ is explicitly defined before it is used and readers are reminded of its meaning at appropriate places.

In the context of signal sampling, it is convenient sometimes to assume unit period ($T = 1$) or unit spacing ($\triangle t = 1$). In either case, the period T, the time-grid spacing $\triangle t$, the frequency-grid spacing $\triangle f$, and the Nyquist frequency f_{max} will each take on a constant numerical value or it will be defined by sample size N only. We obtain two sets of values under the two assumptions, and they are given in Table 2.2. Note that we cannot alter $\triangle t$ independently of $\triangle f$ because their product must satisfy the reciprocity relation: $\triangle f \triangle t = 1/N$. Based on this relation, we may convert the two sets of values to and from each other by scaling $\triangle t$ and $\triangle f$ in the following manner: when $\triangle t$ is scaled by factor N, the inverse factor $1/N$ is used to scale $\triangle f$, so their product remains unchanged after the conversion. (Note that $T = N\triangle t$ and

Table 2.1 Alternate symbols and alternate definitions/assumptions.

Name	Definition/symbol used as consistently as possible in this text	Alternate symbol (alternate definition/ implicit assumption)
A single period (temporal)	$[0, T], [-T/2, T/2]$	$[-T, T]$
A single period (spatial)	$[-L, L]$	$[0, L], [-L/2, L/2]$
Period (length)	$T, 2L$	$2T, L$
Samples per period	$N = 2n+1$ (odd) $N = 2n+2$ (even)	$N = 2n$
Sampled signal	$\{x_0, x_1, \cdots, x_{N-1}\}$	$u_n, x[n], \{f(k)\}$
Sample interval (period)	$\triangle t = T/N$	$t_s, T, \triangle x, 1/(2f_{max})$
Sample point	$t_\ell = \ell \triangle t$	$n\triangle t, nT, n\ (\because \triangle t = 1)$
Sample value	$x_\ell = x(t_\ell)$	$x[n] = x(nT), f(k)$
Sampling rate (frequency)	$\mathbb{R} = 1/\triangle t = N/T$	$N\ (\because T = 1), 2f_{max}$
Fundamental frequency	$f_1 = \triangle f = 1/T$	$f_0, \triangle \omega$
Nyquist frequency	$f_n = n/T \le N/(2T),$ $f_{max}, F/2$	$f_c, f_m, \omega_{max}, F, \Omega,$ $N/2\ (\because T = 1), \Omega/2$
Nyquist interval	$[-f_n, f_n], [-F/2, F/2]$ $[-f_{max}, f_{max}]$	$[-F, F], [-\Omega/2, \Omega/2]$ $[-\Omega, \Omega], [-f_m, f_m]$
Nyquist rate (bandwidth)	$2f_{max} \le 1/\triangle t$	$2f_m, F, 2F, \Omega, 2\Omega$

$f_{max} = \lfloor 1/(2\triangle t) \rfloor$ follow immediately.)

Table 2.2 Constants resulting from assuming unit period or unit spacing.

Symbolic name	Unit period (assume $T = 1$)	Unit spacing (assume $\triangle t = 1$)
T	1	N
$\triangle t$	$1/N$	1
$\triangle f$	1	$1/N$
f_{max}	$N/2$	$1/2$

From our discussion on digital frequency and relations 1—3 in Section 2.1, we recall that $\mathbb{F}_\alpha = f_\alpha \triangle t$; we thus have the equality $\mathbb{F}_\alpha = f_\alpha$ when $\triangle t = 1$, which gives both the same numerical values, but their defining relationship dictates that $\mathbb{F}_\alpha$ is measured by cycles per sample, and f_α is measured by cycles per unit time. For example, when the time is measured by seconds, we have the results in Table 2.3.

Although the results derived using analog frequency with unit time spacing ($\triangle t = 1$) will not be different from those derived using the digital frequency, the explicit incorporation of $\triangle t$ in the latter's definition provides direct means to interpret and apply the results for values of $\triangle t$ other than unity. Therefore in suitable contexts we may use the more convenient digital frequency in the analysis and processing of signals without loss of generality.

Table 2.3 Using analog frequency versus digital frequency.

Symbolic name	Using analog f_α (assume $\triangle t = 1$ sec)	Symbolic name	Using digital $\mathbb{F}_\alpha$ (for arbitrary $\triangle t$)
T	N seconds	$\mathbb{T}$	N samples
$\triangle t$	1 second	$\triangle \ell$	1 sample
$\triangle f$	$1/N$ Hertz	$\triangle \mathbb{F}$	$1/N$ cycles/sample
f_{max}	$1/2$ Hertz	$\mathbb{F}_n$	$1/2$ cycles/sample

2.5   Sampling Period and Alternate Forms of DFT

In the beginning of this chapter we introduced the discrete Fourier transform (DFT) as an interpolating formula which computes the unknown coefficients of a finite Fourier series directly from its sampled values. As a quick recap, recall that in Section 2.1 we obtained the DFT from sampling a band-limited signal $x(t)$ represented by a finite Fourier series of $N = 2n+1$ terms, namely,

$$x(t) = \frac{A_0}{2} + \sum_{r=1}^{n} A_r \cos \frac{2\pi r t}{T} + B_r \sin \frac{2\pi r t}{T},$$

and we stated that since $x(t + T) = t$, we only needed to sample the function over a single period. From interpolating the N equally spaced samples $x_\ell = x(\ell \triangle t)$ for $0 \leq \ell \leq N-1$, we obtained the discrete Fourier transform (DFT) given by Equation (2.7), namely,

$$X_r = \frac{1}{N} \sum_{\ell=0}^{N-1} x_\ell \omega_N^{-r\ell}, \quad \text{where } \omega_N = e^{j2\pi/N}, \ r = 0, 1, \ldots, N-1.$$

The corresponding inverse DFT (or IDFT) was given by Equation (2.6), namely,

$$x_\ell = \sum_{r=0}^{N-1} X_r \omega_N^{r\ell}, \quad \text{where } \omega_N = e^{j2\pi/N}, \ \ell = 0, 1, \ldots, N-1.$$

In this setting, the N samples $\{x_0, x_1, \ldots, x_{N-1}\}$ cover the period $[0, T]$, and the actual sampling time begins at $t = 0$ and ends at $t = T - \triangle t = T - T/N$, because the first sample of the *next* period $[T, 2T]$ will have to be taken at $t = T$. Corresponding to $\{x_0, x_1, \ldots, x_{N-1}\}$, the DFT and IDFT have (discrete) period of N (samples), which is reflected by $x_{\ell+N} = x_\ell$ and $X_{r+N} = X_r$ for $0 \leq r, \ell \leq N-1$.

Now, since a function $x(t)$ with period T is defined over $(-\infty, \infty)$ by periodic extension, the sampling period may begin and end anywhere. (Recall that the period of a cosine function $\cos(\theta)$ is 2π, which can begin at arbitrary θ and ends at $\theta + 2\pi$, including, but not limited to, the standard choice of $[0, 2\pi]$ or $[-\pi, \pi]$.) This observation coupled with the fact that $x_{\ell \pm N} = x_\ell$ and $X_{r \pm N} = X_r$ allows us to obtain the set of N samples corresponding to any period from the N samples collected over $[0, T]$. In particular, we consider the symmetric period $[-T/2, T/2]$; while the samples for interval $[0, T/2]$ are naturally taken from $x_0, x_1, \ldots, x_n$, the next sample x_{n+1} becomes the first sample of the following period; hence, it is also the first sample in the current period $[-T/2, T/2]$. Following this argument, if the sample set $\{x_0, x_1, x_2, \ldots, x_6\}$ covers the period $[0, T]$, the rearranged set $\{x_4, x_5, x_6, x_0, x_1, x_2, x_3\}$ covers the period $[-T/2, T/2]$, which is conventionally labeled as

$\{x_{-3}, x_{-2}, x_{-1}, x_0, x_1, x_2, x_3\}$. Note that the labeling convention abides by the periodicity relation: $x_{-3} = x_{-3+7} = x_4$, $x_{-2} = x_{-2+7} = x_5$, and $x_{-1} = x_{-1+7} = x_6$, because $N = 2n + 1 = 7$ is the period of the sampled sequence. One hidden technical point is that while the samples remain equispaced with the same data spacing $\triangle t = T/N$, the actual sampling time of x_{-n} is $-T/2 + \triangle t/2$, and the last sample x_n is taken at $T/2 - \triangle t/2$. (The first sample of the next period $[T/2, 3T/2]$ will begin at $T/2 + \triangle t/2$, and so on.) The actual placement of samples is shown diagrammatically in Figures 2.8 and 2.9. If we change the variable from t to $\theta = 2\pi t/T$, the samples would be placed in the corresponding period $[0, 2\pi]$ or $[-\pi/2, \pi/2]$ as shown in Figure 2.10.

Figure 2.8 Taking $N = 2n+1$ samples from a single period $[0, T]$.

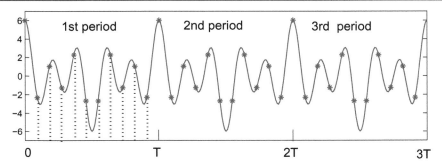

By changing the sampled data set to $\{x_{-n}, \ldots, x_{-1}, x_0, x_1, \ldots, x_n\}$ and following through the derivation analogous to the process described in Section 4.2 of Chapter 4, we obtain another commonly used form of the DFT, namely,

$$(2.8) \qquad X_r = \frac{1}{N} \sum_{\ell=-n}^{n} x_\ell \omega_N^{-r\ell}, \quad \text{where } n = \tfrac{N-1}{2}, \quad \omega_N = e^{j2\pi/N}, \quad -n \le r \le n.$$

The corresponding IDFT is

$$(2.9) \qquad x_\ell = \sum_{r=-n}^{n} X_r \omega_N^{r\ell}, \quad \text{where } n = \tfrac{N-1}{2}, \quad \omega_N = e^{j2\pi/N}, \quad -n \le \ell \le n.$$

Note that the coefficients X_r $(-n \le r \le n)$ computed by (2.8) directly satisfy

$$x(t) = \sum_{r=-n}^{n} X_r e^{j2\pi rt/T} = \frac{A_0}{2} + \sum_{r=1}^{n} A_r \cos \frac{2\pi rt}{T} + B_r \sin \frac{2\pi rt}{T},$$

where $\dfrac{A_0}{2} = X_0$, and $A_r = X_r + X_{-r}$, $B_r = j(X_r - X_{-r})$ for $1 \le r \le n = \tfrac{N-1}{2}$.

Figure 2.9 Rearranging $N = 2n+1$ samples on the time grid.

N = 2n+1 = 7 samples $\{x_0, x_1, x_2, x_3, x_4, x_5, x_6\}$ are taken at $t_k \in [0, T)$, where $t_k = k\Delta t$, k= 0, 1, ..., 6:

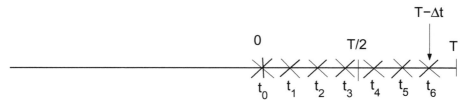

N = 2n+1 = 7 samples $\{x_{-3}, x_{-2}, x_{-1}, x_0, x_1, x_2, x_3\}$ are taken at $t_k \in (-T/2, T/2)$, where $t_k = k\Delta t$, k= -3, ..., 3:

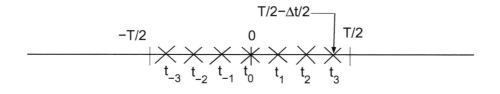

Figure 2.10 The placement of samples after changing variable t to $\theta = 2\pi t/T$.

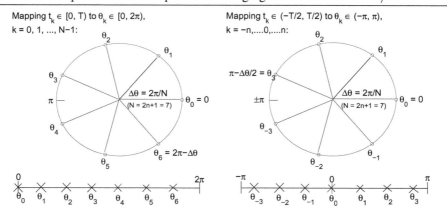

2.6 Sample Size and Alternate Forms of DFT

We have so far linked the DFT sample size N to the number of unknown Fourier coef cients we are seeking, and we have used $N = 2m+1$ by assuming that

$$x(t) = \frac{A_0}{2} + \sum_{r=1}^{m} A_r \cos \frac{2\pi rt}{T} + B_r \sin \frac{2\pi rt}{T}.$$

Now, if we change the sample size from the odd number $N = 2m+1$ to the even number $N = 2m$, we must show that the resulting linear system has only $N = 2m$ unknowns i.e., we must prove that one of the terms in the right-hand side vanishes. This is indeed the case as shown below.

Corresponding to the even sample size $N = 2m$, we have $x_\ell = x(t_\ell)$ with $t_\ell = \ell \triangle t = \ell \left(\frac{T}{2m} \right)$ for $0 \leq \ell \leq 2m - 1$, and the resulting $2m$ linear equations are given by

$$x_\ell = \frac{A_0}{2} + \sum_{r=1}^{m} A_r \cos \frac{2\pi r\ell}{2m} + B_r \sin \frac{2\pi r\ell}{2m}, \quad \ell = 0, 1, \ldots, 2m - 1.$$

Observe that the sine mode corresponding to $r = m$ in the right-hand side is

$$\sin \frac{2\pi m\ell}{2m} = \sin \ell\pi = 0 \text{ for every } \ell,$$

therefore, the term involving B_m vanishes from the right-hand side. By letting $m = n+1$ and $N = 2n+2$, we obtain

$$x_\ell = \frac{A_0}{2} + A_{n+1} \cos \frac{2\pi(n+1)\ell}{N} + \sum_{r=1}^{n} A_r \cos \frac{2\pi r\ell}{N} + B_r \sin \frac{2\pi r\ell}{N}, \quad \ell = 0, 1, \ldots, N - 1.$$

Therefore, we are effectively taking the $2n+2$ discrete-time samples from

(2.10)
$$\tilde{x}(t) = \frac{A_0}{2} + A_{n+1} \cos \frac{2\pi(n+1)t}{T} + \sum_{r=1}^{n} A_r \cos \frac{2\pi rt}{T} + B_r \sin \frac{2\pi rt}{T},$$

which has $2n + 2$ coef cients to be determined. The DFT derived from interpolating the $2n+2$ sampled values of $\tilde{x}(t)$ using (2.10) is given below, with its derivation provided in Chapter 4.

(2.11)
$$X_r = \frac{1}{2n + 2} \sum_{\ell=0}^{2n+1} \tilde{x}_\ell \omega_N^{-r\ell} = \frac{1}{N} \sum_{\ell=0}^{N-1} \tilde{x}_\ell \omega_N^{-r\ell}, \quad r = 0, 1, \cdots, N - 1.$$

This formula is of the same form as the DFT of odd length de ned by (2.7) except that $N = 2n + 2$ and it is tting a different trigonometric polynomial $\tilde{x}(t)$. The corresponding IDFT is

(2.12)
$$\tilde{x}_\ell = \sum_{r=0}^{2n+1} X_r \omega_N^{r\ell} = \sum_{r=0}^{N-1} X_r \omega_N^{r\ell}, \quad \ell = 0, 1, \cdots, N - 1.$$

The $N = 2n+2$ coef cients of $\tilde{x}(t)$ can now be obtained from the computed X_r by applying the following rules:

1. $\frac{A_0}{2} = X_0; A_{\frac{N}{2}} = X_{\frac{N}{2}};$

2. $A_r = X_r + X_{N-r}$, $B_r = j(X_r - X_{N-r})$, for $r = 1, 2, \ldots, \frac{N}{2} - 1$.

Using *even* sample size with the sampling period $[-T/2, T/2]$, the following DFT/IDFT formulas may be obtained. (The direct conversion between Formulas (2.11) and (2.13) is presented in Section 4.2 of Chapter 4.)

(2.13)
$$X_r = \frac{1}{2n+2} \sum_{\ell=-n}^{n+1} \tilde{x}_\ell \omega_N^{-r\ell} = \frac{1}{N} \sum_{\ell=-\frac{N}{2}+1}^{\frac{N}{2}} \tilde{x}_\ell \omega_N^{-r\ell}, \quad -\frac{N}{2} + 1 \leq r \leq \frac{N}{2}.$$

(2.14)
$$\tilde{x}_\ell = \sum_{r=-n}^{n+1} X_r \omega_N^{r\ell} = \sum_{r=-\frac{N}{2}+1}^{\frac{N}{2}} X_r \omega_N^{r\ell}, \quad -\frac{N}{2} + 1 \leq \ell \leq \frac{N}{2}.$$

The $N = 2n+2$ coef cients of $\tilde{x}(t)$ can now be obtained from the computed X_r by applying the following rules:

1. $\dfrac{A_0}{2} = X_0$; $A_{\frac{N}{2}} = X_{\frac{N}{2}}$;

2. $A_r = X_r + X_{-r}$, $B_r = j(X_r - X_{-r})$, for $r = 1, 2, \ldots, \frac{N}{2} - 1$.

Note that when the sample size is an even number $N = 2n+2$, the sampling time for x_0 remains at $t = 0$, the last sample x_{2n+1} in the period $[0, T]$ is taken at $T - \triangle t = T - T/N$ as before; however, the rst sample x_{-n} and the last sample x_{n+1} in the period $[-T/2, T/2]$ are taken from $t = -T/2 + \triangle t$ and $t = T/2$. The actual placement of samples is shown diagrammatically in Figures 2.11, 2.12, and 2.13.

Figure 2.11 Rearranging $N = 2n+2$ samples on the time grid.

N = 2n+2 = 8 samples {x_0, x_1, x_2, x_3, x_4, x_5, x_6, x_7} are taken at $t_k \in [0, T)$, where $t_k = k\Delta t$, k= 0, 1, ..., 7:

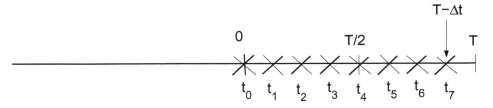

N = 2n+2 = 8 samples {x_{-3}, x_{-2}, x_{-1}, x_0, x_1, x_2, x_3, x_4} are taken at $t_k \in (-T/2, T/2]$, where $t_k = k\Delta t$, k= −3, ..., 4:

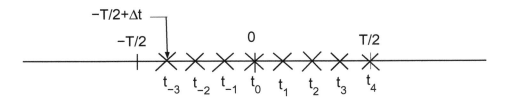

Figure 2.12 The placement of samples after changing variable t to $\theta = 2\pi t/T$.

Mapping $t_k \in [0, T)$ to $\theta_k \in [0, 2\pi)$, k = 0, 1, ..., N − 1:

Mapping $t_k \in (-T/2, T/2]$ to $\theta_k \in (-\pi, \pi]$, k = −n, ..., 0,..., n+1:

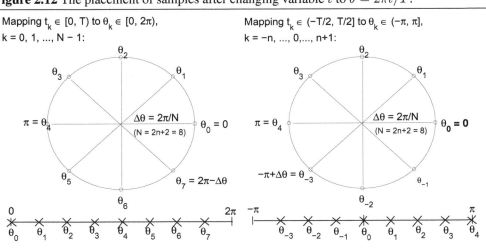

Figure 2.13 Taking $N = 2n+2$ samples from the period $[0, 2\pi]$ or $[-\pi, \pi]$.

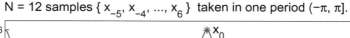

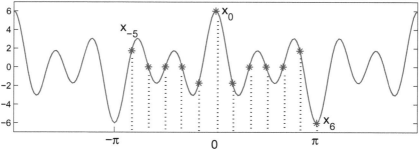

References

1. A. Ambardar. *Analog and Digital Signal Processing*. Brooks/Cole Publishing Company, Paci c Grove, CA, second edition, 1999.

2. W. L. Briggs and V. E. Hensen. *The DFT: An Owner's Manual for the Discrete Fourier Transform*. The Society for Industrial and Applied Mathematics, Philadelphia, PA, 1995.

3. E. Chu and A. George. *Inside the FFT Black Box: Serial and Parallel Fast Fourier Transform Algorithms*, CRC Press, Boca Raton, FL, 2000.

4. R. W. Hamming. *Digital Filters*. Prentice-Hall, Inc., Englewood Cliffs, NJ, third edition, 1989.

5. B. Porat. *A Course in Digital Signal Processing*. John Wiley & Sons, Inc., New York, 1997.

Chapter 3

The Fourier Series

In Chapters 1 and 2 we have been dealing with periodic signals described by a sum of sines and cosines, whereas in general a signal may be described by a mathematical function $f(t)$ which does not represent a sum of sinusoidal terms in its present form. For example, given below is the time-domain description of a periodic triangular wave.

$$f(t) = \begin{cases} t+1, & -1 \leq t \leq 0, \\ -t+1, & 0 < t \leq 1; \end{cases} \quad f(t+2) = f(t).$$

In such cases, the frequency contents of $f(t)$ are revealed by a continuous Fourier series, which must be derived for each individual function using the theory and techniques to be covered in this chapter. The Fourier series may be expressed in the various forms introduced in Chapter 1, and we are simply taking the next step in this chapter to determine its coef cien ts analytically. Since we have initiated the discussion on sampling and reconstruction of functions in Chapter 2, it is not out of place to remark at the outset that the DFT coef cients, which are de ned via the discrete-time samples of $f(t)$, are expected to deviate from the coef cients of the corresponding terms in the Fourier series of $f(t)$, because the DFT coef cients include contributions from all aliased frequencies. The phenomenon of aliasing was brie y discussed in Sections 2.2 and 2.3, and it will be further explored in this chapter.

3.1 Formal Expansions

To expand a general periodic function $f(t)$ into a formal Fourier series (which is also known as a harmonically related trigonometric series), we employ the well-known theorem of Dirichlet, which also gives the suf cien t conditions for the existence of Fourier series.

Theorem 3.1 (Dirichlet s theorem) If $f(t)$ is a real-valued function de ned on $(-\infty, \infty)$ and it satis es the Dirichlet conditions:

(a) $f(t)$ is bounded on any bounded closed subinterval $[a, b]$ of $(-\infty, \infty)$;

(b) $f(t)$ has only a nite number of maxima and minima on any interval $[a, b]$;

(c) $f(t)$ has on $[a, b]$ at most a nite number of discontinuities, each of which is a jump discontinuity;

(d) $f(t)$ is periodic with period T th at is, $f(t + T) = f(t)$;

then for every t at which f is continuous, we have

$$f(t) = \frac{A_0}{2} + \sum_{k=1}^{\infty} A_k \cos \frac{2\pi kt}{T} + B_k \sin \frac{2\pi kt}{T}, \quad \text{where}$$

(3.1)
$$A_k = \frac{2}{T} \int_{-T/2}^{T/2} f(t) \cos \frac{2\pi kt}{T} \, dt, \quad k = 0, 1, 2, \ldots,$$

(3.2)
$$B_k = \frac{2}{T} \int_{-T/2}^{T/2} f(t) \sin \frac{2\pi kt}{T} \, dt, \quad k = 1, 2, \ldots$$

Furthermore, for every t_α at which f has a jump discontinuity, the Fourier series converges to the average of its right- and left-hand limits. That is,

$$\frac{f\left(t_\alpha^+\right) + f\left(t_\alpha^-\right)}{2} = \frac{A_0}{2} + \sum_{k=1}^{\infty} A_k \cos \frac{2\pi kt_\alpha}{T} + B_k \sin \frac{2\pi kt_\alpha}{T}, \quad \text{where}$$

$$f\left(t_\alpha^+\right) \equiv \lim_{t \to t_\alpha^+} f(t), \quad \text{and } f\left(t_\alpha^-\right) \equiv \lim_{t \to t_\alpha^-} f(t).$$

Remark 3.2 If $f(t)$ has a jump discontinuity at t_α, Theorem 3.1 does not require $f(t)$ to be de ned at t_α. For example, the sawtooth function shown below may be de ned by the periodic extension of $f(t) = t$ for either $t \in (0, T]$ or $t \in (0, T)$. In the former case, we have $f(\pm kT) = T$; in the latter case, $f(\pm kT)$ is not de n ed, but one-sided limits exist at the jump discontinuities and Dirichlet s theorem is satis ed .

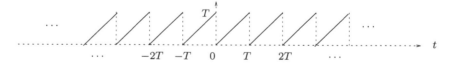

Remark 3.3 Graphically, the function $f(t)$ on any bounded closed interval $[a, b]$ (referred to in Theorem 3.1) may be represented by disjoint arcs of different curves, each de ned by a different formula. In mathematical terms, a function $f(t)$ is said to be *piecewise continuous* in an interval $a \le t \le b$ if there exist n points $a = t_1 < t_2 < t_3 < \cdots < t_n = b$ such that $f(t)$ is continuous in each interval $t_\ell < t < t_{\ell+1}$ and has n ite one-sided limits $f\left(t_\ell^+\right)$ and $f\left(t_{\ell+1}^-\right)$ at the endpoints of each such interval ($\ell = 1, 2, \ldots, n - 1$).

A piecewise continuous $f(t)$:

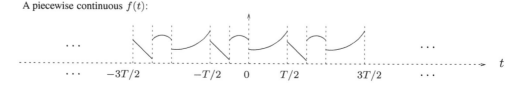

Remark 3.4 For the correct use of mathematical theorems, it is important to know whether a function is continuous on an open interval or a closed interval. We recall that a function $g_\ell(t)$ is said to be continuous throughout a *closed* interval $[t_\ell,\, t_{\ell+1}]$ provided that it is continuous in the *open* interval $(t_\ell,\, t_{\ell+1})$ and also $g_\ell(t_\ell) = g_\ell\left(t_\ell^+\right)$ and $g_\ell(t_{\ell+1}) = g_\ell\left(t_{\ell+1}^-\right)$.

From elementary calculus we recall that if $g_\ell(t)$ is continuous on the closed interval $[t_\ell,\, t_{\ell+1}]$, it must be bounded and (Riemann) integrable on $[t_\ell,\, t_{\ell+1}]$.

Since we may define $g_\ell(t)$ to agree with a piecewise continuous $f(t)$ over the closed interval $[t_\ell,\, t_{\ell+1}]$ except possibly at the endpoints, we have

$$\int_{t_\ell}^{t_{\ell+1}} f(t)\, dt = \int_{t_\ell}^{t_{\ell+1}} g_\ell(t)\, dt.$$

Hence, the piecewise continuous $f(t)$ is integrable on $[a,\, b]$:

$$\int_a^b f(t)\, dt = \sum_{\ell=1}^n \int_{t_\ell}^{t_{\ell+1}} f(t)\, dt = \sum_{\ell=1}^n \int_{t_\ell}^{t_{\ell+1}} g_\ell(t)\, dt.$$

Remark 3.5 Although the coefficient formulas (3.1) and (3.2) are valid for any integrable function $f(t)$, the Fourier series constructed using so-obtained A_k and B_k ($k = 0, 1, \ldots, \infty$) may diverge for some values of t or it may fail to converge to $f(t)$ for infinitely many values of t.

While it is a relatively simple task to derive the expressions (3.1) and (3.2) for the coefficients A_k and B_k ($k = 0, 1, \ldots, \infty$) if we can assume that the Fourier series is convergent and it converges to an integrable function $f(t)$, there is no obvious way to ascertain whether such assumption is valid for an arbitrary harmonically related trigonometric series.

Remark 3.6 A function f is said to have been *normalized* at points of jump discontinuity if $f(t) \stackrel{def}{=} \frac{1}{2}\big(f(t^+) + f(t^-)\big)$ for every t. Since $f(t_c) = f(t_c^+) = f(t_c^-)$ if t_c is a point of continuity, the normalized function agrees with the original $f(t)$ at t_c. Furthermore, changing the value of $f(t)$ at points of jump discontinuity does not change the value of its integral over the interval, nor any of the integrals defining its Fourier coefficients. According to Dirichlet's theorem, the Fourier series expansion of a normalized $f(t)$ converges to itself for every t. Therefore, whenever we *equate* $f(t)$ to its Fourier series expansion by writing

$$f(t) = \frac{A_0}{2} + \sum_{k=1}^{\infty} A_k \cos \frac{2\pi k t}{T} + B_k \sin \frac{2\pi k t}{T}$$

(when not otherwise qualified) in the future, we implicitly assume that $f(t)$ satisfies Dirichlet's conditions and that $f(t)$ has been normalized at points of jump discontinuity.

Examples of normalized functions:

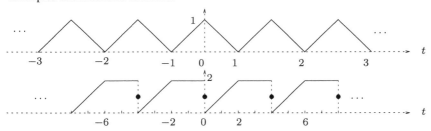

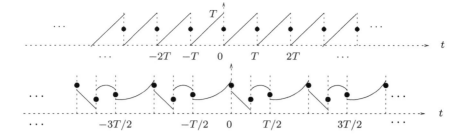

Remark 3.7 The constant term in the Fourier series expansion is conventionally written as $\frac{1}{2}A_0$, so that A_0 can be obtained from the general formula (3.1) for A_k by setting $k = 0$. Formulas (3.1) and (3.2) are known as the Euler or Euler Fourier formulas.

Remark 3.8 For a proof of Dirichlet s theorem, see Philip Franklin s article *A Simple Discussion of the Representation of Functions by Fourier Series* in Selected Papers on Calculus, pp. 357 361, Mathematical Association of America, 1969.

In this chapter we shall prove the same pointwise convergence results (the topic pointwise convergence is covered in Section 3.10.2) for the Fourier series of any piecewise continuous function f under the further assumption that it has nite one-sided derivatives

$$f'\left(t_\ell^+\right) \equiv \lim_{h \to 0} \frac{f\left(t_\ell + h\right) - f\left(t_\ell^+\right)}{h} \text{ and } f'\left(t_\ell^-\right) \equiv \lim_{h \to 0} \frac{f\left(t_\ell^-\right) - f\left(t_\ell - h\right)}{h}$$

at those points where f itself is discontinuous and f' does not exist.

Since the conditions prescribed for f' would be met automatically if both f and f' are piecewise continuous on $[a, b]$ th e function f is then said to be *piecewise smooth* on $[a, b]$, the pointwise convergence of the Fourier series for piecewise smooth functions is guaranteed by the same proof.

3.1.1   Examples

Since the Fourier series coef cients A_k and B_k are de n ed by integrals involving $f(t)$ in Dirichlet s theorem, the analytical formulas for them are available only when the integration can be done analytically and explicitly. An example is given below.

Example 3.9 Find the Fourier series for the given function

$$f(t) = \begin{cases} 2, & -2 \leq t < 0, \\ t, & 0 < t < 2; \end{cases} \quad f(t + 4) = f(t).$$

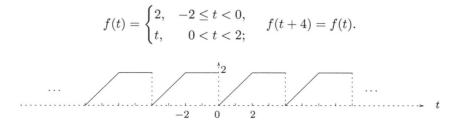

We note that $f(t)$ satis es the conditions of Dirichlet s theorem and that it has a jump discontinuity at $t = 0$ on $[-2, 2]$ because $f(0^-) = 2$ and $f(0^+) = 0$. (Recall that $f(t)$ needs not be de ned at the jump discontinuity, which is the case here.) We nd the coef cients according to Dirichlet s theorem: For $k = 0$, we obtain

$$A_0 = \frac{1}{2} \int_{-2}^{0} 2\, dt + \frac{1}{2} \int_{0}^{2} t\, dt = 3.$$

For $k = 1, 2, \ldots$, we obtain

$$
\begin{aligned}
A_k &= \frac{1}{2} \int_{-2}^{0} 2\cos\frac{k\pi t}{2}\, dt + \frac{1}{2} \int_{0}^{2} t\cos\frac{k\pi t}{2}\, dt \\
&= \left[\frac{2}{k\pi}\sin\frac{k\pi t}{2}\right]_{-2}^{0} + \frac{1}{2}\left[\frac{4}{k^2\pi^2}\cos\frac{k\pi t}{2} + \frac{2t}{k\pi}\sin\frac{k\pi t}{2}\right]_{0}^{2} \\
&= \frac{2}{k^2\pi^2}(\cos k\pi - 1) + \frac{1}{2}\left(\frac{4}{k\pi}\sin k\pi - 0\right) \\
&= \frac{2}{k^2\pi^2}(\cos k\pi - 1),
\end{aligned}
$$

and

$$
\begin{aligned}
B_k &= \frac{1}{2} \int_{-2}^{0} 2\sin\frac{k\pi t}{2}\, dt + \frac{1}{2} \int_{0}^{2} t\sin\frac{k\pi t}{2}\, dt \\
&= \left[-\frac{2}{k\pi}\cos\frac{k\pi t}{2}\right]_{-2}^{0} + \frac{1}{2}\left[\frac{4}{k^2\pi^2}\sin\frac{k\pi t}{2} - \frac{2t}{k\pi}\cos\frac{k\pi t}{2}\right]_{0}^{2} \\
&= -\frac{2}{k\pi} + \frac{2}{k\pi}\cos k\pi + \frac{1}{2}\left(\frac{-4}{k\pi}\cos k\pi - 0\right) \\
&= -\frac{2}{k\pi}.
\end{aligned}
$$

For every t at which $f(t)$ is continuous, we now have

$$
\begin{aligned}
f(t) &= \frac{3}{2} + \sum_{k=1}^{\infty}\left[\frac{2(\cos k\pi - 1)}{k^2\pi^2}\cos\frac{k\pi t}{2} + \left(\frac{-2}{k\pi}\right)\sin\frac{k\pi t}{2}\right] \\
&= \frac{3}{2} - \frac{4}{1^2\pi^2}\cos\frac{\pi t}{2} - \frac{4}{3^2\pi^2}\cos\frac{3\pi t}{2} - \frac{4}{5^2\pi^2}\cos\frac{5\pi t}{2} - \cdots \\
&\quad - \frac{2}{\pi}\sin\frac{\pi t}{2} - \frac{2}{2\pi}\sin\frac{2\pi t}{2} - \frac{2}{3\pi}\sin\frac{3\pi t}{2} - \cdots \\
&= \frac{3}{2} - \frac{4}{\pi^2}\sum_{k=1}^{\infty}\frac{1}{(2k-1)^2}\cos\frac{(2k-1)\pi t}{2} - \frac{2}{\pi}\sum_{k=1}^{\infty}\frac{1}{k}\sin\frac{k\pi t}{2}.
\end{aligned}
$$

We can also verify that the Fourier series indeed converges to the normalized function value at the jump discontinuity at $t = 0$. Note that the Fourier series takes on a much simpler form at $t = 0$ because all cosine and sine functions are replaced by $\cos 0 = 1$ and $\sin 0 = 0$, and the desired relation

$$1 = \frac{f(0^-) + f(0^+)}{2} = \frac{3}{2} - \frac{4}{\pi^2}\sum_{k=1}^{\infty}\frac{1}{(2k-1)^2}$$

can be easily veri ed by substituting the numerical value of the in n ite series, namely,

$$\sum_{k=1}^{\infty}\frac{1}{(2k-1)^2} = 1 + \frac{1}{3^2} + \frac{1}{5^2} + \cdots = \frac{\pi^2}{8},$$

which we shall prove in Example 3.15.

The convergence of the N-term Fourier series is shown in Figure 3.1, where we plot the Fourier series for $N = 8$, 16, 32, 64. Since this particular example involves zero coef cients, we clarify how we count the N terms: as an example, for $N = 8$, the eight Fourier series coef cients used are: $\frac{1}{2}A_0$, A_1, $A_2 = 0$, A_3, $A_4 = 0$, B_1, B_2, and B_3. Accordingly, for $N = 2(n+1) = 8$, 16, 32, 64, the N-term Fourier series uses one DC term, $n+1$ cosine terms (including zero and nonzero terms), and n sine terms. We examine further the graphs of the

Figure 3.1 Illustrating the convergence of the N-term Fourier series.

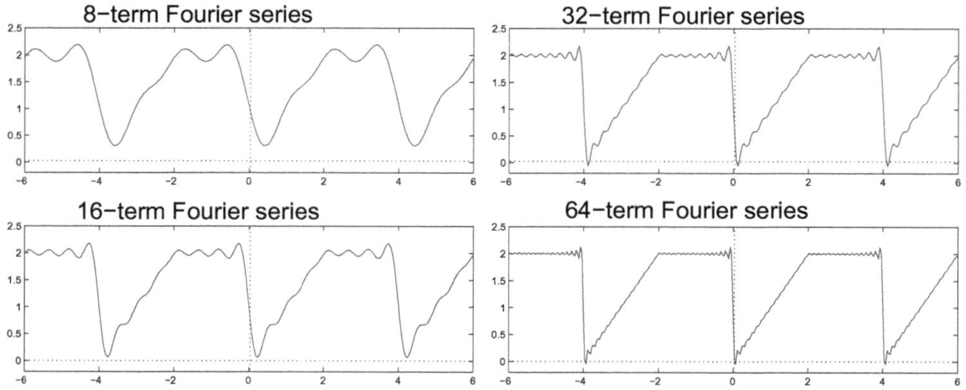

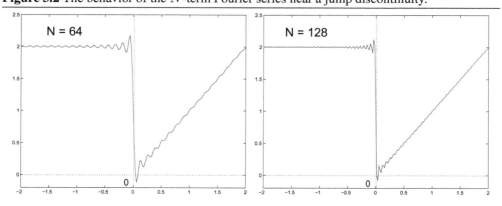

N-term Fourier series near a jump discontinuity in Figure 3.2, where we illustrate the Gibbs effect (to be studied in Section 3.10.4) for $N = 64$ and $N = 128$.

Figure 3.2 The behavior of the N-term Fourier series near a jump discontinuity.

3.2   Time-Limited Functions

If $g(t)$ is only de n ed for a   nite interval $[-T/2, T/2]$, we can construct a periodic function $f(t)$ by repeating $g(t)$ for each period T over $(-\infty, +\infty)$, which is called a *protracted* version (or a periodic extension) of $g(t)$. Because $g(t)$ agrees with $f(t)$ for $t \in [-T/2, T/2]$, the Fourier series of $f(t)$ may be used to represent $g(t)$ in this interval. Note that Dirichlet s theorem applies for every t, so the Fourier series converges to the average of the left- and right-hand limits (of the periodic extension $f(t)$) at jump discontinuities, whether they occur inside or at the ends of the interval on which the time-limited $g(t)$ is defined.

As an example, recall Example 3.9: if we define the time-limited $g(t) = f(t)$ for $t \in [-2, 2]$ in that example, then the Fourier series we obtain for periodic $f(t)$ represents $g(t)$ for $t \in [-2, 2]$.

3.3 Even and Odd Functions

When Dirichlet s theorem is applied to the even and odd functions defined below, their respective Fourier series has only cosine or sine terms. The coefficients of the cosine or sine terms in each case are given in the two theorems following Definition 3.10.

Definition 3.10 A function $f(t)$ is even if and only if $f(-t) = f(t)$ for all t, and it is odd if and only if $f(-t) = -f(t)$ for all t.

By definition, the graph of an even function is symmetric with respect to the y axis if we plot $y = f(t)$ versus t; whereas the graph of an odd function is symmetric with respect to the origin. For example, $f(t) = |t|$ for $t \in [-2, 2]$ is an even function; $f(t) = t$ for $t \in [-2, 2]$ is an odd function; the function $f(t)$ in Example 3.9 is *neither* even *nor* odd.

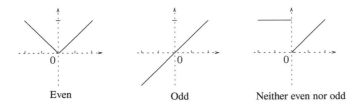

Even	Odd	Neither even nor odd

Noting that the cosine terms are themselves even functions, and the sine terms are themselves odd functions, it comes as no surprise that the expansion of an even function contains only cosine terms, whereas the expansion of an odd function contains only sine terms.

Theorem 3.11 If $f(t)$ is an even function satisfying the conditions of Dirichlet s theorem, the coefficients in the Fourier series of $f(t)$ are given by the formulas

(3.3)
$$A_k = \frac{4}{T} \int_0^{T/2} f(t) \cos \frac{2k\pi t}{T} \, dt, \quad k = 0, 1, 2, \ldots$$
$$B_k = 0, \quad k = 1, 2, \ldots$$

Proof: Using the Euler Fourier formula (3.1), we obtain

$$A_k = \frac{2}{T} \int_{-T/2}^{T/2} f(t) \cos \frac{2\pi kt}{T} dt$$

$$= \frac{2}{T} \left[\int_{-T/2}^{0} f(t) \cos \frac{2\pi kt}{T} dt + \int_{0}^{T/2} f(t) \cos \frac{2\pi kt}{T} dt \right]$$

$$= \frac{2}{T} \left[-\int_{T/2}^{0} f(-s) \cos \frac{2\pi k(-s)}{T} ds + \int_{0}^{T/2} f(t) \cos \frac{2\pi kt}{T} dt \right] \qquad (\text{let } t = -s)$$

$$= \frac{2}{T} \left[\int_{0}^{T/2} f(s) \cos \frac{2\pi ks}{T} ds + \int_{0}^{T/2} f(t) \cos \frac{2\pi kt}{T} dt \right] \qquad (\because f(-s) = f(s))$$

$$= \frac{4}{T} \int_{0}^{T/2} f(t) \cos \frac{2\pi kt}{T} dt. \qquad (\text{let } s = t)$$

Using the Euler Fourier formula (3.2), we obtain

$$B_k = \frac{2}{T} \int_{-T/2}^{T/2} f(t) \sin \frac{2\pi kt}{T} dt$$

$$= \frac{2}{T} \left[\int_{-T/2}^{0} f(t) \sin \frac{2\pi kt}{T} dt + \int_{0}^{T/2} f(t) \sin \frac{2\pi kt}{T} dt \right]$$

$$= \frac{2}{T} \left[-\int_{T/2}^{0} f(-s) \sin \frac{2\pi k(-s)}{T} ds + \int_{0}^{T/2} f(t) \sin \frac{2\pi kt}{T} dt \right] \qquad (\text{let } t = -s)$$

$$= \frac{2}{T} \left[\int_{0}^{T/2} f(s) \sin \frac{2\pi k(-s)}{T} ds + \int_{0}^{T/2} f(t) \sin \frac{2\pi kt}{T} dt \right] \qquad (\because f(-s) = f(s))$$

$$= \frac{2}{T} \left[-\int_{0}^{T/2} f(s) \sin \frac{2\pi ks}{T} ds + \int_{0}^{T/2} f(t) \sin \frac{2\pi kt}{T} dt \right] \qquad (\because \sin(-\theta) = -\sin\theta)$$

$$= 0.$$

∎

Theorem 3.12 If $f(t)$ is an odd function satisfying the conditions of Dirichlet s theorem, the coef cients in the Fourier series of $f(t)$ are given by the formulas

$$A_k = 0, \quad k = 0, 1, 2, \ldots$$

(3.4)

$$B_k = \frac{4}{T} \int_{0}^{T/2} f(t) \sin \frac{2k\pi t}{T} dt, \quad k = 1, 2, \ldots$$

Proof: (Similar to the proof for Theorem 3.11.)

Theorem 3.13 If $f(t)$ is an arbitrary function de n ed over an interval which is symmetric with respect to the origin, it can always be written as the sum of an even function and an odd function.

Proof: By de n ing

(3.5)

$$g_{\text{even}}(t) = \frac{f(t) + f(-t)}{2}, \quad g_{\text{odd}}(t) = \frac{f(t) - f(-t)}{2},$$

we have

$$g_{\text{even}}(t) = g_{\text{even}}(-t), \quad g_{\text{odd}}(t) = -g_{\text{odd}}(-t),$$

and

$$f(t) = g_{\text{even}}(t) + g_{\text{odd}}(t)$$

as desired. ∎

3.4 Half-Range Expansions

Given a time-limited function $f(t)$ de ned over the interval $[0, \widetilde{T}]$, we may construct either

$$g_{\text{even}}(t) = \begin{cases} f(t), & t \in [0, T/2] \\ f(-t), & t \in [-T/2, 0) \end{cases} , \text{ where } T/2 = \widetilde{T},$$

or

$$g_{\text{odd}}(t) = \begin{cases} f(t), & t \in [0, T/2] \\ -f(-t), & t \in [-T/2, 0) \end{cases} , \text{ where } T/2 = \widetilde{T}.$$

Noting that for $t \in [0, T/2] = [0, \widetilde{T}]$,

$$f(t) = g_{\text{even}}(t) = g_{\text{odd}}(t),$$

we may use *either* the cosine series of the even function $g_{\text{even}}(t)$ *or* the sine series of the odd function $g_{\text{odd}}(t)$ to represent the half-range time-limited function $f(t)$ over $[0, \widetilde{T}]$.

Observe that there are (in n itely many) other choices of $g(t)$ which satisfy Dirichlet conditions and agree with $f(t)$ over $[0, \widetilde{T}]$ e.g ., one may simply extend the de n ition of $f(t)$ over the entire interval $[-\widetilde{T}, \widetilde{T}]$. However, when $g(t)$ does not possess the even/odd symmetry properties, the labor of expanding $g(t)$ is doubled because its Fourier series contains both cosine and sine terms.

Example 3.14 Given a time-limited function $f(t) = t - t^2$ for $t \in (0, 1)$,

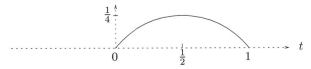

obtain the following three Fourier series expansions of $f(t)$ by treating it as part of an even function, an odd function, and a general function.

$$f(t) = t - t^2 = \begin{cases} \dfrac{1}{6} - \dfrac{1}{\pi^2} \displaystyle\sum_{k=1}^{\infty} \dfrac{\cos 2k\pi t}{k^2}; \\[3mm] \dfrac{8}{\pi^3} \displaystyle\sum_{k=1}^{\infty} \dfrac{\sin(2k-1)\pi t}{(2k-1)^3}; \\[3mm] -\dfrac{1}{3} - \dfrac{4}{\pi^2} \displaystyle\sum_{k=1}^{\infty} (-1)^k \dfrac{\cos k\pi t}{k^2} - \dfrac{2}{\pi} \displaystyle\sum_{k=1}^{\infty} (-1)^k \dfrac{\sin k\pi t}{k}. \end{cases}$$

To obtain the expansion containing only cosine terms, we interpret $f(t)$ for $0 < t < 1$ as part of an *even* function and express $f(t)$ as

$$f(t) = t - t^2 = \frac{A_0}{2} + \sum_{k=1}^{\infty} A_k \cos \frac{2\pi kt}{T} = \frac{A_0}{2} + \sum_{k=1}^{\infty} A_k \cos \pi kt, \quad (\because T/2 = 1)$$

with coef cien ts given by Theorem 3.11:

$$A_k = \frac{4}{T} \int_0^{T/2} (t - t^2) \cos \frac{2\pi kt}{T} \, dt = 2 \int_0^1 (t - t^2) \cos \pi kt \, dt = -\frac{2(1 + \cos k\pi)}{k^2 \pi^2}, \quad k \geq 1;$$

$$A_0 = \frac{4}{T} \int_0^{T/2} (t - t^2) \cos 0 \, dt = 2 \int_0^1 (t - t^2) \, dt = \frac{1}{3}.$$

Because $\cos k\pi = (-1)^k$, the coef cients $A_k = 0$ if k is odd, and $A_k = -\frac{4}{k^2\pi^2}$ if k is even. We thus have

$$f(t) = t - t^2 = \frac{1}{6} - \frac{4}{\pi^2} \left(\frac{\cos 2\pi t}{4} + \frac{\cos 4\pi t}{16} + \frac{\cos 6\pi t}{36} + \cdots + \frac{\cos 2k\pi t}{4k^2} + \cdots \right)$$

$$= \frac{1}{6} - \frac{1}{\pi^2} \sum_{k=1}^{\infty} \frac{\cos 2k\pi t}{k^2}.$$

To obtain the expansion containing only sine terms, we interpret $f(t)$ for $0 < t < 1$ as part of an *odd* function and express $f(t)$ as

$$f(t) = t - t^2 = \sum_{k=1}^{\infty} B_k \sin \frac{2\pi kt}{T} = \sum_{k=1}^{\infty} B_k \sin \pi kt, \quad (\because T/2 = 1)$$

with coef cien ts given by Theorem 3.12:

$$B_k = \frac{4}{T} \int_0^{T/2} (t - t^2) \sin \frac{2\pi kt}{T} \, dt = 2 \int_0^1 (t - t^2) \sin \pi kt \, dt = \frac{4(1 - \cos k\pi)}{k^3 \pi^3}, \quad k \geq 1.$$

Because $\cos k\pi = (-1)^k$, the coef cients $B_k = 0$ if k is even, and $B_k = \frac{8}{k^3\pi^3}$ if k is odd. We thus have

$$f(t) = t - t^2 = \frac{8}{\pi^3} \left(\frac{\sin \pi t}{1} + \frac{\sin 3\pi t}{27} + \frac{\sin 5\pi t}{125} + \cdots + \frac{\sin(2k - 1)\pi t}{(2k - 1)^3} + \cdots \right)$$

$$= \frac{8}{\pi^3} \sum_{k=1}^{\infty} \frac{\sin(2k - 1)\pi t}{(2k - 1)^3}.$$

To obtain the third expansion, we use the de n ition $f(t) = t - t^2$ for $-1 < t < 1$, and apply Dirichlet s theorem to obtain the coef cients in

$$f(t) = t - t^2 = \frac{A_0}{2} + \sum_{k=1}^{\infty} A_k \cos \frac{2\pi kt}{T} + B_k \sin \frac{2\pi kt}{T}, \quad \text{where } T/2 = 1 \text{ as before.}$$

Using Formulas (3.1) and (3.2), we obtain

$$A_k = \frac{2}{T} \int_{-T/2}^{T/2} (t - t^2) \cos \frac{2\pi kt}{T} \, dt = \int_{-1}^1 (t - t^2) \cos \pi kt \, dt = -\frac{4 \cos k\pi}{k^2 \pi^2}, \quad k \geq 1;$$

$$A_0 = \frac{2}{T} \int_{-T/2}^{T/2} (t - t^2) \cos 0 \, dt = \int_{-1}^1 (t - t^2) \, dt = -\frac{2}{3};$$

$$B_k = \frac{2}{T} \int_{-T/2}^{T/2} (t - t^2) \sin \frac{2\pi kt}{T} \, dt = \int_{-1}^1 (t - t^2) \sin \pi kt \, dt = -\frac{2 \cos k\pi}{k\pi}, \quad k \geq 1.$$

Noting that $\cos k\pi = (-1)^k$, we obtain

$$
\begin{aligned}
f(t) = t - t^2 = -\frac{1}{3} - \frac{4}{\pi^2} &\left(-\frac{\cos \pi t}{1} + \frac{\cos 2\pi t}{4} - \frac{\cos 3\pi t}{9} + \cdots \right) \\
- \frac{2}{\pi} &\left(-\frac{\sin \pi t}{1} + \frac{\sin 2\pi t}{2} - \frac{\sin 3\pi t}{3} + \cdots \right) \\
= -\frac{1}{3} - \frac{4}{\pi^2} &\sum_{k=1}^{\infty} (-1)^k \frac{\cos k\pi t}{k^2} - \frac{2}{\pi} \sum_{k=1}^{\infty} (-1)^k \frac{\sin k\pi t}{k}.
\end{aligned}
$$

We remark that although the three series represent the same function $f(t) = t - t^2$ for $0 < t < 1$, they converge to $f(t)$ at different rates (see Figures 3.3, 3.4, 3.5, 3.6, 3.7), which we will investigate further when we study the convergence of Fourier series in Section 3.10 of this chapter.

Figure 3.3 The converging Fourier series of an even function.

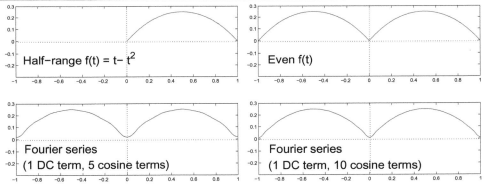

Figure 3.4 The converging Fourier series of an odd function.

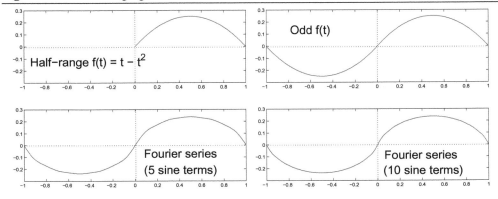

Figure 3.5 De n ing $f(t) = t - t^2$ for the full range: $-1 \le t \le 1$.

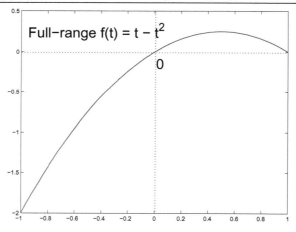

Figure 3.6 The converging Fourier series of $f(t)$ with jump discontinuities.

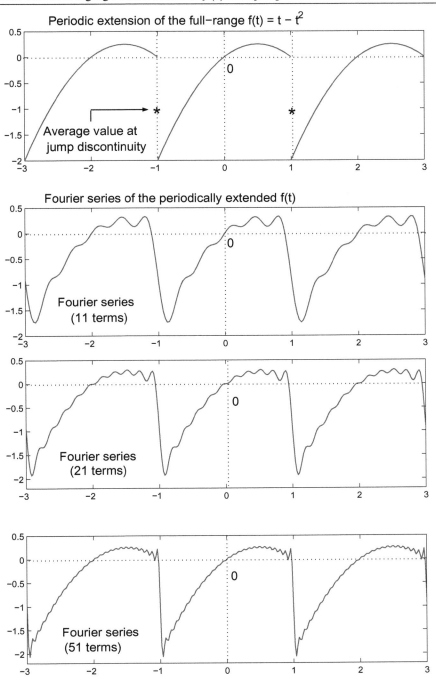

Figure 3.7 The converging Fourier series of $f(t)$ with jump discontinuities.

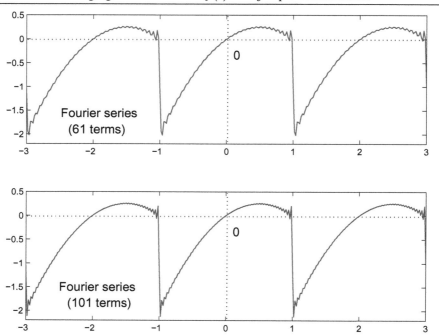

Example 3.15 Using the Fourier series expansions from Example 3.14 to establish the following numerical results:

$$(3.6) \qquad \sum_{k=1}^{\infty} \frac{1}{k^2} = 1 + \frac{1}{2^2} + \frac{1}{3^2} + \frac{1}{4^2} + \frac{1}{5^2} + \cdots = \frac{\pi^2}{6}.$$

$$(3.7) \qquad \sum_{k=1}^{\infty} \frac{(-1)^{k+1}}{k^2} = 1 - \frac{1}{2^2} + \frac{1}{3^2} - \frac{1}{4^2} + \frac{1}{5^2} + \cdots = \frac{\pi^2}{12}.$$

$$(3.8) \qquad \sum_{k=1}^{\infty} \frac{1}{(2k-1)^2} = 1 + \frac{1}{3^2} + \frac{1}{5^2} + \frac{1}{7^2} + \frac{1}{9^2} + \cdots = \frac{\pi^2}{8}.$$

$$(3.9) \qquad \sum_{k=1}^{\infty} \frac{1}{(2k)^2} = \frac{1}{2^2} + \frac{1}{4^2} + \frac{1}{6^2} + \frac{1}{8^2} + \cdots\cdots = \frac{\pi^2}{24}$$

$$(3.10) \qquad \sum_{k=1}^{\infty} \frac{(-1)^{k-1}}{(2k-1)^3} = 1 - \frac{1}{3^3} + \frac{1}{5^3} - \frac{1}{7^3} + \frac{1}{9^3} + \cdots = \frac{\pi^3}{32}.$$

In Example 3.14 we have shown that for the even function

$$g_1(t) = \begin{cases} t - t^2, & 0 < t < 1 \\ -t - t^2, & -1 < t \le 0 \end{cases},$$

the Fourier series expansion is given by

$$g_1(t) = \frac{1}{6} - \frac{1}{\pi^2} \sum_{k=1}^{\infty} \frac{\cos 2k\pi t}{k^2}.$$

Since the even function $g_1(t)$ is continuous at $t = 0$, the Fourier series converges to $g_1(0) = -0 - 0^2 = 0$, and we have

$$0 = g_1(0) = \frac{1}{6} - \frac{1}{\pi^2} \sum_{k=1}^{\infty} \frac{1}{k^2}. \quad (\because \text{ when } t = 0, \cos 2k\pi t = 1 \text{ for every } k)$$

It follows that $\sum_{k=1}^{\infty} \frac{1}{k^2} = \frac{\pi^2}{6}$, and we obtain the desired identity (3.6).

To obtain identity (3.7), we evaluate both $g_1(t)$ and its Fourier series expansion at $t = 0.5$, where $g_1(t)$ is continuous, and we now have

$$\frac{1}{4} = g_1(0.5) = \frac{1}{6} - \frac{1}{\pi^2} \sum_{k=1}^{\infty} \frac{\cos k\pi}{k^2} = \frac{1}{6} + \frac{1}{\pi^2} \sum_{k=1}^{\infty} \frac{(-1)^{k+1}}{k^2}.$$

It follows that $\sum_{k=1}^{\infty} \frac{(-1)^{k+1}}{k^2} = \frac{\pi^2}{4} - \frac{\pi^2}{6} = \frac{\pi^2}{12}$.

We obtain the next identity (3.8) by summing the first two results, (3.6) and (3.7), yielding

$$2 \left[1 + \frac{1}{3^2} + \frac{1}{5^2} + \frac{1}{7^2} + \frac{1}{9^2} + \cdots \right] = \frac{\pi^2}{6} + \frac{\pi^2}{12} = \frac{\pi^2}{4},$$

which gives us (3.8). Similarly, we can obtain identity (3.9) by subtracting (3.7) from (3.6), yielding

$$2 \left[\frac{1}{2^2} + \frac{1}{4^2} + \frac{1}{6^2} + \frac{1}{8^2} + \cdots \right] = \frac{\pi^2}{6} - \frac{\pi^2}{12} = \frac{\pi^2}{12},$$

and the desired result follows immediately.

To show that the last identity (3.10) is true, recall that we have also obtained in Example 3.14 the Fourier series expansion for the odd function

$$g_2(t) = \begin{cases} t - t^2, & 0 < t < 1 \\ t + t^2, & -1 < t \leq 0 \end{cases},$$

and the result was

$$g_2(t) = \frac{8}{\pi^3} \sum_{k=1}^{\infty} \frac{\sin(2k-1)\pi t}{(2k-1)^3}.$$

We again evaluate both $g_2(t)$ and its Fourier series at $t = 0.5$, where $g_2(t)$ is continuous. Noting that at $t = 0.5$, $\sin(2k-1)\pi t = (-1)^{k-1}$ for all $k \geq 1$, we obtain

$$\frac{1}{4} = g_2(0.5) = \frac{8}{\pi^3} \sum_{k=1}^{\infty} \frac{(-1)^{k-1}}{(2k-1)^3},$$

and identity (3.10) follows.

3.5   Fourier Series Using Complex Exponential Modes

In Chapter 1 we show that a Fourier series can also be expressed using complex exponential modes, i.e.,

(3.11)
$$f(t) = \frac{A_0}{2} + \sum_{k=1}^{\infty} A_k \cos \frac{2\pi kt}{T} + B_k \sin \frac{2\pi kt}{T}$$

$$= \sum_{k=-\infty}^{\infty} C_k \, e^{j2\pi kt/T},$$

where
$$C_0 = \frac{A_0}{2}, \quad C_{\pm k} = \frac{A_k \mp jB_k}{2} \quad \text{for } k \geq 1.$$

Using Equations (3.1) and (3.2) from Dirichlet s theorem to evaluate A_k and B_k for $k \geq 1$, we obtain

$$C_{\pm k} = \frac{A_k \mp jB_k}{2}$$

$$= \frac{1}{2} \left[\frac{2}{T} \int_{-T/2}^{T/2} f(t) \left(\cos \frac{2\pi kt}{T} \mp j \sin \frac{2\pi kt}{T} \right) dt \right]$$

$$= \frac{1}{T} \int_{-T/2}^{T/2} f(t) \, e^{\mp j2\pi kt/T} \, dt. \quad (\because e^{\mp j\alpha} = \cos \alpha \mp j \sin \alpha)$$

Note that because

$$C_0 = \frac{A_0}{2} = \frac{1}{2} \left[\frac{2}{T} \int_{-T/2}^{T/2} f(t) \cos 0 \, dt \right] = \frac{1}{T} \int_{-T/2}^{T/2} f(t) \, e^0 \, dt,$$

a single formula expressed as

(3.12)
$$\boxed{C_k = \frac{1}{T} \int_{-T/2}^{T/2} f(t) \, e^{-j2\pi kt/T} \, dt}$$

allows us to obtain C_k for all $k \in (-\infty, \infty)$.

3.6   Complex-Valued Functions

If $u(t) = g(t) + jh(t)$, then the Fourier series of $u(t)$ can be obtained by nding the Fourier series of the real-valued $g(t)$ and the real-valued $h(t)$ separately. That is, if

$$g(t) = \frac{A_0}{2} + \sum_{k=1}^{\infty} A_k \cos \frac{2\pi kt}{T} + B_k \sin \frac{2\pi kt}{T},$$

$$h(t) = \frac{U_0}{2} + \sum_{k=1}^{\infty} U_k \cos \frac{2\pi kt}{T} + V_k \sin \frac{2\pi kt}{T},$$

then we have

$$u(t) = \frac{A_0 + jU_0}{2} + \sum_{k=1}^{\infty} (A_k + jU_k) \cos \frac{2\pi kt}{T} + (B_k + jV_k) \sin \frac{2\pi kt}{T}.$$

3.7 Fourier Series in Other Variables

As indicated in Chapter 1, when the variable of function $f(t)$ with period T is changed from t to $\theta = 2\pi t/T$, we obtain the periodic function $g(\theta) = g(\theta + 2\pi)$ and its Fourier series in terms of $\cos k\theta$ and $\sin k\theta$ or $e^{jk\theta}$:

(3.13)
$$g(\theta) = \frac{A_0}{2} + \sum_{k=1}^{\infty} A_k \cos k\theta + B_k \sin k\theta$$

$$= \sum_{k=-\infty}^{\infty} C_k e^{jk\theta}.$$

Noting that when t varies from $-T/2$ to $T/2$, θ varies from $-\pi$ to π, and that $\theta = 2\pi t/T$ implies $dt = \frac{T}{2\pi}d\theta$ in Equations (3.1) and (3.2), we obtain the Fourier coef cients of $g(\theta)$ from Dirichlet s theorem:

(3.14)
$$A_k = \frac{1}{\pi} \int_{-\pi}^{\pi} g(\theta) \cos k\theta \, d\theta, \quad k = 0, 1, 2, \ldots$$

(3.15)
$$B_k = \frac{1}{\pi} \int_{-\pi}^{\pi} g(\theta) \sin k\theta \, d\theta, \quad k = 1, 2, \ldots.$$

To express C_k, we change the variable in Equation (3.12) to obtain

(3.16)
$$C_k = \frac{1}{2\pi} \int_{-\pi}^{\pi} g(\theta) e^{-jk\theta} \, d\theta, \quad k = \ldots, -1, 0, 1, \ldots$$

3.8 Truncated Fourier Series and Least Squares

Since we can only make use of a nite Fourier series in many applications, it is of practical and theoretical importance to understand how the relationship between a given function $x(t)$ and its Fourier series changes when the latter is truncated after a nite number of terms. To investigate the mathematical connection, we assume that a real-valued function $x(t)$ of period T is approximated by the following trigonometric polynomial of $N = 2n + 1$ terms, i.e.,

$$x(t) \approx \tilde{x}_N(t) = \frac{A_0}{2} + \sum_{k=1}^{n} A_k \cos \frac{2\pi kt}{T} + B_k \sin \frac{2\pi kt}{T}.$$

Using the mathematically equivalent alternate form expressed in variable $\theta = 2\pi t/T$, we may assume that $g(\theta)$ of period 2π is approximated by the same trigonometric polynomial expressed as $\tilde{g}_N(\theta)$:

$$g(\theta) \approx \tilde{g}_N(\theta) = \frac{A_0}{2} + \sum_{k=1}^{n} A_k \cos k\theta + B_k \sin k\theta.$$

It turns out that when the N coef cients are chosen to be the Fourier coef cients de ned according to Dirichlet s Theorem, the discrepancy between the function $g(\theta)$ and the n ite series $\tilde{g}_N(\theta)$ is minimized in the least-squares sense. To prove such direct connection to the least-squares problem, we treat the real coef c ients as unknown variables of the multivariate function (which measures the mean square error of the   t)

$$\Phi(A_0, A_1, B_1, \ldots, A_n, B_n) = \int_{-\pi}^{\pi} [g(\theta) - \tilde{g}_N(\theta)]^2 \, d\theta$$

(3.17)

$$= \int_{-\pi}^{\pi} \left[g(\theta) - \frac{A_0}{2} - \sum_{k=1}^{n} (A_k \cos k\theta + B_k \sin k\theta) \right]^2 \, d\theta.$$

We remark that we have chosen the alternate form in variable θ to simplify the notation some-what. To minimize $\Phi(A_0, A_1, B_1, \ldots, A_n, B_n)$ by standard methods of calculus, we determine A_r and B_r by requiring that the following $N = 2n + 1$ conditions are satis ed:

(3.18) $$\frac{\partial \Phi}{\partial A_r} = 0, \ r = 0, 1, 2, \ldots, n, \ \text{and} \ \frac{\partial \Phi}{\partial B_r} = 0, \ r = 1, 2, \ldots, n.$$

For $r = 1, 2, \ldots, n$, we thus obtain

(3.19) $$\frac{\partial \Phi}{\partial A_r} = -\int_{-\pi}^{\pi} 2 \left[g(\theta) - \tilde{g}_N(\theta) \right] \frac{\partial \tilde{g}_N(\theta)}{\partial A_r} \, d\theta$$

$$= -2 \int_{-\pi}^{\pi} \left[g(\theta) - \frac{A_0}{2} - \sum_{k=1}^{n} (A_k \cos k\theta + B_k \sin k\theta) \right] \cos r\theta \, d\theta = 0,$$

(3.20) $$\frac{\partial \Phi}{\partial B_r} = -\int_{-\pi}^{\pi} 2 \left[g(\theta) - \tilde{g}_N(\theta) \right] \frac{\partial \tilde{g}_N(\theta)}{\partial B_r} \, d\theta$$

$$= -2 \int_{-\pi}^{\pi} \left[g(\theta) - \frac{A_0}{2} - \sum_{k=1}^{n} (A_k \cos k\theta + B_k \sin k\theta) \right] \sin r\theta \, d\theta = 0.$$

To evaluate the integral in (3.19), we make use of the integrated results given by (1.19) and (1.21) on page 15:

$$\int_{-\pi}^{\pi} \cos k\theta \cos r\theta \, d\theta = \begin{cases} 0, & \text{if } k \neq r \\ \pi, & \text{if } k = r \neq 0 \ , \\ 2\pi, & \text{if } k = r = 0 \end{cases} \quad \text{and} \int_{-\pi}^{\pi} \sin k\theta \cos r\theta \, d\theta = 0,$$

and we obtain

(3.21) $$\int_{-\pi}^{\pi} g(\theta) \cos r\theta \, d\theta = A_r \int_{-\pi}^{\pi} \cos^2 r\theta \, d\theta = \pi A_r,$$

Solving for A_r in (3.21), we have

(3.22) $$\boxed{A_r = \frac{1}{\pi} \int_{-\pi}^{\pi} g(\theta) \cos r\theta \, d\theta, \ r = 1, 2, \ldots, n.}$$

Similarly, we evaluate the integral in (3.20) using the identities (which were given by (1.20) and (1.21) on page 15)

$$\int_{-\pi}^{\pi} \sin k\theta \sin r\theta \, d\theta = \begin{cases} 0, & \text{if } k \neq r \\ \pi, & \text{if } k = r \neq 0 \ , \\ 0, & \text{if } k = r = 0 \end{cases} \quad \text{and} \int_{-\pi}^{\pi} \cos k\theta \sin r\theta \, d\theta = 0,$$

and we obtain

$$(3.23) \qquad \int_{-\pi}^{\pi} g(\theta) \sin r\theta \, d\theta = B_r \int_{-\pi}^{\pi} \sin^2 r\theta \, d\theta = \pi B_r.$$

Solving for B_r in (3.23), we have

$$(3.24) \qquad \boxed{B_r = \frac{1}{\pi} \int_{-\pi}^{\pi} g(\theta) \sin r\theta \, d\theta, \quad r = 1, 2, \ldots, n.}$$

For $r = 0$, we require

$$\frac{\partial \Phi}{\partial A_0} = -\int_{-\pi}^{\pi} 2 \left[g(\theta) - \tilde{g}_N(\theta) \right] \frac{\partial \tilde{g}_N(\theta)}{\partial A_0} \, d\theta$$

$$= -2 \int_{-\pi}^{\pi} \left[g(\theta) - \frac{A_0}{2} - \sum_{k=1}^{n} \left(A_k \cos k\theta + B_k \sin k\theta \right) \right] \frac{1}{2} \, d\theta = 0,$$

which is simpli ed to

$$(3.25) \qquad \int_{-\pi}^{\pi} g(\theta) \, d\theta = \frac{A_0}{2} \int_{-\pi}^{\pi} d\theta = \pi A_0.$$

Solving for A_0 in (3.25), we have

$$A_0 = \frac{1}{\pi} \int_{-\pi}^{\pi} g(\theta) \, d\theta = \frac{1}{\pi} \int_{-\pi}^{\pi} g(\theta) \cos 0 \, d\theta,$$

which can be obtained directly from Equation (3.22) by allowing $r = 0$ there.

Since Formulas (3.22) and (3.24) are mathematically equivalent to Formulas (3.1) and (3.2) in Dirichlet s theorem, we have re-derived the Fourier coef cients by nding them as the solution to a least-squares problem.

3.9   Orthogonal Projections and Fourier Series

Orthogonal projections are of fundamental importance in the theory of Fourier series and in many other branches of mathematics. In this section we shall review relevant concepts and results from linear algebra before we apply them to solve the least-squares problem from the preceding section in a more general setting (without involving calculus). We then proceed to study the convergence of Fourier series as well as the Gibbs phenomenon in the next section.

We begin with the de nition which allows us to address the orthogonality of both real-valued and complex-valued functions. (The terms used in the de n ition are explained in the remarks that follow.)

Definition 3.16 If f and g are elements of a linear space V equipped with an inner product $\langle\, ,\, \rangle$, we say f and g are *orthogonal* elements of V whenever $\langle f, g \rangle = 0$.

Remark 3.17 A class of functions, all having the same domain, is said to form a linear space if (i) the sum of any two member functions of the class is also a member; (ii) every scalar multiple of a member function is also a member.

Given below are examples of linear spaces with each being a class of functions de ned on the real line.

(a) All periodic functions with period T.

(b) All functions with the property $f(1) = 0$.

(c) All functions possessing at most a nite number of discontinuities.

(d) All functions with the property $f(2.5) = f(5)$.

Remark 3.18 The class of all functions de ned and continuous in the interval $a \leq t \leq b$ is a linear space, and it is commonly denoted as $C[a, b]$; the class of all functions possessing continuous derivatives of order n in the interval $a \leq t \leq b$ is also a linear space, and it is denoted as $C^n[a, b]$. (Note that only one-sided derivatives are required at the end points, because the behavior of the function outside the domain $a \leq t \leq b$ does not concern us.)

Remark 3.19 An inner product in a linear space is a function of pairs of elements of the space. That is, for each pair f and g in the linear space there is scalar $\langle f, g \rangle$ satisfying the following axioms:

Axiom 1. $\langle f, f \rangle \geq 0$ for every f, and $\langle f, f \rangle = 0$ if and only if f is the zero element.

Axiom 2. $\langle f, g \rangle = \overline{\langle g, f \rangle}$ for every pair f and g.

Axiom 3. $\langle \alpha f + \beta g, h \rangle = \alpha \langle f, h \rangle + \beta \langle g, h \rangle$.

Axiom 4. $\langle f, \alpha g + \beta h \rangle = \overline{\alpha} \langle f, g \rangle + \overline{\beta} \langle f, h \rangle$.

In general, the functions f and g may be real-valued or complex-valued; hence, the scalar $\langle f, g \rangle$ may be real or complex, so are the scalar constants α and β. As usual, the overbar means complex conjugate, which can be omitted when we are only dealing with elements in real linear spaces. Note that $\langle f, f \rangle$ must be real for all f (real or complex) because $\langle f, f \rangle = \overline{\langle f, f \rangle}$ according to axiom 2. Note also that axiom 4 is true whether it is explicitly given as an axiom, because it is implied by axioms 2 and 3:

$$\begin{aligned}
\langle f, \alpha g + \beta h \rangle &= \overline{\langle \alpha g + \beta h, f \rangle} && \text{(by axiom 2)} \\
&= \overline{\alpha \langle g, f \rangle + \beta \langle h, f \rangle} && \text{(by axiom 3)} \\
&= \overline{\alpha} \, \overline{\langle g, f \rangle} + \overline{\beta} \, \overline{\langle h, f \rangle} && \\
&= \overline{\alpha} \langle f, g \rangle + \overline{\beta} \langle f, h \rangle && \text{(by axiom 2)}
\end{aligned}$$

Remark 3.20 For $f, g \in C[a, b]$, an inner product satisfying the four axioms is

$$(3.26) \qquad \langle f, g \rangle = \int_a^b f(t) \, \overline{g(t)} \, dt.$$

If $g(t) = u(t) + jv(t)$, where $u(t)$ and $v(t)$ are real-valued, then its complex conjugate is $\overline{g(t)} = u(t) - jv(t)$; for real-valued $g(t)$, we simply have $\overline{g(t)} = g(t)$.

Remark 3.21 If the linear space V is equipped with an inner product, then for every $f \in V$, we have $\langle f, f \rangle \geq 0$, and its norm may be de ned by

$$(3.27) \qquad \|f\| = \sqrt{\langle f, f \rangle}.$$

Hence, an inner product linear space is also called a *normed* linear space (i.e., a space with a measure associated with it). Using the inner product given by (3.26), we can directly de ne the norm by expressing

$$(3.28) \qquad \|f\|^2 = \int_a^b f(t)\,\overline{f(t)}\,dt = \int_a^b |f(t)|^2\,dt.$$

Note that $|f(t)|^2 = f^2(t)$ if $f(t)$ is real-valued. We can also use the norm $\|f - g\|$ to measure the difference between two functions.

Remark 3.22 The results given by (1.19), (1.20), and (1.21) on page 15 may be used to show that the following pairs of sinusoidal functions are orthogonal over $[-\pi,\ \pi]$:

$$\langle \cos k\theta,\ \cos r\theta \rangle = \int_{-\pi}^{\pi} \cos k\theta \, \cos r\theta \, d\theta = 0, \quad \text{if } k \neq r, \ \ k \geq 0,\ r \geq 0;$$

$$(3.29) \qquad \langle \sin k\theta,\ \sin r\theta \rangle = \int_{-\pi}^{\pi} \sin k\theta \, \sin r\theta \, d\theta = 0, \quad \text{if } k \neq r, \ \ k \geq 1,\ r \geq 1;$$

$$\langle \cos k\theta,\ \sin r\theta \rangle = \int_{-\pi}^{\pi} \cos k\theta \, \sin r\theta \, d\theta = 0, \quad \text{for all } k \geq 0, \text{ and all } r \geq 1.$$

Remark 3.23 Since $f(\theta) = \cos k\theta$ is an even function and $g(\theta) = \sin k\theta$ is an odd function, by Theorems 3.11 and 3.12 on pages 51 and 52 we have

$$(3.30) \qquad \begin{aligned} \int_{-\pi}^{\pi} \cos k\theta \, \cos r\theta \, d\theta &= 2 \int_0^{\pi} \cos k\theta \, \cos r\theta \, d\theta, \\ \int_{-\pi}^{\pi} \sin k\theta \, \sin r\theta \, d\theta &= 2 \int_0^{\pi} \sin k\theta \, \sin r\theta \, d\theta, \end{aligned}$$

and they allow the inner products $\langle \cos k\theta,\ \cos r\theta \rangle$ and $\langle \sin k\theta,\ \sin r\theta \rangle$ de ned over the interval $[0,\ \pi]$ to share the results from (3.29). That is,

$$(3.31) \qquad \begin{aligned} \langle \cos k\theta,\ \cos r\theta \rangle &\stackrel{\text{def}}{=} \int_0^{\pi} \cos k\theta \, \cos r\theta \, d\theta = 0, \quad \text{if } k \neq r, \ \ k \geq 0,\ r \geq 0; \\ \langle \sin k\theta,\ \sin r\theta \rangle &\stackrel{\text{def}}{=} \int_0^{\pi} \sin k\theta \, \sin r\theta \, d\theta = 0, \quad \text{if } k \neq r, \ \ k \geq 1,\ r \geq 1. \end{aligned}$$

Hence, the pairs of sinusoidal functions in (3.31) are also orthogonal over $[0,\ \pi]$.

Theorem 3.24 (Pythagorean theorem) If $\langle f,\ g \rangle = 0$, then $\|f + g\|^2 = \|f\|^2 + \|g\|^2$.

Proof:
$$\begin{aligned} \|f + g\|^2 &= \langle f + g,\ f + g \rangle & \\ &= \langle f,\ f + g \rangle + \langle g,\ f + g \rangle & \text{(by axiom 3)} \\ &= \langle f,\ f \rangle + \langle f,\ g \rangle + \langle g,\ f \rangle + \langle g,\ g \rangle & \text{(by axiom 4)} \\ &= \|f\|^2 + \langle f,\ g \rangle + \overline{\langle f,\ g \rangle} + \|g\|^2 & \text{(by axiom 2)} \\ &= \|f\|^2 + \|g\|^2. & (\because \langle f,\ g \rangle = 0) \end{aligned}$$

■

Theorem 3.25 (Parallelogram theorem) $\|f + g\|^2 + \|f - g\|^2 = 2\|f\|^2 + 2\|g\|^2$.

Proof: Since we have already obtained the expression

$$\|f + g\|^2 = \|f\|^2 + \langle f, g \rangle + \overline{\langle f, g \rangle} + \|g\|^2$$

in the proof of Theorem 3.24, we have, by similar steps,

$$\|f - g\|^2 = \|f\|^2 - \langle f, g \rangle - \overline{\langle f, g \rangle} + \|g\|^2$$

and the result follows.   ∎

Theorem 3.26 The inner product $\langle f, g \rangle$ can be written entirely in terms of the norm in the following form:

(3.32)
$$\langle f, g \rangle = \tfrac{1}{4} \sum_{k=1}^{4} j^k \|f + j^k g\|^2, \text{ where } j = \sqrt{-1},$$
$$= \tfrac{1}{4} \left[\|f + g\|^2 - \|f - g\|^2 + j\|f + jg\|^2 - j\|f - jg\|^2 \right].$$

Proof: We prove this result by reducing the right side to $\langle f, g \rangle$. Note that we have already obtained the inner product expressions for the rst two terms in the proofs of Theorems 3.24 and 3.25, which give us the partial result:

(3.33)
$$\|f + g\|^2 - \|f - g\|^2 = 2\langle f, g \rangle + 2\langle g, f \rangle.$$

We proceed to nd the inner product expression for the third term:

$$\begin{aligned}
j\|f + jg\|^2 &= j\langle f + jg, f + jg \rangle \\
&= j\left[\langle f, f \rangle + \langle f, jg \rangle + \langle jg, f \rangle + \langle jg, jg \rangle \right] \\
&= j\left[\langle f, f \rangle + \overline{j}\langle f, g \rangle + j\langle g, f \rangle + j\overline{j}\langle g, g \rangle \right] \\
&= j\langle f, f \rangle + \langle f, g \rangle - \langle g, f \rangle + j\langle g, g \rangle.
\end{aligned}$$

By similar steps we obtain the inner product expression for the fourth term:

$$j\|f - jg\|^2 = j\langle f, f \rangle - \langle f, g \rangle + \langle g, f \rangle + j\langle g, jg \rangle].$$

The partial result involving the last two terms is

(3.34)
$$j\|f + jg\|^2 - j\|f - jg\|^2 = 2\langle f, g \rangle - 2\langle g, f \rangle.$$

We then obtain, on summing (3.33) and (3.34) as well as including the factor $\tfrac{1}{4}$ on both sides of the equation, the desired result.   ∎

Definition 3.27 A sequence $\phi_1, \phi_2, \phi_3, \ldots$ of elements of a normed linear space V is said to be *orthogonal* if $\langle \phi_k, \phi_r \rangle = 0$ whenever $k \neq r$ and $\|\phi_k\| \neq 0$ for every k. A sequence is said to be *orthonormal* if it is orthogonal and $\|\phi_k\| = 1$ for every k.

Remark 3.28 This de n ition applies to both *finite* or *infinite* sequences.

Remark 3.29 The inner product results from (3.29) show that each of the following sequences is orthogonal over $[-\pi, \pi]$:

(a) $1, \cos\theta, \cos 2\theta, \ldots, \cos n\theta, \ldots$

(b) $\sin\theta, \sin 2\theta, \ldots, \sin n\theta, \ldots$

(c) $1, \cos\theta, \sin\theta, \ldots, \cos n\theta, \sin n\theta, \ldots$.

Theorem 3.30 (Parseval s theorem) If $f(\theta)$ is a real-valued periodic function represented by

$$f(\theta) = \frac{A_0}{2} + \sum_{k=1}^{\infty} A_k \cos k\theta + B_k \sin k\theta = \sum_{k=-\infty}^{\infty} C_k e^{jk\theta},$$

then the power content of $f(\theta)$ in the period 2π is de n ed as the mean-square value

$$(3.35) \qquad \frac{1}{2\pi} \int_{-\pi}^{\pi} f^2(\theta)\, d\theta = \frac{A_0^2}{4} + \frac{1}{2} \sum_{k=1}^{\infty} A_k^2 + B_k^2 = \sum_{k=-\infty}^{\infty} |C_k|^2.$$

Proof: Using the inner product properties given by (3.29) and the integrated results given by (1.19) and (1.20) on page 15, we obtain

$$\frac{1}{2\pi} \int_{-\pi}^{\pi} f^2(\theta)\, d\theta = \frac{1}{2\pi} \int_{-\pi}^{\pi} \left[\frac{A_0}{2} + \sum_{k=1}^{\infty} A_k \cos k\theta + B_k \sin k\theta \right]^2 d\theta$$

$$= \frac{A_0^2}{8\pi} \int_{-\pi}^{\pi} d\theta + \frac{1}{2\pi} \sum_{k=1}^{\infty} \left[A_k^2 \int_{-\pi}^{\pi} \cos^2 k\theta\, d\theta + B_k^2 \int_{-\pi}^{\pi} \sin^2 k\theta\, d\theta \right]$$

$$= \frac{A_0^2}{4} + \frac{1}{2} \sum_{k=1}^{\infty} A_k^2 + B_k^2 \qquad \left(\text{by (1.19) and (1.20)} \right)$$

$$= \sum_{k=-\infty}^{\infty} |C_k|^2. \qquad \left(\because C_0 = \frac{A_0}{2}, \quad |C_{-k}|^2 + |C_k|^2 = \frac{A_k^2 + B_k^2}{2} \right)$$

$\blacksquare$

Remark 3.31 The inner product results from (3.31) show that each of the following sequences is orthogonal over $[0, \pi]$:

(a) $1, \cos\theta, \cos 2\theta, \ldots, \cos n\theta, \ldots$

(b) $\sin\theta, \sin 2\theta, \ldots, \sin n\theta, \ldots$.

Remark 3.32 From identities (1.19) and (1.20) on page 15, we have

$$(3.36) \qquad \| \cos k\theta \|^2 = \langle \cos k\theta, \cos k\theta \rangle = \int_{-\pi}^{\pi} \cos^2 k\theta\, d\theta = \begin{cases} 2\pi & \text{if } k = 0 \\ \pi & \text{if } k > 0 \end{cases};$$

$$\| \sin k\theta \|^2 = \langle \sin k\theta, \sin k\theta \rangle = \int_{-\pi}^{\pi} \sin^2 k\theta\, d\theta = \pi, \quad \text{if } k \neq 0.$$

Using the computed norms to scale the corresponding elements in each orthogonal sequence, we obtain the orthonormal sequences de ned on the interval $[-\pi, \pi]$:

(a) $\frac{1}{\sqrt{2\pi}}, \frac{1}{\sqrt{\pi}} \cos\theta, \frac{1}{\sqrt{\pi}} \cos 2\theta, \ldots, \frac{1}{\sqrt{\pi}} \cos n\theta, \ldots$

(b) $\frac{1}{\sqrt{\pi}}\sin\theta$, $\frac{1}{\sqrt{\pi}}\sin 2\theta$, $\ldots$, $\frac{1}{\sqrt{\pi}}\sin n\theta$, $\ldots$

(c) $\frac{1}{\sqrt{2\pi}}$, $\frac{1}{\sqrt{\pi}}\cos\theta$, $\frac{1}{\sqrt{\pi}}\sin\theta$, $\ldots$, $\frac{1}{\sqrt{\pi}}\cos n\theta$, $\frac{1}{\sqrt{\pi}}\sin n\theta$, $\ldots$.

Example 3.33 Show that the complex trigonometric sequence de ned by the periodic function $\phi_k = \frac{1}{\sqrt{2\pi}}e^{jk\theta}$ for $k \in (-\infty, \infty)$ is an orthonormal sequence.

Proof: We apply the inner product integral (3.26) over the interval $[a, b] = [-\pi, \pi]$, which corresponds to the longest period of ϕ_1, because $e^{jk(\theta+2\pi/k)} = e^{jk\theta}$ for every k.

$$\text{For } k \neq \ell, \quad \langle \phi_k, \phi_\ell \rangle = \frac{1}{2\pi}\int_{-\pi}^{\pi} e^{jk\theta}e^{-j\ell\theta}d\theta = \left.\frac{e^{j(k-\ell)\theta}}{j2(k-\ell)\pi}\right|_{-\pi}^{\pi} = \frac{\sin(k-\ell)\pi}{(k-\ell)\pi} = 0.$$

$$\text{For } k = \ell, \text{ we have } \|\phi_\ell\|^2 = \langle \phi_\ell, \phi_\ell \rangle = \frac{1}{2\pi}\int_{-\pi}^{\pi} e^{j\ell\theta}e^{-j\ell\theta}d\theta = \frac{1}{2\pi}\int_{-\pi}^{\pi} d\theta = 1.$$

∎

3.9.1 The Cauchy–Schwarz inequality

Since the version of Cauchy—Schwarz inequality used in the real vector spaces involves only real-valued dot products, we should point out that complications do arise when we must now work with complex inner product spaces, because $\langle f, g \rangle$ need not be a real number. This fact is re ected by the different requirements in the proof as well as the necessity of using the modulus (absolute value) of $\langle f, g \rangle$ in the left side of the inequality given by Theorem 3.34. Further complication occurs when we use the inner product de ned by (3.26) on piecewise continuous functions, because it becomes necessary to relax the de nition of the inner product (and hence the norm) to permit $\langle g, g \rangle = 0$ (and hence $\|g\| = 0$), even though g is not identically zero. We then have what is called a *pseudo inner product*. For example, if $g(t) = 0$ everywhere in $[a, b]$ except at one point $t = t_a$ and we let $g(t_a) = 1$, then we have $\langle g, g \rangle = \int_a^b |g(t)|^2 = 0$, but $g(t)$ is not identically zero. We therefore present two proofs for Theorem 3.34: the rst proof is simpler but it is not valid in its present form if the part requiring $g \neq$ zero function $\Longrightarrow$ $\langle g, g \rangle \neq 0$ i s dropped from axiom (1); the second proof requires more work but it would still be valid for pseudo inner products.

Theorem 3.34 The Cauchy—Schwarz inequality for every pair f and g from a complex inner product space V is given by

(3.37) $|\langle f, g \rangle|^2 \leq \|f\|^2 \|g\|^2$.

The First Proof: If both f and g are identically zero, then the equality holds because $\langle f, g \rangle = 0$, $\|f\|^2 = 0$, and $\|g\|^2 = 0$. We therefore assume that one element of the pair is not identically zero, and we assume $g \neq$ zero function without loss of generality. According to axiom (1), $\langle g, g \rangle^2 = \|g\|^2 \neq 0$, and we have, for every scalar λ,

$$0 \leq \langle f - \lambda g, f - \lambda g \rangle = \langle f, f - \lambda g \rangle - \lambda \langle g, f - \lambda g \rangle \qquad \text{(by axiom 3)}$$

(3.38)
$$= \langle f, f \rangle - \overline{\lambda}\langle f, g \rangle - \lambda\langle g, f \rangle + \lambda\overline{\lambda}\langle g, g \rangle \quad \text{(by axiom 4)}$$

$$= \|f\|^2 - \overline{\lambda}\langle f, g \rangle - \lambda\overline{\langle f, g \rangle} + \lambda\overline{\lambda}\|g\|^2. \quad \text{(by axiom 2)}$$

The last two terms on the right side may cancel each other if we set

$$\overline{\lambda} = \frac{\overline{\langle f,\, g\rangle}}{\|g\|^2} \quad \text{so that} \quad \lambda\overline{\lambda}\|g\|^2 = \lambda\overline{\langle f,\, g\rangle}.$$

The inequality from (3.38) is thus simpli ed to

$$0 \le \|f\|^2 - \overline{\lambda}\langle f,\, g\rangle = \|f\|^2 - \frac{\overline{\langle f,\, g\rangle}\langle f,\, g\rangle}{\|g\|^2} = \|f\|^2 - \frac{|\langle f,\, g\rangle|^2}{\|g\|^2},$$

or equivalently,

$$0 \le \|f\|^2\|g\|^2 - |\langle f,\, g\rangle|^2,$$

which yields the Cauchy—Schwrz inequality.   ∎

The Second Proof: In this proof we consider two cases based on the value of the inner product $\langle f,\, g\rangle$. If $\langle f,\, g\rangle = 0$, the Cauchy—Schwrz inequality is valid, since the right side of (3.37) cannot be negative. We consider next the case when $\langle f,\, g\rangle \ne 0$. By taking the same initial steps from the r st proof, we again arrive at the inequality

(3.39) $$0 \le \langle f - \lambda g,\, f - \lambda g\rangle = \|f\|^2 - \overline{\lambda}\langle f,\, g\rangle - \lambda\overline{\langle f,\, g\rangle} + \lambda\overline{\lambda}\|g\|^2.$$

In order *not* to make any assumption about the value of $\|g\|^2 = \langle g, g\rangle$, we avoid using any expression involving $\|g\|$ in the denominator. Suppose we now continue by letting $\lambda = \gamma\langle f,\, g\rangle$, where γ is an arbitrary real number so that $\overline{\lambda} = \overline{\gamma}\,\overline{\langle f,\, g\rangle} = \gamma\,\overline{\langle f,\, g\rangle}$, we shall obtain

(3.40)
$$\begin{aligned}
0 &\le \|f\|^2 - \overline{\lambda}\langle f,\, g\rangle - \lambda\overline{\langle f,\, g\rangle} + \lambda\overline{\lambda}\|g\|^2 \\
&= \|f\|^2 - \gamma\,\overline{\langle f,\, g\rangle}\langle f,\, g\rangle - \gamma\,\langle f,\, g\rangle\overline{\langle f,\, g\rangle} + \gamma^2\langle f,\, g\rangle\overline{\langle f,\, g\rangle}\|g\|^2 \\
&= \|f\|^2 - 2\gamma\,|\langle f,\, g\rangle|^2 + \gamma^2|\langle f,\, g\rangle|^2\|g\|^2.
\end{aligned}$$

By expressing the right side as a quadratic expression in the arbitrary real number γ, we obtain

(3.41) $$a\gamma^2 + b\gamma + c \ge 0 \quad \text{for every } \gamma.$$

Because the quadratic formula has real coef cients , namely, $a = |\langle f,\, g\rangle|^2\|g\|^2$, $b = -2|\langle f,\, g\rangle|^2$, and $c = \|f\|^2$, and the formula represents a nonnegative number for all values of the real variable γ, we conclude that the quadratic equation $a\gamma^2 + b\gamma + c = 0$ cannot have two distinct real roots γ_1 and γ_2, since if it did there would be values of γ for which $a\gamma^2 + b\gamma + c < 0$, which is a contradiction. Therefore, the discriminant $b^2 - 4ac$ cannot be positive (otherwise we have two distinct real roots), and we must have

$$b^2 - 4ac = 4|\langle f,\, g\rangle|^4 - 4|\langle f,\, g\rangle|^2\|g\|^2\|f\|^2 \le 0,$$

which yields

$$4|\langle f,\, g\rangle|^4 \le 4|\langle f,\, g\rangle|^2\|g\|^2\|f\|^2.$$

Since $\langle f,\, g\rangle \ne 0$, we can divide both sides by $4|\langle f,\, g\rangle|^2$ to obtain

$$|\langle f,\, g\rangle|^2 \le \|f\|^2\|g\|^2,$$

which completes the proof.   ∎

Example 3.35 Use the Cauchy—Schwarz inequality to show that

$$(3.42) \qquad \left[\int_a^b |h(t)| \, dt \right]^2 \leq (b-a) \int_a^b |h(t)|^2 dt.$$

Proof: To derive this inequality, we apply Theorem 3.34 with the inner product de ned by (3.26) on page 64 to the pair $f(t) = |h(t)|$ and $g(t) = 1$, and we obtain

$$|\langle f, g \rangle|^2 = \left| \int_a^b f(t) \, \overline{g(t)} \, dt \right|^2 = \left| \int_a^b |h(t)| \, dt \right|^2 = \left[\int_a^b |h(t)| \, dt \right]^2,$$

$$\|f\|^2 \|g\|^2 = \left[\int_a^b |h(t)| \, \overline{|h(t)|} \, dt \right] \left[\int_a^b dt \right] = (b-a) \int_a^b |h(t)|^2 dt.$$

The desired result follows because $|\langle f, g \rangle|^2 \leq \|f\|^2 \|g\|^2$. ∎

Example 3.36 Use the result in Examples 3.35 to show that

$$(3.43) \qquad \left| \int_a^b g(t) \, dt - \int_a^b \nu(t) \, dt \right| \leq \sqrt{b-a} \, \|g(t) - \nu(t)\|.$$

Proof: Since the result in Example 3.35 can also be expressed as

$$(3.44) \qquad \int_a^b |h(t)| \, dt \leq \sqrt{b-a} \, \|h(t)\|,$$

by letting $h(t) = g(t) - \nu(t)$ we immediately have

$$\int_a^b |g(t) - \nu(t)| \, dt \leq \sqrt{b-a} \, \|g(t) - \nu(t)\|.$$

The desired inequality follows because

$$\left| \int_a^b g(t) \, dt - \int_a^b \nu(t) \, dt \right| = \left| \int_a^b \big(g(t) - \nu(t) \big) \, dt \right| \leq \int_a^b |g(t) - \nu(t)| \, dt.$$

∎

Example 3.37 Use the Cauchy—Schwarz inequality to show that

$$(3.45) \qquad \int_0^{\pi/2} \sin\theta \cos^{2n}\theta \, d\theta \leq \sqrt{\int_0^{\pi/2} \sin^2\theta \cos^{2n}\theta \, d\theta \int_0^{\pi/2} \cos^{2n}\theta \, d\theta}.$$

Proof: By applying the inner product (3.26) to real functions $f(\theta)$ and $g(\theta)$, the inequality given by Theorem 3.34 may be directly expressed as

$$(3.46) \qquad \left| \int_a^b f(\theta) g(\theta) \, d\theta \right| \leq \sqrt{\int_a^b f^2(\theta) \, d\theta \int_a^b g^2(\theta) \, d\theta}.$$

The desired result is obtained if we let $f(\theta) = \sin\theta \cos^n\theta$ and $g(\theta) = \cos^n\theta$. ∎

3.9.2 The Minkowski inequality

Theorem 3.38 The Minkowski inequality for every pair f and g from a complex inner product space V is given by

(3.47) $$\|f + g\| \leq \|f\| + \|g\|.$$

Proof: Since the inner product $\langle f, g \rangle$ is assumed to be a complex scalar $z = a + jb$, we shall make use of the property $z + \bar{z} = 2a \leq 2|z|$ in the proof. In addition, we also need the Cauchy—Schwarz inequality as shown below.

(3.48)
$$\begin{aligned}
\|f + g\|^2 &= \langle f + g, \, f + g \rangle \\
&= \langle f, \, f \rangle + \langle f, \, g \rangle + \overline{\langle f, \, g \rangle} + \langle g, \, g \rangle && \text{(by axioms 3, 4, 2)} \\
&\leq \|f\|^2 + 2|\langle f, \, g \rangle| + \|g\|^2 && (\because z = \langle f, g \rangle, \, z + \bar{z} \leq 2|z|) \\
&\leq \|f\|^2 + 2\|f\|\,\|g\| + \|g\|^2 && (\because \text{Cauchy—Schwarz inequality}) \\
&= \big(\|f\| + \|g\|\big)^2.
\end{aligned}$$

We complete the proof by taking the positive square roots of both sides. ∎

Example 3.39 Use the Minkowski inequality to show that

(a) $\|f - g\| \leq \|f - h\| + \|h - g\|.$

(b) $\big|\|f\| - \|g\|\big| \leq \|f - g\|.$

Proof: For part (a), we apply inequality (3.47) to $\phi = f - h$ and $\psi = h - g$, and we obtain the desired result:

$$\|f - g\| = \|\phi + \psi\| \leq \|\phi\| + \|\psi\| = \|f - h\| + \|h - g\|.$$

For part (b), we apply inequality (3.47) to $\phi = f - g$ and $\psi = g$, and we obtain

$$\|f\| = \|\phi + \psi\| \leq \|\phi\| + \|\psi\| = \|f - g\| + \|g\|,$$

which yields

(3.49) $$\|f\| - \|g\| \leq \|f - g\|.$$

If we repeat the process with $\phi = g - f$ and $\psi = f$, we shall have

(3.50) $$\|g\| - \|f\| \leq \|g - f\| = \|f - g\|.$$

To obtain the desired result, we simply combine (3.49) and (3.50) into a single inequality, i.e.,

$$\pm\big(\|f\| - \|g\|\big) \leq \|f - g\|, \quad \text{or equivalently,} \quad \big|\|f\| - \|g\|\big| \leq \|f - g\|.$$

∎

3.9.3   Projections

Definition 3.40 Let ϕ be an element of a normed linear space V with $\|\phi\| = 1$. For any $f \in V$ the *projection* of f in the direction of ϕ is denoted by proj $(f : \phi)$ with the de n ition

$$(3.51) \qquad\qquad\qquad \text{proj}(f : \phi) \overset{\text{def}}{=} \langle f, \phi \rangle \phi.$$

Theorem 3.41 The element f is a scalar multiple of ϕ if and only if proj$(f : \phi) = f$.

Proof: If $f = \alpha\phi$ for some scalar α, then we have

$$\text{proj}(f : \phi) = \langle \alpha\phi, \phi \rangle \phi = \alpha\langle \phi, \phi \rangle \phi = \alpha\|\phi\|^2 \phi = \alpha\phi = f. \quad (\because \|\phi\| = 1)$$

Conversely, if proj$(f : \phi) = f$, then by (3.51) we have $\langle f, \phi \rangle \phi = f$, i.e., f is a scalar multiple of ϕ. ∎

Theorem 3.42 If $\psi = f - \text{proj}(f : \phi)$, then the inner product $\langle \psi, \phi \rangle = 0$, and ψ is orthogonal to ϕ.

Proof: By (3.51) we have proj$(f : \phi) = \lambda\phi$ for $\lambda = \langle f, \phi \rangle$. Accordingly,

$$\begin{aligned}
\langle \psi, \phi \rangle &= \langle f - \lambda\phi, \phi \rangle & \\
&= \langle f, \phi \rangle - \lambda\langle \phi, \phi \rangle & \text{(by axiom 3)} \\
&= \langle f, \phi \rangle - \langle f, \phi \rangle\|\phi\|^2 & (\because \lambda = \langle f, \phi \rangle) \\
&= 0. & (\because \|\phi\| = 1 \text{ by de n ition 3.40})
\end{aligned}$$

∎

Definition 3.43 Let V denote a normed linear space, and let Ω_n denote a subspace spanned by an orthonormal sequence $\phi_1, \phi_2, \ldots, \phi_n$ in V. The projection of $f \in V$ into the subspace Ω_n is de n ed by

$$(3.52) \qquad \text{proj}(f : \Omega_n) = \sum_{\ell=1}^{n} \text{proj}(f : \phi_\ell) = \sum_{\ell=1}^{n} \lambda_\ell \phi_\ell, \quad \text{where } \lambda_\ell = \langle f, \phi_\ell \rangle.$$

Example 3.44 Let Ω_{2n+1} be the subspace spanned by the orthonormal sequence

$$\frac{1}{\sqrt{2\pi}}, \ \frac{1}{\sqrt{\pi}}\cos\theta, \ \frac{1}{\sqrt{\pi}}\sin\theta, \ \ldots, \ \frac{1}{\sqrt{\pi}}\cos n\theta, \ \frac{1}{\sqrt{\pi}}\sin n\theta.$$

Recall that these $2n + 1$ elements were shown to have unit norm and be mutually orthogonal with the inner product de ned by the de nite integral over any interval of length 2π, namely,

$$\langle f(\theta), g(\theta) \rangle = \int_{-\pi}^{\pi} f(\theta)\,\overline{g(\theta)}\,d\theta.$$

Using the inner product property from axiom (4),

$$\text{proj}\left(f : \tfrac{1}{\sqrt{\pi}}g\right) = \left\langle f, \tfrac{1}{\sqrt{\pi}}g \right\rangle \tfrac{1}{\sqrt{\pi}}g = \frac{1}{\pi}\langle f, g \rangle g,$$

we express the truncated Fourier series of a periodic function $f(\theta)$ with period 2π as a projection into the subspace Ω_{2n+1} according to de n ition 3.43. That is,

(3.53)
$$
\begin{aligned}
S_{2n+1} &= \text{proj}(f : \Omega_{2n+1}) \\
&= \frac{1}{2\pi}\langle f(\theta), 1 \rangle + \frac{1}{\pi}\sum_{k=1}^{n}\langle f(\theta), \cos k\theta \rangle \cos k\theta + \langle f(\theta), \sin k\theta \rangle \sin k\theta \\
&= \frac{1}{2}A_0 + \sum_{k=1}^{n} A_k \cos k\theta + B_k \sin k\theta,
\end{aligned}
$$

where

(3.54)
$$
A_k = \frac{1}{\pi}\langle f(\theta), \cos k\theta \rangle = \frac{1}{\pi}\int_{-\pi}^{\pi} f(\theta)\cos k\theta \, d\theta, \quad k = 0, 1, 2, \ldots, n;
$$

(3.55)
$$
B_k = \frac{1}{\pi}\langle f(\theta), \sin k\theta \rangle = \frac{1}{\pi}\int_{-\pi}^{\pi} f(\theta)\sin k\theta \, d\theta, \quad k = 1, 2, \ldots, n.
$$

Theorem 3.45 $\text{proj}(f : \Omega_n) = f$ if and only if $f \in \Omega_n$.

Proof: If $\text{proj}(f : \Omega_n) = f$, then by (3.52) f is a linear combination of $\phi_1, \phi_2, \ldots, \phi_n$ from the orthonormal sequence which spans Ω_n, hence $f \in \Omega_n$. Conversely, if $f \in \Omega_n$, then we may express $f = \sum_{k=1}^{n}\alpha_k\phi_k$, where the α_k s are scalars, and the ϕ_k s are from the orthonormal sequence spanning Ω_n, we thus have $\langle \phi_k, \phi_\ell \rangle = 0$ if $k \neq \ell$ and $\langle \phi_\ell, \phi_\ell \rangle = 1$. To obtain $\text{proj}(f : \Omega_n)$ according to (3.52), we compute each coef cient λ_ℓ de ned by the inner product $\langle f, \phi_\ell \rangle$, namely,

$$
\lambda_\ell = \langle f, \phi_\ell \rangle = \left\langle \sum_{k=1}^{n}\alpha_k\phi_k, \phi_\ell \right\rangle = \sum_{k=1}^{n}\alpha_k\langle \phi_k, \phi_\ell \rangle = \alpha_\ell\langle \phi_\ell, \phi_\ell \rangle = \alpha_\ell.
$$

We thus have

$$
\text{proj}(f : \Omega_n) = \sum_{\ell=1}^{n}\lambda_\ell\phi_\ell = \sum_{\ell=1}^{n}\alpha_\ell\phi_\ell = f.
$$

■

Example 3.46 For each $g = \sum_{\ell=1}^{n}\alpha_\ell\phi_\ell$ in Theorem 3.45, show that $\langle g, g \rangle = \sum_{\ell=1}^{n}|\alpha_\ell|^2$.

Proof:
$$
\begin{aligned}
\langle g, g \rangle &= \left\langle \sum_{\ell=1}^{n}\alpha_\ell\phi_\ell, g \right\rangle = \sum_{\ell=1}^{n}\alpha_\ell\langle \phi_\ell, g \rangle = \sum_{\ell=1}^{n}\alpha_\ell\left\langle \phi_\ell, \sum_{k=1}^{n}\alpha_k\phi_k \right\rangle \\
&= \sum_{\ell=1}^{n}\alpha_\ell\sum_{k=1}^{n}\overline{\alpha_k}\langle \phi_\ell, \phi_k \rangle = \sum_{\ell=1}^{n}\alpha_\ell\overline{\alpha_\ell}\langle \phi_\ell, \phi_\ell \rangle = \sum_{\ell=1}^{n}|\alpha_\ell|^2.
\end{aligned}
$$

■

Theorem 3.45 shows that for every $f \in \Omega_n$, we can express it as a linear combination of the elements from the ort honormal basis of Ω_n conveniently, because such an expression is given by $\text{proj}(f : \Omega_n)$ which explicitly de n es each coef cien t to be the inner product of f and an element from the orthonormal sequence.

Example 3.47 Let Ω_{2n+1} again be the subspace spanned by the orthonormal sequence $\frac{1}{\sqrt{2\pi}}, \frac{1}{\sqrt{\pi}}\cos\theta, \frac{1}{\sqrt{\pi}}\sin\theta, \ldots, \frac{1}{\sqrt{\pi}}\cos n\theta, \frac{1}{\sqrt{\pi}}\sin n\theta$. From Theorem 3.45, if $f \in \Omega_{2n+1}$, then $f = \text{proj}(f : \Omega_{2n+1})$. Using the result for $\text{proj}(f : \Omega_{2n+1})$ directly from Example 3.44, we now have

$$\boxed{f(\theta) = \frac{1}{2}A_0 + \sum_{k=1}^{n} A_k \cos k\theta + B_k \sin k\theta,}$$

with A_k and B_k defined by (3.54) and (3.55); hence, the right side represents a Fourier series of $N = 2n + 1$ terms. Note that if we want to apply the result from Example 3.46 to obtain $\|f(\theta)\|^2 = \langle f(\theta), f(\theta) \rangle$, we must use the coefficients with respect to the orthonormal basis. The result is, therefore,

$$\langle f(\theta),\, f(\theta) \rangle = \frac{\pi}{2}|A_0|^2 + \sum_{k=1}^{n} \pi|A_k|^2 + \pi|B_k|^2.$$

Since $\langle f(\theta),\, f(\theta) \rangle = \int_{-\pi}^{\pi} f(\theta)\overline{f(\theta)}\,d\theta = \int_{-\pi}^{\pi} |f(\theta)|^2 d\theta$, the same result is commonly expressed as

$$\boxed{\frac{1}{\pi}\int_{-\pi}^{\pi} |f(\theta)|^2 d\theta = \frac{1}{2}|A_0|^2 + \sum_{k=1}^{n} |A_k|^2 + |B_k|^2.}$$

Theorem 3.48 Let Ω_N be the subspace spanned by the orthonormal sequence $\phi_1, \phi_2, \ldots, \phi_N$. If $\psi = f - \text{proj}(f : \Omega_N)$, then ψ is orthogonal to every element $g_N \in \Omega_N$.

Proof: By definition 3.40, $\text{proj}(f : \Omega_N) = \sum_{k=1}^{N} \lambda_k \phi_k$ for $\lambda_k = \langle f, \phi_k \rangle$. Therefore, for $\ell = 1, 2, \ldots, N$, we have

$$\langle \psi, \phi_\ell \rangle = \left\langle f - \sum_{k=1}^{N} \lambda_k \phi_k, \phi_\ell \right\rangle$$

$$= \langle f, \phi_\ell \rangle - \sum_{k=1}^{N} \lambda_k \langle \phi_k, \phi_\ell \rangle$$

$$= \langle f, \phi_\ell \rangle - \lambda_\ell \langle \phi_\ell, \phi_\ell \rangle \qquad (\because \langle \phi_k, \phi_\ell \rangle = 0,\ k \neq \ell)$$

$$= \langle f, \phi_\ell \rangle - \lambda_\ell \qquad (\because \langle \phi_\ell, \phi_\ell \rangle = \|\phi\|^2 = 1)$$

$$= 0. \qquad (\because \lambda_\ell = \langle f, \phi_\ell \rangle)$$

To show that ψ is orthogonal to every $g_N \in \Omega_N$, we express $g_N = \sum_{\ell=1}^{N} \alpha_\ell \phi_\ell$, and we show that

$$\langle \psi, g_N \rangle = \left\langle \psi, \sum_{\ell=1}^{N} \alpha_\ell \phi_\ell \right\rangle = \sum_{\ell=1}^{N} \langle \psi, \alpha_\ell \phi_\ell \rangle = \sum_{\ell=1}^{N} \overline{\alpha_\ell} \langle \psi, \phi_\ell \rangle = 0. \quad (\because \langle \psi, \phi_\ell \rangle = 0)$$

$\blacksquare$

3.9.4   Least-squares approximation

Theorem 3.45 also shows that if $f \notin \Omega_N$, then $f \neq \text{proj}(f : \Omega_N)$. Therefore, in general, we can only *approximate* an arbitrary periodic function by a *finite* Fourier series. We show next that if Ω_N is a subspace of V, then for every $f \in V$ and every $g_N \in \Omega_N$, the difference

$\|f - g_N\|$ is minimized when $g_N = \text{proj}(f : \Omega_N)$. In other words, we shall show that the best least-squares approximation to f in the subspace Ω_N is given by $g_N = \text{proj}(f : \Omega_N)$.

Theorem 3.49 If Ω_N is a subspace of the normed linear space V and Ω_N is spanned by the orthonormal sequence $\phi_1, \phi_2, \ldots, \phi_N$, then for every $f \in V$, the element $g_N \in \Omega_N$ for which $\|f - g_N\|^2$ is a minimum is $g_N = \text{proj}(f : \Omega_N)$.

Proof: Suppose $\|f - g_N\|^2$ is minimized by $g_N = \sum_{k=1}^{N} \alpha_k \phi_k \in \Omega_N$, where the coefficients α_k are unknowns to be determined. We proceed to evaluate the inner product defining $\|f - g_N\|^2$, and we obtain

$$\langle f - g_N, f - g_N \rangle = \langle f, f \rangle - \langle f, g_N \rangle - \langle g_N, f \rangle + \langle g_N, g_N \rangle$$

$$= \|f\|^2 - \sum_{k=1}^{N} \overline{\alpha}_k \langle f, \phi_k \rangle - \sum_{k=1}^{N} \alpha_k \langle \phi_k, f \rangle + \sum_{k=1}^{N} |\alpha_k|^2$$

$$= \|f\|^2 - \sum_{k=1}^{N} \overline{\alpha}_k \langle f, \phi_k \rangle - \sum_{k=1}^{N} \alpha_k \overline{\langle f, \phi_k \rangle} + \sum_{k=1}^{N} |\alpha_k|^2$$

$$= \|f\|^2 - \sum_{k=1}^{N} \overline{\alpha}_k \lambda_k - \sum_{k=1}^{N} \alpha_k \overline{\lambda}_k + \sum_{k=1}^{N} \overline{\alpha}_k \alpha_k, \quad \text{where } \lambda_k = \langle f, \phi_k \rangle,$$

$$= \|f\|^2 + \sum_{k=1}^{N} \left(\overline{\lambda}_k \lambda_k - \overline{\alpha}_k \lambda_k - \alpha_k \overline{\lambda}_k + \overline{\alpha}_k \alpha_k \right) - \sum_{k=1}^{N} \overline{\lambda}_k \lambda_k$$

$$= \|f\|^2 + \sum_{k=1}^{N} \left(\overline{\lambda}_k - \overline{\alpha}_k \right) \left(\lambda_k - \alpha_k \right) - \sum_{k=1}^{N} |\lambda_k|^2$$

$$= \|f\|^2 + \sum_{k=1}^{N} |\lambda_k - \alpha_k|^2 - \sum_{k=1}^{N} |\lambda_k|^2.$$

To minimize the right side, we focus on the only term involving the unknown α_k s, i.e., the term $\sum_{k=1}^{N} |\lambda_k - \alpha_k|^2$. Because this term cannot be negative, its minimum value is zero, which is reached by setting $\alpha_k = \lambda_k$ for $k = 1, 2, \ldots, N$. Therefore, we minimize $\|f - g_N\|^2$ by choosing

$$g_N = \sum_{k=1}^{N} \alpha_k \phi_k = \sum_{k=1}^{N} \lambda_k \phi_k = \sum_{k=1}^{N} \langle f, \phi_k \rangle \phi_k = \text{proj}(f : \Omega_N),$$

which renders

$$\langle f - g_N, f - g_N \rangle = \|f\|^2 - \sum_{k=1}^{N} |\lambda_k|^2 = \|f\|^2 - \|g_N\|^2, \quad \text{where } \lambda_k = \langle f, \phi_k \rangle.$$

∎

Corollary 3.50 For element f in Theorem 3.49, $g_N = \text{proj}(f : \Omega_N)$ is the best least-squares approximation of f in the subspace Ω_N, and the resulting (minimum) error is given by

$$0 \leq \|f - g_N\|^2 = \|f\|^2 - \|g_N\|^2.$$

Proof: This result was obtained in the proof of Theorem 3.49 when we set $\alpha_k = \lambda_k$ to enable $g_N = \text{proj}(f : \Omega_N)$. ∎

Corollary 3.51 If Ω_N is a subspace of the normed linear space V and Ω_N is spanned by the orthonormal sequence ϕ_1, ϕ_2, ..., ϕ_N, then for every f in V, we have

$$\|f\|^2 \geq \sum_{k=1}^{N} |\langle f, \phi_k \rangle|^2.$$

Proof: Since $\|f\|^2 \geq \|g_N\|^2$ is an immediate result from Corollary 3.50, and $\|g_N\|^2 = \sum_{k=1}^{N} |\langle f, \phi_k \rangle|^2$ was shown in the proof of Theorem 3.49, the desired result follows. ∎

Example 3.52 Let $N = 2n + 1$ and let Ω_N be the subspace spanned by the orthonormal sequence $\frac{1}{\sqrt{2\pi}}$, $\frac{1}{\sqrt{\pi}} \cos\theta$, $\frac{1}{\sqrt{\pi}} \sin\theta$, ..., $\frac{1}{\sqrt{\pi}} \cos n\theta$, $\frac{1}{\sqrt{\pi}} \sin n\theta$. In Example 3.44 we have shown that

(3.56) $$g_N(\theta) = \text{proj}(f : \Omega_N) = \frac{1}{2}A_0 + \sum_{k=1}^{n} A_k \cos k\theta + B_k \sin k\theta,$$

with A_k and B_k de ned by (3.54) and (3.55). As mentioned before, the right side represents a truncated Fourier series of $f(\theta)$. From Theorem 3.49, the N-term Fourier series of $f(\theta)$ is its best least-squares approximation in Ω_N. Using Corollary 3.50 and the result from Example 3.47, the error between $f(\theta)$ and its truncated Fourier series $g_N(\theta)$ is given by

(3.57)
$$0 \leq \|f(\theta) - g_N(\theta)\|^2 = \|f(\theta)\|^2 - \|g_N(\theta)\|^2$$
$$= \int_{-\pi}^{\pi} |f(\theta)|^2 \, d\theta - \left[\frac{\pi}{2}|A_0|^2 + \pi \sum_{k=1}^{n} |A_k|^2 + |B_k|^2 \right];$$

the same result can be represented by the inequality

(3.58) $$\boxed{\frac{1}{\pi} \int_{-\pi}^{\pi} |f(\theta)|^2 d\theta \geq \frac{1}{2}|A_0|^2 + \sum_{k=1}^{n} |A_k|^2 + |B_k|^2,}$$

which is referred to by some authors as Bessel s inequality for nite sum. Observe that this inequality is simply the result of applying Corollary 3.51,

$$\|f\|^2 \geq \sum_{k=1}^{N} |\langle f, \phi_k \rangle|^2$$

to a speci c set of orthonormal functions.

Example 3.53 Find the Fourier series coef cients of $g(\theta)$ and verify that Bessel s inequality (for nite sums) holds for the rst few terms.

$$g(\theta) = \begin{cases} 1, & 0 \leq \theta \leq \pi \\ -1, & -\pi < \theta < 0 \end{cases}.$$

We rst observe that $g(-\theta) = -g(\theta)$, so $g(\theta)$ is an odd function, and its Fourier series coef cients are given by Theorem 3.12 as

$$A_k = 0, \quad k = 0, 1, 2, \ldots ; \quad B_k = \frac{2}{\pi} \int_0^{\pi} g(\theta) \sin k\theta \, d\theta, \quad k = 1, 2, \ldots .$$

Since $g(\theta) = 1$ for $\theta \in [0, \pi]$, we obtain

$$B_k = \frac{2}{\pi} \int_0^\pi \sin k\theta \, d\theta = \frac{2}{k\pi} \left(-\cos k\theta \Big|_0^\pi \right) = \frac{2}{k\pi}(1 - \cos k\pi).$$

Noting that $B_k = 0$ if k is even, and $B_k = \dfrac{4}{k\pi}$ if k is odd, we have

$$g(\theta) = \sum_{r=0}^\infty B_{2r+1} \sin(2r+1)\theta = \frac{4}{\pi} \left(\sin\theta + \frac{\sin 3\theta}{3} + \cdots + \frac{\sin(2r+1)\theta}{2r+1} + \cdots \right).$$

To verify Bessel s inequality for a n ite sum of N terms, we have on the left-hand side

$$\frac{1}{\pi} \int_{-\pi}^\pi g^2(\theta) \, d\theta = \frac{1}{\pi} \int_{-\pi}^\pi d\theta = 2;$$

on the right-hand side we have the partial sum of N nonzero terms, i.e.,

$$S_N = \sum_{r=0}^{N-1} B_{2r+1}^2 = \frac{16}{\pi^2} \sum_{r=0}^{N-1} \frac{1}{(2r+1)^2} = \frac{16}{\pi^2} \left(1 + \frac{1}{9} + \frac{1}{25} + \cdots + \frac{1}{(2N-1)^2} \right).$$

It can now be veri ed that

$$S_1 = \frac{16}{\pi^2} \le 2, \quad S_2 = \frac{160}{9\pi^2} \le 2, \quad S_3 = \frac{4144}{225\pi^2} \le 2, \ \ldots, \text{ etc.}$$

This example also shows that the nonzero Fourier coef cient $B_{2r+1} = \dfrac{4}{(2r+1)\pi} \to 0$ as $k = 2r + 1 \to \infty$.

3.9.5   Bessel's inequality and Riemann's lemma

We revisit the Bessel s inequality (restricted for nite sum at the moment) given by (3.58), and we express the same result as

$$(3.59) \qquad S_n = \frac{1}{2}|A_0|^2 + \sum_{k=1}^n |A_k|^2 + |B_k|^2 \le M,$$

where $M = \frac{1}{\pi} \int_{-\pi}^\pi |f(\theta)|^2 d\theta$ is a constant for given f. Because $S_0, S_1, S_2, \ldots S_n, \ldots$ are partial sums associated with the following in nite series of nonnegative terms:

$$(3.60) \qquad \frac{1}{2}|A_0|^2 + \sum_{k=1}^\infty |A_k|^2 + |B_k|^2,$$

we have $S_0 \le S_1 \le S_2 \le \cdots \le S_n \le \ldots$, and we have found the upper bound M such that every $S_n \le M$; hence, this sequence of partial sums has a limit, and

$$\lim_{n\to\infty} S_n \le M.$$

We have now obtained the full- ed ged Bessel s inequality:

$$(3.61) \qquad \lim_{n\to\infty} S_n = \frac{1}{2}|A_0|^2 + \sum_{k=1}^\infty |A_k|^2 + |B_k|^2 \le \frac{1}{\pi} \int_{-\pi}^\pi |f(\theta)|^2 d\theta;$$

which is, again, a special case of the general result:

$$\sum_{k=1}^{\infty} |\langle f, \phi_k \rangle|^2 \le \|f\|^2,$$

which is valid for any orthonormal sequence $\phi_1, \phi_2, \ldots, \phi_n, \ldots$.

Example 3.54 We continue with Example 3.53, and we may now verify that the full-edged Bessel s inequality holds because

$$\lim_{N \to \infty} S_N = \sum_{r=0}^{\infty} B_{2r+1}^2 = \frac{16}{\pi^2} \sum_{r=0}^{\infty} \frac{1}{(2r+1)^2} = \frac{16}{\pi^2} \left(\frac{\pi^2}{8} \right) = 2,$$

where the numerical result for the sum of the in n ite series was established by identity (3.8) in Example 3.15.

To further analyze the in ite series, we recall the following: (i) an in ite series is said to be convergent if and only if the sequence of its partial sums has a limit; (ii) the terms of a convergent in nite series must tend to zero. Since the in nite series (3.60) was shown to satisfy the condition set out in (i) for convergence, we must have

$$\lim_{k \to \infty} |A_k|^2 = 0, \text{ and } \lim_{k \to \infty} |B_k|^2 = 0,$$

which implies that both A_k and B_k tend to 0 as $k \to \infty$. When the results are given speci cally for the Fourier coef cients, they commonly appear in the following form:

(3.62) $$\lim_{k \to \infty} A_k = \lim_{k \to \infty} \frac{1}{\pi} \int_{-\pi}^{\pi} f(\theta) \cos k\theta \, d\theta = 0;$$

(3.63) $$\lim_{k \to \infty} B_k = \lim_{k \to \infty} \frac{1}{\pi} \int_{-\pi}^{\pi} f(\theta) \sin k\theta \, d\theta = 0.$$

We remark that because $A_k = \frac{1}{\sqrt{\pi}} \langle f, \phi_k \rangle$ with $\phi_k = \frac{1}{\sqrt{\pi}} \cos k\theta$, and $B_k = \frac{1}{\sqrt{\pi}} \langle f, \psi_k \rangle$ with $\psi_k = \frac{1}{\sqrt{\pi}} \sin k\theta$, the above results apply to in nite series with coef cien ts $\lambda_k = \langle f, \phi_k \rangle$ whenever $\phi_1, \phi_2, \phi_3, \ldots$ are mutually orthonormal. (The zero limit is not affected by absorbing an additional constant factor $\frac{1}{\sqrt{\pi}}$ into the Fourier coef cients A_k and B_k.) The latter result is known as Riemann s lemma, which is formally given as Lemma 3.55.

Lemma 3.55 (Riemann s lemma) For a given function f with $\int_{-\pi}^{\pi} |f(\theta)|^2 d\theta < \infty$, and for a given sequence of orthonormal functions $\phi_1, \phi_2, \phi_3, \ldots$,

$$\lim_{k \to \infty} \langle f, \phi_k \rangle = 0.$$

Example 3.56 Assuming that $\int_{-\pi}^{\pi} |g(\theta)|^2 d\theta < \infty$, use Riemann s lemma to prove that

(3.64) $$\lim_{k \to \infty} \frac{1}{\pi} \int_{-\pi}^{\pi} g(\theta) \sin\left(k + \tfrac{1}{2}\right)\theta \, d\theta = 0.$$

Proof: At rst we write out the integral in such a form that Riemann s lemma may be applied.

$$\frac{1}{\pi} \int_{-\pi}^{\pi} g(\theta) \sin\left(k + \tfrac{1}{2}\right)\theta \, d\theta = \frac{1}{\pi} \int_{-\pi}^{\pi} g(\theta) \left(\sin k\theta \cos \frac{\theta}{2} + \cos k\theta \sin \frac{\theta}{2} \right) d\theta$$

$$= \frac{1}{\pi} \int_{-\pi}^{\pi} \left[g(\theta) \sin \frac{\theta}{2} \right] \cos k\theta \, d\theta + \frac{1}{\pi} \int_{-\pi}^{\pi} \left[g(\theta) \cos \frac{\theta}{2} \right] \sin k\theta \, d\theta$$

$$= \frac{1}{\pi} \int_{-\pi}^{\pi} U(\theta) \cos k\theta \, d\theta + \frac{1}{\pi} \int_{-\pi}^{\pi} V(\theta) \sin k\theta \, d\theta,$$

where $U(\theta) = g(\theta)\sin\frac{\theta}{2}$ and $V(\theta) = g(\theta)\cos\frac{\theta}{2}$. By Riemann's lemma, the two integrals on the right side tend to zero:

$$\lim_{k\to\infty} \frac{1}{\pi} \int_{-\pi}^{\pi} U(\theta)\cos k\theta \, d\theta = \frac{1}{\sqrt{\pi}} \lim_{k\to\infty} \left\langle U(\theta), \frac{1}{\sqrt{\pi}}\cos k\theta \right\rangle = 0,$$

$$\lim_{k\to\infty} \frac{1}{\pi} \int_{-\pi}^{\pi} V(\theta)\sin k\theta \, d\theta = \frac{1}{\sqrt{\pi}} \lim_{k\to\infty} \left\langle V(\theta), \frac{1}{\sqrt{\pi}}\sin k\theta \right\rangle = 0.$$

Hence, the integral on the left side must also tend to zero as $k \to \infty$, and we obtain the desired result. ∎

Example 3.57 Assuming that $\int_0^\pi |g(\theta)|^2 d\theta < \infty$, use Riemann's lemma to prove that

$$(3.65) \qquad \lim_{k\to\infty} \frac{1}{\pi} \int_0^\pi g(\theta)\sin\left(k + \tfrac{1}{2}\right)\theta \, d\theta = 0.$$

Proof: By taking the same initial steps in Example 3.56 we rewrite

$$\frac{1}{\pi} \int_0^\pi g(\theta)\sin\left(k + \tfrac{1}{2}\right)\theta \, d\theta = \frac{1}{\pi} \int_0^\pi U(\theta)\cos k\theta \, d\theta + \frac{1}{\pi} \int_0^\pi V(\theta)\sin k\theta \, d\theta,$$

where $U(\theta) = g(\theta)\sin\frac{\theta}{2}$, and $V(\theta) = g(\theta)\cos\frac{\theta}{2}$. Applying Riemann's lemma with the orthonormal cosine or sine sequences defined via the inner products given by (3.31) over the interval $[0, \pi]$, the two integrals on the right side tend to zero:

$$\lim_{k\to\infty} \frac{1}{\pi} \int_0^\pi U(\theta)\cos k\theta \, d\theta = \frac{1}{\sqrt{2\pi}} \lim_{k\to\infty} \left\langle U(\theta), \frac{1}{\sqrt{\frac{\pi}{2}}}\cos k\theta \right\rangle = 0,$$

$$\lim_{k\to\infty} \frac{1}{\pi} \int_0^\pi V(\theta)\sin k\theta \, d\theta = \frac{1}{\sqrt{2\pi}} \lim_{k\to\infty} \left\langle V(\theta), \frac{1}{\sqrt{\frac{\pi}{2}}}\sin k\theta \right\rangle = 0.$$

The desired result follows. ∎

3.10 Convergence of the Fourier Series

3.10.1 Starting with a concrete example

Recall Example 3.14 from Section 3.4, in which we show that the function $f(t) = t - t^2$ $(0 < t < 1)$ has three expansions:

$$f(t) = t - t^2 = \begin{cases} \dfrac{1}{6} - \dfrac{1}{\pi^2} \displaystyle\sum_{k=1}^\infty \dfrac{\cos 2k\pi t}{k^2}; \\[3mm] \dfrac{8}{\pi^3} \displaystyle\sum_{k=1}^\infty \dfrac{\sin(2k-1)\pi t}{(2k-1)^3}; \\[3mm] -\dfrac{1}{3} - \dfrac{4}{\pi^2} \displaystyle\sum_{k=1}^\infty (-1)^k \dfrac{\cos k\pi t}{k^2} - \dfrac{2}{\pi} \displaystyle\sum_{k=1}^\infty (-1)^k \dfrac{\sin k\pi t}{k}. \end{cases}$$

Strictly speaking, the first expansion is the Fourier series of the periodic (even) function $g_1(t)$, $-\infty < t < \infty$, which is formally defined as

$$g_1(t) = \begin{cases} t - t^2, & t \in (0, 1] \\ -t - t^2, & t \in (-1, 0] \end{cases}, \qquad g_1(t + 2) = g_1(t).$$

The graph of $g_1(t)$ is shown in Figure 3.8. Since $g_1(t)$ is continuous at all points, Dirichlet s theorem tells us that

$$g_1(t) = \frac{1}{6} - \sum_{k=1}^{\infty} A_k \cos 2k\pi t, \quad \text{with } A_k = \frac{1}{\pi^2 k^2},$$

is true everywhere. Because the size of the kth cosine term in the series is bounded by the magnitude of its coef cient A_k, the partial sum (from a truncated Fourier series) approaches $g_1(t)$ as fast as the coef cient A_k tends to 0. In this case, when k grows bigger (as more terms are added), the coef cient $A_k \to 0$ as fast as $1/k^2 \to 0$.

Figure 3.8 The graphs of periodic (even) $g_1(t)$ and $g_1'(t)$.

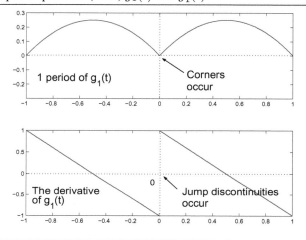

Observe that since

$$g_1'(t) = \begin{cases} 1 - 2t, & t \in (0, 1] \\ -1 - 2t, & t \in (-1, 0] \end{cases}, \quad g_1'(t + 2) = g_1'(t),$$

the rst derivative of $g_1(t)$ has jump discontinuities at $t = 0, \pm 1, \pm 2, \dots$, and $g_1(t)$ is said to have corners at $t = 0, \pm 1, \pm 2, \dots$, which are apparent in the graphs of $g_1(t)$ and $g_1'(t)$ shown in Figure 3.8

We consider next the second expansion, which is, strictly speaking, the Fourier series of the periodic (odd) function $g_2(t)$, $-\infty < t < \infty$, which is formally de ned as

$$g_2(t) = \begin{cases} t - t^2, & t \in (0, 1] \\ t + t^2, & t \in (-1, 0] \end{cases}, \quad g_2(t + 2) = g_2(t).$$

The graph of $g_2(t)$ is shown in Figure 3.9. Since $g_2(t)$ is also continuous at all points, Dirichlet s theorem tells us that

$$g_2(t) = \frac{8}{\pi^3} \sum_{r=1}^{\infty} \frac{\sin(2r - 1)\pi t}{(2r - 1)^3}$$

is true everywhere. Because each sine mode is bounded by the size of its coef cient, the partial sum (from a truncated Fourier series) approaches $g_2(t)$ as fast as the coef cient tends to 0. In this case, when $k = 2r - 1$ grows bigger (as more terms are added), the coefﬁcient $B_k \to 0$ as fast as $1/k^3 \to 0$. Since $1/k^3$ goes to zero faster than $1/k^2$, we conclude that the second expansion converges *faster* than the ﬁrst expansion — the reason (to be formally shown in Section 3.10.3) being that $g_2(t)$ has continuous ﬁrst derivative, which is reﬂected by the fact that no corners appear in the graph of $g_2(t)$ in Figure 3.9. Mathematically,

$$g_2'(t) = \begin{cases} 1 - 2t, & t \in (0, 1] \\ 1 + 2t, & t \in (-1, 0] \end{cases}, \quad g_2'(t + 2) = g_2'(t),$$

and $g_2'(t)$ is indeed continuous everywhere.

Figure 3.9 The graphs of periodic (odd) $g_2(t)$ and $g_2'(t)$.

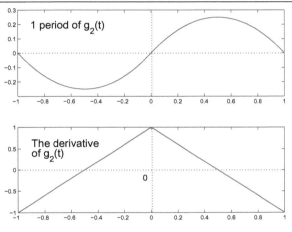

Since both $g_1(t)$ and $g_2(t)$ are everywhere continuous, their expansions converge to the original functions at all points. However, this is no longer the case when we study the convergence of the third expansion which is, strictly speaking, the Fourier series of the periodic function $g_3(t)$, $-\infty < t < \infty$, deﬁned as

$$g_3(t) = t - t^2, \quad t \in (-1, 1], \quad g_3(t + 2) = g_3(t).$$

From the graph of $g_3(t)$ in Figure 3.10, we see that $g_3(t)$ has jump discontinuities at $t = \pm 1, \pm 3, \pm 5, \ldots$. According to Dirichlet's theorem, while the third expansion still converges to $g_3(t)$ at points of continuity, for every t_α at which $g_3(t_\alpha)$ has a jump discontinuity, the Fourier series converges to the average of its right- and left-hand limits. That is, at $t_\alpha = \pm 1, \pm 3, \pm 5, \ldots$,

$$\frac{g_3\left(t_\alpha^+\right) + g_3\left(t_\alpha^-\right)}{2} = -\frac{1}{3} - \frac{4}{\pi^2} \sum_{k=1}^{\infty} (-1)^k \frac{\cos k\pi t_\alpha}{k^2} - \frac{2}{\pi} \sum_{k=1}^{\infty} (-1)^k \frac{\sin k\pi t_\alpha}{k}.$$

Observe that for function $g_3(t)$, the average of the right- and left-hand limits is equal to the constant -1 for all $t_\alpha = \pm 1, \pm 3, \pm 5, \ldots$. Therefore, while the third expansion converges to the original function $g_3(t)$ at all points of continuity, the same expansion converges to the value of -1 at all points of jump discontinuity.

Figure 3.10 The graphs of three periods of $g_3(t)$.

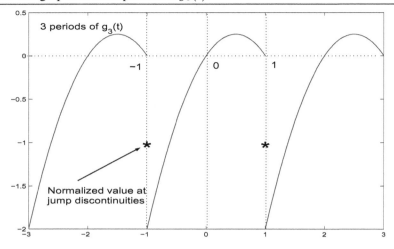

3.10.2   Pointwise convergence—a local property

As remarked earlier, piecewise smooth functions satisfy Dirichlet conditions and we will study the convergence of their Fourier series. Using the functions $g_1(t)$, $g_2(t)$, and $g_3(t)$ shown in Figures 3.8, 3.9, and 3.10 as examples, we see that both corners and jump discontinuities are permitted in a piecewise smooth functions; in addition, as we have remarked earlier, a piecewise smooth function is not required to be de n ed at the points of jump discontinuities.

In the analysis we shall make use of the following properties of the Riemann integral from the theory of calculus:

1. Every continuous function f is bounded and Riemann integrable on $[a, b]$.

2. Every piecewise continuous function g on $[a, b]$ is integrable.

3. Let g and h be integrable functions on $[a, b]$ and let c be a real number, then

   (a) cg is integrable and $\int_a^b cg(t)\,dt = c\int_a^b g(t)\,dt$.

   (b) $g + h$ is integrable and $\int_a^b [g(t) + h(t)]\,dt = \int_a^b g(t)dt + \int_a^b h(t)dt$.

   (c) gh is integrable on $[a, b]$.

   (d) g^2 is integrable on $[a, b]$.

   (e) $|g|$ is integrable on $[a, b]$, and $\left|\int_a^b g(t)\,dt\right| \le \int_a^b |g(t)|\,dt$.

   (f) $|g|^2$ is integrable on $[a, b]$.

   (g) changing the value of function g for a nite number of points does not change the value of $\int_a^b g(t)\,dt$.

We now proceed to prove that the Fourier series of a periodic piecewise smooth function converges to the *normalized* function value at every point. (Recall that g is piecewise smooth on $[a, b]$ if both g and its derivative g' are piecewise continuous on $[a, b]$.) Although the analysis can be carried out using either variable t (with period T) or variable $\theta = 2\pi t/T$ (with period 2π) and the Fourier series can be expressed in a number of mathematically equivalent forms, a function g in variable θ and a corresponding Fourier series expressed using complex exponential modes would be most compact and convenient for the presentation here.

To begin the analysis, let g be an integrable function of period 2π, we denote its truncated Fourier series of $N=2n+1$ terms by

$$\tilde{g}_N(\theta) = \sum_{k=-n}^{n} C_k e^{jk\theta},$$

and we obtain the coef cients from formula (3.16), i.e.,

$$C_k = \frac{1}{2\pi} \int_{-\pi}^{\pi} g(\theta)\, e^{-jk\theta}\, d\theta, \quad -n \le k \le n.$$

In mathematical terms, our objective is to prove that for every $\theta_\ell \in [-\pi, \pi]$,

$$\boxed{\lim_{N\to\infty} \left[\tilde{g}_N(\theta_\ell) - g(\theta_\ell)\right] = 0}$$

where $g(\theta_\ell)$ denotes the rede ned normalized function value at every point θ_ℓ, i.e.,

$$g(\theta_\ell) \overset{\text{def}}{=} \frac{g(\theta_\ell^+) + g(\theta_\ell^-)}{2}.$$

Note that if θ_ℓ is a point of continuity, we have $g(\theta_\ell^+) = g(\theta_\ell^-)$, and the normalized function preserves the values of the original function at all points of continuity. Therefore, the formula we use to change the value of the original function at points of jump discontinuity may be used to rede n e the function value everywhere   it will *not* affect the values of the original function at points of continuity.

To achieve our objective, we need the results from the following two lemmas.

Lemma 3.58 Let $g(\theta)$ be an integrable function of period 2π. Show that the partial sum of the $N=2n+1$ terms from the truncated Fourier series of $g(\theta)$ can be expressed in the integral form

(3.66)
$$\tilde{g}_N(\theta) = \frac{1}{2\pi} \int_{-\pi}^{\pi} g(\theta + \lambda)\frac{\sin\left(n + \frac{1}{2}\right)\lambda}{\sin\frac{\lambda}{2}}\, d\lambda,$$

or equivalently,

(3.67)
$$\tilde{g}_N(\theta) = \frac{1}{2\pi} \int_{-\pi}^{\pi} g(\theta - \lambda)\frac{\sin\left(n + \frac{1}{2}\right)\lambda}{\sin\frac{\lambda}{2}}\, d\lambda.$$

Proof:

$$\tilde{g}_N(\theta) = \sum_{k=-n}^{n} C_k e^{jk\theta} = \frac{1}{2\pi} \sum_{k=-n}^{n} \left[\int_{-\pi}^{\pi} g(\tau) e^{-jk\tau} \, d\tau \right] e^{jk\theta} \qquad \text{(by de n ition of } C_k\text{)}$$

$$= \frac{1}{2\pi} \sum_{k=-n}^{n} \left[\int_{-\pi+\theta}^{\pi+\theta} g(\tau) e^{-jk\tau} \, d\tau \right] e^{jk\theta} \qquad (\because g(\tau) \text{ has period } 2\pi)$$

$$= \frac{1}{2\pi} \sum_{k=-n}^{n} \left[\int_{-\pi}^{\pi} g(\theta+\lambda) e^{-jk(\theta+\lambda)} \, d\lambda \right] e^{jk\theta} \qquad (\text{let } \tau = \theta + \lambda)$$

$$= \frac{1}{2\pi} \sum_{k=-n}^{n} \left[\int_{-\pi}^{\pi} g(\theta+\lambda) e^{-jk(\theta+\lambda)} e^{jk\theta} \, d\lambda \right]$$

$$= \frac{1}{2\pi} \sum_{k=-n}^{n} \left[\int_{-\pi}^{\pi} g(\theta+\lambda) e^{-jk\lambda} \, d\lambda \right]$$

$$= \frac{1}{2\pi} \int_{-\pi}^{\pi} g(\theta+\lambda) \left[\sum_{k=-n}^{n} e^{-jk\lambda} \right] d\lambda.$$

To obtain the integral form (3.66), we are now required to show that

$$(3.68) \qquad \sum_{k=-n}^{n} e^{-jk\lambda} = \frac{\sin\left(n+\frac{1}{2}\right)\lambda}{\sin\frac{\lambda}{2}}. \qquad \text{(limit of the right side exists as } \lambda \to 0)$$

We note that the left side is a power series expressed in $z = e^{-j\lambda}$, and we could make use of the following result:

$$(3.69) \qquad \left(z^{-\frac{1}{2}} - z^{\frac{1}{2}}\right) \sum_{k=-n}^{n} z^k = \sum_{k=-n}^{n} z^{k-\frac{1}{2}} - \sum_{k=-n}^{n} z^{k+\frac{1}{2}} = z^{-n-\frac{1}{2}} - z^{n+\frac{1}{2}},$$

which allows us to express the power series

$$\sum_{k=-n}^{n} z^k = \frac{z^{-n-\frac{1}{2}} - z^{n+\frac{1}{2}}}{z^{-\frac{1}{2}} - z^{\frac{1}{2}}} \qquad \text{if } z^{-\frac{1}{2}} \neq z^{\frac{1}{2}}.$$

Letting $z = e^{-j\lambda}$, $-\pi \leq \lambda \leq \pi$, we obtain the desired result:

$$\sum_{k=-n}^{n} e^{-jk\lambda} = \frac{e^{j(n+\frac{1}{2})\lambda} - e^{-j(n+\frac{1}{2})\lambda}}{e^{j\frac{\lambda}{2}} - e^{-j\frac{\lambda}{2}}} = \frac{\sin\left(n+\frac{1}{2}\right)\lambda}{\sin\frac{\lambda}{2}}. \qquad \text{(limit is } 2n+1 \text{ as } \lambda \to 0)$$

Note that we have used Euler s identity, namely, $e^{j\alpha} - e^{-j\alpha} = 2j\sin\alpha$, in the last step.

To convert (3.66) to the equivalent integral form (3.67), we change variable λ in (3.66) to $\mu = -\lambda$, and we obtain

$$\tilde{g}_N(\theta) = -\frac{1}{2\pi} \int_{+\pi}^{-\pi} g(\theta-\mu) \frac{-\sin\left(n+\frac{1}{2}\right)\mu}{-\sin\frac{\mu}{2}} \, d\mu \qquad (\because \mu = -\lambda, \ \lambda = -\mu, \ d\lambda = -d\mu)$$

$$= \frac{1}{2\pi} \int_{-\pi}^{+\pi} g(\theta-\mu) \frac{\sin\left(n+\frac{1}{2}\right)\mu}{\sin\frac{\mu}{2}} \, d\mu$$

$$= \frac{1}{2\pi} \int_{-\pi}^{\pi} g(\theta-\lambda) \frac{\sin\left(n+\frac{1}{2}\right)\lambda}{\sin\frac{\lambda}{2}} \, d\lambda. \qquad \text{(change dummy variable } \mu \text{ back to } \lambda)$$

∎

Lemma 3.59 The value of the function $g(\theta)$ de ned for any point $\theta = \theta_\ell$ can be expressed as an integral consistent with the partial sum $\tilde{g}_N(\theta)$ from Lemma 3.58. Such an expression is speci cally constructed as

$$g(\theta_\ell) = \frac{1}{2\pi} \int_{-\pi}^{\pi} g(\theta_\ell) \frac{\sin\left(n + \frac{1}{2}\right)\lambda}{\sin\frac{\lambda}{2}} \, d\lambda.$$

Proof: Since θ_ℓ is a parameter independent of λ, we are asked to prove

$$g(\theta_\ell) = g(\theta_\ell) \left[\frac{1}{2\pi} \int_{-\pi}^{\pi} \frac{\sin\left(n + \frac{1}{2}\right)\lambda}{\sin\frac{\lambda}{2}} \, d\lambda \right],$$

which is equivalent to showing that

$$\int_{-\pi}^{\pi} \frac{\sin\left(n + \frac{1}{2}\right)\lambda}{\sin\frac{\lambda}{2}} \, d\lambda = 2\pi.$$

This result can be easily obtained if we integrate both sides of equation (3.68), i.e,

$$\int_{-\pi}^{\pi} \frac{\sin\left(n + \frac{1}{2}\right)\lambda}{\sin\frac{\lambda}{2}} \, d\lambda = \int_{-\pi}^{\pi} \sum_{k=-n}^{n} e^{-jk\lambda} \, d\lambda$$

$$= \int_{-\pi}^{\pi} \left[1 + \sum_{k=1}^{n} \left(e^{-jk\lambda} + e^{jk\lambda} \right) \right] d\lambda$$

$$= \int_{-\pi}^{\pi} 1 \, d\lambda + \int_{-\pi}^{\pi} \left[\sum_{k=1}^{n} 2\cos k\lambda \right] d\lambda \qquad \text{(by Euler s formula)}$$

$$= 2\pi + \sum_{k=1}^{n} \left[\int_{-\pi}^{\pi} 2\cos k\lambda \, d\lambda \right] \qquad \text{(integrate term by term)}$$

$$= 2\pi + \sum_{k=1}^{n} \frac{4\sin k\pi}{k}$$

$$= 2\pi. \qquad (\sin k\pi = 0, \ 1 \le k \le n)$$

∎

Using the integral representation of $\tilde{g}_N(\theta_\ell)$ and $g(\theta_\ell)$ from the last two lemmas, we can now prove the pointwise convergence theorem.

Theorem 3.60 If $g(\theta)$ is a piecewise smooth function of period 2π, its Fourier series converges to the normalized function value $g(\theta_\ell) = \frac{1}{2}\left[g(\theta_\ell^+) + g(\theta_\ell^-)\right]$ at every point θ_ℓ.

Proof: Our objective is to show that for every $\theta_\ell \in [-\pi, \pi]$,

$$\lim_{N \to \infty} \left[\tilde{g}_N(\theta_\ell) - \frac{g(\theta_\ell^+) + g(\theta_\ell^-)}{2} \right] = 0.$$

On substituting the integral forms from Lemmas 3.58 and 3.59 for $\tilde{g}_N(\theta_\ell)$ and $g(\theta_\ell)$, we may express $\frac{1}{2}\left[\tilde{g}_N(\theta_\ell) - g(\theta_\ell) \right]$ in two forms:

(3.70) $\qquad \frac{1}{2}\left[\tilde{g}_N(\theta_\ell) - g(\theta_\ell) \right] = \frac{1}{4\pi} \int_{-\pi}^{\pi} \left[g(\theta_\ell + \lambda) - g(\theta_\ell) \right] \frac{\sin\left(n + \frac{1}{2}\right)\lambda}{\sin\frac{\lambda}{2}} \, d\lambda$

(3.71) $\qquad \frac{1}{2}\left[\tilde{g}_N(\theta_\ell) - g(\theta_\ell) \right] = \frac{1}{4\pi} \int_{-\pi}^{\pi} \left[g(\theta_\ell - \lambda) - g(\theta_\ell) \right] \frac{\sin\left(n + \frac{1}{2}\right)\lambda}{\sin\frac{\lambda}{2}} \, d\lambda.$

We then obtain, on summing (3.70) and (3.71),

(3.72)
$$\tilde{g}_N(\theta_\ell) - g(\theta_\ell) = \frac{1}{4\pi} \int_{-\pi}^{\pi} \left\{ \left[g(\theta_\ell + \lambda) - 2g(\theta_\ell) + g(\theta_\ell - \lambda) \right] \frac{\sin\left(n + \frac{1}{2}\right)\lambda}{\sin\frac{\lambda}{2}} \right\} d\lambda$$

$$= \frac{1}{2\pi} \int_{0}^{\pi} \left\{ \left[g(\theta_\ell + \lambda) - 2g(\theta_\ell) + g(\theta_\ell - \lambda) \right] \frac{\sin\left(n + \frac{1}{2}\right)\lambda}{\sin\frac{\lambda}{2}} \right\} d\lambda.$$

In the derivation above it is valid to express $\int_{-\pi}^{\pi} \{F(\lambda)\} \, d\lambda = 2 \int_{0}^{\pi} \{F(\lambda)\} \, d\lambda$ because the integrand $F(\lambda)$ on the right side of (3.72) is an even function, i.e., $F(\lambda) = F(-\lambda)$ for $\lambda \in [0, \pi]$. Since $g(\theta_\ell) = \frac{1}{2}\left[g(\theta_\ell^+) + g(\theta_\ell^-) \right]$ at any point θ_ℓ, we substitute this into both sides of (3.72), and we obtain

(3.73)
$$\tilde{g}_N(\theta_\ell) - \frac{g(\theta_\ell^+) + g(\theta_\ell^-)}{2}$$

$$= \frac{1}{2\pi} \int_{0}^{\pi} \left[g(\theta_\ell + \lambda) - g(\theta_\ell^+) - g(\theta_\ell^-) + g(\theta_\ell - \lambda) \right] \frac{\sin\left(n + \frac{1}{2}\right)\lambda}{\sin\frac{\lambda}{2}} \, d\lambda.$$

$$= \frac{1}{2\pi} \int_{0}^{\pi} \left[g(\theta_\ell + \lambda) - g(\theta_\ell^+) \right] \frac{\sin\left(n + \frac{1}{2}\right)\lambda}{\sin\frac{\lambda}{2}} \, d\lambda$$

$$+ \frac{1}{2\pi} \int_{0}^{\pi} \left[g(\theta_\ell - \lambda) - g(\theta_\ell^-) \right] \frac{\sin\left(n + \frac{1}{2}\right)\lambda}{\sin\frac{\lambda}{2}} \, d\lambda.$$

To prove that the limit of the left side tends to zero as $N = 2n + 1 \to \infty$, we must show that both of the integrals on the right side tend to zero as $n \to \infty$. We show next that the first integral tends to zero. Because we can rewrite

$$I_1 = \frac{1}{2\pi} \int_{0}^{\pi} \left[g(\theta_\ell + \lambda) - g(\theta_\ell^+) \right] \frac{\sin\left(n + \frac{1}{2}\right)\lambda}{\sin\frac{\lambda}{2}} \, d\lambda$$

$$= \frac{1}{2\pi} \int_{0}^{\pi} \left[\frac{g(\theta_\ell + \lambda) - g(\theta_\ell^+)}{\lambda} \right] \left[\frac{\lambda}{\sin\frac{\lambda}{2}} \right] \sin\left(n + \frac{1}{2}\right)\lambda \, d\lambda,$$

we shall let

$$U(\lambda) = \frac{g(\theta_\ell + \lambda) - g(\theta_\ell^+)}{\lambda}, \quad V(\lambda) = \frac{\lambda}{\sin\frac{\lambda}{2}},$$

and our objective now is to show that $G(\lambda) = U(\lambda)V(\lambda)$ is integrable on $[0, \pi]$. With this result we can immediately apply the identity (3.65) from Example 3.57 (an application of Riemann s lemma) to obtain

$$\lim_{n \to \infty} \frac{1}{\pi} \int_{0}^{\pi} G(\lambda) \sin\left(n + \frac{1}{2}\right)\lambda \, d\lambda = 0$$

and it follows that the integral I_1 tends to zero as $n \to \infty$.

Since the product of two integrable functions is integrable, we examine $U(\lambda)$ and $V(\lambda)$ separately. We see that $U(\lambda)$ is unde ned at $\lambda = 0$; for $U(\lambda)$ to be piecewise continuous (and thus integrable on $[0, \pi]$), we need to show that it has a n ite limit as $\lambda \to 0^+$. This is indeed the case, because

$$\lim_{\lambda \to 0^+} U(\lambda) = \lim_{\lambda \to 0^+} \frac{g(\theta_\ell + \lambda) - g(\theta_\ell^+)}{\lambda} = g'(\theta_\ell^+),$$

and we know that the piecewise smooth function $g(\theta)$ has a   nite one-sided derivative every-where.

For $V(\lambda)$ we encounter the same dif culty at $\lambda = 0$, and we need to show that it has a n ite limit as $\lambda \to 0^{+}$. By L Hospital s rule, we obtain

$$\lim_{\lambda \to 0} V(\lambda) = \lim_{\lambda \to 0} \frac{\lambda}{\sin \frac{\lambda}{2}} = \lim_{\lambda \to 0} \frac{2}{\cos \frac{\lambda}{2}} = 2,$$

so $V(\lambda)$ is piecewise continuous and thus integrable on $[0, \pi]$.

The second integral in (3.73) can be shown to tend to zero in a similar manner. With both integrals on the right side of (3.73) tending to zero, we obtain the desired result for every $\theta_\ell \in [-\pi, \pi]$,

$$\lim_{N \to \infty} \left[\tilde{g}_N(\theta_\ell) - \frac{g(\theta_\ell^{+}) + g(\theta_\ell^{-})}{2} \right] = 0.$$

∎

It is worth noting that we may use the same proof for the following theorem, where the Fourier series may not converge at every point because we do not require g to be piecewise smooth.

Theorem 3.61 If $g(\theta)$ is an integrable function of period 2π, its Fourier series converges to $\frac{1}{2}\left[g(\theta_\ell^{+}) + g(\theta_\ell^{-})\right]$ at any point θ_ℓ where g has both a right-sided and a left-sided derivative.

Corollary 3.62 If $g(\theta)$ is an integrable function of period 2π, its Fourier series converges to $g(\theta_\ell)$ at any point where g is differentiable.

Proof: If g is differentiable at θ_ℓ, then

(i) g is continuous at θ_ℓ and we have $g(\theta_\ell) = g(\theta_\ell^{+}) = g(\theta_\ell^{-})$;

(ii) g satis es Theorem 3.61, and its Fourier series converges to $\frac{1}{2}\left[g(\theta_\ell^{+}) + g(\theta_\ell^{-})\right]$, which must equal $g(\theta_\ell)$ based on the result from (i).

∎

3.10.3 The rate of convergence—a global property

Given a piecewise smooth function $g(\theta)$ of period 2π, we have shown that its Fourier series converges at every point $\theta_\ell \in [-\pi, \pi]$ in Theorem 3.60. The pointwise convergence of the Fourier series is a local property, because the number of terms required for a partial sum to get closed to a limit at a particular point varies with the location of the point θ_ℓ i.e., the local rate of convergence varies from point to point, which is the cause of the Gibbs phenomenon to be discussed in the next subsection. In this subsection we study the convergence of the Fourier series in the global sense; that is, we examine how fast the coef cients A_k and B_k tend to zero as $k \to \infty$ th is provides a way to measure how fast a converging series tends to its limit knowing only that all basis functions $\cos k\theta$ and $\sin k\theta$ are bounded by unity in size. As demonstrated by Example 3.14 in Sections 3.4 and 3.10.1, it is possible to obtain more than one Fourier series expansion when a time-limited function is extended into different periodic functions (even, odd, or neither), and what affected the convergence rate is the continuity of the nth derivative ($n \geq 0$) of the extended function. This mathematical connection can now be formally established as shown below.

Let $g(\theta)$ be a piecewise smooth function of period 2π (whether $g(\theta)$ is given or it results from a periodic extension does not affect our analysis.) We consider the formal Fourier series expansion of $g(\theta)$ given by

$$\tilde{g}(\theta) = \frac{A_0}{2} + \sum_{k=1}^{\infty} A_k \cos k\theta + B_k \sin k\theta,$$

with coef cien ts from (3.14) and (3.15), namely,

$$A_k = \frac{1}{\pi} \int_{-\pi}^{\pi} g(\theta) \cos k\theta \, d\theta, \quad k = 0, 1, 2, \ldots$$

$$B_k = \frac{1}{\pi} \int_{-\pi}^{\pi} g(\theta) \sin k\theta \, d\theta, \quad k = 1, 2, \ldots .$$

Since both $g(\theta)$ and $g'(\theta)$ are presumed to be piecewise continuous, we obtain each coef-cien t by evaluating the integral on the right side over $\mathcal{M}$ subintervals (pieces)

$$A_k = \frac{1}{\pi} \sum_{m=1}^{\mathcal{M}} \int_{\theta_{m-1}}^{\theta_m} g(\theta) \cos k\theta \, d\theta, \quad B_k = \frac{1}{\pi} \sum_{m=1}^{\mathcal{M}} \int_{\theta_{m-1}}^{\theta_m} g(\theta) \sin k\theta \, d\theta,$$

where $\theta_0 = -\pi$, $\theta_{\mathcal{M}} = \pi$, and the other θ_m s mark the ends of each subinterval (piece) where the *potential* jump discontinuities in the function or in its derivative occur. Note that $g(\theta)$ has corners wherever $g'(\theta)$ has jump discontinuities.

To apply the technique of integration by parts to the integral $\int_a^b g(\theta) \cos k\theta \, d\theta$, we let $\mathbf{u} = g(\theta)$, $\mathbf{dv} = \cos k\theta \, d\theta$, and we obtain

$$\int_a^b g(\theta) \cos k\theta \, d\theta = \mathbf{uv}\Big|_a^b - \mathbf{v}\,\mathbf{du} = \frac{1}{k} g(\theta) \sin k\theta \Big|_a^b - \frac{1}{k} \int_a^b \sin k\theta \, g'(\theta) \, d\theta.$$

Using the above result with $a = \theta_{m-1}$ and $b = \theta_m$, we obtain

$$A_k = \left[\frac{1}{k\pi} \sum_{m=1}^{\mathcal{M}} g(\theta) \sin k\theta \Big|_{\theta_{m-1}}^{\theta_m} \right] - \frac{1}{k\pi} \sum_{m=1}^{\mathcal{M}} \int_{\theta_{m-1}}^{\theta_m} g'(\theta) \sin k\theta \, d\theta.$$

By expanding the sum of the integrated terms, we have (for an example with $\mathcal{M} = 3$)

$$\frac{1}{k\pi} \sum_{m=1}^{3} g(\theta) \sin k\theta \Big|_{\theta_{m-1}}^{\theta_m} = \frac{1}{k\pi} \Big[g(\theta_1^-) \sin k\theta_1 - g(\theta_0^+) \sin k\theta_0$$

$$+ g(\theta_2^-) \sin k\theta_2 - g(\theta_1^+) \sin k\theta_1$$

$$+ g(\theta_3^-) \sin k\theta_3 - g(\theta_2^+) \sin k\theta_2 \Big]$$

$$= \begin{cases} 0 & \text{if } g(\theta) \text{ is everywhere continuous,} \\ \dfrac{\alpha_k}{k} \leq \dfrac{c}{k} & \text{if } g(\theta) \text{ has jump discontinuities,} \end{cases}$$

where c is a constant independent of k, because $g(\theta_m^+)$ and $g(\theta_m^-)$ do not vary with k, and $\sin k\theta_m$ is bounded by 1 in size. Note that $g(\theta_3^-) = g(\pi^-)$, and $g(\theta_0^+) = g(-\pi^+) = g(\pi^+)$ because $g(\theta)$ has a period of 2π. When $g(\theta)$ is everywhere continuous, we have $g(\theta_3^-) = g(\theta_0^+)$ and $g(\theta_m^-) = g(\theta_m^+)$ at all other end points, so the six integrated terms sum to zero. We can then repeat the integration by parts process to evaluate the new integral that remains and obtain

$$A_k = -\frac{1}{k\pi} \sum_{m=1}^{\mathcal{M}} \int_{\theta_{m-1}}^{\theta_m} g'(\theta) \sin k\theta \, d\theta$$

$$= \left[\frac{1}{k^2\pi} \sum_{m=1}^{\mathcal{M}} g'(\theta) \cos k\theta \Big|_{\theta_{m-1}}^{\theta_m}\right] - \frac{1}{k^2\pi} \sum_{m=1}^{\mathcal{M}} \int_{\theta_{m-1}}^{\theta_m} g''(\theta) \cos k\theta \, d\theta.$$

By the same argument we will again have the integrated terms summed to zero if the first derivative $g'(\theta)$ is everywhere continuous, and we are left with a new integral which now involves factor $1/k^2$ and $g''(\theta)$ in the integrand. This process can be repeated until the integrated terms involving the factor $1/k^{n+1}$ and the nth derivative $g^{(n)}(\theta)$ cannot cancel out because $g^{(n)}(\theta)$ is discontinuous somewhere. A summary of our findings follows.

If $g(\theta)$ has jump discontinuities, we expect the coefficients A_k to be of order $1/k$, because the integrated terms involving the factor $1/k$ do not cancel out in the analysis. As a result, as $k \to \infty$, the coefficients A_k approach zero at a rate proportional to $1/k$. On the other hand, if $g(\theta)$ is everywhere continuous, then the coefficients A_k tend to zero *at least* as fast as $1/k^2$. If, in addition, the first derivative $g'(\theta)$ is discontinuous somewhere, then A_k s tend to zero at a rate proportional to $1/k^2$.

In general, if piecewise smooth $g(\theta)$ and its first $n-1$ derivatives are everywhere continuous, then as $k \to \infty$, the coefficients A_k tend to zero *at least* as fast as $1/k^{n+1}$. If, in addition, the nth derivative is discontinuous somewhere, then the A_k s tend to zero at a rate proportional to $1/k^{n+1}$. By essentially identical analysis the preceding statements are true for coefficients B_k.

Now that we complete the analysis, it would be timely and useful to revisit Example 3.14 in Section 3.10.1, in which we examined and compared the different convergence rates of three Fourier series.

3.10.4 The Gibbs phenomenon

We shall study the Gibbs phenomenon using Example 3.53 from Section 3.9.4, in which we have shown that for the periodic square wave function

$$g(\theta) = \begin{cases} 1, & 0 < \theta \le \pi \\ -1, & -\pi < \theta \le 0 \end{cases}, \quad g(\theta + 2\pi) = g(\theta),$$

the partial sum of the first N nonzero terms of its Fourier series is given by

$$(3.74) \qquad \tilde{g}_N(\theta) = \frac{4}{\pi}\left[\sin\theta + \frac{\sin 3\theta}{3} + \cdots + \frac{\sin(2N-1)\theta}{2N-1}\right] = \frac{4}{\pi} \sum_{k=1}^{N} \frac{\sin(2k-1)\theta}{2k-1}.$$

Observe that $g(\theta)$ has jump discontinuities of size 2 at $\theta = 0, \pm\pi, \pm 2\pi, \ldots$, and that the graphs of $\tilde{g}_5(\theta)$, $\tilde{g}_7(\theta)$, $\tilde{g}_9(\theta)$, and $\tilde{g}_{11}(\theta)$ in Figure 3.11 show undying ripples moving toward (and staying) at these discontinuities. This peculiar effect is known as the Gibbs phenomenon which we can neither reduce nor eliminate by including more terms from the Fourier series, because it is caused by the *nonuniform* pointwise convergence of the (infinite) Fourier series near the jump discontinuities.

Figure 3.11 Gibbs phenomenon and nite Fourier series of the square wave.

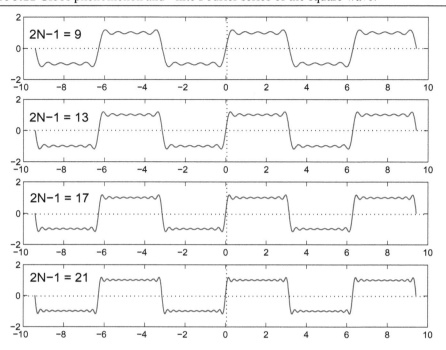

Since the sizes of the ripples are determined by the local maxima and minima (corresponding to the peaks and valleys in the graph) of $\tilde{g}_N(\theta)$, we proceed to nd the local maxima/minima by rst solving the nonlinear equation $\tilde{g}'_N(\theta) = 0$, where

(3.75)
$$
\begin{aligned}
\tilde{g}'_N(\theta) &= \frac{4}{\pi}\sum_{k=1}^{N}\cos(2k-1)\theta = \frac{4}{\pi}\left[\frac{1}{\sin\theta}\sum_{k=1}^{N}\sin\theta\cos(2k-1)\theta\right] \\
&= \frac{4}{\pi}\left[\frac{1}{\sin\theta}\sum_{k=1}^{N}\frac{\sin(2k)\theta - \sin(2k-2)\theta}{2}\right] \\
&= \frac{2}{\pi}\frac{\sin(2N)\theta}{\sin\theta}.
\end{aligned}
$$

On solving $\tilde{g}'_N(\theta) = 0$ within the half period $(0,\,\pi)$ we obtain zeros at $\theta_r = r\pi/(2N)$, $r = 1,\,2,\,\ldots,\,2N-1$. To *estimate* the local maximum or minimum values of $\tilde{g}_N(\theta)$ at the θ_r s as $N \to \infty$, we may express it as an integral obtained directly from its derivative $\tilde{g}'_N(\theta)$ according to the Fundamental Theorem of Calculus:

(3.76)
$$
\tilde{g}_N(\theta) = \tilde{g}_N(\theta) - \tilde{g}_N(0) = \int_0^\theta \tilde{g}'_N(\lambda)\,d\lambda = \frac{2}{\pi}\int_0^\theta \frac{\sin(2N)\lambda}{\sin\lambda}\,d\lambda.
$$

At $\theta = \dfrac{r\pi}{2N}$, $r = 1, 2, \ldots, 2N - 1$, we have

$$\tilde{g}_N\left(\frac{r\pi}{2N}\right) = \frac{2}{\pi}\int_0^{r\pi/(2N)} \frac{\sin(2N)\lambda}{\sin\lambda}\, d\lambda$$

$$= \frac{2}{\pi}\int_0^{r\pi} \frac{\sin\mu}{\sin\left(\frac{\mu}{2N}\right)} \frac{1}{2N}\, d\mu \qquad \text{(change variable: } \mu = 2N\lambda\text{)}$$

(3.77)

$$= \frac{2}{\pi}\int_0^{r\pi} \frac{\sin\mu}{\mu} \frac{\frac{\mu}{2N}}{\sin\left(\frac{\mu}{2N}\right)}\, d\mu$$

$$= \frac{2}{\pi}\int_0^{r\pi} \frac{\sin\lambda}{\lambda}\left[\frac{\sin\left(\frac{\lambda}{2N}\right)}{\frac{\lambda}{2N}}\right]^{-1}\, d\lambda. \qquad \text{(change } \mu \text{ to } \lambda\text{)}$$

As $N \to \infty$ we have, by L Hospital s rule,

$$\lim_{N\to\infty} \frac{\sin\left(\frac{\lambda}{2N}\right)}{\frac{\lambda}{2N}} = \lim_{\nu\to 0} \frac{\sin\nu}{\nu} = \lim_{\nu\to 0} \frac{\cos\nu}{1} = 1,$$

which allows us to obtain the limit value of (3.77):

(3.78) $$\lim_{N\to\infty} \tilde{g}_N\left(\frac{r\pi}{2N}\right) = \lim_{\nu\to 0} \frac{2}{\pi}\int_0^{r\pi} \frac{\sin\lambda}{\lambda}\left[\frac{\sin\nu}{\nu}\right]^{-1}\, d\lambda = \frac{2}{\pi}\int_0^{r\pi} \frac{\sin\lambda}{\lambda}\, d\lambda,$$

where the numerical values of the sine-integral function,

(3.79) $$\mathbf{Si}\,(\theta) \overset{\text{def}}{=} \int_0^\theta \frac{\sin\lambda}{\lambda}\, d\lambda,$$

can be found in standard mathematical tables. At $\theta_1 = \pi/(2N)$ for $N = 100$, we have the rst local maximum

$$\tilde{g}_{100}(\theta_1) = \tilde{g}_{100}(0.016) \approx \frac{2}{\pi}\,\mathbf{Si}\,(\pi) \approx \frac{2}{\pi}(1.8516) = 1.1788 > g(\theta_1).$$

Since $g(\theta_1) = 1$ and the jump size is 2 at the discontinuity at $\theta = 0$, the size of the overshoot relative to the jump is measured by $(1.1788 - 1)/2 \approx 9\%$, which is based on the limiting value so it can not be further reduced or eliminated by letting N approach ∞. At $\theta_2 = 2\pi/(2N)$ for $N = 100$, we have the next local minimum

$$\tilde{g}_{100}(\theta_2) = \tilde{g}_{100}(0.0314) \approx \frac{2}{\pi}\,\mathbf{Si}\,(2\pi) \approx \frac{2}{\pi}(1.4182) = 0.90285 < g(\theta_2).$$

Since $g(\theta_2) = 1$ and the jumpsize is 2 at the nearest discontinuity at $\theta = 0$, the size of the unders hoot relative to the jump is measured by $(1 - 0.90285)/2 \approx 5\%$. Graphically, the overshoots and undershoots of $\tilde{g}_N(\theta)$ are compressed/pinched into a spike (of the same magnitude) at the nearest jump as $N \to \infty$. Mathematically, the pointwise convergence ensured by Theorem 3.1 is *not* compromised because the nonin nitesimal overshoot occurs over an interval whose length approaches zero as $N \to \infty$.

3.10.5 The Dirichlet kernel perspective

In this subsection we shall try to understand the Gibbs phenomenon from a perspective which can be applied to a wide range of functions, including but not limited to the square wave function studied exclusively in the last subsection. We begin by recalling Lemma 3.58, where we proved that given an integrable function $g(\theta)$ of period 2π, we may express the $N = (2n+1)$-term partial sum of its Fourier series in the integral form

$$(3.80) \qquad \tilde{g}_N(\theta) = \frac{1}{2\pi} \int_{-\pi}^{\pi} g(\theta - \lambda) \frac{\sin(n + \frac{1}{2})\lambda}{\sin \frac{\lambda}{2}} \, d\lambda.$$

By de n ing the Dirichlet integrating kernel

$$(3.81) \qquad \mathbf{D_n}(\lambda) \equiv \sum_{k=-n}^{n} e^{-jk\lambda} = \frac{\sin(n + \frac{1}{2})\lambda}{\sin \frac{\lambda}{2}}, \qquad \text{(recall formula (3.68))}$$

we may express (3.80) as

$$(3.82) \qquad \tilde{g}_N(\theta) = \frac{1}{2\pi} \int_{-\pi}^{\pi} g(\theta - \lambda) \, \mathbf{D_n}(\lambda) \, d\lambda,$$

where both $g(\theta)$ and $\mathbf{D_n}(\lambda)$ are periodic with period 2π th is property combined with the change of variable allows us to derive one more useful expression for $\tilde{g}_N(\theta)$:

$$\begin{aligned}
\tilde{g}_N(\theta) &= \frac{1}{2\pi} \int_{-\pi}^{\pi} g(\theta - \lambda) \, \mathbf{D_n}(\lambda) \, d\lambda \\
&= \frac{1}{2\pi} \int_{\theta-\pi}^{\theta+\pi} g(\mu) \, \mathbf{D_n}(\theta - \mu) \, d\mu \qquad \text{(change variable: } \mu = \theta - \lambda\text{)} \\
&= \frac{1}{2\pi} \int_{-\pi}^{\pi} g(\mu) \, \mathbf{D_n}(\theta - \mu) \, d\mu. \qquad (\because g \text{ and } \mathbf{D_n} \text{ are periodic})
\end{aligned}$$

We have thus proved that

$$(3.83) \qquad \boxed{\tilde{g}_N(\theta) = \frac{1}{2\pi} \int_{-\pi}^{\pi} g(\theta - \lambda) \, \mathbf{D_n}(\lambda) \, d\lambda = \frac{1}{2\pi} \int_{-\pi}^{\pi} g(\lambda) \, \mathbf{D_n}(\theta - \lambda) \, d\lambda.}$$

The Dirichlet kernel is an important tool in mathematical analysis and applications we   shall discuss rst its alternate forms and main properties before we use it to explain the Gibbs effect.

1. The Dirichlet kernel in common use may be de ned by any one of the following formulas:

$$(3.84) \qquad \mathbf{D_n}(\lambda) \equiv \sum_{k=-n}^{n} e^{-jk\lambda} = \frac{\sin(n + \frac{1}{2})\lambda}{\sin \frac{\lambda}{2}}.$$

   When the same de n itions are expressed in terms of $N = 2n+1$ instead of n, we shall use a different notation to avoid confusion:

$$(3.85) \qquad \mathbf{D}(N, \lambda) \equiv \sum_{k=-\frac{N-1}{2}}^{\frac{N-1}{2}} e^{-jk\lambda} = \frac{\sin(0.5N\lambda)}{\sin(0.5\lambda)}.$$

2. The Dirichlet kernel $\mathbf{D_n}(\lambda)$ is periodic and it is an even function as shown by the graphs for $n = 8, 12, 16, 20$ in Figure 3.12 below. Over the period $(-\pi, \pi)$ the function $\mathbf{D_n}(\lambda)$ has zeros at $\lambda_{\pm r} = \pm 2r\pi/(2n+1)$ for $r = 1, 2, \ldots, n$. The area between $\lambda_1 = 2\pi/(2n+1)$ and $\lambda_{-1} = -2\pi/(2n+1)$ is called the m ainlobe of the kernel; the sidelobes are areas between adjacent zeros on each side The graph of $\mathbf{D_n}(\lambda)$ for $n = 8$ is shown with more details in Figure 3.13.

Figure 3.12 The Dirichlet kernel $\mathbf{D_n}(\lambda)$ for $n = 8, 12, 16, 20$.

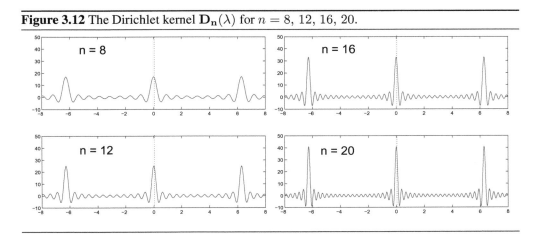

Figure 3.13 One period of the Dirichlet kernel $\mathbf{D_n}(\lambda)$ for $n = 8$.

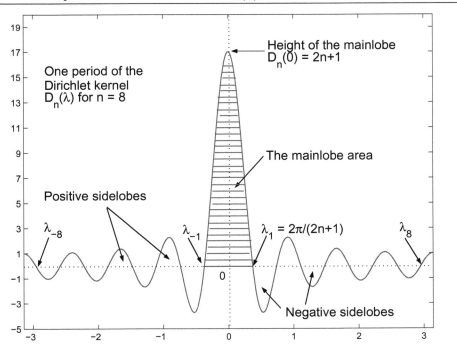

3. The height of the mainlobe is given by $\max \mathbf{D_n}(\lambda) = \mathbf{D_n}(0) = 2n+1$, or equivalently, $\mathbf{D}(N, 0) = N$. There is also a local maximum or minimum between every pair of adjacent zeros, which occurs approximately at the midpoint $\lambda = \pm(2r+1)\pi/(2n+1)$. The peak of the highest sidelobe (in absolute value) occurs at $\lambda = \pm 3\pi/(2n+1)$ and its value is approximately $(2n+1)/(1.5\pi)$, which is obtained by taking the absolute value of the limit of $\mathbf{D_n}(\lambda)$ at $\lambda = \pm 3\pi/(2n+1)$:

$$\lim_{n \to \infty} \mathbf{D_n}\left(\frac{\pm 3\pi}{2n+1}\right) = \lim_{n \to \infty} \frac{\sin(\pm 1.5\pi)}{\sin\left(\frac{\pm 1.5\pi}{2n+1}\right)} \approx \frac{\mp 1}{\left(\frac{\pm 1.5\pi}{2n+1}\right)} = \frac{-(2n+1)}{1.5\pi},$$

where we have made use of the approximation $\sin \theta \approx \theta$ as θ tends to 0.

4. The ratio of the highest sidelobe to the mainlobe is therefore $1/1.5\pi$. When this value represents the signal-to-noise ratio in signal ltering applications, it is customarily expressed in decibels (tenths of a bel), abbreviated dB, and it equals -13.5 dB $= 20 \log_{10}(1/1.5\pi)$ according to the de n ition which equates ± 20 dB to a ten-fold increase or decrease in the peak values:

$$\text{decibel units} \equiv 20 \log_{10}|\text{ratio}|.$$

Observe that the height (in absolute value) of the sidelobes is decreasing on either side, and each negative sidelobe is followed by a positive sidelobe. At each endpoint of the period from $-\pi$ to π, we have $\mathbf{D_n}(\pm \pi) = -1$ (if n is odd) or 1 (if n is even).

5. Recall that in the proof of Lemma 3.59 we have shown that the total signed area

(3.86) $$\int_{-\pi}^{\pi} \mathbf{D_n}(\lambda)\, d\lambda = \int_{-\pi}^{\pi} \mathbf{D}(N, \lambda)\, d\lambda = 2\pi.$$

Observe from the graphs that the total area represented by the sidelobes must be negative, because each negative area is larger than the next positive area. From this we may infer that the area of the mainlobe is greater than 2π.

6. Because of the symmetry $\mathbf{D_n}(\lambda) = \mathbf{D_n}(-\lambda)$, the graph of $\mathbf{D_n}(\theta_\ell - \lambda)$ can be obtained by centering the graph of $\mathbf{D_n}(\lambda)$ at θ_ℓ.

We may now apply the Dirichlet kernel to the example from Section 3.10.4: recall that $g(\theta)$ is the square wave function with $g(\theta) = 1$ over $(0, \pi]$ and $g(\theta) = -1$ over $(-\pi, 0]$. Since 100 terms of nonzero B_{2k+1} correspond to $n = 200$ and $N = 2n + 1 = 401$ (including terms with zero A_k and zero B_{2k} coef cien ts), we can now use the Dirichlet kernel $\mathbf{D}_{200}(\lambda)$ to evaluate the approximating partial sum $\tilde{g}_{401}(\theta)$ at the jump at $\theta = 0$:

$$\tilde{g}_{401}(0) = \frac{1}{2\pi} \int_{-\pi}^{\pi} g(\lambda)\, \mathbf{D}_{200}(0 - \lambda)\, d\lambda, \quad \text{where } \mathbf{D_n}(-\lambda) = \mathbf{D_n}(\lambda),$$

$$= \frac{1}{2\pi} \left[\int_0^{\pi} (+1)\, \mathbf{D}_{200}(\lambda)\, d\lambda + \int_{-\pi}^0 (-1)\, \mathbf{D}_{200}(\lambda)\, d\lambda \right]$$

$$= 0,$$

which is the (expected) midpoint between $g(0^+) = 1$ and $g(0^-) = -1$; the same formula can also be used to evaluate $\tilde{g}_{401}(\theta)$ in the neighborhood $(0, \epsilon]$ of the jump at $\theta = 0$:

$$\tilde{g}_{401}(\epsilon) = \frac{1}{2\pi} \int_{-\pi}^{\pi} g(\lambda) \, \mathbf{D}_{200}(\epsilon - \lambda) \, d\lambda$$

$$= \frac{1}{2\pi} \left[\int_0^{\pi} \mathbf{D}_{200}(\epsilon - \lambda) \, d\lambda - \int_{-\pi}^0 \mathbf{D}_{200}(\epsilon - \lambda) \, d\lambda \right]$$

$$\approx \frac{1}{2\pi} \int_{-\epsilon}^{\epsilon} \mathbf{D}_{200}(\lambda) \, d\lambda.$$

Now, if $\epsilon \approx \lambda_1 = 2\pi/401 \approx 0.016$ (as before), we know $\tilde{g}_{401}(0.016) > 1$ because the area inside the mainlobe must be greater than 2π so that the total signed area equals 2π. Since $\lambda_1 = 2\pi/(2n+1)$ tends to the origin as $n \to \infty$, the mainlobe becomes taller and narrower, whereas its area remains greater than 2π. This explains the lack of disappearance of the overshoot at the jump as $n \to \infty$. To obtain the numerical values for the areas of the mainlobe and sidelobes, we shall again make use of the $\mathbf{Si}(\lambda)$ function, which is related to the integral of the Dirichlet kernel $\mathbf{D}(N, \lambda)$ as shown in the following lemma:

Lemma 3.63

(3.87)
$$\lim_{N \to \infty} \frac{1}{2\pi} \int_{-2r\pi/N}^{2r\pi/N} \mathbf{D}(N, \lambda) \, d\lambda = \frac{2}{\pi} \mathbf{Si}(r\pi).$$

Proof:

$$\lim_{N \to \infty} \frac{1}{2\pi} \int_{-2r\pi/N}^{2r\pi/N} \mathbf{D}(N, \lambda) \, d\lambda = \lim_{N \to \infty} \frac{1}{\pi} \int_0^{2r\pi/N} \mathbf{D}(N, \lambda) \, d\lambda$$

$$= \lim_{N \to \infty} \frac{1}{\pi} \int_0^{2r\pi/N} \frac{\sin(0.5N\lambda)}{\sin(0.5\lambda)} \, d\lambda$$

$$= \lim_{N \to \infty} \frac{2}{\pi} \int_0^{r\pi} \frac{\sin \mu}{N \sin \frac{\mu}{N}} \, d\mu \qquad (\text{let } \mu = 0.5N\lambda)$$

$$= \frac{2}{\pi} \int_0^{r\pi} \frac{\sin \mu}{N \left(\frac{\mu}{N}\right)} \, d\mu \qquad (\because \sin \theta \to \theta \text{ as } \theta \to 0)$$

$$= \frac{2}{\pi} \int_0^{r\pi} \frac{\sin \mu}{\mu} \, d\mu$$

$$= \frac{2}{\pi} \mathbf{Si}(r\pi).$$

∎

3.10.6 Eliminating the Gibbs effect by the Cesaro sum

The Gibbs effect can be eliminated if we use the arithmetic mean of the successive partial sums from the Fourier series of $g(\theta)$ to smooth the approximation. That is, instead of using the partial sum $\tilde{g}_N(\theta)$, we take the average of all partial sums from $\tilde{g}_1(\theta)$ to $\tilde{g}_N(\theta)$ in succession, the result is called the Cesaro sum, which is formally defined as

(3.88)
$$\tilde{f}_N(\theta) \equiv \frac{1}{n+1} \sum_{\ell=0}^{n} \tilde{g}_{2\ell+1}(\theta) = \frac{1}{n+1} \left(\tilde{g}_1(\theta) + \tilde{g}_3(\theta) + \cdots + \tilde{g}_{2n+1}(\theta) \right),$$

where $N = 2n+1$, and $\tilde{g}_{2\ell+1}(\theta) = \sum_{k=-\ell}^{\ell} C_\ell e^{j\ell\theta}$. We assume that $g(\theta)$ is discontinuous with finite jumps as before. Our objective is to show that the Cesaro sum $\tilde{f}_N(\theta)$ does not exhibit the Gibbs effect suffered by the partial sum $\tilde{g}_N(\theta)$.

To relate the Cesaro sum $\tilde{f}_N(\theta)$ directly to the general truncated Fourier series $\tilde{g}_N(\theta) = \sum_{k=-n}^{n} C_k e^{jk\theta}$, we express

$$
\begin{aligned}
\tilde{f}_N(\theta) &\equiv \frac{1}{n+1} \sum_{\ell=0}^{n} \tilde{g}_{2\ell+1}(\theta) \\
&= \frac{1}{n+1} \Big[(n+1)C_0 + nC_{-1}e^{-j\theta} + nC_1 e^{j\theta} + \cdots + C_n e^{jn\theta} \Big] \\
&= C_0 + \sum_{\ell=1}^{n} \frac{n+1-\ell}{n+1} \left(C_{-\ell} e^{-j\ell\theta} + C_\ell^{j\ell\theta} \right) \\
&= \sum_{k=-n}^{n} \left(\alpha_k C_k \right) e^{jk\theta},
\end{aligned}
$$

(3.89)

where each Fourier coefficient C_k in the partial sum $\tilde{g}_N(\theta)$ has been modified by a factor α_k defined by

(3.90)
$$
\alpha_k = \frac{n+1-|k|}{n+1}, \quad -n \leq k \leq n.
$$

Observe that the α_k factors are always positive and their values decay linearly from $\alpha_0 = 1$ to $\alpha_n = \dfrac{1}{n+1}$ as $|k|$ increases from 0 to n.

To relate the Cesaro sum $\tilde{f}_N(\theta)$ directly to the original function $g(\theta)$, we shall prove the following lemma:

Lemma 3.64 The Cesaro sums of an integrable function $g(\theta)$ of period 2π can be expressed in the integral form

(3.91)
$$
\tilde{f}_N(\theta) = \frac{1}{2\pi} \int_{-\pi}^{\pi} g(\theta - \lambda) \mathbf{F_n}(\lambda) \, d\lambda,
$$

where $\mathbf{F_n}(\lambda)$ is called the Fejer kernel, which is the arithmetic mean of the $n+1$ successive Dirichlet kernels:

(3.92)
$$
\mathbf{F_n}(\lambda) = \frac{1}{n+1} \sum_{\ell=0}^{n} \mathbf{D}_\ell(\lambda) = \frac{\sin^2(n+1)\frac{\lambda}{2}}{(n+1)\sin^2\frac{\lambda}{2}}.
$$

Proof:

$$\tilde{f}_N(\theta) \equiv \frac{1}{n+1} \sum_{\ell=0}^{n} \tilde{g}_{2\ell+1}(\theta)$$

$$= \frac{1}{n+1} \sum_{\ell=0}^{n} \left[\frac{1}{2\pi} \int_{-\pi}^{\pi} g(\theta - \lambda) \mathbf{D}_\ell(\lambda) \, d\lambda \right] \qquad \text{(by Lemma 3.58)}$$

$$= \frac{1}{2\pi} \int_{-\pi}^{\pi} g(\theta - \lambda) \left[\frac{1}{n+1} \sum_{\ell=0}^{n} \mathbf{D}_\ell(\lambda) \right] d\lambda$$

$$= \frac{1}{2\pi} \int_{-\pi}^{\pi} g(\theta - \lambda) \left[\sum_{\ell=0}^{n} \frac{\sin\left(\ell + \frac{1}{2}\right)\lambda}{(n+1)\sin\frac{\lambda}{2}} \right] d\lambda \qquad \text{(by de n ition of } \mathbf{D}_\ell(\lambda))$$

$$= \frac{1}{2\pi} \int_{-\pi}^{\pi} g(\theta - \lambda) \left[\frac{1}{(n+1)\sin\frac{\lambda}{2}} \sum_{\ell=1}^{n+1} \sin(2\ell - 1)\frac{\lambda}{2} \right] d\lambda$$

$$= \frac{1}{2\pi} \int_{-\pi}^{\pi} g(\theta - \lambda) \left[\frac{\sin^2(n+1)\frac{\lambda}{2}}{(n+1)\sin^2\frac{\lambda}{2}} \right] d\lambda. \qquad \text{(by result from Example 1.4)}$$

$$= \frac{1}{2\pi} \int_{-\pi}^{\pi} g(\theta - \lambda) \mathbf{F_n}(\lambda) \, d\lambda.$$

∎

The Fejer kernel $\mathbf{F_n}(\lambda)$ is periodic with period 2π (one period is shown in Figure 3.14), and it has the following properties:

Property 1. For every n, we have $\left|\mathbf{F_n}(\lambda)\right| = \mathbf{F_n}(\lambda) \geq 0$ for every λ.

Property 2. $\mathbf{F_n}(\lambda) = 0$ for $\lambda = \pm 2r\pi/(n+1)$, where $1 \leq r \leq n$. At $\lambda = 0$, we have

$$(3.93) \qquad \mathbf{F_n}(0) = \frac{1}{n+1} \sum_{\ell=0}^{n} \mathbf{D}_\ell(0) = \frac{1}{n+1} \sum_{\ell=0}^{n} (2\ell + 1) = n + 1.$$

Property 3. Because the signed area of the Dirichlet kernel $\mathbf{D}_\ell(\lambda)$ is 2π for every ℓ, we immediately have

$$(3.94) \qquad \int_{-\pi}^{\pi} \left|\mathbf{F_n}(\lambda)\right| d\lambda = \int_{-\pi}^{\pi} \mathbf{F_n}(\lambda) \, d\lambda = \frac{1}{n+1} \sum_{\ell=0}^{n} \int_{-\pi}^{\pi} \mathbf{D}_\ell(\lambda) \, d\lambda = 2\pi$$

for every n. By contrast $\left|\mathbf{D_n}(\lambda)\right| \neq \mathbf{D_n}(\lambda)$, and

$$\int_{-\pi}^{\pi} \left|\mathbf{D_n}(\lambda)\right| d\lambda > 2\pi.$$

Theorem 3.65 The Cesaro sums of an integrable function g of period 2π are bounded by the maximum value of g. That is, if $|g(\theta)| \leq M$ for every θ, then

$$\left|\tilde{f}_N(\theta)\right| = \left| \frac{1}{2\pi} \int_{-\pi}^{\pi} g(\theta - \lambda) \mathbf{F_n}(\lambda) \, d\lambda \right| \leq M.$$

Figure 3.14 One period of the Fejer kernel $\mathbf{F_n}(\lambda)$ for $n = 8$.

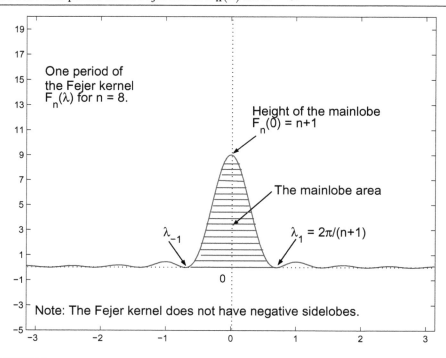

Proof:

$$\left| \tilde{f}_N(\theta) \right| = \left| \frac{1}{2\pi} \int_{-\pi}^{\pi} g(\theta - \lambda)\, \mathbf{F_n}(\lambda)\, d\lambda \right| \qquad \text{(by Lemma 3.64)}$$

$$\leq \frac{1}{2\pi} \int_{-\pi}^{\pi} \left| g(\theta - \lambda) \right| \left| \mathbf{F_n}(\lambda) \right| d\lambda$$

$$\leq \frac{M}{2\pi} \int_{-\pi}^{\pi} \left| \mathbf{F_n}(\lambda) \right| d\lambda$$

$$= M. \qquad\qquad\qquad \big(\text{by property 3 of } \mathbf{F_n}(\lambda)\big)$$

∎

Because for every $N = 2n+1$, the Cesaro sum $\tilde{f}_N(\theta)$ is bounded by the maximum value of $g(\theta)$ according to Theorem 3.65, it cannot overshoot the function and the Gibbs phenomenon will not occur with Cesaro sums. (There is no undershoot because the Fejer kernel has no negative sidelobes.)

As illustrated in Figure 3.15, the computed Cesaro sums of the square wave converge without suffering from the Gibbs effect. As indicated inside each plot in Figure 3.15, the same result is obtained by *either* modifying the Fourier series coef cient C_k or nding the average of the indicated number of partial sums. For example, in the rst plot, the seven partial sums used would be $\tilde{g}_1(\theta),\, \tilde{g}_3(\theta),\, \tilde{g}_5(\theta),\, \ldots,\, \tilde{g}_{13}(\theta)$. Because the Fourier series of the square wave has only nonzero odd-indexed sine terms, $\tilde{g}_1(\theta)$ involves one nonzero sine term, $\tilde{g}_3(\theta)$ involves two nonzero sine terms, and the last partial sum $\tilde{g}_{13}(\theta)$ involves seven nonzero sine terms with the highest index being 13.

Figure 3.15 Illustrating the convergence of the Cesaro sums of the square wave.

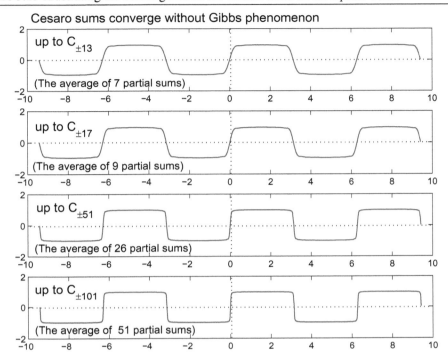

3.10.7 Reducing the Gibbs effect by Lanczos smoothing

The Lanczos method smooths the partial sum $\tilde{g}_N(\theta)$ at each θ by integration instead of summation. That is, we replace the partial sum by the averaged value computed by the de nite integral

$$(3.95) \qquad \tilde{h}_N(\theta) = \frac{1}{2\tau} \int_{\theta-\tau}^{\theta+\tau} \tilde{g}_N(\lambda)\, d\lambda,$$

where $\tau = \pi/n$, and the interval centered at θ has length 2τ, which is the period of the last term $e^{\pm jn\theta} = \cos n\theta \pm j \sin n\theta$ in the unmodi ed partial sum

$$\tilde{g}_N(\theta) = \sum_{k=-n}^{n} C_k\, e^{jk\theta}.$$

It was observed by Lanczos that $2\tau = 2\pi/n$ approximates the period of the ripples produced by the Gibbs effect. By carrying out the integration, we obtain

$$
\tilde{h}_N(\theta) = \frac{1}{2\tau} \int_{\theta-\tau}^{\theta+\tau} \tilde{g}_N(\lambda)\, d\lambda = \frac{1}{2\tau} \int_{\theta-\tau}^{\theta+\tau} \sum_{k=-n}^{n} C_k\, e^{jk\lambda}\, d\lambda
$$

$$
= \frac{1}{2\tau} \sum_{k=-n}^{n} C_k \int_{\theta-\tau}^{\theta+\tau} e^{jk\lambda}\, d\lambda
$$

(3.96)
$$
= \sum_{k=-n}^{n} C_k \left[\frac{e^{jk(\theta+\tau)} - e^{jk(\theta-\tau)}}{2jk\tau} \right]
$$

$$
= \sum_{k=-n}^{n} C_k\, e^{jk\theta} \left[\frac{\sin(k\tau)}{k\tau} \right] \qquad \text{(by Euler s formula)}
$$

$$
= \sum_{k=-n}^{n} \left[C_k\, \sigma_k \right] e^{jk\theta},
$$

which shows the effect on the Fourier coef cient: each C_k is modi ed by the Lanczos sigma factor

(3.97)
$$
\sigma_k \equiv \frac{\sin(k\pi/n)}{k\pi/n}.
$$

The convergence of the Fourier series after its coef cients are modi ed by the Lanczos sigma factor is illustrated in Figure 3.16.

3.10.8   The modification of Fourier series coefficients

The truncation of a Fourier series and the modi cation of its coef cients can both be understood as the result of applying a *spectral* (or *frequency-domain*) *window* (in contrast to the time-domain window treated in Chapter 8) to the Fourier coef cients of

$$
\tilde{g}_\infty(\theta) = \sum_{k=-\infty}^{\infty} C_k e^{jk\theta}.
$$

The spectral window used to obtain the partial sum $\tilde{g}_{2n+1}(\theta)$ is given by

(3.98)
$$
d_k = \begin{cases} 1, & \text{for } -n \le k \le n; \\ 0, & \text{otherwise.} \end{cases}
$$

The truncated spectrum is the *pointwise* product of the two sequences $\{C_k\}$ and $\{d_k\}$, which results in

$$
\tilde{g}_{2n+1}(\theta) = \sum_{k=-\infty}^{\infty} \left(d_k C_k \right) e^{jk\theta} = \sum_{k=-n}^{n} C_k e^{jk\theta}.
$$

Since d_k s are interpreted as the Fourier series coef cients of the window function $w(\theta)$, using the result from (3.81) we have

$$
w(\theta) = \sum_{k=-\infty}^{\infty} d_k e^{jk\theta} = \sum_{k=-n}^{n} e^{jk\theta} = \sum_{k=-n}^{n} e^{-jk\theta} = \frac{\sin\left(n + \frac{1}{2}\right)\theta}{\sin\frac{\theta}{2}} = \mathbf{D_n}(\theta).
$$

Figure 3.16 Fourier series with coefficients modified by the Lanczos sigma factor.

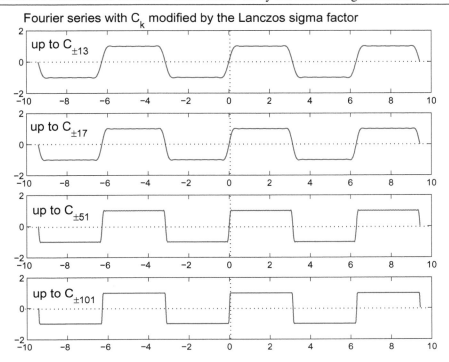

Following Lemma 3.58, we have

$$\tilde{g}_{2n+1}(\theta) = \frac{1}{2\pi} \int_{-\pi}^{\pi} g(\theta - \lambda)\, \mathbf{D_n}(\lambda)\, d\lambda,$$

where the right side defines the *periodic convolution* of the signal function $g(\theta)$ and the window function $w(\theta)$. (The subject of convolution is formally treated in Chapter 6. Readers are referred to Section 6.4 for discussion on periodic convolution and Fourier series.)

The spectral window defined by nonzero $d_k = 1$ for $-n \leq k \leq n$ is called the N-point ($N = 2n + 1$) rectangular frequency-domain window for obvious reason (see Figure 3.17). Because its corresponding window function in the time domain is the Dirichlet kernel $\mathbf{D_n}(\theta)$, the truncation of the Fourier expansion (of a function with jump discontinuities) by a rectangular spectral window causes the Gibbs effect as explained in Section 3.10.5.

Following (3.90), the spectral window corresponding to the Fejer kernel is given by

$$(3.99) \qquad \alpha_k = \begin{cases} \dfrac{n + 1 - |k|}{n + 1}, & \text{for } -n \leq k \leq n; \\ 0, & \text{otherwise.} \end{cases}$$

The N nonzero α_k s define an N-point ($N = 2n + 1$) triangular frequency-domain window (see Figure 3.17). Using the result from (3.91), the modified partial sum $\tilde{f}_{2n+1}(\theta)$ can be expressed as the periodic convolution of the signal function $g(\theta)$ and the window function $w(\theta)$ defined by the Fejer kernel $\mathbf{F_n}(\theta)$. That is,

$$\tilde{f}_{2n+1}(\theta) = \frac{1}{2\pi} \int_{-\pi}^{\pi} g(\theta - \lambda)\, \mathbf{F_n}(\lambda)\, d\lambda.$$

Figure 3.17 The three N-point frequency-domain windows for $N = 2n+1 = 11$.

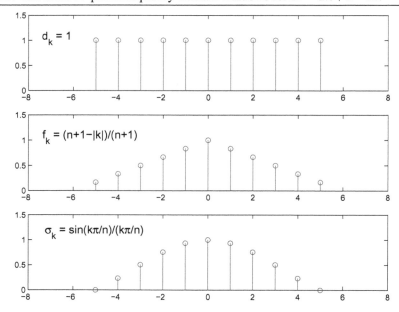

Following (3.97), the spectral window for Lanczos smoothing is de ned by the sigma factors

$$(3.100) \qquad \sigma_k = \begin{cases} \dfrac{\sin(k\pi/n)}{(k\pi/n)}, & \text{for } -n \le k \le n; \\ 0, & \text{otherwise.} \end{cases}$$

The graph of the Lanczos window for $N = 2n + 1 = 11$ is also given in Figure 3.17. Our
analysis in Sections 3.10.6 and 3.10.7 shows that the Gibbs effect can be eliminated or reduced
by applying the two *tapered* frequency-domain windows.

3.11   Accounting for Aliased Frequencies in DFT

We now provide the mathematical argument behind our prior discussion on al iasing in Sections 2.2 and 2.3. Recall that the DFT coef cients are de ned by formula (2.7), which is
restated below for easy reference.

$$(3.101) \qquad X_r = \frac{1}{N} \sum_{\ell=0}^{N-1} g_\ell \omega_N^{-r\ell}, \quad \omega_N \stackrel{\text{def}}{=} e^{j2\pi/N}, \quad r = 0, 1, \ldots, N-1,$$

where $g_\ell = g(\ell \triangle t)$, with $\triangle t = T/N$, are the N equally spaced samples of $g(t)$ over the period
$[0, T)$. To link the DFT coef cients X_r to the complex Fourier series coef cients of $g(t)$, we
simply evaluate the Fourier series of $g(t)$ for its sample values. That is, we evaluate

$$g(t) = \sum_{k=-\infty}^{\infty} C_k \, e^{j2\pi kt/T}$$

at $t = \ell \triangle t = \ell T/N$, and we obtain

$$(3.102) \qquad g_\ell = \sum_{k=-\infty}^{\infty} C_k e^{j2\pi k\ell/N} = \sum_{k=-\infty}^{\infty} C_k \omega_N^{k\ell}, \;\; 0 \le \ell \le N-1.$$

Using the right-hand side of Equation (3.102) to replace g_ℓ in Equation (3.101), we obtain

$$X_r = \frac{1}{N} \sum_{\ell=0}^{N-1} \left[\sum_{k=-\infty}^{\infty} C_k \omega_N^{k\ell} \right] \omega_N^{-r\ell}, \;\; \text{for } r = 0, 1, \ldots, N-1$$

$$= \frac{1}{N} \sum_{\ell=0}^{N-1} \sum_{k=-\infty}^{\infty} C_k \omega_N^{(k-r)\ell}$$

$$= \sum_{k=-\infty}^{\infty} C_k \left[\frac{1}{N} \sum_{\ell=0}^{N-1} \omega_N^{(k-r)\ell} \right].$$

To further simplify the result, we apply the properties of ω_N to show that

$$C_k \left[\frac{1}{N} \sum_{\ell=0}^{N-1} \omega_N^{(k-r)\ell} \right] = \begin{cases} C_{r+mN}, & \text{if } k = r + mN, \\ 0, & \text{if } k \ne r + mN. \end{cases}$$

In the first case $k = r + mN$, we have $(k - r) = mN$; hence, $\omega_N^{(k-r)\ell} = \left(\omega_N^N \right)^{m\ell} = 1$ for every ℓ the sum of the N terms is N, and the result given above follows immediately. In the second case $k \ne r + mN$, we have $(k - r) = q \ne mN$. Therefore $\omega_N^q \ne 1$, and we can sum the geometric series of ω_N^q, which results in zero as shown below.

$$\sum_{\ell=0}^{N-1} \omega_N^{(k-r)\ell} = \sum_{\ell=0}^{N-1} \left(\omega_N^q \right)^\ell = \frac{1 - \left(\omega_N^q \right)^N}{1 - \omega_N^q} = \frac{1 - \left(\omega_N^N \right)^q}{1 - \omega_N^q} = 0. \qquad (\because \omega_N^N = 1)$$

We thus obtain

$$X_r = \sum_{m=-\infty}^{\infty} C_{r+mN}, \;\; r = 0, 1, \ldots, N-1,$$

which reveals how the frequencies aliased into the Nyquist interval by the sampling process are accounted for in the resulting DFT coefficients. It is interesting to note that the contributions from the aliased frequencies effectively make the DFT coefficients a *periodic* sequence with period N, because X_r and X_{r+kN} (for every k) are represented by the sum of the same set of Fourier series coefficients.

To complete the story, let $\tilde{g}(t)$ denote the function reconstructed from the $N = 2n+1$ DFT coefficients, and we express

$$\tilde{g}(t) = \frac{\tilde{A}_0}{2} + \sum_{r=1}^{n} \tilde{A}_r \cos \frac{2\pi rt}{T} + \tilde{B}_r \sin \frac{2\pi rt}{T},$$

with the following remarks:

1. The reconstructed $\tilde{g}(t)$ is periodic with commensurate frequencies $f_r = \frac{r}{T}$ for $1 \le r \le n$, and we have $\tilde{g}(t + T) = \tilde{g}(t)$.

2. The reconstructed $\tilde{g}(t)$ is band-limited to the Nyquist interval $[-f_n, f_n] = \left[-\frac{n}{T}, \frac{n}{T} \right]$. (Recall that the Nyquist interval is imposed solely by the sampling interval $\triangle t = T/N$, and sampling is an irreversible process.)

3. The amplitudes of the sine and cosine components of $\tilde{g}(t)$ are distorted by the aliased frequencies as shown below.

$$\frac{\tilde{A}_0}{2} = X_0 = \sum_{m=-\infty}^{\infty} C_{mN},$$

$$\tilde{A}_r = X_r + X_{N-r} = \sum_{m=-\infty}^{\infty} \left(C_{r+mN} + C_{N-r+mN}\right) = \sum_{m=-\infty}^{\infty} \left(C_{r+mN} + C_{-r+mN}\right),$$

$$\tilde{B}_r = j\left(X_r - X_{N-r}\right) = \sum_{m=-\infty}^{\infty} j\left(C_{r+mN}^{\cdot} - C_{-r+mN}\right).$$

4. In contrast, the original continuous-time signal $g(t)$ is not assumed to be band-limited, and we express $g(t)$ by its continuous Fourier series as

$$g(t) = \frac{A_0}{2} + \sum_{k=1}^{\infty} A_k \cos \frac{2\pi k t}{T} + B_k \sin \frac{2\pi k t}{T},$$

where $\dfrac{A_0}{2} = C_0$, $A_k = C_k + C_{-k}$, and $B_k = j(C_k - C_{-k})$ for every k.

5. Although the reconstructed $\tilde{g}(t)$ has the appearance of a truncated Fourier series, it is not equal to the truncated Fourier series of $g(t)$ t he corresponding components have different amplitudes due to aliasing.

6. The faster the Fourier series coef cients converge to zero, the less impact the aliased frequencies have on the reconstructed signal. Our earlier investigation on the convergence rate reveals that the Fourier series coef cients of $g(t)$ converges at a higher rate if the derivate $g'(t)$ exists th is links the phenomenon of aliasing to the differentiability of $g(t)$.

7. To have undistorted $\tilde{A}_r = A_r$ and $\tilde{B}_r = B_r$ for $0 \leq r \leq n$, we need to remove the components with frequencies higher than f_n from the original signal $g(t)$ *before* sampling this is where the an ti-aliasing lter (discussed in Section 2.3) comes in. Filtering is the topic of Chapter 10.

3.11.1   Sampling functions with jump discontinuities

Here is another hidden technical point: Since the Fourier series converges to the average of the right-hand and left-hand limits at points of jump discontinuity, if any sampling point $t_\ell = \ell \triangle t$ happens to coincide with a point of jump discontinuity, then f_ℓ must be assigned the average limit value (to which its Fourier series converges) *regardless of* whether $f(t)$ is de ned at $t = t_\ell$ or not. Note that the points of jump discontinuity can occur inside or at the end points of $[0, T]$; in either case the rule above must be followed in determining the sample values for the DFT computation.

Example 3.66 In this example we show the signals reconstructed from N computed DFT coef cients . For $N = 8$, 16, 32, and 64, we obtain the DFT coef cients by transforming N

equally spaced samples taken from the chosen period $[0, 4)$ of the piecewise smooth function given in Example 3.9:

$$f(t) = \begin{cases} t, & 0 < t < 2, \\ 2, & 2 \le t < 4; \end{cases} \qquad f(t+4) = f(t).$$

Note that because there is a jump discontinuity at $t_0 = 0$, the data sample f_0 is assigned the average limit value 1 , which is explicitly shown here when $N = 8$ samples of $f(t)$ are taken from the period $[0, 4)$:

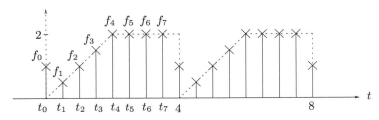

$$\{f_0, f_1, f_2, f_3, f_4, f_5, f_6, f_7\} = \{1, 0.5, 1, 1.5, 2, 2, 2, 2\}.$$

For $N = 8$, 16, and 32, the computed DFT coefﬁcients $\{X_0, X_1, \ldots, X_{N-1}\}$ are recorded in Table 3.1, which is MATLAB output (displayed in format short) from running the DFT code dft.m provided in Section 4.7 in Chapter 4. The function reconstructed using $N = 2n+2$ DFT coefﬁcients can be expressed as

$$(3.103) \qquad \tilde{f}(t) = \frac{\tilde{A}_0}{2} + \tilde{A}_{n+1} \cos \frac{2\pi(n+1)t}{T} + \sum_{r=1}^{n} \left(\tilde{A}_r \cos \frac{2\pi rt}{T} + \tilde{B}_r \sin \frac{2\pi rt}{T} \right),$$

where $T = 4$ because the N samples are taken from one period of $f(t)$, $\frac{1}{2}\tilde{A}_0 = X_0$, $\tilde{A}_{n+1} = X_{n+1}$, $\tilde{A}_r = X_r + X_{N-r}$, and $\tilde{B}_r = j(X_r - X_{N-r})$ for $1 \le r \le n$. For $N = 8$, 16, 32, and 64, the graphs of the reconstructed $\tilde{f}(t)$ are shown in Figure 3.18.

Figure 3.18 Graphs of $\tilde{f}(t)$ reconstructed using N computed DFT coefﬁcients.

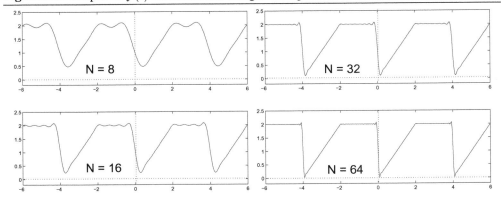

Table 3.1 The DFT coefficients computed in Example 3.66 ($N = 8, 16, 32$).

r	X_r ($N = 8$)	X_r ($N = 16$)	X_r ($N = 32$)
0	1.5000	1.5000	1.5000
1	$-0.2134 + j0.3018$	$-0.2053 + j0.3142$	$-0.2033 + j0.3173$
2	$-0.0000 + j0.1250$	$-0.0000 + j0.1509$	$-0.0000 + j0.1571$
3	$-0.0366 + j0.0518$	$-0.0253 + j0.0935$	$-0.0232 + j0.1030$
4	$-0.0000 - j0.0000$	$-0.0000 + j0.0625$	$-0.0000 + j0.0754$
5	$-0.0366 - j0.0518$	$-0.0113 + j0.0418$	$-0.0088 + j0.0585$
6	$-0.0000 - j0.1250$	$-0.0000 + j0.0259$	$-0.0000 + j0.0468$
7	$-0.2134 - j0.3018$	$-0.0081 + j0.0124$	$-0.0049 + j0.0381$
8		$-0.0000 + j0.0000$	$-0.0000 + j0.0312$
9		$-0.0081 - j0.0124$	$-0.0033 + j0.0256$
10		$-0.0000 - j0.0259$	$-0.0000 + j0.0209$
11		$-0.0113 - j0.0418$	$-0.0025 + j0.0167$
12		$-0.0000 - j0.0625$	$-0.0000 + j0.0129$
13		$-0.0253 - j0.0935$	$-0.0021 + j0.0095$
14		$-0.0000 - j0.1509$	$-0.0000 + j0.0062$
15		$-0.2053 - j0.3242$	$-0.0020 + j0.0031$
16			$-0.0000 + j0.0000$
17			$-0.0020 - j0.0031$
18			$-0.0000 - j0.0062$
19			$-0.0021 - j0.0095$
20			$-0.0000 - j0.0129$
21			$-0.0025 - j0.0167$
22			$-0.0000 - j0.0209$
23			$-0.0033 - j0.0256$
24			$-0.0000 - j0.0312$
25			$-0.0049 - j0.0381$
26			$-0.0000 - j0.0468$
27			$-0.0088 - j0.0585$
28			$-0.0000 - j0.0754$
29			$-0.0232 - j0.1030$
30			$-0.0000 - j0.1571$
31			$-0.2033 - j0.3173$

Remark: We will learn in Chapter 4 that κN sample values of the reconstructed $\tilde{f}(t)$ can be obtained by applying the inverse DFT to the sequence formed by the N computed DFT coefficients and $(\kappa - 1)N$ zeros, provided that κ is an integer and the zeros are appropriately inserted. In other words, the evaluation of the reconstructed $\tilde{f}(t)$ at equally spaced κN data points amounts to the inverse transform of κN zero-padded DFT coefficients. The process of zero padding the DFT is discussed in full detail in Section 4.6.2 in Chapter 4.

References

1. W. L. Briggs and V. E. Hensen. *The DFT: An Owner's Manual for the Discrete Fourier Transform*. The Society for Industrial and Applied Mathematics, Philadelphia, PA, 1995.

2. H. F. Davis. *Fourier Series and Orthogonal Functions*. Allyn and Bacon, Inc., Boston, MA, 1963.

3. J. C. Goswami and A. K. Chan. *Fundamentals of Wavelets*. John Wiley & Sons, Inc., New York, 1999.

4. R. W. Hamming. *Digital Filters*. Prentice-Hall, Inc., Englewood Cliffs, NJ, third edition, 1989.

5. Y. Nievergelt. *Wavelets Made Easy*. Birkhauser, Cambridge, MA, 1999.

6. H. J. Weaver. *Applications of Discrete and Continuous Fourier Analysis*. John Wiley & Sons, Inc., New York, 1983.

7. C. R. Wylie. *Advanced Engineering Mathematics*. McGraw-Hall Book Company, New York, fourth edition, 1975.

Chapter 4

DFT and Sampled Signals

We have seen the theoretical relationship between the DFT coef cien ts and the Fourier series coef cients of a periodic signal $x(t)$ in the previous chapter, and that relation was established by assuming, on the one hand, that $x(t)$ can be represented by a Fourier series

$$x(t) = \sum_{k=-\infty}^{\infty} C_k \, e^{j2\pi kt/T}, \quad C_k = \frac{1}{T} \int_{-T/2}^{T/2} f(t) \, e^{-j2\pi kt/T} \, dt;$$

and assuming, on the other hand, that the N discrete-time samples $\{x_\ell\}$ transformed by the DFT were equally spaced over a single period T o f the signal $x(t)$; i.e., we have $N\triangle t = T$.

However, because DFT is a numerical formula which we apply only to the sampled function values, the samples transformed by the DFT in practice are likely observations of an unknown signal or phenomenon. For example, one use (among many important applications) of the DFT is to analyze the frequency contents of an *unknown* signal $f(t)$, and we must be prepared to account for the distortions caused by the potential mi smatch between the period (N samples imply a period of $N\triangle t$) irreversibly imposed by using the DFT and the (unknown) true period of $f(t)$. To understand and deal with such problems and their consequences, we propose the following:

1. We will derive the DFT formulas to show why the samples are supposed to be taken over a single period of the envelope function in the r st place. (The derivation was omitted when DFT was r st introduced in Chapter 2.)

2. We will sample known functions for irregular intervals (longer or shorter than the known periods) to create mismatched periods for our experiment, so we can study the possible consequences.

4.1   Deriving the DFT and IDFT Formulas

As indicated in Sections 2.5 and 2.6 in Chapter 2, there is more than one DFT formula depending on the chosen sampling period and sample size. Following a similar derivation for the odd-size DFT given in our earlier book on fast Fourier transform algorithms [13], we derive the DFT formula for even sample size $N = 2n + 2$ over the period $[0, T]$. In addition, we will show that the resulting formula can be converted to its alternate form for the symmetric period $[-T/2, T/2]$.

We indicated in Section 2.6 that an even-size DFT can be derived from using the trigonometric polynomial

$$(4.1) \qquad p(t) = a_0 + a_{n+1} \cos \frac{2\pi(n+1)t}{T} + \sum_{r=1}^{n} a_r \cos \frac{2\pi r t}{T} + b_r \sin \frac{2\pi r t}{T}$$

to interpolate the samples of a periodic function $x(t)$. In anticipation of the desired change of variable from t to $\theta = 2\pi t/T$ in the derivation, we perform the variable change at the outset by directly using

$$(4.2) \qquad p(\theta) = a_0 + a_{n+1} \cos(n+1)\theta + \sum_{r=1}^{n} a_r \cos r\theta + b_r \sin r\theta$$

to interpolate the $N = 2n + 2$ equally spaced samples of $x(\theta)$ over the period $[0, 2\pi]$. The samples are denoted as $x_\ell = x(\theta_\ell)$ for $0 \le \ell \le 2n+1$, where $\theta_\ell = \ell \triangle \theta$ with $\triangle \theta = 2\pi/N = \pi/(n+1)$. For $N = 2n+2 = 8$, we show the mapping of $\{t_0, t_1, \ldots, t_7\}$ to $\{\theta_0, \theta_1, \ldots, \theta_7\}$ in Figure 4.1.

Figure 4.1 Mapping $t_\ell \in [0, T)$ to $\theta_\ell = 2\pi t_\ell/T \in [0, 2\pi)$ for $0 \le \ell \le 2n+1$.

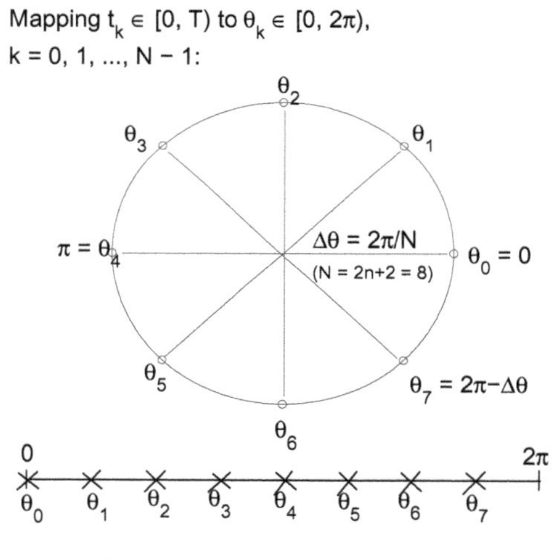

In order to explicitly display all of the terms (for clarity) during the derivation without loss of generality, we consider a particular size $N = 2n + 2 = 6$ with $n = 2$. Since $p(\theta)$ interpolates every x_ℓ, we must have $x_\ell = p(\theta_\ell)$, i.e.,

$$(4.3) \qquad x_\ell = p(\theta_\ell) = a_0 + a_1 \cos \theta_\ell + b_1 \sin \theta_\ell + a_2 \cos 2\theta_\ell + b_2 \sin 2\theta_\ell + a_3 \cos 3\theta_\ell.$$

Corresponding to $\ell = 0, 1, \ldots, 5$, we have a system of six equations

$$\begin{bmatrix} 1 & \cos\theta_0 & \sin\theta_0 & \cos 2\theta_0 & \sin 2\theta_0 & \cos 3\theta_0 \\ 1 & \cos\theta_1 & \sin\theta_1 & \cos 2\theta_1 & \sin 2\theta_1 & \cos 3\theta_1 \\ 1 & \cos\theta_2 & \sin\theta_2 & \cos 2\theta_2 & \sin 2\theta_2 & \cos 3\theta_2 \\ 1 & \cos\theta_3 & \sin\theta_3 & \cos 2\theta_3 & \sin 2\theta_3 & \cos 3\theta_3 \\ 1 & \cos\theta_4 & \sin\theta_4 & \cos 2\theta_4 & \sin 2\theta_4 & \cos 3\theta_4 \\ 1 & \cos\theta_5 & \sin\theta_5 & \cos 2\theta_5 & \sin 2\theta_5 & \cos 3\theta_5 \end{bmatrix} \begin{bmatrix} a_0 \\ a_1 \\ b_1 \\ a_2 \\ b_2 \\ a_3 \end{bmatrix} = \begin{bmatrix} x_0 \\ x_1 \\ x_2 \\ x_3 \\ x_4 \\ x_5 \end{bmatrix}.$$

Using Euler s formula $e^{j\theta} = \cos\theta + j\sin\theta$, we may now express

$$\cos r\theta_\ell = \frac{e^{jr\theta_\ell} + e^{-jr\theta_\ell}}{2} \quad \text{and} \quad \sin r\theta_\ell = \frac{e^{jr\theta_\ell} - e^{-jr\theta_\ell}}{2j}.$$

Note that when $N = 2n + 2$, we have $\theta_\ell = \ell\Delta\theta = \ell\pi/(n+1)$ and $(n+1)\theta_\ell = \ell\pi$. Hence, for the special case $r = n + 1$, we have

$$e^{j(n+1)\theta_\ell} = e^{j\ell\pi} = e^{-j\ell\pi} = e^{-j(n+1)\theta_\ell},$$

which implies

$$\cos(n+1)\theta_\ell = e^{j(n+1)\theta_\ell}.$$

Using these complex exponentials to express the cosine and sine values in (4.3) yields

$$x_\ell = p(\theta_\ell) = \left(\frac{a_2 + jb_2}{2}\right)e^{-j2\theta_\ell} + \left(\frac{a_1 + jb_1}{2}\right)e^{-j\theta_\ell} + a_0$$
$$+ \left(\frac{a_1 - jb_1}{2}\right)e^{j\theta_\ell} + \left(\frac{a_2 - jb_2}{2}\right)e^{j2\theta_\ell} + a_3 e^{j3\theta_\ell}.$$

Noting that $e^{\pm jr\theta_\ell} = \left(e^{j\theta_\ell}\right)^{\pm r}$; we may use the power $\pm r$ a s index and rename the coef - cients of $e^{\pm jr\theta_\ell}$ as $X_{\pm r}$, we thus obtain

(4.4) $\quad x_\ell = p(\theta_\ell) = X_{-2}e^{-j2\theta_\ell} + X_{-1}e^{-j\theta_\ell} + X_0 + X_1 e^{j\theta_\ell} + X_2 e^{j2\theta_\ell} + X_3 e^{j3\theta_\ell}.$

To further simplify the right-hand side, recall that $\theta_\ell = \ell\Delta\theta$, so $\theta_1 = \Delta\theta = 2\pi/N$, and $e^{j\theta_\ell} = e^{j\ell\theta_1} = \left(e^{j\theta_1}\right)^\ell = \omega^\ell$ if we de n e

$$\boxed{\omega = e^{j\theta_1} = e^{j2\pi/N}, \quad N = 2n + 2.}$$

Equation (4.4) can now be written as

(4.5) $\quad x_\ell = p(\theta_\ell) = X_{-2}\omega^{-2\ell} + X_{-1}\omega^{-\ell} + X_0 + X_1\omega^\ell + X_2\omega^{2\ell} + X_3\omega^{3\ell},$

where $\omega = e^{j\pi/3}$ for $N = 6$.

We further note that $\omega = e^{j2\pi/N}$ is the Nth primitive root of unity it is easy to verify that $\omega^N = 1$ and $\omega^{-r} = \omega^{-r+N}$. By changing $\left(\omega^{-r}\right)^\ell$ in the above equation to the equivalent $\left(\omega^{N-r}\right)^\ell$, where $r = 1, 2$ and $N = 6$, we obtain

(4.6) $\quad x_\ell = p(\theta_\ell) = X_0 + X_1\omega^\ell + X_2\omega^{2\ell} + X_3\omega^{3\ell} + X_4\omega^{4\ell} + X_5\omega^{5\ell}.$

Corresponding to $\ell = 0, 1, \ldots, 5$, we now have a system of six equations with unknowns being X_r for $r = 0, 1, \ldots, 5$:

(4.7)
$$\begin{bmatrix} 1 & 1 & 1 & 1 & 1 & 1 \\ 1 & \omega & \omega^2 & \omega^3 & \omega^4 & \omega^5 \\ 1 & \omega^2 & \omega^4 & \omega^6 & \omega^8 & \omega^{10} \\ 1 & \omega^3 & \omega^6 & \omega^9 & \omega^{12} & \omega^{15} \\ 1 & \omega^4 & \omega^8 & \omega^{12} & \omega^{16} & \omega^{20} \\ 1 & \omega^5 & \omega^{10} & \omega^{15} & \omega^{20} & \omega^{25} \end{bmatrix} \begin{bmatrix} X_0 \\ X_1 \\ X_2 \\ X_3 \\ X_4 \\ X_5 \end{bmatrix} = \begin{bmatrix} x_0 \\ x_1 \\ x_2 \\ x_3 \\ x_4 \\ x_5 \end{bmatrix}.$$

This can be written as a matrix equation $\boldsymbol{MX} = \boldsymbol{x}$, and we shall obtain the scalar DFT formula (2.11) for each X_r by solving this matrix equation analytically. To accomplish that, three additional steps are required:

Step 1. For $\omega \equiv e^{j2\pi/N} = \cos(2\pi/N) + j\sin(2\pi/N)$, we need to prove the following properties (which are required in Step 2):

(a) $\omega^{-1} = \overline{\omega}$, $\omega^{\pm N} = 1$, and $\omega^{\pm N/2} = -1$.

(b) $\omega^{\pm \ell \pm N} = \omega^{\pm \ell}$.

(c) $\sum_{k=0}^{N-1} \omega^k = 1 + \omega + \omega^2 + \cdots + \omega^{N-1} = 0$.

(d) For $1 \le \rho, q \le N$, $\sum_{k=0}^{N-1} \omega^{k(\rho-q)} = \begin{cases} 0 & \text{if } \rho \ne q, \\ N & \text{if } \rho = q. \end{cases}$

(e) If $\omega_N \equiv e^{j2\pi/N}$, prove $\omega_N^2 = \omega_{N/2}$. (Note: the notation ω_N is used when we need to refer to ω for more than one value of N at the same time.)

Proof: For part **(a)**, we apply the de n ition of ω to obtain

$$\omega^{-1} = e^{-j2\pi/N} = \cos(2\pi/N) - j\sin(2\pi/N) = \overline{\omega}.$$
$$\omega^{\pm N} = \left(e^{j2\pi/N}\right)^{\pm N} = e^{\pm j2\pi} = \cos 2\pi \pm j\sin 2\pi = 1.$$
$$\omega^{\pm N/2} = \left(e^{j2\pi/N}\right)^{\pm N/2} = e^{\pm j\pi} = \cos \pi \pm j\sin \pi = -1.$$

For part **(b)**, using $\omega^{\pm N} = 1$ from (a), we immediately have

$$\omega^{\pm \ell \pm N} = \omega^{\pm \ell} \omega^{\pm N} = \omega^{\pm \ell}.$$

For part **(c)**, we use the closed-form expression for the geometric series to obtain

$$\sum_{k=0}^{N-1} \omega^k = 1 + \omega + \omega^2 + \cdots + \omega^{N-1} = \frac{1 - \omega^N}{1 - \omega} = 0. \qquad (\because \omega^N = 1 \text{ from (a)})$$

For part **(d)**, let $m = \rho - q$. The condition $1 \le \rho, q \le N$ implies $0 \le m \le N - 1$. If $\rho \ne q$, then $m \ne 0$, and we use again the closed-form expression for the geometric series to obtain

$$\sum_{k=0}^{N-1} \omega^{k(\rho-q)} = \sum_{k=0}^{N-1} \left(\omega^m\right)^k = \frac{1 - \left(\omega^m\right)^N}{1 - \omega^m} = \frac{1 - \left(\omega^N\right)^m}{1 - \omega^m} = 0. \qquad (\because \omega^N = 1)$$

If $\rho = q$, then we have

$$\sum_{k=0}^{N-1} \omega^{k(\rho-q)} = \sum_{k=0}^{N-1} \omega^0 = \sum_{k=0}^{N-1} 1 = N.$$

For part **(e)**, we use the de n ition of ω_N to obtain

$$\omega_N^2 = \left(e^{j2\pi/N}\right)^2 = e^{j4\pi/N} = e^{j2\pi/(N/2)} = \omega_{N/2}.$$

■

Step 2. The Fourier matrix M is de n ed by $M(\rho, q) = \omega^{(\rho-1)(q-1)}$ for $1 \le \rho, q \le N$. With the properties of ω now available from Step 1, we may prove that the Fourier matrix M is invertible, and its inverse is $\frac{1}{N}\overline{M}$. (We have used $\overline{M}$ to denote the complex conjugate of M.)

Remarks: For $N = 6$, the Fourier matrix M appeared in (4.7), and it is a simple task to verify that the elements in the ρth $(1 \leq \rho \leq N)$ row are $M(\rho, r) = \omega^{(\rho-1)(r-1)}$ for $1 \leq r \leq N$; and the elements in the qth $(1 \leq q \leq N)$ column are $M(r, q) = \omega^{(r-1)(q-1)}$ for $1 \leq r \leq N$.

Proof: To obtain $M^{-1} = \frac{1}{N}\overline{M}$, we may form the product $D = M\overline{M}$ and show that $D = NI$, where I is the identity matrix. That is, for $1 \leq \rho, q \leq N$, we must show

$$D(\rho, q) = \sum_{r=1}^{N} M(\rho, r)\,\overline{M}(r, q) = \begin{cases} 0 & \text{if } \rho \neq q, \\ N & \text{if } \rho = q. \end{cases}$$

We proceed to prove the desired result below.

$$\begin{aligned} D(\rho, q) = \sum_{r=1}^{N} M(\rho, r)\,\overline{M}(r, q) &= \sum_{r=1}^{N} \omega^{(\rho-1)(r-1)}\,\overline{\omega}^{(r-1)(q-1)} \\ &= \sum_{r=1}^{N} \omega^{(\rho-1)(r-1)}\,\omega^{-(r-1)(q-1)} \quad (\because \overline{\omega} = \omega^{-1}) \\ &= \sum_{r=1}^{N} \omega^{(r-1)(\rho-q)} \\ &= \sum_{k=0}^{N-1} \omega^{k(\rho-q)} \quad (\because k = r - 1) \\ &= \begin{cases} 0 & \text{if } \rho \neq q, \\ N & \text{if } \rho = q. \end{cases} \quad (\text{from Step 1(d)}) \end{aligned}$$

To demonstrate the result that we have just proved, we display D for $N = 6$:

$$D = M\overline{M} = \begin{bmatrix} 6 & 0 & 0 & 0 & 0 & 0 \\ 0 & 6 & 0 & 0 & 0 & 0 \\ 0 & 0 & 6 & 0 & 0 & 0 \\ 0 & 0 & 0 & 6 & 0 & 0 \\ 0 & 0 & 0 & 0 & 6 & 0 \\ 0 & 0 & 0 & 0 & 0 & 6 \end{bmatrix} = 6 \begin{bmatrix} 1 & 0 & 0 & 0 & 0 & 0 \\ 0 & 1 & 0 & 0 & 0 & 0 \\ 0 & 0 & 1 & 0 & 0 & 0 \\ 0 & 0 & 0 & 1 & 0 & 0 \\ 0 & 0 & 0 & 0 & 1 & 0 \\ 0 & 0 & 0 & 0 & 0 & 1 \end{bmatrix}.$$

Therefore, we have obtained

$$\frac{1}{N} M\overline{M} = \frac{1}{N} D = I,$$

which yields

$$M^{-1} = \frac{1}{N}\overline{M}.$$

$\blacksquare$

Step 3. Solve the matrix equation $MX = x$ by inverting the matrix M, i.e.,

$$X = M^{-1}x = \frac{1}{N}\overline{M}x.$$

For $N = 6$, we obtain

(4.8)
$$
\begin{bmatrix} X_0 \\ X_1 \\ X_2 \\ X_3 \\ X_4 \\ X_5 \end{bmatrix} = \frac{1}{6} \begin{bmatrix} 1 & 1 & 1 & 1 & 1 & 1 \\ 1 & \omega^{-1} & \omega^{-2} & \omega^{-3} & \omega^{-4} & \omega^{-5} \\ 1 & \omega^{-2} & \omega^{-4} & \omega^{-6} & \omega^{-8} & \omega^{-10} \\ 1 & \omega^{-3} & \omega^{-6} & \omega^{-9} & \omega^{-12} & \omega^{-15} \\ 1 & \omega^{-4} & \omega^{-8} & \omega^{-12} & \omega^{-16} & \omega^{-20} \\ 1 & \omega^{-5} & \omega^{-10} & \omega^{-15} & \omega^{-20} & \omega^{-25} \end{bmatrix} \begin{bmatrix} x_0 \\ x_1 \\ x_2 \\ x_3 \\ x_4 \\ x_5 \end{bmatrix}.
$$

Since $X = M^{-1}x$ expresses X as a matrix-vector product, we can now express each element of X by the corresponding scalar equation, which is the formula for the DFT:

(4.9)
$$
X_r = \frac{1}{N} \sum_{\ell=0}^{N-1} x_\ell \omega_N^{-r\ell} = \frac{1}{2n+2} \sum_{\ell=0}^{2n+1} x_\ell \omega_N^{-r\ell}, \ \ 0 \le r \le 2n+1.
$$

Similarly, because $x = MX$, the inverse DFT formula (IDFT) has the form

(4.10)
$$
x_\ell = \sum_{r=0}^{N-1} X_r \omega_N^{r\ell} = \sum_{r=0}^{2n+1} X_r \omega_N^{r\ell}, \ \ 0 \le \ell \le 2n+1.
$$

4.2   Direct Conversion Between Alternate Forms

We indicated in Section 2.6 that if the given $N = 2n + 2$ samples $\tilde{x}(t_\ell)$ are equally spaced over $[-T/2, T/2]$, we may obtain the alternate DFT/IDFT formulas.

(4.11)
$$
\tilde{X}_r = \frac{1}{N} \sum_{\ell=-\frac{N}{2}+1}^{\frac{N}{2}} \tilde{x}_\ell \omega_N^{-r\ell} = \frac{1}{2n+2} \sum_{\ell=-n}^{n+1} \tilde{x}_\ell \omega_N^{-r\ell}, \ \ -n \le r \le n+1;
$$

(4.12)
$$
\tilde{x}_\ell = \sum_{r=-\frac{N}{2}+1}^{\frac{N}{2}} \tilde{X}_r \omega_N^{r\ell} = \sum_{r=-n}^{n+1} X_r \omega_N^{r\ell}, \ \ -n \le \ell \le n+1.
$$

To convert the DFT formula given by (4.9) to its alternate form given by (4.11), we recall the simple fact behind the derivation of a DFT formula: the periodic $x(t)$ was uniformly sampled for a single period of length T, so the sample sequence $\{x_\ell\}$ over $[0, T]$ is related to the sample sequence $\{\tilde{x}_\ell\}$ over $[-T/2, T/2]$ through the periodicity of $x(t)$. For the even sample size $N = 2n + 2$, we have

(4.13)
$$
x_\ell = \begin{cases} \tilde{x}_\ell & \text{for } 0 \le \ell \le n+1; \\ \tilde{x}_{\ell-N} & \text{for } n+2 \le \ell \le 2n+1. \end{cases}
$$

Using this fact in the DFT formula (4.9), we obtain (4.11) as shown below.

$$
\begin{aligned}
X_r &= \frac{1}{N} \sum_{\ell=0}^{2n+1} x_\ell \omega_N^{-r\ell} = \frac{1}{N} \sum_{\ell=0}^{n+1} x_\ell \omega_N^{-r\ell} + \frac{1}{N} \sum_{\ell=n+2}^{2n+1} x_\ell \omega_N^{-r\ell} \\
&= \frac{1}{N} \sum_{\ell=0}^{n+1} \tilde{x}_\ell \omega_N^{-r\ell} + \frac{1}{N} \sum_{\ell=n+2}^{2n+1} \tilde{x}_{\ell-N} \omega_N^{-r\ell} && \text{(from (4.13))} \\
&= \frac{1}{N} \sum_{\ell=0}^{n+1} \tilde{x}_\ell \omega_N^{-r\ell} + \frac{1}{N} \sum_{m=-n}^{-1} \tilde{x}_m \omega_N^{-r(m+N)} && \text{(let } m = \ell - N) \\
&= \frac{1}{N} \sum_{\ell=0}^{n+1} \tilde{x}_\ell \omega_N^{-r\ell} + \frac{1}{N} \sum_{m=-n}^{-1} \tilde{x}_m \omega_N^{-rm} && (\because \omega_N^{-N} = 1) \\
&= \frac{1}{N} \sum_{\ell=0}^{n+1} \tilde{x}_\ell \omega_N^{-r\ell} + \frac{1}{N} \sum_{\ell=-n}^{-1} \tilde{x}_\ell \omega_N^{-r\ell} && \text{(denote } m \text{ by } \ell) \\
&= \frac{1}{N} \sum_{\ell=-n}^{n+1} \tilde{x}_\ell \omega_N^{-r\ell}. && \text{(combine partial sums)}
\end{aligned}
$$

Observe that the right-hand side represents $\tilde{X}_r$ if $0 \leq r \leq n+1$. For $n+2 \leq r \leq 2n+1$, the right-hand side represents $\tilde{X}_{r-N}$ as shown below.

$$
\begin{aligned}
X_r &= \frac{1}{N} \sum_{\ell=-n}^{n+1} \tilde{x}_\ell \omega_N^{-r\ell} && \text{(for } n+2 \leq r \leq 2n+1) \\
&= \frac{1}{N} \sum_{\ell=-n}^{n+1} \tilde{x}_\ell \omega_N^{-(r-N)\ell} && (\because \omega_N^{-r+N} = \omega_N^{-r}) \\
&= \tilde{X}_{r-N}. && (\because -n \leq r - N \leq -1)
\end{aligned}
$$

These results reveal that the relationship between the two sets of DFT coefficients mirrors the relationship between the two sample sequences:

$$
(4.14) \qquad X_r = \begin{cases} \tilde{X}_r & \text{for } 0 \leq r \leq n+1; \\ \tilde{X}_{r-N} & \text{for } n+2 \leq r \leq 2n+1. \end{cases}
$$

This is expected because $X_{r\pm N} = X_r$ from either formula; i.e., $\{X_r\}$ is a periodic sequence with period N.

As to converting formula (4.11) back to (4.9), we simply make use of the same fact in the opposite direction:

$$
(4.15) \qquad \tilde{x}_\ell = \begin{cases} x_\ell & \text{for } 0 \leq \ell \leq n+1; \\ x_{\ell+N} & \text{for } -n \leq \ell \leq -1. \end{cases}
$$

The corresponding IDFT formulas can be converted to each other by the same process.

4.3   DFT of Concatenated Sample Sequences

If we concatenate the N-sample sequence $\{x_0, x_1, \ldots, x_{N-1}\}$ to itself, then we obtain an M-sample sequence $\{y_0, y_1, \ldots, y_{M-1}\}$, where $M = 2N$, and

$$y_\ell = \begin{cases} x_\ell & \text{for } 0 \leq \ell \leq N-1; \\ x_{\ell-N} & \text{for } N \leq \ell \leq 2N-1. \end{cases}$$

Suppose that the N DFT coefficients computed from the N-sample sequence are $\{X_0, X_1, \ldots, X_{N-1}\}$; how are they related to the $2N$ DFT coefficients $\{Y_0, Y_1, \ldots, Y_{2N-1}\}$ computed from the concatenated $2N$-sample sequence? To answer this question, we apply the definition of DFT to compute Y_{2r} and Y_{2r+1} for $r = 0, 1, \ldots, N-1$:

$$\begin{aligned}
Y_{2r} &= \frac{1}{M} \sum_{\ell=0}^{M-1} y_\ell \omega_M^{-2r\ell} \\
&= \frac{1}{2N} \sum_{\ell=0}^{N-1} x_\ell \omega_{2N}^{-2r\ell} + \frac{1}{2N} \sum_{\ell=N}^{2N-1} x_{\ell-N} \omega_{2N}^{-2r\ell} \\
&= \frac{1}{2N} \sum_{\ell=0}^{N-1} x_\ell \omega_{2N}^{-2r\ell} + \frac{1}{2N} \sum_{k=0}^{N-1} x_k \omega_{2N}^{-2r(k+N)} && (\text{ let } k = \ell - N) \\
&= \frac{1}{2N} \sum_{\ell=0}^{N-1} x_\ell \left(\omega_{2N}^{-2r\ell} + \omega_{2N}^{-2r(\ell+N)} \right) && (\text{ rename } k \text{ to be } \ell) \\
&= \frac{1}{2N} \sum_{\ell=0}^{N-1} x_\ell \left(\omega_{2N}^{-2r\ell} + \omega_{2N}^{-2r\ell} \right) && \left(\because \left(\omega_{2N}^{-2N} \right)^r = 1 \right) \\
&= \frac{1}{N} \sum_{\ell=0}^{N-1} x_\ell \omega_N^{-r\ell} \\
&= X_r.
\end{aligned}$$

Through the same steps we obtain

$$\begin{aligned}
Y_{2r+1} &= \frac{1}{M} \sum_{\ell=0}^{M-1} y_\ell \omega_M^{-(2r+1)\ell} \\
&= \frac{1}{2N} \sum_{\ell=0}^{N-1} x_\ell \omega_{2N}^{-(2r+1)\ell} + \frac{1}{2N} \sum_{\ell=N}^{2N-1} x_{\ell-N} \omega_{2N}^{-(2r+1)\ell} \\
&= \frac{1}{2N} \sum_{\ell=0}^{N-1} x_\ell \omega_{2N}^{-(2r+1)\ell} + \frac{1}{2N} \sum_{k=0}^{N-1} x_k \omega_{2N}^{-(2r+1)(k+N)} && (\text{ let } k = \ell - N) \\
&= \frac{1}{2N} \sum_{\ell=0}^{N-1} x_\ell \left(\omega_{2N}^{-(2r+1)\ell} + \omega_{2N}^{-(2r+1)(\ell+N)} \right) && (\text{ rename } k \text{ to be } \ell) \\
&= \frac{1}{2N} \sum_{\ell=0}^{N-1} x_\ell \left(\omega_{2N}^{-(2r+1)\ell} - \omega_{2N}^{-(2r+1)\ell} \right) && \left(\because \left(\omega_{2N}^{-N} \right)^{2r+1} = -1 \right) \\
&= 0.
\end{aligned}$$

This establishes the relationship between the DFT coef cien ts $\{Y_k\}$ and $\{X_k\}$ as

$$
(4.16) \qquad Y_k = \begin{cases} X_{k/2} & \text{if } k = 2r \\ 0 & \text{if } k \neq 2r \end{cases}, \quad k = 0, 1, \dots, 2N - 1.
$$

This result can be extended to multiple sequences: if the qN-sample sequence $\{y_\ell\}$ is obtained by concatenating the N-sample sequence $\{x_\ell\}$, where $q \geq 2$ is a positive integer, then the DFT coef cients $\{Y_k\}$ and $\{X_k\}$ are related as

$$
(4.17) \qquad Y_k = \begin{cases} X_{k/q} & \text{if } k = qr \\ 0 & \text{if } k \neq qr \end{cases}, \quad k = 0, 1, \dots, qN - 1.
$$

The derivation given above for the case $q = 2$ can be adapted for the case $q > 2$ in an obvious way: instead of combining only two subsequences, we combine the $q > 2$ subsequences resulted from applying the DFT de n ition to the concatenated sequence $\{y_\ell\}$ in a similar manner.

4.4 DFT Coefficients of a Commensurate Sum

In preparation for the discussion forthcoming in this section, we assume that readers are familiar with the contents of Chapter 2 and the following sections from Chapter 1:

- Section 1.6 Per iodicity and Commensurate Frequencies.

- Section 1.8 Expres sing Single Component Signals.

- Section 1.9.1 Expre ssing sequence of discrete-time samples.

- Section 1.9.2 Per iodicity of sinusoidal sequences.

In this section we shall relate the DFT coef cien ts of a commensurate sum to the DFT coef cients of its components. Recall from Section 1.6 that a commensurate $y(t)$ is periodic with its fundamental frequency being the GCD of the individual frequencies and its common period being the LCM of the individual periods. For example, when $f_k = k/T$, the fundamental frequency is $f_1 = 1/T$, and the composite function

$$
y(t) = C_0 + \sum_{k=1}^{n} C_k \cos\left(\frac{2\pi k t}{T} - \phi_k\right)
$$

is commensurate and periodic with common period T, i.e., $y(t + T) = y(t)$.

4.4.1 DFT coefficients of single-component signals

We consider sampling a single-component signal

$$
x(t) = C_k \cos(2\pi f_k t - \phi_k), \text{ where } f_k = \frac{k}{T},
$$

at intervals of $\triangle t = T/N$, where N is chosen to satisfy the Nyquist condition $1/\triangle t > 2k/T$ so that aliasing will not occur. Note that because $T = N\triangle t$, the Nyquist condition is equally de n ed by $N > 2k$. From $f_k = k/T$, we know that $x(t)$ completes k cycles as t varies from 0 to $T = N\triangle t$; i.e., the N samples span k periods of $x(t)$.

For single-component signals, we may obtain the IDFT formula for x_ℓ directly: for $\ell = 0, 1, \ldots, N-1$ with $N > 2k$ (ensured by satisfying the Nyquist condition),

$$
\begin{aligned}
x_\ell \equiv x(\ell\triangle t) &= C_k \cos(2\pi f_k \ell \triangle t - \phi_k) \\
&= C_k \cos(2\pi \ell k / N - \phi_k) && (\because f_k \triangle t = k/N) \\
&= C_k \cos(k\ell\theta_1 - \phi_k) && (\text{let } \theta_1 = 2\pi/N) \\
&= \left(\tfrac{1}{2}C_k e^{-j\phi_k}\right)e^{jk\ell\theta_1} + \left(\tfrac{1}{2}C_k e^{j\phi_k}\right)e^{-jk\ell\theta_1} && (\text{by Euler s formula}) \\
&= X_k \omega_N^{k\ell} + X_{-k} \omega_N^{-k\ell} && (\because \omega_N = e^{j\theta_1}) \\
&= X_k \omega_N^{k\ell} + X_{N-k} \omega_N^{(N-k)\ell}. && (\because \omega_N^{-k} = \omega_N^{N-k})
\end{aligned}
$$

(4.18)

Accordingly, there are only two nonzero DFT coef cients with indexes k and $N - k$ in the IDFT formula in this case, i.e,

(4.19)
$$
x_\ell = \sum_{r=0}^{N-1} X_r \omega_N^{r\ell} = X_k \omega_N^{k\ell} + X_{N-k} \omega_N^{(N-k)\ell},
$$

where $2k < N$, $X_k = \tfrac{1}{2}C_k e^{-j\phi_k}$, $X_{N-k} = \tfrac{1}{2}C_k e^{j\phi_k}$.

Remarks: If we change variable to $\theta = 2\pi t/T$, then $\cos(2\pi kt/T - \phi_k) = \cos(k\theta - \phi_k)$, and we would sample the mathematically equivalent

(4.20)
$$
x(\theta) = C_k \cos(k\theta - \phi_k)
$$

at intervals of

$$
\triangle\theta = 2\pi\triangle t/(N\triangle t) = 2\pi/N = \theta_1,
$$

which leads to exactly the same formulation already used in (4.18):

$$
x_\ell = x(\ell\triangle\theta) = C_k \cos(k\ell\theta_1 - \phi_k),
$$

and we arrive at the same results given by (4.18) and (4.19).

However, formula (4.20) has its own role to play in signal reconstruction: while the continuous function $x(\theta)$ can be reconstructed from the DFT coef cients $X_{\pm k}$ alone, the reconstruction of the analog signal $x(t)$ requires the actual value of $T = N\triangle t$ ($\because f_k = k/T$), which may or may not be available depending on whether the sampling rate or interval is known or not, *although this does not prevent us from constructing an analog signal at any desired output frequency* r ecall the following comments from Section 1.9.1: by simply adjusting $\triangle t$ at the time of output, the same set of digital samples may be converted to analog signals with different frequencies.

Example 4.1 In this example we study the sampling and reconstruction of a single-component signal $x(t) = 3.2 \cos(1.5\pi t - \pi/4)$ for two cases:

(i) The Nyquist condition is satis ed.

(ii) The Nyquist condition is *not* satis ed .

At r st we identify the physical frequency of the analog signal by expressing

$$x(t) = 3.2 \cos\left(2\pi f_k t - \pi/4\right) \text{ with } f_k = \frac{3}{4} = \frac{k}{T} = \frac{k}{N\triangle t}.$$

For case (i), we choose the sample size $N = 8$ over $T = 4$ seconds, so the condition $N > 2k = 6$ is satis ed , and the result from using $k = 3$ and $N = 8$ in (4.19) is the IDFT formula:

$$(4.21) \qquad x_\ell = X_3 \omega_8^{3\ell} + X_5 \omega_8^{5\ell} = 1.6 e^{-j\pi/4} \omega_8^{3\ell} + 1.6 e^{j\pi/4} \omega_8^{5\ell}.$$

We can now reconstruct the function $x(\theta)$ from the DFT coef cients obtained from (4.21):

$$(4.22) \qquad x(\theta) = C_k \cos\left(k\theta - \phi_k\right)$$
$$= 3.2 \cos(3\theta - \pi/4). \qquad (\because k = 3,\ X_k = \tfrac{1}{2}C_k e^{-j\phi_k} = 1.6 e^{-j\pi/4})$$

Since $\theta = 2\pi t/T$, we can rewrite (4.22) as

$$x(t) = 3.2 \cos\left(2\pi f_k t - \pi/4\right) \text{ with } f_k = \frac{k}{T} = \frac{3}{N\triangle t} = \frac{3}{8\triangle t}.$$

Since we have chosen $N = 8$ over $T = 4$ sec, the sampling interval is $\triangle t = 0.5$ sec, and we obtain $f_k = 3/(8\triangle t) = 0.75$ Hz, with which we get back the original signal $x(t) = 3.2 \cos(1.5\pi t - \pi/4)$. Hence, we can reconstruct the analog signal $x(t)$ if we know the sampling rate used to obtain $\{x_\ell\}$ in the rst place; more importantly, we can output $x(t)$ at any physical frequency by setting and adjusting the sampling interval $\triangle t$ as desired.

For case (ii), we choose the sample size $N = 4$ over $T = 4$ seconds, thus $N \not> 2k = 6$ and we expect to see the effect of aliasing. We can determine the aliased frequency using different methods   although they lead to the same result, we gain valuable insight about the methods themselves.

Method I. Since $\triangle t = T/N = 4/4 = 1$ second, by sampling the given signal $x(t) = 3.2 \cos(1.5\pi t - \pi/4)$ at intervals of $\triangle t$ we obtain

$$(4.23) \qquad
\begin{aligned}
x_\ell = x(\ell\triangle t) &= 3.2 \cos\left(1.5\pi\ell\triangle t - \pi/4\right), \qquad \ell = 0,1,2,3, \\
&= 3.2 \cos\left(1.5\pi\ell - \pi/4\right) \qquad (\because \triangle t = 1 \text{ second}) \\
&= 3.2 \cos\left(1.5\pi\ell - \pi/4 - 2\pi\ell\right) \quad (\because \cos\left(\theta \pm 2\pi\ell\right) = \cos\theta) \\
&= 3.2 \cos\left(-0.5\pi\ell - \pi/4\right) \\
&= 3.2 \cos\left(0.5\pi\ell + \pi/4\right) \qquad \text{(note the phase reversal)} \\
&= 3.2 \cos\left(2\pi f_a\ell + \pi/4\right), \qquad \text{where } f_a = 0.25 \text{ Hz.}
\end{aligned}$$

Since $f_{\max} = 1/(2\triangle t) = 0.5$ Hz, we see that the higher frequency $f_k = 0.75$ Hz present in the original signal has been aliased into an equivalent lower frequency $f_a = 0.25$ Hz inside the Nyquist interval $[-f_{\max},\ f_{\max}] = [-0.5,\ 0.5]$ in this case.

Remarks: Recall that the sampling rate $\mathbb{R} = 1/\triangle t$ (samples per second or Hz). For this example we have $\triangle t = 1$ and $\mathbb{R} = 1$, so

$$\tfrac{1}{2}\mathbb{R} < f_k = 0.75 \text{ Hz} < \mathbb{R}.$$

We shall learn in Chapter 7 that the Fourier transform of a sampled sequence is periodic with period $\mathbb{R}$, hence the aliased frequency $\tilde{f}_a \in [-\frac{1}{2}\mathbb{R}, \frac{1}{2}\mathbb{R}]$ can be computed directly as

$$(4.24) \qquad \tilde{f}_a = f_k - \mathbb{R} = 0.75 - 1 = -0.25 \text{ Hz},$$

where the negative frequency is interpreted as phase reversal (discussed previously in Section 1.9), which occurs when we turn $\tilde{f}_a = -0.25$ Hz into positive $f_a = -\tilde{f}_a = 0.25$ Hz through the trigonometric identity $\cos(-\theta) = \cos\theta$. Note that the phase reversal can be avoided when the following relation holds.

$$(4.25) \qquad \mathbb{R} < f_k < \frac{3}{2}\mathbb{R}, \text{ because } 0 < f_k - \mathbb{R} < \frac{1}{2}\mathbb{R}.$$

Method II. To show that we obtain the same aliased frequency from the DFT coefficients, let us adapt (4.18) for $N < 2k$ we begin by repeating the first part of (4.18):

$$(4.26) \qquad \begin{aligned} x_\ell &\equiv x(\ell\triangle t), & \ell &= 0, 1, \ldots, N-1, \\ &= C_k \cos(2\pi f_k \ell \triangle t - \phi_k) \\ &= C_k \cos(2\pi \ell k/N - \phi_k) & (\because f_k \triangle t = k/N) \\ &= C_k \cos(k\ell\theta_1 - \phi_k) & (\text{let } \theta_1 = 2\pi/N) \\ &= \left(\tfrac{1}{2}C_k e^{-j\phi_k}\right)e^{jk\ell\theta_1} + \left(\tfrac{1}{2}C_k e^{j\phi_k}\right)e^{-jk\ell\theta_1} & (\text{by Euler's formula}) \\ &= \left(\tfrac{1}{2}C_k e^{-j\phi_k}\right)\omega_N^{k\ell} + \left(\tfrac{1}{2}C_k e^{j\phi_k}\right)\omega_N^{-k\ell}. & (\because \omega_N = e^{j\theta_1}) \end{aligned}$$

At this point, we can apply formula (4.26) to a problem with known numerical values for C_k, ϕ_k, k, and N. However, because we choose $N < 2k$, we must use the properties $\omega_N^{\pm mN} = 1$ to reduce k to $r < N/2$ so that ω_N^r and ω_N^{N-r} correspond to the terms in the IDFT formula.

For our example, we continue from the last line in formula (4.26) with $C_k = 3.2$, $\phi_k = \pi/4$, $k = 3$, and $N = 4$:

$$(4.27) \qquad \begin{aligned} x_\ell &= 1.6e^{-j\pi/4}\omega_4^{3\ell} + 1.6e^{j\pi/4}\omega_4^{-3\ell}, & \ell &= 0, 1, 2, 3, \\ &= 1.6e^{-j\pi/4}\omega_4^{-\ell} + 1.6e^{j\pi/4}\omega_4^{\ell} & (\because \omega_4^{\pm 4} = 1, \therefore \omega_4^3 = \omega_4^{-1}, \omega_4^{-3} = \omega_4) \\ &= X_{-1}\omega_4^{-\ell} + X_1\omega_4^{\ell} & (\text{note } r = 1 \text{ and } r < N/2) \\ &= X_3\omega_4^{3\ell} + X_1\omega_4^{\ell}. \end{aligned}$$

Hence the two nonzero DFT coefficients are

$$X_1 = 1.6e^{j\pi/4}, \ X_3 = 1.6e^{-j\pi/4}.$$

We can now reconstruct the function $y(\theta)$ using the DFT coefficients:

$$(4.28) \qquad \begin{aligned} y(\theta) &= C_r \cos(r\theta - \phi_r), & \text{where } r < N/2, \\ &= 3.2\cos(\theta + \pi/4). & (\because r = 1, X_r = \tfrac{1}{2}C_r e^{-j\phi_r} = 1.6e^{j\pi/4}) \end{aligned}$$

Since $\theta = 2\pi t/T = 2\pi t/(N\triangle t) = 2\pi t/4 = 0.5\pi t$, the reconstructed analog signal

$$(4.29) \qquad y(t) = 3.2\cos(0.5\pi t + \pi/4)$$

contains aliased frequency $f_a = 0.25$ Hz. Note again the phase reversal from $\phi_k = \pi/4$ in $x(t)$ or $x(\theta)$ to $\phi_r = -\pi/4$ in the reconstructed $y(\theta)$ or $y(t)$.

4.4.2 Making direct use of the digital frequencies

Recall that using the digital frequency de ned by

$$\mathbb{F}_k = f_k \triangle t = \frac{k}{T} \triangle t = \frac{k}{N},$$

we may conveniently express

$$x_\ell = C_k \cos(2\pi \mathbb{F}_k \ell - \phi_k),$$

where $\mathbb{F}_k < \frac{1}{2}$, because $2k < N$ when the Nyquist condition is satis ed. Note that $\mathbb{F}_k = k/N$ directly conveys the following information:

1. If k and N do not share a common factor, then the period of the N-sample sequence $\{x_\ell\}$ is indeed N samples, and they span k periods of its envelope function, which leads to two nonzero DFT coef cients indexed by k and $N - k$.

2. If k and N share a common factor q, then we have

$$\mathbb{F}_k = \frac{k}{N} = \frac{q\tilde{k}}{q\tilde{N}} = \frac{\tilde{k}}{\tilde{N}},$$

which tells us that

$$x_{\tilde{N}} = C_k \cos(2\pi\tilde{k} - \phi_k)$$

and the period of $\{x_\ell\}$ is $\tilde{N} = N/q$ samples. The fact that we have computed the DFT coef cients based on $N = q\tilde{N}$ samples simply means that the DFT coef cien ts are positioned at index $k = q\tilde{k}$ and $N - k$, which are their rightful places when the sample size is $N = q\tilde{N}$.

Let us now turn to the composite signal: suppose that we sample

$$y(t) = C_0 + \sum_{k=1}^{n} C_k \cos\left(\frac{2\pi kt}{T} - \phi_k\right)$$

to obtain the N-sample sequence (with $N > 2n$)

$$y_\ell = C_0 + \sum_{k=1}^{n} C_k \cos(2\pi \mathbb{F}_k \ell - \phi_k);$$

we immediately see that $\mathbb{F}_k = k/N$ if $T = N\triangle t$ for $k = 1, 2, \ldots, n$, and the DFT coef cients of $\{y_\ell\}$ are exactly the union of the DFT coef cients of each component.

Recall that the period of composite $y(t)$ is determined by its fundamental frequency $f_1 = 1/T$. It is now clear that when we sample the composite signal $y(t)$ for a single period, we have in fact sampled its components for multiple periods. Since the N samples span k periods for the kth component, it has the effect of putting the DFT coef cients from different components in their separate positions so they will not interfere with each other. It turns out that sampling a signal (whether composite or single-component) for integer number of periods is the key to avoid the l eakage of frequencies, which is the subject of Section 4.5.

Example 4.2 To determine the appropriate sampling rate and duration for the signal $x(t) = 10\cos(10\pi t)$, we rewrite $x(t) = 10\cos(2\pi f_\alpha t)$ with $f_\alpha = 5$ Hz, and we obtain the digital frequency $\mathbb{F}_\alpha = f_\alpha \triangle t = 5\triangle t$. To satisfy the Nyquist condition, we must have

$$\mathbb{F}_\alpha = 5\triangle t \le \frac{1}{2},$$

which immediately leads to $\triangle t \le 0.1$ sec. By choosing $\triangle t = 0.05$ sec, we obtain $\mathbb{F}_\alpha = 5\triangle t = 0.25$, which enables us to express $\mathbb{F}_\alpha$ as a rational fraction

$$\mathbb{F}_\alpha = \frac{1}{4} = \frac{k}{N}.$$

Therefore, by sampling $x(t)$ at intervals of 0.05 sec, the $N = 4$ samples will span $k = 1$ period of $x(t)$, and the sampling duration is $N\triangle t = 4 \times 0.05 = 0.2$ sec. The sampling rate is $\mathbb{R} = 1/\triangle t = 1/0.05 = 20$ (samples per second or Hertz).

From the resulting discrete-time sinusoid

$$x_\ell = 10\cos(2\pi \mathbb{F}_\alpha \ell) = 10\cos(2\pi k\ell/N), \quad \text{where } k = 1, N = 4,$$

we obtain the two nonzero DFT coef cients

$$X_1 = \tfrac{1}{2}(10) = 5, \text{ and } X_3 = \tfrac{1}{2}(10) = 5.$$

This result can be veri ed by actually computing

$$x_\ell = 10\cos(2\pi \mathbb{F}_\alpha \ell) = 10\cos(0.5\pi \ell), \quad \ell = 0, 1, 2, 3,$$

to obtain the sequence

$$\{x_0, x_1, x_2, x_3\} = \{10, 0, -10, 0\}$$

and use the DFT formula to obtain

$$
\begin{bmatrix} X_0 \\ X_1 \\ X_2 \\ X_3 \end{bmatrix}
= \frac{1}{4}
\begin{bmatrix}
1 & 1 & 1 & 1 \\
1 & \omega_4^{-1} & \omega_4^{-2} & \omega_4^{-3} \\
1 & \omega_4^{-2} & \omega_4^{-4} & \omega_4^{-6} \\
1 & \omega_4^{-3} & \omega_4^{-6} & \omega_4^{-9}
\end{bmatrix}
\begin{bmatrix} 10 \\ 0 \\ -10 \\ 0 \end{bmatrix}
= \frac{1}{4}
\begin{bmatrix}
1 & 1 & 1 & 1 \\
1 & -j & -1 & j \\
1 & -1 & 1 & -1 \\
1 & j & -1 & -j
\end{bmatrix}
\begin{bmatrix} 10 \\ 0 \\ -10 \\ 0 \end{bmatrix}
= \begin{bmatrix} 0 \\ 5 \\ 0 \\ 5 \end{bmatrix}
$$

Observe that we have simpli ed the DFT matrix using the value

$$\omega_4 = e^{j2\pi/4} = \cos(0.5\pi) + j\sin(0.5\pi) = j$$

and the de n ition $j^2 = -1$.

Example 4.3 Using the experience gained from the last example, we can now determine the appropriate sampling rate and duration for $y(t) = 3.2\cos(1.5\pi t - \pi/4)$ in two simple steps:

Step 1. We require $\mathbb{F}_\beta = 0.75\triangle t \le \dfrac{1}{2}$, which leads to $\triangle t \le \dfrac{2}{3}$.

Step 2. We choose $\triangle t = 0.2$ to obtain $\mathbb{F}_\beta = 0.75 \times 0.2 = 0.15 = \dfrac{3}{20} = \dfrac{k}{N}$.

Therefore, the sampling duration is $N\triangle t = 20 \times 0.2 = 4$ (sec), and the $N = 20$ samples span $k = 3$ periods of $y(t)$. The sampling rate is $\mathbb{R} = 1/\triangle t = 1/0.2 = 5$ (Hertz). The two nonzero DFT coef cients of $\{y_\ell\}$ so sampled are indexed by $k = 3$ and $N - k = 17$:

$$Y_3 = 1.6e^{-j\pi/4} = \tfrac{4\sqrt{2}}{5} - j\tfrac{4\sqrt{2}}{5}, \text{ and } Y_{17} = 1.6e^{j\pi/4} = \tfrac{4\sqrt{2}}{5} + j\tfrac{4\sqrt{2}}{5}.$$

Example 4.4 We now repeat Step 2 in **Example 4.3** with $\triangle t = 0.5$, and we obtain

$$\mathbb{F}_\beta = 0.75 \times 0.5 = 0.375 = \frac{3}{8} = \frac{k}{N}.$$

The sampling duration is $N\triangle t = 8 \times 0.5 = 4$ (sec), and we have $N = 8$ samples spanning over $k = 3$ periods of $y(t)$. The sampling rate is now reduced to $\mathbb{R} = 1/0.5 = 2$ (Hertz). The two nonzero DFT coefficients of the eight-sample sequence $\{y_\ell\}$ are indexed by $k = 3$ and $N - k = 5$:

$$Y_3 = 1.6e^{-j\pi/4} = \tfrac{4\sqrt{2}}{5} - j\,\tfrac{4\sqrt{2}}{5}, \quad \text{and } Y_5 = 1.6e^{j\pi/4} = \tfrac{4\sqrt{2}}{5} + j\,\tfrac{4\sqrt{2}}{5}.$$

4.4.3 Common period of sampled composite signals

Suppose the composite signal

$$x(t) = C_\alpha \cos(2\pi f_\alpha t) + C_\beta e^{j2\pi f_\beta t}$$

has been sampled above the Nyquist rate at intervals of $\triangle t$ seconds (which is the reciprocal of the sampling rate of $\mathbb{R}$ Hertz), and we express the sample sequence as

$$x_\ell = C_\alpha \cos(2\pi \mathbb{F}_\alpha \ell) + C_\beta e^{j2\pi \mathbb{F}_\beta \ell}, \quad \ell = 0, 1, \ldots,$$

where the digital frequencies $\mathbb{F}_\alpha = f_\alpha \triangle t < \frac{1}{2}$ and $\mathbb{F}_\beta = f_\beta \triangle t < \frac{1}{2}$. To determine the periodicity of the sample sequence $\{x_\ell\}$, we take the following steps:

Step 1: Express the digital frequency of each component signal as a rational fraction. For the given signal, let us assume

$$\mathbb{F}_\alpha = \frac{k_1}{N_1} \text{ and } \mathbb{F}_\beta = \frac{k_2}{N_2}.$$

Step 2: Find the common period $N = \text{LCM}(N_1, N_2)$.

We now verify that $x_{\ell+N} = x_\ell$. To proceed, we assume $N = m_1 N_1 = m_2 N_2$ (because N is the least common multiple of N_1 and N_2), and we obtain

$$C_\alpha \cos\big(2\pi \mathbb{F}_\alpha(\ell + N)\big) = C_\alpha \cos(2\pi \mathbb{F}_\alpha \ell + 2\pi k_1 m_1) = C_\alpha \cos(2\pi \mathbb{F}_\alpha \ell),$$

and

$$C_\beta e^{j2\pi \mathbb{F}_\beta(\ell+N)} = C_\beta e^{j2\pi \mathbb{F}_\beta \ell} e^{j2\pi k_2 m_2} = C_\beta e^{j2\pi \mathbb{F}_\beta \ell}.$$

The desired result follows immediately:

$$\begin{aligned} x_{\ell+N} &= C_\alpha \cos\big(2\pi \mathbb{F}_\alpha(\ell + N)\big) + C_\beta e^{j2\pi \mathbb{F}_\beta(\ell+N)} \\ &= C_\alpha \cos(2\pi \mathbb{F}_\alpha \ell) + C_\beta e^{j2\pi \mathbb{F}_\beta \ell} \\ &= x_\ell. \end{aligned}$$

If we denote the fundamental frequency of the composite envelope function $x(t)$ by f_o, then f_o is the greatest common divisor of the individual frequencies (recall Section 1.6):

$$f_o = \text{GCD}(f_\alpha, f_\beta),$$

and the period of the composite signal $x(t)$ is $T_o = 1/f_o$ sec. Since the sampling duration is $T_N = N\triangle t$ seconds in total, we have sampled $\Lambda = T_N/T_o$ periods of $x(t)$ for the N samples.

It also turns out that we may directly obtain $\Lambda = \text{GCD}(k_1, k_2)$, because

$$\frac{1}{T_o} = f_o = \text{GCD}(f_\alpha, f_\beta) = \text{GCD}\left(\frac{k_1}{N_1\triangle t}, \frac{k_2}{N_2\triangle t}\right),$$

and we obtain

$$\frac{T_N}{T_o} = T_N \times f_o = T_N \times \text{GCD}\left(\frac{k_1 m_1}{N\triangle t}, \frac{k_2 m_2}{N\triangle t}\right) \qquad (\because N = N_1 m_1 = N_2 m_2)$$

$$= T_N \times \frac{1}{N\triangle t} \times \text{GCD}(k_1 m_1, k_2 m_2)$$

$$= \text{GCD}(k_1 m_1, k_2 m_2) \qquad (\because T_N = N\triangle t)$$

$$= \text{GCD}(k_1, k_2). \qquad (\because \text{GCD}(m_1, m_2) = 1)$$

Example 4.5 The function

$$y(t) = 4.5 \cos\left(2\pi f_\alpha t\right) + 7.2 \cos\left(2\pi f_\beta t\right) = 4.5 \cos\left(1.2\pi t\right) + 7.2 \cos\left(1.8\pi t\right)$$

was used in **Examples 1.1** and **1.2** in Section 1.6 to demonstrate a commensurate sum with fundamental frequency $f_o = 0.3$ Hz. Suppose $y(t)$ has been sampled at intervals of $\triangle t = 0.5$ (sec) so that $y_\ell = 4.5 \cos\left(2\pi \mathbb{F}_\alpha \ell\right) + 7.2 \cos\left(2\pi \mathbb{F}_\beta \ell\right)$, where

$$\mathbb{F}_\alpha = f_\alpha \triangle t = 0.6 \times 0.5 = \frac{3}{10} = \frac{k_1}{N_1}; \quad \mathbb{F}_\beta = f_\beta \triangle t = 0.9 \times 0.5 = \frac{9}{20} = \frac{k_2}{N_2}.$$

We can now determine the period of the sequence $\{y_\ell\}$, which is $N = \text{LCM}(N_1, N_2) = \text{LCM}(10, 20) = 20$ (samples), and they span over $\Lambda = \text{GCD}(k_1, k_2) = \text{GCD}(3, 9) = 3$ periods of the original signal $y(t)$. Since the total time for taking 20 samples is $T_N = N\triangle t = 20 \times 0.5 = 10$ seconds, we get the same result from computing $\Lambda = T_N \times f_o = 10 \times 0.3 = 3$ periods.

Applying the DFT formula to the $N = 20$ sample sequence $\{y_0, y_1, \ldots, y_{19}\}$, we obtain four nonzero DFT coef cients indexed by $\tilde{k}_1 = k_1 m_1 = 6$, $N - \tilde{k}_1 = 14$, $\tilde{k}_2 = k_2 m_2 = 9$, and $N - \tilde{k}_2 = 11$, i.e.,

$$Y_6 = \tfrac{1}{2}C_\alpha = 2.25, \quad Y_{-6} = \tfrac{1}{2}C_\alpha = 2.25 = Y_{14};$$

$$Y_9 = \tfrac{1}{2}C_\beta = 3.6, \quad Y_{-9} = \tfrac{1}{2}C_\beta = 3.6 = Y_{11}.$$

The original function $y(\theta)$ can be reconstructed from

$$y(\theta) = \sum_{r=-\frac{N}{2}+1}^{\frac{N}{2}} Y_r e^{jr\theta} = \sum_{r=-9}^{10} Y_r e^{jr\theta}$$

$$= 3.6 e^{-j9\theta} + 2.25 e^{-j6\theta} + 2.25 e^{j6\theta} + 3.6 e^{j9\theta}$$

$$= 2.25\left(e^{j6\theta} + e^{-j6\theta}\right) + 3.6\left(e^{j9\theta} + e^{-j9\theta}\right)$$

$$= 4.5 \cos\left(6\theta\right) + 7.2 \cos\left(9\theta\right).$$

Observe that $y\left(\theta + \tfrac{2}{3}\pi\right) = 4.5(6\theta + 4\pi) + 7.2(9\theta + 6\pi) = y(\theta)$. To convert the variable from θ back to t, it is important to recall that the DFT formula was derived by assuming that the

N samples are equally spaced over $T = T_N = N\triangle t$ seconds, so $\theta = 2\pi t/T = 2\pi t/T_N = 2\pi t/(10) = 0.2\pi t$, and we are able to reconstruct the original signal

$$y(t) = 4.5\cos(1.2\pi t) + 7.2\cos(1.8\pi t).$$

Remark 1: For the chosen $\triangle t = 0.5$ sec (or sampling rate $\mathbb{R} = 2$ Hz), the common period $N = 20$ represents the smallest number of samples we must take so that the period $T_N = N\triangle t$ imposed by the DFT is an integer multiple of the fundamental period T_o of the original signal $y(t)$ in this case, we have $T_N = 3T_o$. The sampling of $y(t)$ based on these choices is illustrated in the top plot in Figure 4.2.

Figure 4.2 Sampling $y(t)$ at 2 Hz (for three periods) and 3 Hz (for one period).

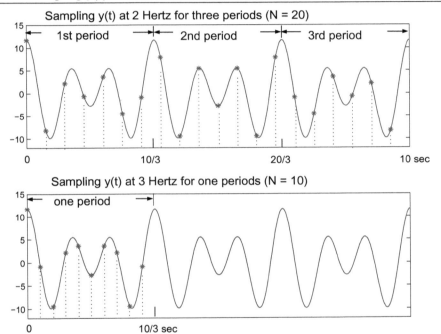

Remark 2: The sampling rate may vary as long as (i) the Nyquist condition is satis ed and (ii) $T_N = N\triangle t = mT_o$, where $m \geq 1$ is a positive integer. For this example, using $\triangle t = \frac{1}{3}$ will result in $\mathbb{F}_\alpha = \frac{1}{5}$ and $\mathbb{F}_\beta = \frac{3}{10}$. Since $N = \text{LCM}(5, 10) = 10$ and $\Lambda = \text{GCD}(1, 3) = 1$, the ten samples are equally spaced over a single period of $y(t)$, and we verify that $T_N = N\triangle t = 10 \times \frac{1}{3} = T_o$. Applying the DFT to the ten samples, we obtain the four nonzero DFT coef cients: $Y_2 = 2.25$, $Y_{-2} = 2.25 = Y_8$, $Y_3 = 3.6$, and $Y_{-3} = 3.6 = Y_7$. The reconstructed $y(\theta)$ is now given by

$$y(\theta) = 4.5\cos(2\theta) + 7.2\cos(3\theta).$$

Observe that $y(\theta + 2\pi) = y(\theta)$. Letting $\theta = 2\pi t/T_N = 0.6\pi t$, we again recover

$$y(t) = 4.5\cos(1.2\pi t) + 7.2\cos(1.8\pi t).$$

The sampling of $y(t)$ based on these new choices is illustrated in the second plot in Figure 4.2. The signal fully reconstructed based on the twenty computed DFT coefficients $\{Y_0, Y_1, \ldots, Y_{19}\} = \{0, 0, \ldots, 0, Y_6, 0, 0, Y_9, 0, Y_{11}, 0, 0, Y_{14}, 0, 0, \ldots, 0\}$ from Table 4.1 is shown in Figure 4.3.

Table 4.1 Numerical values of M DFT coefficients when $T_M = T_o$ and $T_M = 3T_o$.

	$M = 10$ (one period)			$M = 20$ (three periods)		
r	DFT Y_r	A_r	B_r	DFT Y_r	A_r	B_r
0	0	0	—	0	0	—
1	0	0	0	0	0	0
2	2.25	4.5	0	0	0	0
3	3.60	7.2	0	0	0	0
4	0	0	0	0	0	0
5	0	0	—	0	0	0
6	0			2.25	4.5	0
7	3.60			0	0	0
8	2.25			0	0	0
9	0			3.6	7.2	0
10				0	0	—
11				3.6		
12				0		
13				0		
14				2.25		
15				0		
16				0		
17				0		
18				0		
19				0		

4.5 Frequency Distortion by Leakage

In the context of **Example 4.5** from the previous section, the term leakage refers to the consequent distortion of frequency contents when the total number of samples M is neither equal to $N = \text{LCM}(N_1, N_2)$ nor equal to an integer multiple of N. We study again the function used in the cited example:

$$y(t) = 4.5 \cos\left(2\pi f_\alpha t\right) + 7.2 \cos\left(2\pi f_\beta t\right) = 4.5 \cos\left(1.2\pi t\right) + 7.2 \cos\left(1.8\pi t\right).$$

For sampling rate $\mathbb{R} = 2$ Hz the digital frequencies of $\{y_\ell\}$ were determined to be

$$\mathbb{F}_\alpha = \frac{k_1}{N_1} = \frac{3}{10}, \quad \mathbb{F}_\beta = \frac{k_2}{N_2} = \frac{9}{20},$$

with $N = \text{LCM}(N_1, N_2) = 20$. Suppose we have sampled $y(t)$ at $\mathbb{R} = 2$ Hz to obtain a total of $M = 10$ samples, then $M = 0.5N$, and the sampling duration $T_M = M\triangle t = M/\mathbb{R} = 5$ seconds. Since the ratio $T_M/T_o = T_M \times f_o = 5 \times 0.3 = 1.5$, the function $y(t)$ is now sampled

Figure 4.3 Signal reconstructed using computed DFT coefﬁcients from Table 4.1.

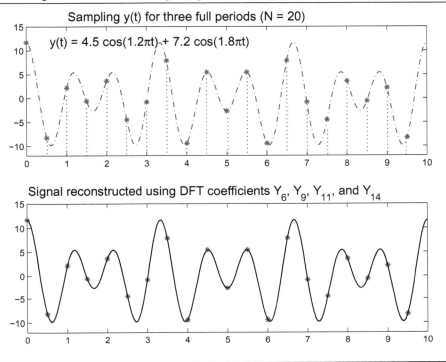

Figure 4.4 Sampling $y(t)$ at 2 Hz for 1.5 periods.

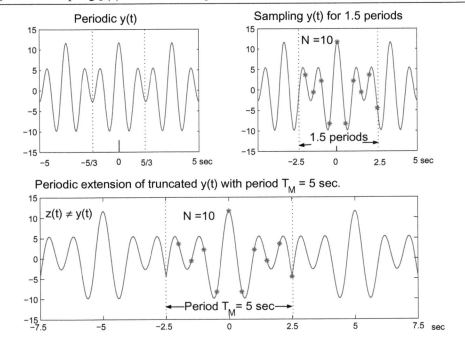

for 1.5 periods instead of an integer number of periods. The sampling of $y(t)$ based on these choices is illustrated in Figure 4.4.

To investigate the consequence of applying the DFT to the M-sample sequence with $M = 10$, we express the digital frequencies as

$$\mathbb{F}_\alpha = \frac{3}{10} = \frac{3}{M}, \quad \mathbb{F}_\beta = \frac{9}{20} = \frac{4.5}{10} = \frac{4.5}{M},$$

and we obtain the analytical expression

$$\begin{aligned}
y_\ell &= 4.5 \cos\left(2\pi \tfrac{3}{M} \ell\right) + 7.2 \cos\left(2\pi \tfrac{4.5}{M} \ell\right), & \ell = 0, 1, \ldots, M-1, \\
&= 4.5 \cos\left(3\ell\theta_1\right) + 7.2 \cos\left(4.5\ell\theta_1\right) & \left(\text{let } \theta_1 = \tfrac{2\pi}{M}\right) \\
&= 2.25\left(e^{j3\ell\theta_1} + e^{-j3\ell\theta_1}\right) + 3.6\left(e^{j4.5\ell\theta_1} + e^{-j4.5\ell\theta_1}\right) & \text{(by Euler s formula)} \\
&= 2.25\,\omega_M^{3\ell} + 2.25\,\omega_M^{-3\ell} + 3.6\,\omega_M^{4.5\ell} + 3.6\,\omega_M^{-4.5\ell} & \left(\text{let } \omega_M = e^{j\theta_1}\right) \\
&= 2.25\,\omega_M^{3\ell} + 2.25\,\omega_M^{7\ell} + 3.6\,\omega_M^{4.5\ell} + 3.6\,\omega_M^{5.5\ell}. & \left(\because \omega_M^{M \pm r} = \omega_M^{\pm r}\right)
\end{aligned}$$

(4.30)

Comparing the right-hand side of (4.30) with the M-term IDFT formula

$$y_\ell = \sum_{r=0}^{M-1} Y_r \omega_M^{r\ell}, \quad \text{where } \omega_M \overset{\text{def}}{=} e^{j2\pi/M},$$

clearly we won t nd terms to match $\omega_M^{4.5\ell}$ and $\omega_M^{5.5\ell}$, because the exponent r in the IDFT formula must be an integer. This happens because the corresponding component in the original signal $y(t)$ is *nonharmonic* with respect to the interval $T_M = 5$ seconds chosen for the DFT computation:

$$7.2 \cos\left(1.8\pi t\right) = 7.2 \cos\left(2\pi \frac{4.5}{5} t\right) = 7.2 \cos\left(2\pi \frac{4.5}{T_M} t\right),$$

which shows a frequency $f_\beta = 4.5/T_M$ instead of $f_k = k/T_M$ required by a harmonic component. We will show how the frequency content of a nonharmonic component spreads across the entire DFT spectrum. Because of such leakage from the nonharmonic component, while the two terms $2.25\,\omega_M^{3\ell}$ and $2.25\,\omega_M^{7\ell}$ from the harmonic component $4.5 \cos\left(1.2\pi t\right)$ contribute to the corresponding terms in the IDFT formula, they are no longer solely responsible for the values of Y_3 and Y_7 the effect of leakage on the entire DFT spectrum is studied in the following sections.

4.5.1   Fourier series expansion of a nonharmonic component

To determine the ten DFT coef cients of the nonharmonic component

$$z(t) = 7.2 \cos\left(1.8\pi t\right) = 7.2 \cos\left(2\pi \frac{4.5}{T_M} t\right),$$

which was sampled at $\mathbb{R} = 2$ Hz for $T_M = 5$ seconds in the last example, we can proceed analytically by nding, at rst, the Fourier series coef cients of $z(t)$ using formula (3.12) from

Chapter 3:

$$
\begin{aligned}
C_k &= \frac{1}{T_M} \int_{-T_M/2}^{T_M/2} z(t)\, e^{-j2\pi kt/T_M}\, dt \\
&= \frac{7.2}{5} \int_{-5/2}^{5/2} \cos(1.8\pi t)\, e^{-j2\pi kt/5}\, dt \qquad\qquad (\because T_M = 5) \\
&= \frac{7.2}{5} \int_{-5/2}^{5/2} \left[\frac{e^{j1.8\pi t} + e^{-j1.8\pi t}}{2} \right] e^{-j2\pi kt/5}\, dt \qquad \text{(by Euler s formula)} \\
&= \frac{3.6}{5} \int_{-5/2}^{5/2} e^{j2\pi(4.5-k)t/5} + e^{j2\pi(-4.5-k)t/5}\, dt.
\end{aligned}
$$

To evaluate the de n ite integral on the right-hand side, we make use of the following result:

$$
\frac{1}{2\tau} \int_{-\tau}^{\tau} e^{j\alpha t}\, dt = \frac{1}{2\tau} \left.\frac{e^{j\alpha t}}{j\alpha}\right|_{-\tau}^{\tau} = \frac{1}{\alpha\tau} \left[\frac{e^{j\alpha\tau} - e^{-j\alpha\tau}}{2j} \right] = \frac{\sin \alpha\tau}{\alpha\tau}.
$$

Letting $\tau = 5/2$, $\alpha = 2\pi(4.5 - k)/5$, and $\beta = 2\pi(-4.5 - k)/5$, we immediately obtain

$$
\begin{aligned}
C_k &= \frac{3.6}{2\tau} \int_{-\tau}^{\tau} e^{j\alpha t} + e^{j\beta t}\, dt \\
&= 3.6 \left[\frac{\sin \alpha\tau}{\alpha\tau} + \frac{\sin \beta\tau}{\beta\tau} \right] \\
&= 3.6 \left[\frac{\sin(4.5 - k)\pi}{(4.5 - k)\pi} + \frac{\sin(-4.5 - k)\pi}{(-4.5 - k)\pi} \right] \\
&= 3.6 \left[\frac{\sin(4.5 - k)\pi}{(4.5 - k)\pi} + \frac{\sin(4.5 + k)\pi}{(4.5 + k)\pi} \right].
\end{aligned}
$$

We can further simplify the right-hand side using

$$
\sin(\gamma\pi \mp k\pi) = \sin \gamma\pi \cos k\pi \mp \cos 4.5\pi \sin k\pi = (-1)^k \sin \gamma\pi
$$

with $\gamma = 4.5$, and we obtain

$$
\begin{aligned}
C_k &= 3.6(-1)^k \sin 4.5\pi \left[\frac{1}{(4.5 - k)\pi} + \frac{1}{(4.5 + k)\pi} \right] \\
&= 3.6(-1)^k \left[\frac{9}{(4.5^2 - k^2)\pi} \right]. \qquad\qquad (\because \sin 4.5\pi = 1)
\end{aligned}
$$

Note that $C_k \neq 0$ for every integer $k \in (-\infty, \infty)$. Because the factor $(4.5^2 - k^2)$ occurs in the denominator, $C_{\pm 4}$ has the largest magnitude when $(4.5^2 - k^2)$ takes on the smallest value with $k = \pm 4$.

4.5.2 Aliased DFT coefficients of a nonharmonic component

Using the results from the last section, we can express the component $z(t)$ (which is non-harmonic with respect to the sampling duration T_M chosen for the DFT computation) by its

Fourier series expansion, i.e.,

$$z(t) = 7.2 \cos{(1.8\pi t)} = 7.2 \cos{\left(2\pi \frac{4.5}{T_M} t\right)} = \sum_{k=-\infty}^{\infty} C_k \, e^{j2\pi kt/T_M}, \quad \text{where}$$

$$T_M = 5, \quad C_k = \frac{3.6 \times 9 \times (-1)^k}{(4.5^2 - k^2)\pi}.$$

As discussed previously in Section 3.11, the Fourier series coefficients C_k of $z(t)$ can be directly linked to the DFT coefficients Z_r defined by

$$Z_r = \frac{1}{M} \sum_{\ell=0}^{M-1} z_\ell \omega_M^{-r\ell}, \quad \omega_M = e^{j2\pi/M}, \quad r = 0, 1, \ldots, M-1,$$

where $z_\ell = z(\ell \triangle t)$ are the M equally spaced samples of $z(t)$ over the imposed period $[0, T_M]$. The relationship

$$Z_r = \sum_{m=-\infty}^{\infty} C_{r+mM}, \quad r = 0, 1, \ldots, M-1$$

derived in Section 3.11 explains how all of the DFT coefficients of sampled $z(t)$ are affected by the infinite number of nonzero Fourier series coefficients.

Recall that in our example we were analyzing the DFT coefficients of sampled composite signal

$$y(t) = 4.5 \cos{(1.2\pi t)} + 7.2 \cos{(1.8\pi t)},$$

for which we have the IDFT formula

$$y_\ell = 2.25 \, \omega_M^{3\ell} + 2.25 \, \omega_M^{7\ell} + \sum_{r=0}^{M-1} Z_r \omega_M^{r\ell} = \sum_{r=0}^{M-1} Y_r \omega_M^{r\ell}, \quad \ell = 0, 1, \ldots, M-1;$$

hence, the DFT coefficients Y_r can now be expressed as

$$Y_r = \begin{cases} Z_3 + 2.25 & \text{if } r = 3 \\ Z_7 + 2.25 & \text{if } r = 7 \\ Z_r & \text{if } r \neq 3 \,\&\, r \neq 7 \end{cases}$$

for $r = 0, 1, \ldots, M-1$ (recall $M = 10$ in our example), where

$$Z_r = \sum_{m=-\infty}^{\infty} C_{r+mM}$$

$$= \sum_{m=-\infty}^{\infty} \frac{3.6 \times 9 \times (-1)^{r+mM}}{(4.5^2 - (r+mM)^2)\pi}$$

$$= \sum_{m=-\infty}^{\infty} \frac{32.4(-1)^r}{20.25\pi - (r+10M)^2\pi}. \qquad (\because M = 10, \, (-1)^{mM} = 1)$$

To approximate the DFT coefficients Y_r using this analytical formula, we may evaluate

$$Z_r = \sum_{m=-K}^{K} C_{r+mM},$$

using suf ciently large nite K. For $M = 10$, using $K = 500$, we obtain $Y_0 \approx 0.1143$, $Y_1 \approx -0.1267$, $Y_2 \approx 0.1768$, $Y_3 \approx 1.9032$, $Y_4 \approx 1.5667$, $Y_5 \approx 4.5456$, $Y_6 \approx 1.5667$, $Y_7 \approx 1.9032$, $Y_8 \approx 0.1768$, and $Y_9 \approx -0.1267$. Therefore, the entire DFT spectrum is affected by leakage from the nonharmonic frequency.

4.5.3 Demonstrating leakage by numerical experiments

In the last section we used the composite signal from Example 4.5 to study frequency distortion by leakage analytically; we shall now do the same numerically. We have sampled the composite signal

$$y(t) = 4.5 \cos(1.2\pi t) + 7.2 \cos(1.8\pi t)$$

at intervals of $\triangle t = 0.5$ for $M = 10$ samples over the duration $T_M = 1.5 T_o = 5$ seconds, and we apply the DFT formula to compute $\{Y_r\}$ directly from the M-sample sequence $\{y_\ell\}$. The computed DFT coef cients $\{Y_0, Y_1, \ldots, Y_9\}$ are given in Table 4.2. The reconstructed signal

$$
\begin{aligned}
(4.31) \quad z(t) &= \sum_{r=0}^{9} Y_r \, e^{j2\pi rt/T_M} \\
&= \frac{A_0}{2} + A_5 \cos \frac{20\pi t}{T_M} + \sum_{r=1}^{4} A_r \cos \frac{2\pi rt}{T_M} + B_r \sin \frac{2\pi rt}{T_M} \\
&= \frac{A_0}{2} + A_5 \cos(4\pi t) + \sum_{r=1}^{4} A_r \cos(0.4\pi rt) + B_r \sin(0.4\pi rt)
\end{aligned}
$$

contains $M = 10$ terms with coef cien ts A_r and B_r explicitly given in Table 4.2. Note that $B_r = 0$ for every r because the function in the sampled interval remains to be an even function; however, we now have $A_r \neq 0$ for $r = 0, 1, 2, 3, 4, 5$. Clearly the reconstructed signal $z(t) \neq 4.5 \cos(1.2\pi t) + 7.2 \cos(1.8\pi t)$, and we illustrate how $z(t)$ deviates from $y(t)$ in Figure 4.5. In Figure 4.6 we show that leakage can be reduced by increasing the number of samples from $N = 10$ to $N = 20$.

4.5.4 Mismatching periodic extensions

In this section we offer another useful perspective on the cause of frequency leakage. Suppose we have sampled a signal $x(t)$ with period T to obtain M equally spaced samples over $[0, T_M]$ for spectrum analysis, and T_M is not an integer multiple of T. By carrying out the DFT on the M samples denoted by $z_\ell \overset{def}{=} x_\ell$ for $\ell = 0, 1, \ldots, M-1$, we obtain

$$(4.32) \qquad Z_r = \frac{1}{M} \sum_{\ell=0}^{M-1} z_\ell \omega_M^{-r\ell} \qquad (r = 0, 1, \ldots, M-1);$$

by carrying out the IDFT on computed Z_r, we recover

$$(4.33) \qquad z_\ell = \sum_{r=0}^{M-1} Z_r \omega_M^{r\ell} \qquad (\ell = 0, 1, \ldots, M-1).$$

Observe that the periodicity of the M-sample sequence, expressed as $z_{\ell+M} = z_\ell$, is imposed by the IDFT computation prescribed by (4.33), which reinforces the fact that the M samples are taken by the DFT to represent a single period of some unknown signal $z(t)$; hence, the

Table 4.2 Numerical values of M distorted DFT coef cients when $T_M = 1.5T_o$.

r	$M = 10\ (T_M = 1.5T_o)$ DFT Y_r	A_r	B_r	$M = 20\ (T_M = 1.5T_o)$ DFT Y_r	A_r	B_r
0	0.1140368	0.2280736	—	0.4215058	0.8430116	—
1	−0.1264108	−0.2528215	0	−0.4474678	−0.8949357	0
2	0.1765708	0.3531417	0	0.5448626	1.0897252	0
3	1.9034602	3.8069204	0	1.4256876	2.8513752	0
4	1.5664108	3.1328215	0	2.3303310	4.6606621	0
5	4.5459011	4.5459011	—	2.2729506	4.5459011	0
6	1.5664108			−0.7639203	−1.5278405	0
7	1.9034602			0.4777726	0.9555452	0
8	0.1765708			−0.3682918	−0.7365835	0
9	−0.1264108			0.3210571	0.6421141	0
10				−0.3074691	−0.3074691	—
11				0.3210571		
12				−0.3682918		
13				0.4777726		
14				−0.7639203		
15				2.2729506		
16				2.3303310		
17				1.4256876		
18				0.5448626		
19				−0.4474678		

Figure 4.5 Signal reconstructed using $M = 10$ DFT coef cients from Table 4.2.

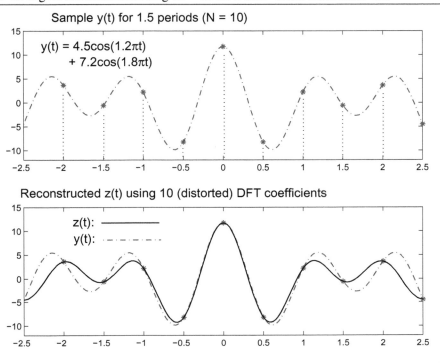

Figure 4.6 Signal reconstructed using $M = 20$ DFT coef cients from Table 4.2.

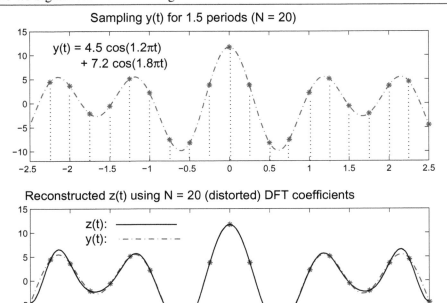

reconstructed signal must satisfy $z(t+T_M) = z(t)$. Clearly, $x_{\ell+M} \neq x_\ell$ and $x(t+T_M) \neq x(t)$ when T_M/T is not an integer, *so the signal $z(t)$ analyzed by the DFT is not the original signal $x(t)$.* By treating $z(t)$ as a time-limited function over $[0, T_M]$ at rst, its periodic extension gives us the protracted version of $z(t)$ with period T_M. That is,

$$z(t) \stackrel{def}{=} \begin{cases} x(t), & t \in [0, T_M], \\ x(t - mT_M), & |t| > T_M, \ t - mT_M \in [0, T_M], \quad m \text{ is an integer.} \end{cases}$$

Since the Fourier series expansion of the T_M-periodic $z(t)$ is different from that of the T-periodic $x(t)$, so are the corresponding DFT coef cients; i.e., $Z_r \neq X_r$. The leakage error in the computed Z_r can thus be attributed to truncating $x(t)$ at the wrong place, and for this reason the leakage error is also referred to as the truncation error or nite sample error in the literature.

Understanding frequency leakage from this perspective will be useful when we connect the error in the computed spectrum to the application of windows in Chapter 8.

4.5.5   Minimizing leakage in practice

While in theory we can eliminate leakage by sampling a given signal for an integer number of periods, this cannot be easily accomplished in practice when we attempt to analyze samples from an unknown signal becaus e we would not know its period. The usual strategies employed to minimize leakage are to (i) experiment with increasing sampling rate and sampling duration if this option is available (in general, a higher sampling rate and a longer sampling duration help minimize the effects of aliasing and leakage); (ii) run the DFT on increasing number of samples until there is little change in the computed spectrum (for example, we have demonstrated how leakage can be reduced by increasing the number of samples from $N = 10$ in Figure 4.5 to $N = 20$ in Figure 4.6); and (iii) use tapered windows to truncate the sample sequence before running the DFT   this is a topic covered in Chapter 8.

4.6   The Effects of Zero Padding

4.6.1   Zero padding the signal

There are two commonly cited reasons for extending the sample sequence by adding zeros. (i) Some FFT computer programs require the user to input exactly 2^n samples. (When this condition is not met, some program will automatically append zeros to the input data so the data length is extended to the next power of two.) We remark that the radix-2 FFT is simply one fast method of computing the DFT coef cients   it does not alter the mathematical de ni-tion or properties of the DFT. Although FFT algorithms for other lengths (including arbitrary prime length) have been developed, their implementations may not be available in every FFT package. (ii) When the DFT spectrum is too sparse for us to visualize a continuous analog spectrum $X(f)$, one may wish to decrease the spectral spacing $\triangle f$ on the frequency grid. Re-call that $\triangle f = 1/(N\triangle t)$; hence, $\triangle f$ can be reduced if we enlarge N by adding zeros. (The continuous analog spectrum $X(f)$ is called the Fourier transform of $x(t)$, which is formally treated in Chapter 5.)

How does zero padding affect the DFT spectrum? In the rst case, the zero-padding of input data to the next power of two $\tilde{M} = 2^s \geq N$ may result in $\tilde{M} = \alpha N$, where α is not an integer. In such case the zero-padded input data of length $\tilde{M} = \alpha N$ may be interpreted as the result obtained by truncating zero-padded input data sequence of length $M = qN$, where $q > a$ is an integer the case we study next, and the frequency distortion caused by improper truncation of the data sequence in the time-domain will be studied in Chapter 8.

We develop next the zero-padding strategy which allows us to obtain the additional values needed for visualizing a continuous spectrum *without* distorting the original DFT spectrum; i.e., the original DFT coef cients can be recovered from the new results. To accomplisth that, we need to append zeros to an N-sample sequence $\{x_\ell\}$ to extend its length to $M = qN$, where q is an integer and $q \geq 2$, and when we assume $N = 2n+2$ and $M = 2m+2$, the DFT of the zero-padded M-sample sequence $\{z_\ell\}$ is given by

(4.34)
$$Z_r = \frac{1}{M} \sum_{\ell=-m}^{m+1} z_\ell \omega_M^{-r\ell}, \quad \omega_M \stackrel{def}{=} e^{j2\pi/M} \quad (-m \leq r \leq M-m+1)$$

$$= \frac{1}{M} \sum_{\ell=-n}^{n+1} z_\ell \omega_M^{-r\ell}. \qquad (\because z_\ell = 0 \text{ for } \ell < -n \text{ or } \ell > n+1))$$

Note that neighboring Z_r and Z_{r+1} are now separated by

$$\triangle \tilde{f} = \frac{1}{M\triangle t} = \frac{1}{q}\left[\frac{1}{N\triangle t}\right] = \frac{1}{q}\triangle f.$$

Letting $r = q \times k$ for $-n \leq k \leq n-1$ in the de n ing formula (4.34) for Z_r, we obtain

(4.35)
$$Z_{q \times k} = \frac{1}{M} \sum_{\ell=-n}^{n+1} z_\ell \omega_M^{-qk\ell} = \frac{1}{qN} \sum_{\ell=-n}^{n+1} z_\ell \omega_{qN}^{-qk\ell} \quad (\because M = qN)$$

$$= \frac{1}{qN} \sum_{\ell=-n}^{n+1} z_\ell \omega_N^{-k\ell} \quad (\because \omega_{qN}^q = e^{j2\pi/N} = \omega_N)$$

$$= \frac{1}{q}\left[\frac{1}{N} \sum_{\ell=-n}^{n+1} x_\ell \omega_N^{-k\ell}\right] \quad (\because z_\ell = x_\ell, \ -n \leq \ell \leq n+1)$$

$$= \frac{1}{q} X_k. \qquad (\text{by de n ition of DFT})$$

Hence we can recover $X_k = q Z_{q \times k}$ for $-n \leq q \leq n+1$. (An example is shown later in Figures 4.7, 4.8, 4.9, and Table 4.3.)

How do we interpret the $q-1$ values of Z_r between $Z_{q \times k}$ and $Z_{q \times (k+1)}$? Since we can obtain the de ning formula of $\frac{1}{q}X_k$ by evaluating

(4.36)
$$Z(\theta) = \frac{1}{q}\left[\frac{1}{N} \sum_{\ell=-n}^{n+1} x_\ell e^{-j\ell\theta}\right] = \frac{1}{M} \sum_{\ell=-n}^{n+1} z_\ell e^{-j\ell\theta}$$

at $\theta = k(2\pi/N)$, and we can also obtain the de ning formula (4.34) of Z_r by evaluating the same $Z(\theta)$ at $\theta = r(2\pi/M)$, the Z_r values between $\frac{1}{q}X_n$ and $\frac{1}{q}X_{n+1}$ are simply additional interpolating frequency points supplied by the same function $Z(\theta)$.

Therefore, by zero padding the signal, we effortlessly obtain additional values of $Z(\theta)$ so that we can plot a visually denser spectrum. The IDFT of $\{q Z_r\}$ returns the original N samples plus $M - N$ zeros no information is gained or lost by zero padding. This explains why zero padding in the time domain leads to interpolation in the frequency domain.

To directly and explicitly demonstrate Z_r s interpolation of the DFT, we begin with the DFT (of zero-padded sequence) de ned by Equation (4.34) and proceed as shown below.

$$
\begin{aligned}
Z_r &= \frac{1}{qN} \sum_{\ell=-n}^{n+1} x_\ell\, \omega_{qN}^{-r\ell}, && \text{where } N = 2n+2,\ M = qN, \\
&= \frac{1}{M} \sum_{\ell=-n}^{n+1} \left[\sum_{k=-n}^{n+1} X_k \omega_N^{k\ell} \right] \omega_M^{-r\ell} && \text{(by IDFT de nition)} \\
&= \frac{1}{M} \sum_{k=-n}^{n+1} X_k \left[\sum_{\ell=-n}^{n+1} \omega_N^{k\ell} \omega_M^{-r\ell} \right] \\
&= \frac{1}{M} \sum_{k=-n}^{n+1} X_k \left[\sum_{\ell=-n}^{n+1} \omega_N^{\ell(k-r/q)} \right] && (\because \omega_M^q = \omega_{qN}^q = e^{j2\pi/N} = \omega_N) \\
&= \frac{1}{M} \sum_{k=-n}^{n+1} X_k L_k \big(k - r/q\big),
\end{aligned}
$$

(4.37)

where

$$
L_k\big(k - r/q\big) = \sum_{\ell=-n}^{n+1} \omega_N^{\ell(k-r/q)} = \begin{cases} N & \text{if } m = k - r/q = 0; \\ 0 & \text{if } m = k - r/q \neq 0 \text{ is an integer;} \\ \dfrac{1 - e^{j2\pi\lambda_k}}{1 - \omega_N^{\lambda_k}} & \text{if } \lambda_k = k - r/q \neq 0 \text{ is not an integer.} \end{cases}
$$

Note that $\omega_N^{\lambda_k N} = e^{j2\pi\lambda_k} \neq 1$ when λ_k is not an integer. Assuming that $q = M/N$ is an integer as before, we may now use (4.37) to show

$$
Z_{q\times\mu} = \frac{1}{M} \sum_{k=-n}^{n+1} X_k L_k\big(k - \tfrac{q\times\mu}{q}\big) = \frac{1}{M} X_\mu L_\mu\big(\mu - \tfrac{q\times\mu}{q}\big) = \frac{N}{M} X_\mu = \frac{1}{q} X_\mu;
$$

for Z_r with r not being an integer multiple of q, we have the interpolated value according to (4.37):

$$
Z_r = \frac{1}{M} \sum_{k=-n}^{n+1} X_k L_k(k - r/q) = \frac{1}{M} \sum_{k=-n}^{n+1} X_k \left[\frac{1 - e^{j2\pi\lambda_k}}{1 - e^{j2\pi\lambda_k/N}} \right],
$$

where $\lambda_k = k - r/q$ is *not* an integer.

When the zero-padded data length M is not an integer multiple of the original length N, then $q = M/N$ is not an integer; while Z_r is still de ned by Equation (4.34), we no longer have the conventional DFT $\{X_\mu\}$ appear as a subset of the new DFT $\{Z_r\}$.

Example 4.6 In Figures 4.8 and 4.9 we demonstrate the effect of zero padding using the Gaussian function

$$
x(t) = e^{-at^2}\ (a > 0), \quad t \in (-\infty, \infty),
$$

and its Fourier transform

$$
X(f) = \sqrt{\pi/a}\, e^{-\pi^2 f^2/a}, \quad f \in (-\infty, \infty);
$$

the latter represents the continuous analog spectrum of the nonperiodic time-domain function $x(t)$ in the frequency domain. (The Fourier transform pair involving the Gaussian function is derived in Example 5.2 in Chapter 5.) The graphs of $x(t)$ and $X(f)$ are shown for $t \in [-5, 5]$ in Figure 4.7. To obtain the DFT coefficients in Figure 4.8, we have performed the following steps:

Step 1. Take $N = 10$ equally spaced samples

$$\{x_{-4},\, x_{-3},\, x_{-2},\, x_{-1},\, x_0,\, x_1,\, x_2,\, x_3,\, x_4,\, x_5\}$$

from the interval $-5 < t \le 5$ as identified in the first plot in Figure 4.8.

Step 2. Compute $N = 10$ DFT coefficients using the formula

$$X_r = \frac{1}{N} \sum_{\ell=-4}^{5} x_\ell\, \omega_N^{-r\ell}, \quad -4 \le r \le 5.$$

The computed X_r s (scaled by $T = 10$) are identified in the second plot in Figure 4.8, and their numerical values are recorded in Table 4.3.

Remark: In this step, if needed, we may rearrange the data to obtain

$$\{\tilde{x}_0,\, \tilde{x}_1,\, \tilde{x}_2,\, \tilde{x}_3,\, \tilde{x}_4,\, \tilde{x}_5,\, \tilde{x}_6,\, \tilde{x}_7,\, \tilde{x}_8,\, \tilde{x}_9\}$$
$$=\{x_0,\, x_1,\, x_2,\, x_3,\, x_4,\, x_5,\, x_{-4},\, x_{-3},\, x_{-2},\, x_{-1}\},$$

and compute the $N = 10$ DFT coefficients by the alternate formula:

$$X_r = \frac{1}{N} \sum_{\ell=0}^{9} \tilde{x}_\ell\, \omega_N^{-r\ell}, \quad 0 \le r \le 9.$$

Using the relationship $X_r = X_{r-N}$, we can convert the computed X_r, $6 \le r \le 9$, back to X_{-4}, X_{-3}, X_{-2}, and X_{-1}.

This is a useful strategy in computing practice when only one of the two equivalent DFT formulas is implemented by an available FFT computer program.

To obtain the extra ten DFT coefficients in Figure 4.9, we have performed the following steps:

Step 1. Take $N = 10$ equally spaced samples

$$\{x_{-4}, x_{-3}, x_{-2}, x_{-1}, x_0, x_1, x_2, x_3, x_4, x_5\}$$

from the interval $-5 < t \le 5$ as identified in the first plot in Figure 4.9.

Step 2. Zero-pad the 10-sample sequence $\{x_\ell\}$ by appending ten more zeros. That is, we double the sample length from N to qN with $q = 2$. As shown below, the ten zeros are split up and appended to both ends of the given sequence. The resulting 20-sample sequence $\{z_\ell\}$ is

$$\{z_{-9},\, z_{-8},\, \ldots,\, z_{-1},\, z_0,\, z_1,\, \ldots,\, z_{10}\}$$
$$= \{0, 0, 0, 0, 0,\, x_{-4},\, x_{-3},\, x_{-2},\, x_{-1},\, x_0,\, x_1,\, x_2,\, x_3,\, x_4,\, x_5,\, 0, 0, 0, 0, 0\}.$$

Step 3. Compute $M = 20$ DFT coefficients using the formula

$$Z_r = \frac{1}{M} \sum_{\ell=-9}^{10} z_\ell \, \omega_M^{-r\ell}, \quad -9 \leq r \leq 10.$$

The computed Z_r s (scaled by $qT = 20$) are identified in the second plot in Figure 4.9. The numerical values of Z_r s are recorded in Table 4.3, so they can be compared with the ten previously computed X_r s directly.

Remark: In this step, if needed, we may rearrange the data to obtain

$$\{\tilde{z}_0, \tilde{z}_1, \ldots, \tilde{z}_{10}, \tilde{z}_{11}, \ldots, \tilde{z}_{19}\} = \{z_0, z_1, \ldots, z_{10}, z_{-9}, \ldots, z_{-1}\},$$

and compute the $M = 20$ DFT coefficients by the alternate formula:

$$Z_r = \frac{1}{M} \sum_{\ell=0}^{19} \tilde{z}_\ell \, \omega_M^{-r\ell}, \quad 0 \leq r \leq 19.$$

Using the relationship $Z_r = Z_{r-M}$, we can convert the computed Z_r, $11 \leq r \leq 19$, back to $Z_{-9}, Z_{-8}, \ldots, Z_{-1}$.

We mention again that this is a useful strategy in computing practice when only one of the two equivalent DFT formulas is implemented by an available FFT computer program.

Figure 4.7 The Gaussian function $x(t)$ and its Fourier transform $X(f)$.

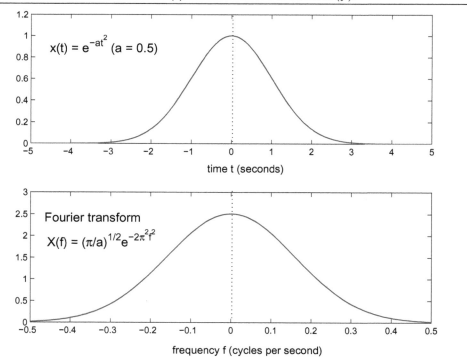

Figure 4.8 Computing ten DFT coef cients from ten signal samples.

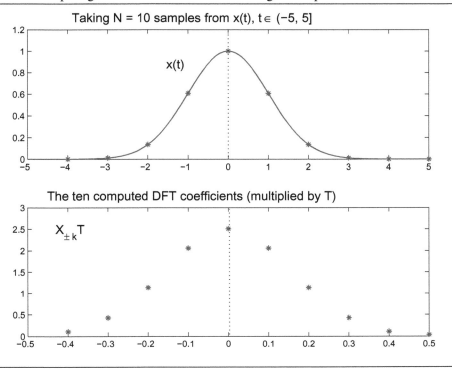

Figure 4.9 Computing twenty DFT coef cients by zero padding ten signal samples.

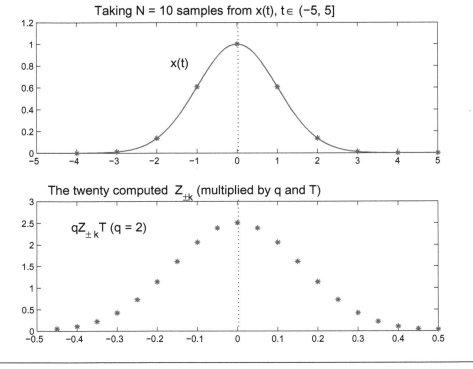

Table 4.3 Numerical values of the DFT coef cie nts plotted in Figures 4.8 and 4.9.

$\pm r$	$X_{\pm r}T$	$\pm r$	$qZ_{\pm r}T\ (q=2)$
N=10 (T=10)		M=qN=20 (T=10)	
		-9	0.0524351
-4	0.1085813	-8	0.1085813
		-7	0.2239264
-3	0.4243447	-6	0.4243447
		-5	0.7300004
-2	1.1381158	-4	1.1381158
		-3	1.6077044
-1	2.0576168	-2	2.0576168
		-1	2.3859337
0	2.5066245	0	2.5066245
		1	2.3859337
1	2.0576168	2	2.0576168
		3	1.6077044
2	1.1381158	4	1.1381158
		5	0.7300004
3	0.4243447	6	0.4243447
		7	0.2239264
4	0.1085813	8	0.1085813
		9	0.0524351
5	0.0360585	10	0.0360585

4.6.2 Zero padding the DFT

In view of the symmetry in the formulation of DFT and IDFT, it is expected that zero padding the DFT will lead to signal interpolation in the time domain. This is indeed the case provided that we preserve the DFT property $|Z_r| = |Z_{M-r}|$ (for $r > 0$) in the M-point zero-padded $\{Z_r\}$, because this would have held when the DFT and IDFT are derived directly from any M-point sequence. To show how to transplant this property from the original N-point DFT to the zero-padded M-point DFT, we shall use a concrete example to help with the explanation that follows. In the example we are required to extend the DFT length from $N = 2n+2 = 6$ to $M = 2m+2 = 10$, and the questions are: Where should we put the zeros? What other changes should we make to ensure $|Z_r| = |Z_{M-r}|$ in the zero-padded new DFT?

We first review some key relationships embedded in the derivation of the DFT. Recall that DFT was derived to give us the $N = 2n+2$ coefficients of the interpolating trigonometric polynomial when the latter is expressed in complex exponential modes:

$$(4.38) \qquad x(t) = \sum_{r=-n}^{n+1} X_r e^{j2\pi rt/T}.$$

By evaluating (4.38) at N equally spaced signal samples x_ℓ over $[0, T]$, we obtain the system of equations:

$$(4.39) \qquad x_\ell = \sum_{r=-n}^{n+1} X_r \omega_N^{r\ell}, \quad \omega_N = e^{j2\pi/N}, \qquad 0 \le \ell \le N-1.$$

For $N = 2n + 2 = 6$, the coefficients in (4.39) are

$$\{X_{-2}, X_{-1}, X_0, X_1, X_2, X_3\},$$

and we have $|X_r| = |X_{-r}|$ for $1 \le r \le 2$; because it is through defining $X_{\pm r} = \frac{1}{2}(A_r \mp jB_r)$ we convert $x(t)$ from the expression using the pure cosine and sine modes to the one using complex exponential modes. By making use of the fact $\omega_N^{-r} = \omega_N^{N-r}$, we relabel the terms $X_{-r}\omega_N^{-r\ell}$ in (4.39) as $X_{N-r}\omega_N^{(N-r)\ell}$ to obtain the IDFT

$$(4.40) \qquad x_\ell = \sum_{r=0}^{N-1} X_r \omega_N^{r\ell}, \quad 0 \le \ell \le N-1,$$

which then leads to the DFT formula for X_r ($1 \le r \le N-1$). Hence the DFT coefficients in our example for $N = 6$ are

$$\{X_0, X_1, X_2, X_3, X_4, X_5\} = \{X_0, X_1, X_2, X_3, X_{-2}, X_{-1}\},$$

and $|X_r| = |X_{-r}|$ for $1 \le r \le n$ in (4.39) is translated to $|X_r| = |X_{N-r}|$ for $1 \le r \le n$ in (4.40).

To figure out where to insert the zeros in the N-point DFT $\{X_r\}$, we start with the equations defined by (4.39) (or its equivalent for odd N), because any additional terms in the DFT must *originate* from (4.39). Depending on N being even or odd, we take the following steps to arrive at the definition of zero-padded DFT $\{Z_r\}$:

Step 1. (Skip this step if N is odd.) Starting with (4.39), we split the term with coefficient X_{n+1} (which is X_3 in the example) into two halves so that we can extend the property

$|X_r| = |X_{-r}|$ for $1 \leq r \leq n$ to $|\hat{X}_r| = |\hat{X}_{-r}|$ over the extended symmetric range for $1 \leq r \leq n + 1$ as shown below.

$$x_\ell = \sum_{r=-n}^{n+1} X_r \omega_N^{r\ell} \qquad (\text{ start with } (4.39))$$

$$= \tfrac{1}{2} X_{n+1} \omega_N^{(n+1)\ell} + \left[\sum_{r=-n}^{n} X_r \omega_N^{r\ell} \right] + \tfrac{1}{2} X_{n+1} \omega_N^{(n+1)\ell}$$

(4.41)

$$= \tfrac{1}{2} X_{n+1} \omega_N^{-(n+1)\ell} + \left[\sum_{r=-n}^{n} X_r \omega_N^{r\ell} \right] + \tfrac{1}{2} X_{n+1} \omega_N^{(n+1)\ell}$$

$$(\because N = 2n+2, \ \omega_N^{n+1} = e^{j\pi} = e^{-j\pi} = \omega_N^{-n-1})$$

$$= \sum_{r=-n-1}^{n+1} \hat{X}_r \omega_N^{r\ell},$$

where

(4.42)
$$\hat{X}_r = \begin{cases} \tfrac{1}{2} X_{n+1} & \text{if } r = -(n+1), \\ X_r & \text{if } -n \leq r \leq n, \\ \tfrac{1}{2} X_{n+1} & \text{if } r = n+1. \end{cases}$$

For $N = 2n + 2 = 6$ in our example, we split X_3 to obtain

$$\{\hat{X}_{-3}, \ \hat{X}_{-2}, \ \hat{X}_{-1}, \ \hat{X}_0, \ \hat{X}_1, \ \hat{X}_2, \ \hat{X}_3\} = \{\tfrac{1}{2} X_3, \ X_{-2}, \ X_{-1}, \ X_0, \ X_1, \ X_2, \ \tfrac{1}{2} X_3\}.$$

Observe that $|\hat{X}_r| = |\hat{X}_{-r}|$ for $1 \leq r \leq 3$.

Step 2(a). (For even N) We now add $M - N - 1 = 2(m - n) - 1$ zeros to the $(N+1)$-point sequence $\{\hat{X}_r\}$ in the following manner: add $m - n - 1$ zeros before $\hat{X}_{-(n+1)}$, and add $m - n$ zeros after $\hat{X}_{n+1}$. For $M = 10$ and $N = 6$, we have $M - N - 1 = 3$ in our example, so we obtain

$$\{Z_{-4}, \ Z_{-3}, \ Z_{-2}, \ Z_{-1}, \ Z_0, \ Z_1, \ Z_2, \ Z_3, \ Z_4, \ Z_5\}$$
$$= \{0, \ \hat{X}_{-3}, \ \hat{X}_{-2}, \ \hat{X}_{-1}, \ \hat{X}_0, \ \hat{X}_1, \ \hat{X}_2, \ \hat{X}_3, \ 0, \ 0\}$$
$$= \{0, \ \tfrac{1}{2} X_3, \ X_{-2}, \ X_{-1}, \ X_0, \ X_1, \ X_2, \ \tfrac{1}{2} X_3, \ 0, \ 0\}, \quad (\text{from } (4.42))$$

where $|Z_r| = |Z_{-r}|$ holds for $1 \leq r \leq 4$. Observe that $|Z_3| = |Z_{-3}|$ because we split the X_3 term, which is no longer the last term in the zero-padded 10-point sequence. Because the condition $|Z_{-r}| = |Z_r|$ for $1 \leq r \leq m$ is satis ed , we may now interpret Z_r $(-m \leq r \leq m + 1)$ as the coef cien ts of

(4.43)
$$z_\ell = \sum_{r=-m}^{m+1} Z_r \omega_M^{r\ell}, \quad \omega_M = e^{j2\pi/M}, \qquad 0 \leq \ell \leq M - 1.$$

Step 2(b). (For odd N) When $N = 2n + 1$, we already have equal number of X_r and X_{-r}, so the term splitting in Step 1 is *not* needed. Since $M - N$ is an even number when $M = 2m + 1$ and $N = 2n + 1$ are both odd, we add exactly $m - n$ zeros before X_{-n} and exactly $m - n$ zeros after X_n.

Step 3(a). (For even N) The property $\omega_M^{-r} = \omega_M^{M-r}$ enables us to relabel the terms $Z_{-r}\omega_M^{-r\ell}$ in (4.43) as $Z_{M-r}\omega_M^{(M-r)\ell}$ to obtain the M-point IDFT

(4.44)
$$z_\ell = \sum_{r=0}^{M-1} Z_r \omega_M^{r\ell}, \quad \omega_M = e^{j2\pi/M}, \quad 0 \le \ell \le M-1.$$

To complete our example, the zero-padded DFT contains

$$\begin{aligned}
\{Z_r\} &= \{Z_0, Z_1, Z_2, Z_3, Z_4, Z_5, Z_6, Z_7, Z_8, Z_9\} \\
&= \{Z_0, Z_1, Z_2, Z_3, Z_4, Z_5, Z_{-4}, Z_{-3}, Z_{-2}, Z_{-1}\} \\
&= \{X_0, X_1, X_2, \tfrac{1}{2}X_3, 0, 0, 0, \tfrac{1}{2}X_3, X_{-2}, X_{-1}\} \quad (\text{from step 2(a)}) \\
&= \{X_0, X_1, X_2, \tfrac{1}{2}X_3, 0, 0, 0, \tfrac{1}{2}X_3, X_4, X_5\}. \quad (\because X_{-r} = X_{N-r})
\end{aligned}$$

We have thus arrived at the following de␣n␣ition of Z_r ($0 \le r \le M-1$) in terms of given X_r ($0 \le r \le N-1$) with $M = 2m+2$ and $N = 2n+2$:

(4.45)
$$Z_r = \begin{cases} X_r, & 0 \le r \le n, \\ \tfrac{1}{2}X_{n+1}, & r = n+1, \\ 0, & n+2 \le r \le M-n-2, \\ \tfrac{1}{2}X_{n+1}, & r = M-n-1, \\ X_{r-M+N}, & M-n \le r \le M-1. \end{cases}$$

The process we use to arrive at the de␣n␣ition of the new DFT $\{Z_r\}$ ensures that $|Z_r| = |Z_{M-r}|$ holds for $1 \le r \le m$.

Step 3(b). (For odd N) Repeating the process in 3(a) on the zero-padded sequence from step 2(b) mandates that $M-N$ zeros should be inserted between X_n and X_{n+1} when $N = 2n+1$.

With the zero-padded DFT properly de␣␣ned, we can now demonstrate how the computed M-sample sequence $\{z_\ell\}$ interpolates the N-sample sequence $\{x_d\}$ in the time domain. Assuming that $q = M/N$ is an integer, we show next that the computed $\{z_\ell\}$ contains $\{x_d\}$ th␣ e value of z_ℓ agrees with the value of $x_{\ell/q}$ when ℓ is an integer multiple of q, i.e., $z_\ell = x_{\ell/q}$ for $0 \le \ell/q \le N-1$.

We proceed by computing z_ℓ from the zero-padded $\{Z_r\}$ de␣␣ned in (4.45) using the M-point IDFT, where ℓ is an integer multiple of q, and we assume $q = M/N = (2m+2)/(2n+2)$ is an integer:

$$z_\ell = \sum_{r=0}^{M-1} Z_r \, \omega_M^{r\ell}, \qquad\qquad (0 \le \ell \le M-1)$$

(4.46)
$$= \sum_{r=0}^{n} X_r \, \omega_M^{r\ell} + \left[\tfrac{1}{2}X_{n+1} \left(\omega_M^{(n+1)\ell} + \omega_M^{(M-n-1)\ell} \right) \right] + \sum_{r=M-n}^{M-1} X_{r-M+N} \, \omega_M^{r\ell}$$

$$= \sum_{r=0}^{n} X_r \, \omega_M^{r\ell} + X_{n+1} e^{j\pi\ell/q} + \sum_{r=M-n}^{M-1} X_{r-M+N} \, \omega_M^{r\ell}.$$

Note that in deriving this intermediate result we have made use of the fact that $\omega_M^M = 1$ and $\omega_M^{\pm(n+1)\ell} = e^{\pm j2\pi(n+1)\ell/M} = e^{\pm j\pi\ell/q}$, where $q = M/N$. Since ℓ/q is an integer, we have

$e^{j\pi\ell/q} = e^{-j\pi\ell/q}$, and the result follows. We continue to simplify formula (4.46):

$$z_\ell = \sum_{r=0}^{n+1} X_r \omega_M^{r\ell} + \sum_{r=M-n}^{M-1} X_{r-M+N}\, \omega_M^{r\ell} \qquad (\because X_{n+1}\omega_M^{(n+1)\ell} = X_{n+1}e^{j\pi\ell/q})$$

$$= \sum_{r=0}^{n+1} X_r \omega_M^{r\ell} + \sum_{d=n+2}^{N-1} X_d\, \omega_M^{(d+M-N)\ell} \qquad (\text{de ne } d = r - M + N)$$

$$= \sum_{r=0}^{n+1} X_r \omega_M^{r\ell} + \sum_{r=n+2}^{N-1} X_r\, \omega_M^{r\ell}\omega_M^{M\ell}\omega_M^{-N\ell} \qquad (\text{relabel } d \text{ as } r\,;\ \text{note } \omega_M^{M\ell} = 1)$$

$$(4.47)\qquad = \sum_{r=0}^{n+1}\left[\frac{1}{N}\sum_{k=0}^{N-1} x_k\omega_N^{-kr}\right]\omega_M^{r\ell} + \sum_{r=n+2}^{N-1}\left[\frac{1}{N}\sum_{k=0}^{N-1} x_k\omega_N^{-kr}\right]\omega_M^{r\ell}\omega_M^{-N\ell}$$

$$= \frac{1}{N}\sum_{k=0}^{N-1} x_k\left[\sum_{r=0}^{n+1}\omega_N^{-kr}\omega_M^{r\ell}\right] + \frac{1}{N}\sum_{k=0}^{N-1} x_k\left[\sum_{r=n+2}^{N-1}\omega_N^{-kr}\omega_M^{r\ell}\right]\omega_M^{-N\ell}$$

$$= \frac{1}{N}\sum_{k=0}^{N-1} x_k\left[\sum_{r=0}^{n+1}\omega_N^{-r(k-\ell/q)} + \omega_N^{-N\ell/q}\sum_{r=n+2}^{N-1}\omega_N^{-r(k-\ell/q)}\right] \qquad (\because \omega_M^{q} = \omega_N)$$

$$= \frac{1}{N}\sum_{k=0}^{N-1} x_k\sum_{r=0}^{N-1}\omega_N^{-r(k-\ell/q)} \qquad (\because \omega_N^{-N\ell/q} = 1 \text{ when } \ell/q \text{ is an integer})$$

$$= \frac{1}{N}\sum_{k=0}^{N-1} x_k\, T_k(k - \ell/q),$$

where
$$T_k(k-\ell/q) = \sum_{r=0}^{N-1}\omega_N^{-r(k-\ell/q)}$$
$$= \begin{cases} N & \text{if } k-\ell/q = 0, \\ 0 & \text{otherwise.} \end{cases} \qquad (\because k-\ell/q \text{ is a nonzero integer})$$

Since ℓ is an integer multiple of q, we express the result from (4.46) with $\ell = q \times d$, we immediately obtain

$$z_\ell = z_{q\times d} = \frac{1}{N}\sum_{k=0}^{N-1} x_k T_k\left(k - \frac{q\times d}{q}\right) = \frac{1}{N} x_d T_d\left(d - \frac{q\times d}{q}\right) = x_d = x_{\ell/q}$$

for $0 \le \ell/q \le N-1$.

Example 4.7 To demonstrate the effect of zero padding the DFT, we make use of the DFT coef cients from Table 3.1, which were computed in Example 3.66 (in Section 3.11.1, Chapter 3) from data taken from the function

$$f(t) = \begin{cases} t, & 0 < t < 2, \\ 2, & 2 \le t < 4; \end{cases} \qquad f(t+4) = f(t).$$

For $N=8$ and 16, we display the zero-padded DFT coef cient sequences of length $2N=16$ and 32 in Table 4.4. Observe that the steps given in this section for padding zeros have been followed: for $N=8$, the coef cien t X_4 is split and seven zeros are inserted; for $N=16$, X_8 is split and 15 zeros are inserted. In Figure 4.10, we explicitly contrast the signal sample values obtained by inverse transforming the N DFT coefficients (*without* zero-padding) with those obtained from inverse transforming the $2N$ zero-padded DFT coefficients.

Note that for the two cases $N = 8$ and $N = 16$ we demonstrate in this example, doubling the sample sizes to $2N$ serves our purpose because our objective is to explicitly show where the extra values occur. To plot a smooth graph of the function $\tilde{f}(t)$ reconstructed from N DFT coefficients in Example 3.66, one may extend N to κN by padding zeros, where κ can take on an integer value as large as needed. (A reminder: Depending on whether N is even or odd, the zeros are padded in different manners as explained earlier in this section.)

Table 4.4 Zero pad the DFT coefficients computed in Example 3.66 ($N = 8$, 16).

r	Zero-padded X_r ($2N = 16$)	Zero-padded X_r ($2N = 32$)
0	1.5000000	1.5000000
1	$-0.2133884 + j0.3017767$	$-0.2052667 + j0.3142087$
2	$-0.0000000 + j0.1250000$	$0.0000000 + j0.1508884$
3	$-0.0366117 + j0.0517767$	$-0.0253112 + j0.0935379$
4	$0 - j0.0000000$	$0.0000000 + j0.0625000$
5	0	$-0.0113005 + j0.0417612$
6	0	$0.0000000 + j0.0258884$
7	0	$-0.0081216 + j0.0124320$
8	0	$0 + j0.0000000$
9	0	0
10	0	0
11	0	0
12	$0 - j0.0000000$	0
13	$-0.0366117 - j0.0517767$	0
14	$-0.0000000 - j0.1250000$	0
15	$-0.2133884 - j0.3017767$	0
16		0
17		0
18		0
19		0
20		0
21		0
22		0
23		0
24		$0 + j0.0000000$
25		$-0.0081216 - j0.0124320$
26		$0.0000000 - j0.0258884$
27		$-0.0113005 - j0.0417612$
28		$0.0000000 - j0.0625000$
29		$-0.0253112 - j0.0935379$
30		$0.0000000 - j0.1508884$
31		$-0.2052667 - j0.3242087$

Figure 4.10 The effect of zero padding the DFT as done in Table 4.4.

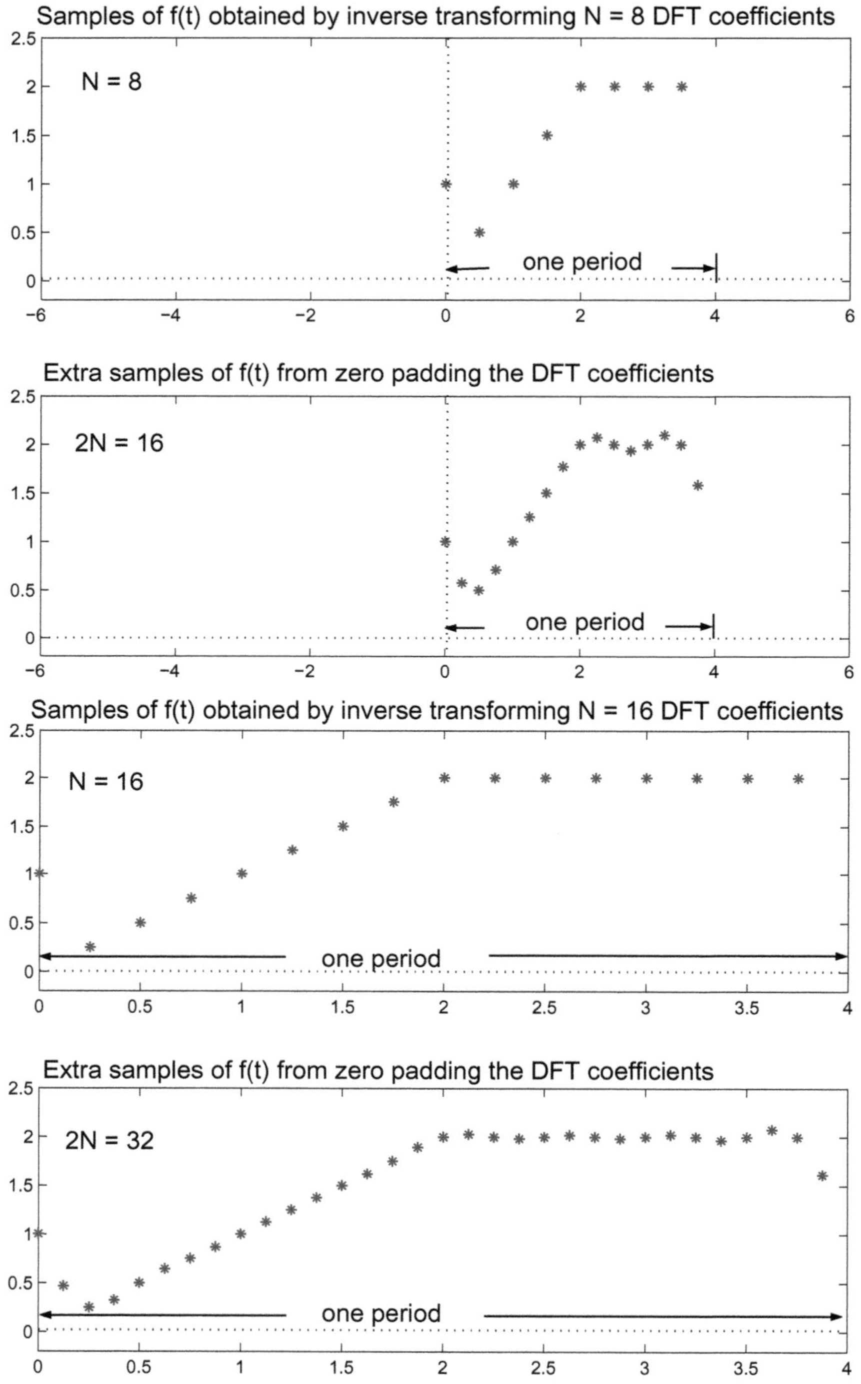

Samples of f(t) obtained by inverse transforming N = 8 DFT coefficients
N = 8
one period
Extra samples of f(t) from zero padding the DFT coefficients
2N = 16
one period
Samples of f(t) obtained by inverse transforming N = 16 DFT coefficients
N = 16
one period
Extra samples of f(t) from zero padding the DFT coefficients
2N = 32
one period

4.7 Computing DFT Defining Formulas Per Se

While there is no doubt that appropriate FFT algorithms should be used to compute the DFT at all times, it remains useful to learn to program and compute the DFT as matrix-vector products according to its various de ning formulas for the following reasons:

1. The various FFT algorithms are tailored to DFT of speci c lengths. While there is a mismatch in lengths, zero-padding commonly occurs, and the DFT computed from the zero-padded signal may not contain the original DFT as we discussed in Section 4.6.

   Therefore, when test problems of *small* sizes are used for debugging or aiding theoretical understanding, it is often useful to compute the exact DFT as matrix-vector products.

2. There are a number of different ways to formulate the DFT depending on the sampling period and sample size as discussed in Sections 2.5 and 2.6. While each formula can be computed as a matrix-vector product, the FFT programs available will not compute every formula directly and exactly. Even when the FFT program which computes the desired DFT formula of the given length is available, the FFT output may not be in the desired order.

   We can eliminate such uncertainties by checking the output of a selected FFT program against the desired DFT matrix-vector product on small test problems.

3. Since the DFT formulas are numerical formulas at the core of digital signal processing (DSP), there are DFT-like formulas which cannot be computed by the FFT e.g., part of the Chirp-FFT codes implements DFT-like formula per se; learning to program the DFT in an environment suitable for DSP applications can only help with our future tasks.

Since we use MATLAB®[1] for all numerical computation in this book, we want to show how to and how *not* to compute the product of a DFT matrix and a vector in MATLAB. At the same time we also want to use the very specially structured DFT matrix as a vehicle to bring out those programming techniques which set MATLAB apart from other high-level procedural programming languages such as C, Fortran, and Pascal.

4.7.1 Programming DFT in MATLAB

For comparison and contrast, we show multiple MATLAB implementations of the same formula, and we also show how to simulate a C, Fortran, or Pascal program in MATLAB. The various MATLAB programs and timing results demonstrate that very signi cant reduction in execution time may be achieved by using built-in functions, high-level matrix operators, and aggregated data structures.

In order to *connect* the MATLAB code to the mathematical equations, the following variable names are chosen to denote the mathematical symbols indicated. Since $\overline{M}(r, \ell) \equiv \omega_N^{r\ell}$ (note that ω_N is rede n ed as $\overline{\omega}_N = \omega_N^{-1}$ so the negative power can be omitted in coding the DFT matrix), it is important to note that $\overline{M}(r, \ell)$ is stored in M_dft(r+1,ell+1), because the index range of arrays in MATLAB begins at one instead of zero.

Suppose that matrix A and vector x have been entered, the MATLAB command for computing their product is simply y=A*x. Therefore, to compute the DFT of sequence x with period N according to the matrix equation $X = \frac{1}{N}\overline{M}x$, one only needs to generate the DFT

---

[1]MATLAB is a registered trademark of The MathWorks, Inc.

Table 4.5 Variable names in MATLAB code.

Variable types	Mathematical expressions	MATLAB expressions
matrix	Ω or $\Omega_{N \times N}$	`Omega or Omega(1:N, 1:N)`
	$\overline{M}$ or $\overline{M}_{N \times N}$	`M_dft or M_dft(1:N, 1:N)`
matrix element	$\Omega_{r,\ell}, 0 \leq r, \ell \leq N - 1.$	`Omega(r+1, ell+1)`
	$\overline{M}_{r,\ell}, 0 \leq r, \ell \leq N - 1.$	`M_dft(r+1, ell+1)`
matrix row	$\overline{M}_{r*}$ or $\overline{M}[r,*]$	`M_dft(r+1, :)`
	Ω_{r*} or $\Omega[r,*]$	`Omega(r+1, :)`
matrix column	$\overline{M}_{*\ell}$ or $\overline{M}[*,\ell]$	`M_dft(:, ell+1)`
	$\Omega_{*\ell}$ or $\Omega[*,\ell]$	`Omega(:, ell+1)`
scalar	ω or $\omega_N = e^{-j2\pi/N}$	`w = exp(-j*2*pi/N)`

matrix $\overline{M}$, which is de ned by $\overline{M}_{r,\ell} \equiv \omega_N^{r\ell}$, with $0 \leq r, \ell \leq N - 1$. In MATLAB, this can be done in more than one way four programs are discussed below.

1. Simulating C, Fortran, or Pascal Code in MATLAB: (Not recommended)

```
function M_dft = dft1_matrix(N)
%
% Input  N:     order of the DFT matrix
% Output M_dft: the DFT matrix (without division by N)
%
w = exp(-j*2*pi/N);    % j is MATLAB constant for sqrt(-1)
for r = 0:N-1          % access matrix elements row by row
   for ell = 0:N-1
      power = r*ell;
      M_dft(r+1,ell+1) = w^power;   % compute each scalar
   end                              % element
end
```

For DFT matrices of order $N = 100, 200, 400, 512,$ and 800, the execution times and total op counts are given in Table 4.6. (Note that MATLAB 5.3 built-in function `flops.m` is used to obtain the op counts reported here, but this function is no longer avaiable in the current version 7.4 of MATLAB.)

2. Using MATLAB's matrix building functions and operators: We review the needed functions and operators before the code of functions `dft2_matrix.m` and `dft2b_matrix.m` are presented.

• MATLAB command `A=ones(M,N)` generates an $M \times N$ matrix of all ones. If $M = N$, one would commonly use `A=ones(N)` instead of `A=ones(N,N)`, although the latter is also correct. Since MATLAB supports the product of a scalar and a matrix, the command `Omega=w*ones(N)` generates a constant matrix with all entries $\Omega_{r,\ell} = \omega$.

• MATLAB command `v=0:N-1` generates a row vector $v = [0, 1, \cdots, N - 1]$. To get a column vector, simply use MATLAB s matrix transpose operator as in the command `u=v'`.

Table 4.6 Testing function dft1_matrix.m using MATLAB 5.3 and 7.4.

	MATLAB 5.3			MATLAB 7.4
	M-File Timings	Total Flop		M- le Timings
N	(CPU 1.3 GHz)	Counts	N	(CPU 3.2 GHz)
100	0.13 sec	990,246	100	0.03 sec
200	0.52 sec	4,697,686	200	0.16 sec
400	3.39 sec	21,715,446	400	1.73 sec
512	7.03 sec	37,285,462	512	3.39 sec
800	25.13 sec	98,477,476	800	12.13 sec

- MATLAB commands `v=0:N-1` and `Power=v'*v` compute the outer product given below.

$$\text{Power} = \mathtt{v'*v} = \begin{bmatrix} 0 & 0 & 0 & 0 & \cdots & 0 \\ 0 & 1 & 2 & 3 & \cdots & N-1 \\ 0 & 2 & 4 & 6 & \cdots & 2(N-1) \\ 0 & 3 & 6 & 9 & \cdots & 3(N-1) \\ \multicolumn{6}{c}{\dotfill} \\ 0 & (N-1) & 2(N-1) & 3(N-1) & \cdots & (N-1)^2 \end{bmatrix}.$$

- MATLAB command `C=A.^B` computes elements $C_{r,\ell} = A_{r,\ell}^{B_{r,\ell}}$. This implies $C_{r,\ell} = \omega^k$ if $A_{r,\ell} = \omega$ and $B_{r,\ell} = k$. Therefore, the DFT matrix may be generated by the command `M_dft=(w*ones(N)).^Power`. For example, the DFT matrix of order $N = 4$ may be generated as

$$\mathtt{M\_dft = (w*ones(N)).\string^Power} = \begin{bmatrix} \omega & \omega & \omega & \omega \\ \omega & \omega & \omega & \omega \\ \omega & \omega & \omega & \omega \\ \omega & \omega & \omega & \omega \end{bmatrix} \cdot \begin{bmatrix} 0 & 0 & 0 & 0 \\ 0 & 1 & 2 & 3 \\ 0 & 2 & 4 & 6 \\ 0 & 3 & 6 & 9 \end{bmatrix}$$

$$= \begin{bmatrix} \omega^0 & \omega^0 & \omega^0 & \omega^0 \\ \omega^0 & \omega^1 & \omega^2 & \omega^3 \\ \omega^0 & \omega^2 & \omega^4 & \omega^6 \\ \omega^0 & \omega^3 & \omega^6 & \omega^9 \end{bmatrix}.$$

To avoid storing the matrix of all ones and to eliminate the redundant computation of `w*1` entailed by the command `w*ones(N)`, the MATLAB command `M_dft=w.^Power` could be used because the operator `.^` supports mixed-mode operands.

$$\mathtt{M\_dft = w.\string^Power} = \omega \cdot \begin{bmatrix} 0 & 0 & 0 & 0 \\ 0 & 1 & 2 & 3 \\ 0 & 2 & 4 & 6 \\ 0 & 3 & 6 & 9 \end{bmatrix} = \begin{bmatrix} \omega^0 & \omega^0 & \omega^0 & \omega^0 \\ \omega^0 & \omega^1 & \omega^2 & \omega^3 \\ \omega^0 & \omega^2 & \omega^4 & \omega^6 \\ \omega^0 & \omega^3 & \omega^6 & \omega^9 \end{bmatrix}.$$

Using these matrix building functions and operators, we construct the function dft2_matrix.m below. Observe that the code works with the contents of the vector or matrix directly one need not address the individual elements according to their positions in the arrays. Speci cally, comparing with dft1_matrix.m, we no longer address the elements `M_dft(r+1,ell+1)` explicitly in the code of `dft2_matrix.m`.

```
function M_dft = dft2_matrix(N)
%
% Input N:        order of the DFT matrix
% Output M_dft: the DFT matrix (without division by N)
%
w = exp(-j*2*pi/N);   % j is MATLAB constant for sqrt(-1)
v = 0:N-1;            % content of v is [0,1, ..., N-1]
Power = v'*v;        % store outer product in matrix Power
M_dft = w.^Power;    % compute DFT matrix w/o division by N
```

Since the inverse DFT matrix is the complex conjugate of the (symmetric) DFT matrix, the command M_inv = conj(dft2_matrix(N)) produces the inverse. Alternatively, one may choose to output both matrices as shown in the modi ed listing below. Note that we have added the second output argument M_inv.

```
function [M_dft, M_inv] = dft2b_matrix(N)
%
% Input N:        order of the DFT matrix
% Output M_dft: the DFT matrix (without division by N)
%          M_inv: the inverse DFT matrix
%
w = exp(-j*2*pi/N);   % j is MATLAB constant for sqrt(-1)
v = 0:N-1;            % content of v is [0,1, ..., N-1]
Power = v'*v;        % store outer product in matrix Power
M_dft = w.^Power;    % compute DFT matrix w/o division by N
M_inv = conj(M_dft); % compute the inverse DFT matrix
```

For DFT matrices of order $N = 100, 200, 400, 512,$ and 800, the execution times and total op counts are given in Table 4.7. Comparing these results with those in Table 4.6, we see the dramatic decrease in execution times, although both functions perform essentially the same sequence of arithmetic operations.

Table 4.7 Testing function dft2_matrix.m using MATLAB 5.3 and 7.4.

	MATLAB 5.3			MATLAB 7.4	
N	M-File Timings (CPU 1.3 GHz)	Total Flop Counts	N	M- le Timings (CPU 3.2 GHz)	
100	0.06 sec	980,146	100	0.03 sec	
200	0.19 sec	4,657,486	200	0.08 sec	
400	0.83 sec	21,555,046	400	0.36 sec	
512	1.45 sec	37,022,806	512	0.61 sec	
800	3.65 sec	97,836,676	800	1.58 sec	

3. **Reducing redundant arithmetic in building the DFT matrix:** Inside both of the functions dft1_matrix.m and dft2_matrix.m, each individual matrix element $\overline{M}_{r,\ell} = \omega_N^{r\ell}$ is computed by literally raising ω_N to the power of $r \times \ell$. By noting that

$$\omega_N^{r(\ell+1)} = \omega_N^{r\ell} * \omega_N^{r},$$

and that

$$\text{M\_dft}(r+1,\text{ell}+2) = \omega_N^{r(\ell+1)},$$

$$\text{M\_dft}(r+1,\text{ell}+1) = \omega_N^{r\ell}, \text{ and } \text{M\_dft}(r+1,2) = \omega_N^{r},$$

we obtain

$$\text{M\_dft}(r+1,\text{ell}+2) = \text{M\_dft}(r+1,\text{ell}+1) * \text{M\_dft}(r+1,2).$$

Accordingly, to compute the entire column M_dft(:,ell+2), we simply apply the componentwise operator .* to multiply column M_dft(:,ell+1) and column M_dft(:,2) element by element. That is, using the DFT matrix of order 4 as an example, the command

$$\text{M\_dft}(:,\text{ell}+2) = \text{M\_dft}(:,\text{ell}+1).* \text{M\_dft}(:,2)$$

computes the third and the fourth column as shown below:

$$\text{M\_dft}(:,2).*\text{M\_dft}(:,2) = \begin{bmatrix} \omega^0 \\ \omega^1 \\ \omega^2 \\ \omega^3 \end{bmatrix} .* \begin{bmatrix} \omega^0 \\ \omega^1 \\ \omega^2 \\ \omega^3 \end{bmatrix} = \begin{bmatrix} \omega^0 \\ \omega^2 \\ \omega^4 \\ \omega^6 \end{bmatrix} = \text{M\_dft}(:,3).$$

$$\text{M\_dft}(:,3).*\text{M\_dft}(:,2) = \begin{bmatrix} \omega^0 \\ \omega^2 \\ \omega^4 \\ \omega^6 \end{bmatrix} .* \begin{bmatrix} \omega^0 \\ \omega^1 \\ \omega^2 \\ \omega^3 \end{bmatrix} = \begin{bmatrix} \omega^0 \\ \omega^3 \\ \omega^6 \\ \omega^9 \end{bmatrix} = \text{M\_dft}(:,4).$$

Thus, the cost for computing the N elements in each column is N (complex) multiplications. To generate column 2, which is needed for computing all subsequent columns, we tailor the command C=A.^B introduced earlier to a special case. Here A is a scalar instead of a matrix, and B and C are vectors instead of matrices. Thus, the same command now computes vector element $C_\ell = A^{B_\ell}$. That is, the MATLAB commands

```
w = exp(-j*2*pi/4);   f = 0:3;   v = w.^f'
```

compute

$$v = \omega.^{\begin{bmatrix} 0 \\ 1 \\ 2 \\ 3 \end{bmatrix}} = \begin{bmatrix} \omega^0 \\ \omega^1 \\ \omega^2 \\ \omega^3 \end{bmatrix}, \text{ where } \omega = e^{-j2\pi/4}.$$

The code of function dft3_matrix is given below.

```
function M_dft = dft3_matrix(N)
%
% Input  N:     order of the DFT matrix
% Output M_dft: the DFT matrix (without division by N)
%
M_dft(:,1)=ones(N,1);  % generate the first column
If N > 1
   w = exp(-j*2*pi/N); % j is MATLAB constant for sqrt(-1)
```

```
v = (0:N-1)';           % column vector v=[0,1,2,...,N-1]'
M_dft(:,2) = w.^v;      % generate the second column
for ell = 2:N-1
    M_dft(:,ell+1) = M_dft(:,ell).*M_dft(:,2);
end
end
```

For DFT matrices of order $N = 100, 200, 400, 512$, and 800, the execution times and total op counts are given in Table 4.8. Comparing these results with those in Tables 4.6 and 4.7, we see the dramatic reduction in total op counts (as expected), but such reduction does not necessarily reduce execution times! On the contrary, the execution times of function dft3_matrix are significantly longer than function dft2_matrix. This example shows that we must be aware of such potential tradeoff when we apply the conventional wisdom in op count reduction.

Table 4.8 Testing function dft3_matrix.m using MATLAB 5.3 and 7.4.

	MATLAB 5.3			MATLAB 7.4	
N	M-File Timings (CPU 1.3 GHz)	Total Flop Counts	N	M-le Timings (CPU 3.2 GHz)	
100	0.02 sec	64,259	100	0.004 sec	
200	0.12 sec	250,287	200	0.065 sec	
400	1.62 sec	984,143	400	1.188 sec	
512	4.03 sec	1,605,603	512	2.625 sec	
800	16.70 sec	3,895,455	800	10.078 sec	

4. **Generating DFT matrix one column at a time:** If the matrix columns can be generated and used one by one in forming the matrix-vector product, we will not need to store the entire matrix, and can reduce storage from $\Theta(N^2)$ to $\Theta(N)$. To accomplish this goal, we need to view the matrix-vector product as a linear combination of the matrix columns. For example, we compute $X = \frac{1}{N}\overline{M}x$, where

$$\overline{M}x = \begin{bmatrix} 1 & 1 & 1 & 1 \\ 1 & \omega^1 & \omega^2 & \omega^3 \\ 1 & \omega^2 & 1 & \omega^2 \\ 1 & \omega^3 & \omega^2 & \omega^1 \end{bmatrix} \begin{bmatrix} x_0 \\ x_1 \\ x_2 \\ x_3 \end{bmatrix} = x_0 \begin{bmatrix} 1 \\ 1 \\ 1 \\ 1 \end{bmatrix} + x_1 \begin{bmatrix} 1 \\ \omega^1 \\ \omega^2 \\ \omega^3 \end{bmatrix} + x_2 \begin{bmatrix} 1 \\ \omega^2 \\ 1 \\ \omega^2 \end{bmatrix} + x_3 \begin{bmatrix} 1 \\ \omega^3 \\ \omega^2 \\ \omega^1 \end{bmatrix}.$$

Note that such computation can be done literally in MATLAB s interactive environment by entering the following command:

```
X = x(1)*M_dft(:,1)+x(2)*M_dft(:,2)+x(3)*M_dft(:,3)+x(4)*M_dft(:,4);
X = X/N;
```

where x(k) denotes the kth element in vector x, and M_dft(:,k) denotes the kth column in matrix M_dft. Therefore, in order *not* to store all columns, we must n d an *efficient* way to generate the columns one at a time. We show next how this can be done for the DFT matrix de ned by $\overline{M}_{r,\ell} = \omega_N^{r\ell}$ for $0 \leq r, \ell \leq N-1$.

Observe rst that since $\omega_N^N = 1$, one has $\omega_N^{k+mN} = \omega_N^k$ for $0 \le k \le N-1$, which implies that $\omega_N^{r\ell} = \omega_N^q$ with $q = r\ell \bmod N$. Since $0 \le q \le N-1$, ω_N^q is the $(q+1)$st element in the second column (or row) of the DFT matrix $\overline{M}$. This relationship is demonstrated below using the DFT matrix of order $N = 4$, in which $\omega \equiv \omega_4 = e^{-j2\pi/4}$.

$$
\texttt{M\_dft} = \begin{bmatrix} 1 & 1 & 1 & 1 \\ 1 & \omega^1 & \omega^2 & \omega^3 \\ 1 & \omega^2 & \omega^4 & \omega^6 \\ 1 & \omega^3 & \omega^6 & \omega^9 \end{bmatrix} = \begin{bmatrix} \omega^0 & \omega^0 & \omega^{0 \bmod 4} & \omega^{0 \bmod 4} \\ \omega^0 & \omega^1 & \omega^{2 \bmod 4} & \omega^{3 \bmod 4} \\ \omega^0 & \omega^2 & \omega^{4 \bmod 4} & \omega^{6 \bmod 4} \\ \omega^0 & \omega^3 & \omega^{6 \bmod 4} & \omega^{9 \bmod 4} \end{bmatrix} = \begin{bmatrix} \omega^0 & \omega^0 & \omega^0 & \omega^0 \\ \omega^0 & \omega^1 & \omega^2 & \omega^3 \\ \omega^0 & \omega^2 & \omega^0 & \omega^2 \\ \omega^0 & \omega^3 & \omega^2 & \omega^1 \end{bmatrix}.
$$

Clearly, the N distinct values of $\omega_N^{r\ell}$ are contained in a single vector column 2 of the matrix `M_dft`, and the other columns are formed by a subset of elements from column 2. In MATLAB, when the elements of `u= [b b a f g c c]` are to be chosen from another vector `v= [a b c d e f g]`, one could use the command `u=v([2 2 1 6 7 3 3])`, in which the integer index vector contains the original positions of the desired elements in vector `v`. It can be easily veri ed that `u(1)=v(2)=b`, `u(2)=v(2)=b`, `u(3)=v(1)=a`, and so on.

Note that the MATLAB command

```
u = v([2 2 1 6 7 3 3])
```

is equivalent to the command

```
u = [v(2) v(2) v(1) v(6) v(7) v(3) v(3)]).
```

This is a convenient programming technique when the relationship between the contents of vectors u and v is re ected by the index array, assuming that such an array can be easily established. We next show that this is indeed the case when building other columns in the DFT matrix from column 2.

Recall that for correctness in MATLAB code, we must store $\overline{M}_{r,\ell} = \omega_N^{r\ell}$ in `M_dft(r+1, ell+1)`, because $0 \le r, \ell \le N-1$, whereas array indices in MATLAB begin at one instead of zero. By noting that

$$\texttt{M\_dft(r+1, ell+1)} = \omega_N^{r\ell} = \omega_N^{r\ell \bmod N} = \omega_N^q, \quad \text{where } \ell \ge 2,$$

and

$$\texttt{M\_dft(q+1, 2)} = \omega_N^q,$$

we obtain the value of `M_dft(r+1,ell+1)` by entering these MATLAB commands:

```
q = rem(r*ell,N);   M_dft(r+1, ell+1) = M_dft(q+1, 2).
```

Note that `rem.m` is the MATLAB built-in function for computing the quantity $q = r\ell \bmod N$. Applying this relationship to the DFT matrix of order 4, we see that from the second column of matrix `M_dft`,

$$
\texttt{M\_dft}(:,2) = v = \begin{bmatrix} \omega^0 \\ \omega^1 \\ \omega^2 \\ \omega^3 \end{bmatrix},
$$

one obtains the third and fourth columns of matrix M_dft as shown below.

$$\texttt{M\_dft(:,3)} = \texttt{v(}\begin{bmatrix}1 & 3 & 1 & 3\end{bmatrix}\texttt{)} = \texttt{v(}\begin{bmatrix}1 & 3 & 1 & 3\end{bmatrix}'\texttt{)} = \begin{bmatrix} v(1) \\ v(3) \\ v(1) \\ v(3) \end{bmatrix} = \begin{bmatrix} \omega^0 \\ \omega^2 \\ \omega^0 \\ \omega^2 \end{bmatrix},$$

and

$$\texttt{M\_dft(:,4)} = \texttt{v(}\begin{bmatrix}1 & 4 & 3 & 2\end{bmatrix}\texttt{)} = \texttt{v(}\begin{bmatrix}1 & 4 & 3 & 2\end{bmatrix}'\texttt{)} = \begin{bmatrix} v(1) \\ v(4) \\ v(3) \\ v(2) \end{bmatrix} = \begin{bmatrix} \omega^0 \\ \omega^3 \\ \omega^2 \\ \omega^1 \end{bmatrix}.$$

Note that the index vector, i.e., the integer vector [1 3 1 3] in the MATLAB expression v([1 3 1 3)]), can be either a row vector or a column vector, and the result is always a column vector if v is a column vector. As expected, if either index vector is used with a row vector, the result will always be a row vector. Therefore, one should never need to transpose an index vector for such usage in a MATLAB program.

Another unique feature of MATLAB is that many of its built-in functions accept both scalar and array input arguments. This is so for function rem.m in the DFT computation. For example, suppose v is an integer column vector of length $N = 4$, the command p=rem(v,N) produces a column vector p with $p(r) \equiv v(r) \bmod N$:

$$p = \texttt{rem(v',N)} = \begin{bmatrix} \texttt{rem(v(1),N)} \\ \texttt{rem(v(2),N)} \\ \texttt{rem(v(3),N)} \\ \texttt{rem(v(4),N)} \end{bmatrix} = \begin{bmatrix} v(1) \bmod N \\ v(2) \bmod N \\ v(3) \bmod N \\ v(4) \bmod N \end{bmatrix}.$$

Finally, recall that MATLAB s binary operators accept operands in mixed-modes. For example, let p be a column vector of length $N = 4$, the MATLAB command p+1 adds the scalar 1 to each element of vector p. That is,

$$\texttt{p + 1} = \begin{bmatrix} p(1) \\ p(2) \\ p(3) \\ p(4) \end{bmatrix} + 1 = \begin{bmatrix} p(1) + 1 \\ p(2) + 1 \\ p(3) + 1 \\ p(4) + 1 \end{bmatrix}.$$

As expected, if q is a row vector, the MATLAB expression q+1 produces a row vector with $q(r) = q(r) + 1$.

The programming techniques introduced above are now used to build function dft.m.

```
function Xout = dft(x)
%
% Input  x:    One period of the sequence to be transformed.
% Output Xout: Xout = (1/N)*M_dft*x
%
% Reference: Introduction to Scientific Computing by
%            Charles F. Van Loan (p. 181)
```

```
%
N = length(x);            % determine period N
Xout = x(1)*ones(N,1);    % M_dft(:,1) is a column of all ones
if N > 1
    w = exp(-j*2*pi/N);   % j is MATLAB constant for sqrt(-1)
    v = (0:N-1)';         % column vector v=[0,1,...,N-1]'
    Mdf2 = w.^v;          % compute 2nd column M_dft(:,2)
    for ell = 1:N-1       % start with column ell+1 = 2
        q = rem(v*ell, N); % compute row vector q=r*ell mod N
                          % for building index vector q+1
        Xout = Xout + x(ell+1)*Mdf2(q+1);% accumulate partial
                                     % sums: add one column
    end                            % each time
end
Xout = Xout/N;            % including division by N in Xout
```

For sequences of randomly generated complex numbers x of length $N = 100, 200, 400, 512$, and 800, the execution times and total op counts are given in Table 4.9. Note that because only one column of the DFT matrix is actually computed, and the matrix is not stored in its entirety, function dft requires $\Theta(N)$ storage. However, for transforming x of length N, it is still a $\Theta\left(N^2\right)$ algorithm in time. Comparing with the results in Tables 4.6, 4.7, and 4.8, function dft is the fastest $\Theta\left(N^2\right)$ algorithm for transforming x note the huge reduction of execution time: for $N = 800$, the M- le timing is 0.34 seconds, which compares with 25.13 seconds, 3.65 seconds, and 16.70 seconds reported in Tables 4.6 to 4.8.

Table 4.9 Testing function dft.m using MATLAB 5.3 and 7.4.

	MATLAB 5.3			MATLAB 7.4
	M-File Timings	Total Flop		M- le Timings
N	(CPU 1.3 GHz)	Counts	N	(CPU 3.2 GHz)
100	0.02 sec	135,360	100	0.003 sec
200	0.04 sec	532,488	200	0.008 sec
400	0.11 sec	2,108,544	400	0.029 sec
512	0.15 sec	3,446,304	512	0.047 sec
800	0.34 sec	8,384,256	800	0.114 sec

References

1. A. Ambardar. *Analog and Digital Signal Processing*. Brooks/Cole Publishing Company, Paci c Grove, CA, second edition, 1999.

2. W. L. Briggs and V. E. Hensen. *The DFT: An Owner's Manual for the Discrete Fourier Transform*. The Society for Industrial and Applied Mathematics, Philadelphia, PA, 1995.

3. E. Chu and A. George. *Inside the FFT Black Box: Serial and Parallel Fast Fourier Transform Algorithms*. CRC Press, Boca Raton, FL, 2000.

4. B. Porat. *A Course in Digital Signal Processing*. John Wiley & Sons, Inc., New York, 1997.

5. W. H. Press, S. A. Teukolsky, W. T. Vetterling, and B. P. Flannery. *Numerical Recipes in C++: The Art of Scientific Computing*. Cambridge University Press, Cambridge, UK, second edition, 2001.

6. C. F. Van Loan. *Computational Frameworks for the Fast Fourier Transform*. The Society for Industrial and Applied Mathematics, Philadelphia, PA, 1992.

Chapter 5

Sampling and Reconstruction of Functions—Part II

In Chapter 2 we introduce a number of fundamental concepts in function sampling through the Fourier series representation of a periodic band-limited function. In the real world we encounter many signals which are not periodic and they are not time limited. Although it is useful to think of such a signal as a periodic signal with period $T = \infty$ at times, the unmodi ed Fourier series which represents a periodic signal with nite period is no longer a proper representation.

In this chapter we consider sampling a signal $x(t)$ which is not required to be periodic and it is not required to be time-limited either. The frequency-domain contents of $x(t)$ are de ned by its Fourier transform (to be derived in this chapter)

$$(5.1) \qquad X(f) = \int_{-\infty}^{\infty} x(t)\, e^{-j2\pi ft}\, dt,$$

where the independent variable f represents the continuously varying frequency, and the frequency domain is the entire real axis. If $X(f)$ is a real-valued function, its frequency-domain plot is the graph of $X(f)$ versus f; if $X(f)$ is complex-valued, its frequency-domain plot consists of the graph of $\mathrm{Re}\big(X(f)\big)$ versus f and the graph of $\mathrm{Im}\big(X(f)\big)$ versus f. The frequency contents may be formally represented by $\big\{(f, \mathrm{Re}(X(f)), \mathrm{Im}(X(f)))\big\}$: an in nite set made up of a continuous spectrum of the frequency f.

Under conditions which are to be discussed, the analytical form of the original function in the time domain can be recovered from the inverse Fourier transform (to be derived in this chapter) of $X(f)$, i.e.,

$$(5.2) \qquad x(t) = \int_{-\infty}^{\infty} X(f)\, e^{j2\pi ft}\, df.$$

The right-hand side of this equation is called the Fourier integral representation of $x(t)$. Note that there is a noncountable number of frequencies in the Fourier integral because each real number f corresponds to one frequency.

For a function $x(t)$ which is de n ed for every t, the suf cient condition for the existence of a Fourier transform $X(f)$ is given by

$$(5.3) \qquad \int_{-\infty}^{\infty} |x(t)|\, dt < \infty.$$

That is, the function $x(t)$ is absolutely integrable. Since this is not a necessary condition, there are functions which have Fourier transforms even though they are not absolutely integrable. For example, a Fourier transform can be formally de ned for a periodic function $x^p(t)$ (see Chapter 7) with the help of the generalized functions (see Chapter 6.)

Observe that the Fourier transform $X(f)$ de ned by the integral (5.1) remains unchanged if we let the absolutely integrable function $x(t)$ take on different values at any nite number of points. This means that if $x(t)$ is not continuous, then it is possible that two different functions share the same transform $X(f)$. Therefore, the function $x(t)$ cannot be uniquely determined by inverse transforming $X(f)$ unless it is continuous or it satis es two more conditions: (i) $x(t)$ has only a nite number of maxima and minima on any nite interval; (ii) $x(t)$ has on any n ite interval at most a nite number of discontinuities, each of which is a jump discontinuity. In the latter case, the inverse Fourier transform of $X(f)$ produces $\tilde{x}(t)$, which agrees with $x(t)$ at every t at which $x(t)$ is continuous, and $\tilde{x}(t)$ equals the average of the left-hand limit $x(t_\alpha^-)$ and the right-hand limit $x(t_\alpha^+)$ at every t_α at which a jump discontinuity occurs. The proof for the existence of the inverse Fourier transform may be found in texts treating the theory of Fourier integrals [3, 22, 51, 54].

We shall begin this chapter by deriving the sampling theorem for nonperiodic band-limited functions in Section 5.1. The theorem determines an appropriate choice of sampling rate so that the original (unknown) function $x(t)$ can be reconstructed analytically. This process is directly connected to the inverse Fourier transform of a frequency-limited $X(f)$ as de ned by (5.2). The Fourier transform pair de ned by (5.1) and (5.2) is derived next in Section 5.2. The frequency contents represented by the Fourier transform are then examined, the properties of the Fourier transform are derived, examples of Fourier transform pair are given, and the relationship between Fourier series coef cients and sampled Fourier transforms of time-limited and almost time-limited $x(t)$ is established in the sections that follow.

5.1   Sampling Nonperiodic Band-Limited Functions

The nonperiodic $x(t)$ is said to be *band limited* up to the maximum frequency f_{max} if its Fourier transform $X(f)$ is zero outside the interval $[-F/2, F/2] = [-f_{max}, f_{max}]$, which is called the Nyquist interval as introduced in Chapter 2. (Note that $F = 2f_{max}$.) By assuming that $X(f)$ is a real-valued function with a nite range of independent variable f, we de ne a periodic $X^p(f)$ which is the protracted version of the frequency-limited $X(f)$; i.e., $X^p(f) \equiv X(f)$ for $f \in [-F/2, F/2]$ and $X^p(f \pm F) = X^p(f)$ holds for arbitrary f. The sampling theorem for nonperiodic band-limited functions makes use of the Fourier integral representation of $x(t)$ as well as the Fourier series representation of $X^p(f)$ (which is a periodic extension of $X(f)$). To proceed, we assume that the following two conditions have been met: (i) $x(t)$ is suf ciently well-behaved so that its Fourier transform exists; (ii) $X^p(f)$ is real-valued and it satis es the Dirichlet conditions given by Theorem 3.1 (in Chapter 3) so it possesses a Fourier series.

5.1.1 Fourier series of frequency-limited $X(f)$

Since $X^p(f)$ is periodic with a finite period $F = 2f_{max}$, its Fourier series representation can be written (using complex exponential modes) as

$$(5.4) \qquad X^p(f) \ = \ \sum_{\ell=-\infty}^{\infty} C_\ell \, e^{j2\pi\ell f/F}, \text{ where}$$

$$(5.5) \qquad C_\ell \ = \ \frac{1}{F} \int_{-F/2}^{F/2} X^p(f) \, e^{-j2\pi\ell f/F} \, df.$$

Since $X^p(f) = X(f)$ for $f \in [-F/2, F/2]$, we immediately obtain

$$(5.6) \qquad X(f) \ = \ \sum_{\ell=-\infty}^{\infty} C_\ell \, e^{j2\pi\ell f/F}, \text{ where}$$

$$(5.7) \qquad C_\ell \ = \ \frac{1}{F} \int_{-F/2}^{F/2} X(f) \, e^{-j2\pi\ell f/F} \, df.$$

So the frequency-limited $X(f)$ has a Fourier series representation in the frequency domain. (Note that there is an infinite number of terms in the Fourier series above.)

5.1.2 Inverse Fourier transform of frequency-limited $X(f)$

To construct $x(t)$ analytically, we begin with Equation (5.2), which expresses $x(t)$ as the inverse Fourier transform of $X(f)$, and recall that frequency-limited $X(f) = 0$ outside the interval $[-F/2, F/2]$; hence,

$$(5.8) \qquad x(t) = \int_{-\infty}^{\infty} X(f) \, e^{j2\pi ft} \, df = \int_{-F/2}^{F/2} X(f) \, e^{j2\pi ft} \, df.$$

Since $x(t)$ is not time-limited, the temporal variable t may take on any value. If we let $t = -\ell/F$,

$$(5.9) \qquad x\left(-\frac{\ell}{F}\right) = \int_{-F/2}^{F/2} X(f) \, e^{-j2\pi\ell f/F} \, df.$$

Note that the integral on the right-hand side of the last equation is identical to the integral that occurs in equation (5.7) which defines the Fourier series coefficient C_ℓ. Thus, the coefficients C_ℓ are now connected to the discrete-time samples of $x(t)$ at $t = -\ell\Delta t$ if we define $\Delta t = 1/F$. That is, for every integer ℓ,

$$(5.10) \qquad C_\ell = \frac{1}{F} x\left(-\frac{\ell}{F}\right) = \frac{x(-\ell\Delta t)}{F}, \text{ where } F = 2f_{max}.$$

The Fourier series coefficients of $X(f)$ are therefore isomorphic to the values of its inverse transform $x(t)$ sampled at intervals of $\Delta t = 1/F$. The corresponding sampling rate is given by $1/\Delta t = F = 2f_{max}$, which is the Nyquist rate defined in Chapter 2. Since f_{max} (cycles per unit time) is the highest frequency present in the band-limited signal $x(t)$, a sampling rate of $2f_{max}$ (samples per unit time) results in two samples per cycle for the highest frequency present, which is the lowest sampling rate we can possibly use to avoid aliasing.

5.1.3   Recovering the signal analytically

To reconstruct $x(t)$ from the equally spaced discrete-time samples, we combine the inverse Fourier transform with the Fourier series representation of $X(f)$ so that

(5.11)

$$
\begin{aligned}
x(t) &= \int_{-\infty}^{\infty} X(f)\, e^{j2\pi ft}\, df \quad \text{(by the de\,n\,ition of inverse Fourier transform)} \\
&= \int_{-F/2}^{F/2} X(f)\, e^{j2\pi ft}\, df, \qquad \because X(f) = 0 \text{ when } f < -F/2 \text{ or } f > F/2 \\
&= \int_{-F/2}^{F/2} \left(\sum_{\ell=-\infty}^{\infty} C_\ell e^{j2\pi \ell f/F} \right) e^{j2\pi ft}\, df \quad \text{(use Fourier series representation)} \\
&= \int_{-F/2}^{F/2} \left(\sum_{\ell=-\infty}^{\infty} \frac{x(-\ell\triangle t)}{F} e^{j2\pi \ell f \triangle t} \right) e^{j2\pi ft}\, df, \quad \because C_\ell = \frac{x(-\ell\triangle t)}{F}, \ \triangle t = \frac{1}{F} \\
&= \sum_{\ell=-\infty}^{\infty} \frac{x(-\ell\triangle t)}{F} \int_{-F/2}^{F/2} e^{j2\pi (t+\ell\triangle t)f}\, df \\
&= \sum_{\ell=-\infty}^{\infty} \frac{x(-\ell\triangle t)}{F} \left(\frac{e^{j\pi(t+\ell\triangle t)F} - e^{-j\pi(t+\ell\triangle t)F}}{j2\pi(t+\ell\triangle t)} \right) \\
&= \sum_{\ell=-\infty}^{\infty} x(-\ell\triangle t) \frac{\sin\big(\pi F(t+\ell\triangle t)\big)}{\pi F(t+\ell\triangle t)}, \qquad \because \frac{e^{j\theta} - e^{-j\theta}}{2j} = \sin\theta \\
&= \sum_{\ell=-\infty}^{\infty} x(\ell\triangle t) \frac{\sin\big(\pi F(t-\ell\triangle t)\big)}{\pi F(t-\ell\triangle t)} \\
&= \sum_{\ell=-\infty}^{\infty} x(\ell\triangle t)\, \text{sinc}\big(F(t-\ell\triangle t)\big), \qquad \because \text{sinc}\,(\lambda) \equiv \frac{\sin\pi\lambda}{\pi\lambda}.
\end{aligned}
$$

Note that each component function is a **sinc** function de ned by $\text{sinc}(\lambda) = \sin(\pi\lambda)/(\pi\lambda)$. Since $\text{sinc}(\lambda) = 0/0$ at $\lambda = 0$, we apply L Hospital s rule to evaluate the limit as $\lambda \to 0$:

$$
\lim_{\lambda \to 0} \text{sinc}(\lambda) = \lim_{\lambda \to 0} \frac{\sin \pi\lambda}{\pi\lambda} = \lim_{\lambda \to 0} \frac{\pi \cos \pi\lambda}{\pi} = \cos 0 = 1,
$$

and we de ne $\text{sinc}(0) = 1$ based on the limit obtained. Observe that $\text{sinc}(\lambda) = 0$ for $\lambda = \pm 1, \pm 2, \pm 3, \ldots$. Observe further that the function $\text{sinc}(\lambda) = \sin(\pi\lambda)/(\pi\lambda)$ is not periodic, because when the numerator $\sin(\lambda)$ does repeat over each 2π interval, the denominator will never be the same, and the function $\text{sinc}(\lambda)$ will never repeat itself.

The results we have derived are summarized in the sampling theorem given below.

Theorem 5.1 (Sampling theorem) If the signal $x(t)$ is known to be band-limited with bandwidth $F = 2f_{max}$, then $x(t)$ can be sampled at the Nyquist rate $1/\triangle t = 2f_{max}$, and we can determine a unique $x(t)$ by interpolating the sequence of samples according to the formula

$$
x(t) = \sum_{\ell=-\infty}^{\infty} x(\ell\triangle t) \frac{\sin\big(\pi F(t-\ell\triangle t)\big)}{\pi F(t-\ell\triangle t)}.
$$

5.1.4 Further discussion of the sampling theorem

To clearly show the nature of the interpolating formula and the properties of the individual **sinc** function, we let $t_\ell = \ell \triangle t$ and de ne

$$L_\ell(t) = \frac{\sin(\pi F(t - t_\ell))}{\pi F(t - t_\ell)}.$$

Now we express the same interpolating formula as

$$p(t) = \sum_{\ell=-\infty}^{\infty} x(t_\ell) L_\ell(t),$$

and we show next that at each Nyquist sample point $t_k = k \triangle t$, we have

$$L_\ell(t_k) = \begin{cases} 1, & \text{if } \ell = k; \\ 0, & \text{if } \ell \neq k. \end{cases}$$

Note that when the running index $\ell = k$, because $L_k(t_k) = \sin 0/0 = 0/0$ is in an indeterminate form, we establish $L_k(t) = 1$ in the limit as $t \to t_k$ by applying L Hospital s rule:

$$\lim_{t \to t_k} L_k(t) = \lim_{t \to t_k} \frac{\pi F \cos(\pi F(t - t_k))}{\pi F} = \cos 0 = 1.$$

When $\ell \neq k$, we have $(t_k - t_\ell) = (k - \ell)\triangle t = (k - \ell)/F$, yielding

$$L_\ell(t_k) = \frac{\sin(\pi F(t_k - t_\ell))}{\pi F(t_k - t_\ell)} = \frac{\sin(\pi(k - \ell))}{\pi(k - \ell)} = 0.$$

Since $L_k(t_k) = 1$ and $L_k(t_m) = 0$ for every $m \neq k$, the **sinc** function denoted by $L_k(t)$ crosses the t-axis at all Nyquist sample points except for the one at $t_k = k \triangle t$. The graphs of $L_{-3}(t)$, $L_0(t)$, and $L_1(t)$ are shown in Figure 5.1.

Accordingly, for every integer $k \in (-\infty, \infty)$, we have

$$\begin{aligned} p(t_k) &= \sum_{\ell=-\infty}^{\infty} x(t_\ell) L_\ell(t_k) \\ &= x(t_k) L_k(t_k) \qquad (\because L_\ell(t_k) = 0 \text{ if } \ell \neq k) \\ &= x(t_k), \qquad\qquad (\because L_k(t_k) = 1) \end{aligned}$$

which means that $p(t)$ takes on the sample value $x(t_k)$ at $t = t_k$ for every k. Therefore, the interpolating formula $p(t)$ passes all sample values as long as they are spaced $\triangle t = 1/F = 1/(2f_{max})$ apart as required.

While the sampling theorem gives us insight and guidance to the sampling process, we are still faced with the following dilemma: in order to sample the signal at the Nyquist rate, we need to know the bandwidth *a priori*. Strictly speaking, the bandwidth is not known unless we have already obtained the complete set of values of $X(f)$ th is is not possible when $x(t)$ is an unknown function, which we wish to sample in order to recover its frequency contents. In practice this dilemma is usually resolved by sampling the signal frequently enough so that its graph resembles the physical signal. Of course, for signals with known bandwidth (e.g., speech or voice information of certain quality), we would be able to sample at or above Nyquist rate to suit our processing needs. There are other legitimate concerns:

Figure 5.1 The graphs of $L_\ell(t)$ for $\ell = -3,\ 0,\ 1$.

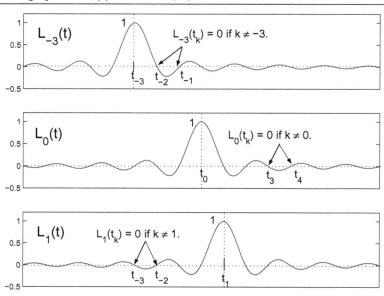

1. To use the interpolating formula given in the theorem, we need an in nite number of samples extended from $-\infty$ to ∞.

2. In theory we can interpolate $x(t)$ between sample points. However, this interpolating formula is very expensive to compute, because it involves trigonometric function in every term.

3. As mentioned earlier we actually need two nonzero samples per cycle for the highest frequency present. Therefore, when this cannot be guaranteed by the sampling process, we must sample at a rate higher than the minimum Nyquist rate required by the sampling theorem.

4. Many signals in the real world are not band limited. In fact, the uncertainty principle of quantum mechanics [21, 27, 37] does not permit a signal to be arbitrarily narrow in both time and frequency. For example, several rectangular pulses and their Fourier transforms are shown in Figure 6.1 in Chapter 6: we see that the narrower the pulse becomes in the time domain, the wider its transform spreads out in the frequency domain.

We will return to address these issues after we learn more about the Fourier transform.

5.2   Deriving the Fourier Transform Pair

We derive the Fourier transform of $x(t)$ by thinking of it as a periodic signal with period $T = \infty$ in the time domain. We then investigate how the Fourier series representation of a periodic function should be modi ed when the period T approaches ∞.

We begin with the Fourier series of a periodic signal $x^p(t)$, which is expressed using the complex exponential modes below.

$$(5.12) \qquad x^P(t) \quad = \quad \sum_{\ell=-\infty}^{\infty} C_\ell \, e^{j2\pi\ell t/T}, \text{ where}$$

$$(5.13) \qquad C_\ell \quad = \quad \frac{1}{T} \int_{-T/2}^{T/2} x^P(t) \, e^{-j2\pi\ell t/T} \, dt.$$

We are interested in representing the nonperiodic signal

$$x(t) = \lim_{T\to\infty} x^P(t) = \lim_{T\to\infty} \sum_{\ell=-\infty}^{\infty} C_\ell \, e^{j2\pi\ell t/T}.$$

To have a closed-form representation of the right-hand side, we modify the Fourier series and the formula for the Fourier coef cients in the following manner:

1. When $T \to \infty$, the frequency spacing $\triangle f = 1/T$ becomes in nitesimal, and we may replace it by df when turning the summation into an integral in the limit.

2. When $T \to \infty$, $\triangle f = 1/T \to 0$, the set of discrete frequencies $f_k = k/T = k\triangle f$ turns into a noncountable set of continuous frequencies $f \in (-\infty, \infty)$ each   frequency is represented by a real number f.

3. With regard to the formula for the Fourier coef cient C_ℓ, we preserve the closed-form integral by evaluating

$$\lim_{T\to\infty} TC_\ell = \lim_{T\to\infty} \int_{-T/2}^{T/2} x^P(t) \, e^{-j2\pi\ell t/T} \, dt$$

$$= \int_{-\infty}^{\infty} x(t) \, e^{-j2\pi ft} \, dt \quad \text{(This is the Fourier transform of } x(t).)$$

$$= X(f).$$

Note that in evaluating the limit as $T \to \infty$, we have replaced $x^P(t)$ with $x(t)$ in the integrand, and we replace the particular discrete frequency $\ell/T = \ell\triangle f$ in the exponent with a real number f. The outcome is the Fourier transform of $x(t)$.

4. With these changes in place, we return to modify the Fourier series representation itself:

$$x(t) = \lim_{T\to\infty} \sum_{\ell=-\infty}^{\infty} C_\ell \, e^{j2\pi\ell t/T}$$

$$= \lim_{T\to\infty} \sum_{\ell=-\infty}^{\infty} (TC_\ell) \, e^{j2\pi(\ell/T)t} \left(\frac{1}{T}\right)$$

$$= \int_{-\infty}^{\infty} X(f) \, e^{j2\pi ft} \, df. \quad \text{(This is the inverse Fourier transform of } X(f).)$$

The two functions $x(t)$ and $X(f)$ are de ned in terms of each other, and they form the Fourier transform pair. It will be convenient on many occasions to denote the Fourier transform of $x(t)$ by $\mathcal{F}\{x(t)\}$, so we have

$$\mathcal{F}\{x(t)\} = X(f) = \int_{-\infty}^{\infty} x(t) \, e^{-j2\pi ft} \, dt.$$

Similarly, the inverse Fourier transform of $X(f)$ may be denoted by $\mathcal{F}^{-1}\{X(f)\}$, and we have

$$\mathcal{F}^{-1}\{X(f)\} = x(t) = \int_{-\infty}^{\infty} X(f)\, e^{j2\pi ft}\, df.$$

Observe that by evaluating the integral forms of $x(t)$ and $X(f)$ at the central ordinates, i.e., at $t = 0$ and $f = 0$ respectively, we obtain another useful relationship:

(5.14)
$$\boxed{X(0) = \int_{-\infty}^{\infty} x(t)\, dt; \quad x(0) = \int_{-\infty}^{\infty} X(f)\, df.}$$

The operators $\mathcal{F}$ and $\mathcal{F}^{-1}$ may be used to express the relationship between transforms of different functions conveniently. For example, we may use the compact notation

$$\mathcal{F}\{x(t - t_a)\} = X(f)\, e^{-j2\pi ft_a}$$

to express the relation between the Fourier transform of the shifted function $x(t - t_a)$ and that of the original function $x(t)$. (The relation itself needs to be proved and that will be shown in Section 5.6.)

5.3   The Sine and Cosine Frequency Contents

In the last section we derived the Fourier transform from the Fourier series using complex exponential modes, and we saw some connection between $X(f) = \mathcal{F}\{x(t)\}$ and the Fourier series coefficients . Recall that the Fourier series has alternate forms; in particular, it may be expressed using sine and cosine modes. Since the different sets of coefficients used by alternate forms of the Fourier series are directly convertible from each other, we expect to obtain the sine and cosine frequency contents of $x(t)$ from its Fourier transform $X(f)$ too. We begin our derivation by seeking an alternate form of the Fourier integral representation of $x(t)$.

$$\begin{aligned}
x(t) &= \int_{-\infty}^{\infty} X(f)\, e^{j2\pi ft}\, df \\
&= \int_{-\infty}^{\infty} X(f)\big(\cos(2\pi ft) + j\sin(2\pi ft)\big)\, df \\
&= \int_{0}^{\infty} \Big(X(f)\cos(2\pi ft) + X(-f)\cos\big(2\pi(-f)t\big) \\
&\qquad\qquad + jX(f)\sin(2\pi ft) + jX(-f)\sin\big(2\pi(-f)t\big)\Big)\, df \\
&= \int_{0}^{\infty} \big(X(f) + X(-f)\big)\cos(2\pi ft)\, df + \int_{0}^{\infty} j\big(X(f) - X(-f)\big)\sin(2\pi ft)\, df \\
&= \int_{0}^{\infty} X_{\cos}(f)\cos(2\pi ft)\, df + \int_{0}^{\infty} X_{\sin}(f)\sin(2\pi ft)\, df,
\end{aligned}$$

where we have obtained the cosine and sine contents in terms of $X(f)$ and $X(-f)$, namely,

$$X_{\cos}(f) = X(f) + X(-f),$$

and

$$X_{\sin}(f) = j\big(X(f) - X(-f)\big).$$

By combining the integrals de n ing $X(f)$ and $X(-f)$, we may also express $X_{\cos}(f)$ and $X_{\sin}(f)$ each as an integral transform of $x(t)$. In the derivation below, we apply Euler s formulas

$$\cos(\theta) = \frac{e^{j\theta} + e^{-j\theta}}{2} \quad \text{and} \quad \sin(\theta) = \frac{e^{j\theta} - e^{-j\theta}}{2j}$$

with $\theta = 2\pi ft$, and we obtain

$$
\begin{aligned}
X_{\cos}(f) &= X(f) + X(-f) \\
&= \int_{-\infty}^{\infty} x(t)\, e^{-j2\pi ft}\, dt + \int_{-\infty}^{\infty} x(t)\, e^{-j2\pi(-f)t}\, dt \\
&= \int_{-\infty}^{\infty} x(t)\left(e^{-j2\pi ft} + e^{j2\pi ft}\right) dt \\
&= 2\int_{-\infty}^{\infty} x(t)\cos(2\pi ft)\, dt \quad \text{(by Euler s formula)}
\end{aligned}
$$

and

$$
\begin{aligned}
X_{\sin}(f) &= j\left(X(f) - X(-f)\right) \\
&= j\int_{-\infty}^{\infty} x(t)\left(e^{-j2\pi ft} - e^{j2\pi ft}\right) dt \\
&= 2\int_{-\infty}^{\infty} x(t)\sin(2\pi ft)\, dt. \quad \text{(by Euler s formula)}
\end{aligned}
$$

We now have two integral representations for $x(t)$:

$$
\begin{aligned}
x(t) &= \int_{-\infty}^{\infty} X(f)\, e^{j2\pi ft}\, df \\
&= \int_{0}^{\infty} X_{\cos}(f)\cos(2\pi ft)\, df + \int_{0}^{\infty} X_{\sin}(f)\sin(2\pi ft)\, df.
\end{aligned}
$$

Note that for $f \in [0, \infty)$, the complete set of the values of $X_{\cos}(f)$ de nes the cosine frequency contents of $x(t)$, and the complete set of the values of $X_{\sin}(f)$ de nes the sine frequency contents of $x(t)$.

5.4 Tabulating Two Sets of Fundamental Formulas

We have seen earlier that the interplay between the Fourier series and the Fourier transform is critical in developing the sampling theorem, and we will continue to rely on the connection between the two in future development. Since there are commonly used alternate forms for both, we tabulate the fundamental formulas and their alternates for easy reference in Table 5.1.

5.5 Connections with Time/Frequency Restrictions

Our derivations in this chapter and Chapter 3 show that under suitable conditions, a given signal $x(t)$ may be represented by a Fourier integral, a Fourier series, or an interpolating formula. These alternate forms are valid and connected under the conditions identi ed in Table 5.2, in which we use the shorthand notation $x(t) \Longleftrightarrow X(f)$ t o denote a Fourier transform pair.

Table 5.1 Two sets of fundamental formulas in Fourier analysis.

Fourier series and its coef cien ts of periodic real-valued function $x^p(t)$	Fourier transform and its inverse of nonperiodic real-valued function $x(t)$
$x^p(t) = \sum_{k=-\infty}^{\infty} C_k e^{j2\pi kt/T}$ $C_k = \frac{1}{T} \int_{-T/2}^{T/2} x^p(t)\, e^{-j2\pi kt/T}\, dt$	$x(t) = \int_{-\infty}^{\infty} X(f)\, e^{j2\pi ft} df$ $X(f) = \int_{-\infty}^{\infty} x(t)\, e^{-j2\pi ft} dt$
$x^p(t) = \frac{1}{2}A_0 + \sum_{k=1}^{\infty} A_k \cos 2\pi \frac{k}{T} t$ $\qquad + \sum_{k=1}^{\infty} B_k \sin 2\pi \frac{k}{T} t$ $A_k = \frac{2}{T} \int_{-T/2}^{T/2} x^p(t) \cos 2\pi \frac{k}{T} t\, dt,$ $\qquad k = 0, 1, 2, \ldots$ $B_k = \frac{2}{T} \int_{-T/2}^{T/2} x^p(t) \sin 2\pi \frac{k}{T} t\, dt,$ $\qquad k = 1, 2, \ldots$	$x(t) = \int_0^{\infty} X_{\cos}(f) \cos(2\pi ft) df$ $\qquad + \int_0^{\infty} X_{\sin}(f) \sin(2\pi ft) df$ $X_{\cos}(f) = 2 \int_{-\infty}^{\infty} x(t) \cos(2\pi ft) dt$ $X_{\sin}(f) = 2 \int_{-\infty}^{\infty} x(t) \sin(2\pi ft) dt$
$A_k = C_k + C_{-k} = 2\mathrm{Re}(C_k)$ $B_k = j(C_k - C_{-k}) = -2\mathrm{Im}(C_k)$	$X_{\cos}(f) = X(f) + X(-f)$ $X_{\sin}(f) = j(X(f) - X(-f))$
$C_{\pm k} = \frac{1}{2}(A_k \mp jB_k)$	$X(\pm f) = \frac{1}{2}(X_{\cos}(f) \mp jX_{\sin}(f))$

Table 5.2 Connections with time/frequency restrictions.

Generic pair $x(t) \Longleftrightarrow X(f)$ $t \in (-\infty, \infty)$ $f \in (-\infty, \infty)$	$x(t) = \int_{-\infty}^{\infty} X(f)\, e^{j2\pi ft} df,$ $X(f) = \int_{-\infty}^{\infty} x(t)\, e^{-j2\pi ft} dt.$	Assumptions: $x(t)$ is real-valued; $X(f)$ is either real-valued $\qquad$ or complex-valued
Time-limited pair $x(t) \Longleftrightarrow X(f)$ $t \in [-T/2, T/2]$ $f \in (-\infty, \infty)$	$x(t) = \int_{-\infty}^{\infty} X(f)\, e^{j2\pi ft} df,$ $X(f) = \int_{-T/2}^{T/2} x(t)\, e^{-j2\pi ft} dt.$	Fourier series connection: $x(t) = \sum_{k=-\infty}^{\infty} C_k e^{j2\pi kt/T},$ for $t \in [-T/2, T/2]$, where coef cient $C_k = \frac{1}{T} X\left(\frac{k}{T}\right),$ for $k = \ldots, -1, 0, 1, \ldots$
Band-limited pair $x(t) \Longleftrightarrow X(f)$ $t \in (-\infty, \infty)$ $f \in [-F/2, F/2]$	$x(t) = \int_{-F/2}^{F/2} X(f)\, e^{j2\pi ft} df,$ $X(f) = \int_{-\infty}^{\infty} x(t)\, e^{-j2\pi ft} dt.$	Sampling theorem connection (assume $X(f)$ is real-valued): $x(t) = \sum_{k=-\infty}^{\infty} x(t_k) \frac{\sin \pi F(t-t_k)}{\pi F(t-t_k)}$ where $t_k = k\triangle t$, $\triangle t = 1/F$, for $k = \ldots, -1, 0, 1, \ldots$

5.5.1 Examples of Fourier transform pair

Example 5.1 The Fourier transform of the decaying exponential function

$$x(t) = \begin{cases} e^{-at}, & \text{for } t \in [0, \infty) \quad (a > 0); \\ 0, & \text{for } t \in (-\infty, 0) \end{cases}$$

can be easily obtained from the de n ition

$$X(f) = \int_{-\infty}^{\infty} x(t) \, e^{-j2\pi ft} dt = \int_{0}^{\infty} e^{-at} e^{-j2\pi ft} dt$$

$$= \int_{0}^{\infty} e^{-(a+j2\pi f)t} dt$$

$$= \frac{1}{a + j2\pi f}$$

$$= \frac{a}{a^2 + 4\pi^2 f^2} - j\frac{2\pi f}{a^2 + 4\pi^2 f^2}.$$

Since $X(f)$ is complex-valued, the frequency-domain plot of $x(t)$ consists of the graph of $\mathrm{Re}(X(f))$ and the graph of $\mathrm{Im}(X(f))$ versus f for every $f \in (-\infty, \infty)$, which are shown in Figure 5.2.

Figure 5.2 Time-domain and frequency-domain plots of $x(t) = e^{-at}$.

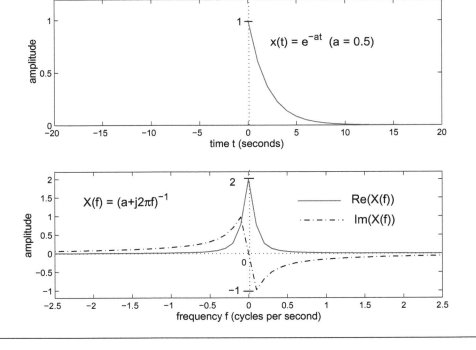

Example 5.2 The Gaussian function

$$x(t) = e^{-at^2} \quad (a > 0)$$

is defined for every $t \in (-\infty, \infty)$. We find its Fourier transform by evaluating the integral

$$
\begin{aligned}
X(f) &= \int_{-\infty}^{\infty} \{e^{-at^2}\} e^{-j2\pi ft} dt \\
&= \int_{-\infty}^{\infty} e^{-a(t^2 + j2\pi ft/a)} dt \\
&= \int_{-\infty}^{\infty} e^{-a[(t+j\pi f/a)^2 + \pi^2 f^2/a^2]} dt \\
&= e^{-\pi^2 f^2/a} \int_{-\infty}^{\infty} e^{-a(t+j\pi f/a)^2} dt \\
&= e^{-\pi^2 f^2/a} \int_{-\infty}^{\infty} e^{-a\lambda^2} d\lambda && \text{(change variable to } \lambda = t + j\pi f/a; \ d\lambda = dt) \\
&= \frac{1}{\sqrt{a}} e^{-\pi^2 f^2/a} \int_{-\infty}^{\infty} e^{-u^2} du && \text{(change variable to } u = \sqrt{a}\lambda; \ du = \sqrt{a}\, d\lambda) \\
&= \sqrt{\pi/a}\, e^{-\pi^2 f^2/a}. && \text{(using the known result } \int_{-\infty}^{\infty} e^{-u^2} du = \sqrt{\pi})
\end{aligned}
$$

We thus have

$$
X(f) = \mathcal{F}\{e^{-at^2}\} = \sqrt{\pi/a}\, e^{-\pi^2 f^2/a} \quad \text{for every } f \in (-\infty, \infty).
$$

Observe that $X(f)$ is real-valued and it is also a Gaussian function (see Figure 5.3). In this example, $x(t)$ is neither time limited nor band limited. Furthermore, when the scalar constant $a = \pi$, we have self-reciprocity: $\mathcal{F}\{x(t)\} = x(f)$.

Example 5.3 (Figure 5.4 Time-Limited Pair) We consider the rectangular pulse function (also known as the square wave or boxcar function), which is assumed to have a pulse width of $2t_0$ and it is defined as

$$
x_{\text{rect}}(t) = \begin{cases} \dfrac{1}{2t_0}, & \text{for } t \in [-t_0, t_0]; \\ 0, & \text{for } |t| > t_0. \end{cases}
$$

Its Fourier transform is

$$
\begin{aligned}
X(f) &= \int_{-\infty}^{\infty} x_{\text{rect}}(t)\, e^{-j2\pi ft} dt \\
&= \frac{1}{2t_0} \int_{-t_0}^{t_0} e^{-j2\pi ft} dt \\
&= \frac{e^{-j2\pi ft_0} - e^{j2\pi ft_0}}{-j4\pi ft_0} \\
&= \frac{\sin(2\pi ft_0)}{2\pi ft_0} && \text{(by Euler's formula)} \\
&= \text{sinc}(2ft_0), && \because \text{sinc}(x) \equiv \sin(\pi x)/(\pi x).
\end{aligned}
$$

Thus the time-limited rectangular pulse and the frequency-domain **sinc** function form a Fourier transform pair (see Figure 5.4). By letting $t_0 = 1/2$, we obtain the pair

$$
z_{\text{rect}}(t) = \begin{cases} 1, & t \in [-1/2, 1/2]; \\ 0, & |t| > 1/2. \end{cases} \quad \Longleftrightarrow \quad Z(f) = \text{sinc}(f) = \begin{cases} \sin(\pi f)/(\pi f), & f \neq 0; \\ 1, & f = 0. \end{cases}
$$

Figure 5.3 Gaussian function and its real-valued Fourier transform.

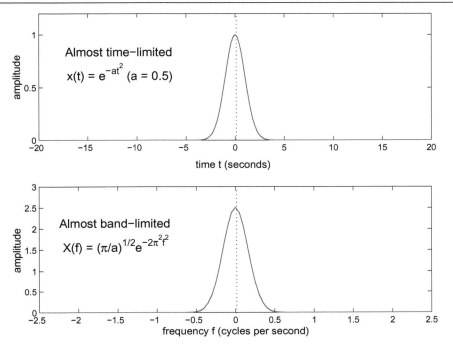

Figure 5.4 Time-limited rectangular pulse and its Fourier transform.

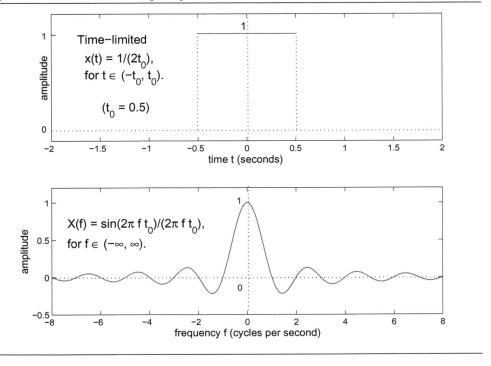

Note that $Z(f)$ is de n ed for every $f \in (-\infty, \infty)$.

To demonstrate the Fourier series connection, we give the Fourier series representation of $x_{\text{rect}}^p(t)$, which is the periodic extension of $x_{\text{rect}}(t)$, and the two functions agree over the period $[-T/2, T/2]$. The periodic pulse function $x_{\text{rect}}^p(t)$ with period T is de ned as

$$x_{\text{rect}}^p(t) = \begin{cases} \dfrac{1}{2t_0}, & \text{for } t \in [-t_0, t_0]; \\ 0, & \text{for } t_0 < |t| \leq T/2. \end{cases}$$

It is worth noting that because there are discontinuities at $t = \pm t_0$, the Fourier series obtained for $x_{\text{rect}}^p(t)$ actually converges to the normalized function shown in Figure 5.5:

$$\hat{x}_{\text{rect}}^p(t) = \begin{cases} \dfrac{1}{2t_0}, & \text{for } t \in (-t_0, t_0); \\ \dfrac{1}{4t_0}, & \text{for } t = \pm t_0; \\ 0, & \text{for } t_0 < |t| \leq T/2. \end{cases}$$

This is so because the Fourier series converges to the average of the left- and right-hand limits at the point of discontinuity.

Figure 5.5 Connecting Fourier series coef cients to Fourier transform.

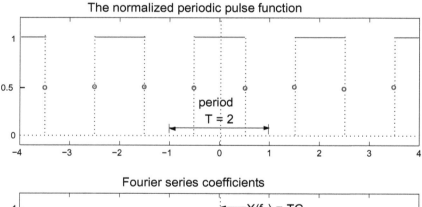

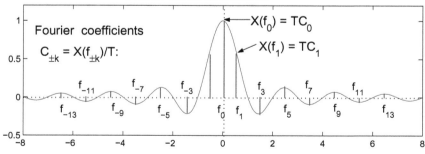

We can now nd the Fourier series of $x_{\text{rect}}^p(t)$, which converges to $\hat{x}_{\text{rect}}^p(t)$. Using the complex exponential modes, we write

$$\hat{x}_{\text{rect}}^p(t) = \sum_{k=-\infty}^{\infty} C_k e^{j2\pi kt/T},$$

where the coef cients are computed from $x_{\text{rect}}^p(t)$:

$$
\begin{aligned}
C_k &= \frac{1}{T} \int_{-T/2}^{T/2} x_{\text{rect}}^p(t)\, e^{-j2\pi kt/T} \, dt \\
&= \frac{1}{T} \int_{-t_0}^{t_0} x_{\text{rect}}(t)\, e^{-j2\pi kt/T} \, dt \\
&= \frac{1}{T} \left(\frac{1}{2t_0} \int_{-t_0}^{t_0} e^{-j2\pi kt/T} \, dt \right) \\
&= \frac{1}{T} \text{sinc}(2kt_0/T).
\end{aligned}
$$

Comparing the equation de ning C_k with the Fourier transform of the pulse function, we immediately have

$$
C_k = \frac{1}{T} \text{sinc}(2f_k t_0) = \frac{1}{T} X(f_k), \quad f_k = k/T.
$$

That is, we may obtain the Fourier series coef cient C_k by evaluating the Fourier transform $X(f) = \text{sinc}(2ft_0)$ at $f = f_k = k/T$, and scale the result by $1/T$, for all $k = \ldots, -1, 0, 1, \ldots$. This connection is illustrated in Figure 5.5.

Example 5.4 (Figure 5.6 Band-Limited Pair) We now consider the Fourier transform of the time-domain function $\text{sinc}(2f_c t)$, which turns out to be the rectangular pulse function

$$
X_{\text{rect}}(f) = \begin{cases} \dfrac{1}{2f_c}, & \text{for } f \in [-f_c, f_c]; \\ 0, & \text{for } |f| > f_c. \end{cases}
$$

We show this by nding the inverse Fourier transform of $X_{\text{rect}}(f)$:

$$
\begin{aligned}
\mathcal{F}^{-1}\{X_{\text{rect}}(f)\} &= \int_{-\infty}^{\infty} X_{\text{rect}}(f)\, e^{j2\pi ft} \, df \\
&= \frac{1}{2f_c} \int_{-f_c}^{f_c} e^{j2\pi ft} \, df \\
&= \frac{e^{j2\pi f_c t} - e^{-j2\pi f_c t}}{j4\pi f_c t} \\
&= \frac{\sin(2\pi f_c t)}{2\pi f_c t} \qquad \text{(by Euler s formula)} \\
&= \text{sinc}(2f_c t).
\end{aligned}
$$

We have thus shown that the time-domain function $x(t) \equiv \text{sinc}(2f_c t)$ is band limited with bandwidth $2f_c$ (see Figure 5.6). Note that by letting $f_c = 1/2$, we obtain the pair

$$
z(t) = \text{sinc}(t) = \begin{cases} \sin(\pi t)/(\pi t), & t \neq 0; \\ 1, & t = 0. \end{cases} \quad \Longleftrightarrow \quad Z_{\text{rect}}(f) = \begin{cases} 1, & f \in [-1/2, 1/2]; \\ 0, & |f| > 1/2. \end{cases}
$$

5.6   Fourier Transform Properties

The Fourier transform is a linear operation and obeys superposition. Its mathematical properties are summarized in Table 5.3, and their derivations follow the summary. While most of

Figure 5.6 A band-limited Fourier transform pair.

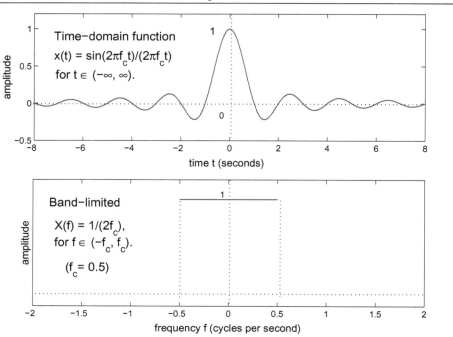

the derivations are fairly straightforward, the experience and technical details are useful when these properties need to be re-derived for alternate forms of Fourier transform. (For example, the Fourier transform may be de ned using angular frequency $\omega = 2\pi f$ instead of the rotational frequency f. We will follow up on this later.) These properties are operational and their utilities are demonstrated by examples.

5.6.1   Deriving the properties

1. Linearity

$$\mathcal{F}\{\alpha x(t) + \beta y(t)\} = \int_{-\infty}^{\infty} \left(\alpha x(t) + \beta y(t)\right) e^{-j2\pi ft} dt$$

$$= \alpha \int_{-\infty}^{\infty} x(t) e^{-j2\pi ft} dt + \beta \int_{-\infty}^{\infty} y(t) e^{-j2\pi ft} dt$$

$$= \alpha X(f) + \beta Y(f).$$

Table 5.3 Fourier transform properties.

1. Linearity	$\mathcal{F}\{\alpha x(t) + \beta y(t)\} = \alpha X(f) + \beta Y(f)$
2. Time shift	$\mathcal{F}\{x(t - t_a)\} = X(f) e^{-j2\pi f t_a}$
3. Frequency shift	$\mathcal{F}\{x(t) e^{j2\pi f_a t}\} = X(f - f_a)$
4. Modulation	$\mathcal{F}\{x(t) \cos 2\pi f_a t\} = \frac{1}{2}(X(f + f_a) + X(f - f_a))$
	$\mathcal{F}\{x(t) \sin 2\pi f_a t\} = \frac{j}{2}(X(f + f_a) - X(f - f_a))$
5. Time scaling	$\mathcal{F}\{x(\alpha t)\} = \frac{1}{\|\alpha\|} X\left(\frac{f}{\alpha}\right), \alpha \neq 0.$
6. Folding	$\mathcal{F}\{x(-t)\} = X(-f)$
7. Transform of a transform	$\mathcal{F}\{\mathcal{F}\{x(t)\}\} = \mathcal{F}\{X(f)\} = x(-t)$
8. Transform of the derivative	$\mathcal{F}\{x'(t)\} = j2\pi f X(f)$
9. Derivative of the transform	$X'(f) = -j2\pi \mathcal{F}\{t x(t)\}$

2. The Time-Shift Property

$$
\begin{aligned}
\mathcal{F}\{x(t - t_a)\} &= \int_{-\infty}^{\infty} x(t - t_a) e^{-j2\pi f t} dt \\
&= \int_{-\infty}^{\infty} x(s) e^{-j2\pi f(s + t_a)} ds \qquad (\because s = t - t_a, \; t = s + t_a, \; dt = ds) \\
&= \int_{-\infty}^{\infty} x(s) e^{-j2\pi f s} e^{-j2\pi f t_a} ds \\
&= \left[\int_{-\infty}^{\infty} x(s) e^{-j2\pi f s} ds\right] e^{-j2\pi f t_a} \\
&= X(f) e^{-j2\pi f t_a}.
\end{aligned}
$$

3. The Frequency-Shift Property

$$
\begin{aligned}
\mathcal{F}\{x(t) e^{j2\pi f_a t}\} &= \int_{-\infty}^{\infty} \left[x(t) e^{j2\pi f_a t}\right] e^{-j2\pi f t} dt \\
&= \int_{-\infty}^{\infty} x(t) e^{-j2\pi (f - f_a) t} dt \\
&= X(f - f_a).
\end{aligned}
$$

4. The Modulation Property

$$\because X(f + f_a) + X(f - f_a) = \mathcal{F}\{x(t)\, e^{-j2\pi f_a t}\} + \mathcal{F}\{x(t)\, e^{j2\pi f_a t}\} \quad \text{(by frequency shift)}$$
$$= \mathcal{F}\{x(t)[e^{-j2\pi f_a t} + e^{j2\pi f_a t}]\} \quad \text{(by linearity)}$$
$$= 2\mathcal{F}\{x(t)\cos 2\pi f_a t\}, \quad \text{(by Euler s formula)}$$
$$\therefore \mathcal{F}\{x(t)\cos 2\pi f_a t\} = \frac{1}{2}\left(X(f + f_a) + X(f - f_a)\right).$$

Similarly,

$$\because X(f + f_a) - X(f - f_a) = \mathcal{F}\{x(t)\, e^{-j2\pi f_a t}\} - \mathcal{F}\{x(t)\, e^{j2\pi f_a t}\} \quad \text{(by frequency shift)}$$
$$= \mathcal{F}\{x(t)[e^{-j2\pi f_a t} - e^{j2\pi f_a t}]\} \quad \text{(by linearity)}$$
$$= -2j\mathcal{F}\{x(t)\sin 2\pi f_a t\}, \quad \text{(by Euler s formula)}$$
$$\therefore \mathcal{F}\{x(t)\sin 2\pi f_a t\} = \frac{j}{2}\left(X(f + f_a) - X(f - f_a)\right).$$

5. The Time-Scaling Property

Case (i) When $\alpha > 0$, we have

$$\mathcal{F}\{x(\alpha t)\} = \int_{-\infty}^{\infty} x(\alpha t)\, e^{-j2\pi f t}\, dt$$
$$= \int_{-\infty}^{\infty} x(s)\, e^{-j2\pi f s/\alpha} \frac{1}{\alpha}\, ds \qquad \left(\because s = \alpha t,\ t = \frac{s}{\alpha},\ dt = \frac{1}{\alpha}\, ds\right)$$
$$= \frac{1}{\alpha}\left[\int_{-\infty}^{\infty} x(s)\, e^{-j2\pi\left(\frac{f}{\alpha}\right)s}\, ds\right]$$
$$= \frac{1}{\alpha} X\left(\frac{f}{\alpha}\right).$$

Case (ii) When $\alpha < 0$, the new variable $s = \alpha t$ changes sign. Thus when $t \to -\infty$, $s \to \infty$; when $t \to \infty$, $s \to -\infty$. We have taken into account the sign change in the lower and upper limits of the integral in the derivation below.

$$\mathcal{F}\{x(\alpha t)\} = \int_{-\infty}^{\infty} x(\alpha t)\, e^{-j2\pi f t}\, dt$$
$$= \int_{\infty}^{-\infty} x(s)\, e^{-j2\pi f s/\alpha} \frac{1}{\alpha}\, ds \qquad (\because s = \alpha t \text{ changes sign})$$
$$= \frac{1}{\alpha}\left[(-1)\int_{-\infty}^{\infty} x(s)\, e^{-j2\pi\left(\frac{f}{\alpha}\right)s}\, ds\right]$$
$$= -\frac{1}{\alpha}\left[\int_{-\infty}^{\infty} x(s)\, e^{-j2\pi\left(\frac{f}{\alpha}\right)s}\, ds\right]$$
$$= \frac{1}{|\alpha|} X\left(\frac{f}{\alpha}\right). \qquad (\because -\alpha = |\alpha| \text{ when } \alpha < 0)$$

Combining the two cases, we have

$$\mathcal{F}\{x(\alpha t)\} = \frac{1}{|\alpha|} X\left(\frac{f}{\alpha}\right),\ \alpha \neq 0.$$

6. The Folding Property This is a direct result of the time-scaling property: let $\alpha = -1$ in the time scaling formula above, we immediately obtain the folding property.

7. Transform of a Transform

$$\because x(t) = \mathcal{F}^{-1}\{X(f)\} = \int_{-\infty}^{\infty} X(f)\, e^{j2\pi ft} df$$

$$\therefore x(-t) = \int_{-\infty}^{\infty} X(f)\, e^{-j2\pi tf} df = \mathcal{F}\{X(f)\} = \mathcal{F}\{\mathcal{F}\{x(t)\}\}.$$

8. Transform of the Derivative

$$\mathcal{F}\{x'(t)\} = \int_{-\infty}^{\infty} x'(t)\, e^{-j2\pi ft}\, dt \qquad \text{(by de n ition of Fourier transform)}$$

$$= x(t)\, e^{-j2\pi ft}\Big|_{-\infty}^{\infty}$$

$$\quad - \int_{-\infty}^{\infty} x(t)\left\{-j2\pi f e^{-j2\pi ft}\right\} dt \quad \text{(via integration by parts)}$$

$$= j2\pi f \int_{-\infty}^{\infty} x(t)\, e^{-j2\pi ft}\, dt \qquad \text{(assume } x(t) \to 0 \text{ as } |t| \to \infty)$$

$$= j2\pi f X(f).$$

9. Derivative of the transform

$$\because \mathcal{F}^{-1}\{X'(f)\} = \int_{-\infty}^{\infty} X'(f)\, e^{j2\pi ft} df \qquad \text{(inverse Fourier transform)}$$

$$= X(f)\, e^{j2\pi ft}\Big|_{-\infty}^{\infty}$$

$$\quad - \int_{-\infty}^{\infty} X(f)\left\{j2\pi t\, e^{j2\pi ft}\right\} df \quad \text{(via integration by parts)}$$

$$= -j2\pi t \int_{-\infty}^{\infty} X(f)\, e^{j2\pi ft} df \qquad \text{(assume } X(f) \to 0 \text{ as } |f| \to \infty)$$

$$= -j2\pi t\, x(t).$$

$$\therefore X'(f) = \mathcal{F}\{-j2\pi t\, x(t)\} = -j2\pi \mathcal{F}\{t x(t)\}.$$

5.6.2 Utilities of the properties

We demonstrate how to use these properties to nd Fourier transform pairs by several examples.

Example 5.5 (Figure 5.7) Given the Fourier transform pair (from Example 5.3)

$$x_{\text{rect}}(t) = \begin{cases} \dfrac{1}{2t_0}, & \text{for } t \in [-t_0, t_0]; \\ 0, & \text{for } |t| > t_0. \end{cases} \iff X(f) = \text{sinc}(2ft_0) = \frac{\sin(2\pi ft_0)}{2\pi ft_0},$$

we apply the time-shift property and obtain the following pair:

$$x_{\text{rect}}(t - \alpha) = \begin{cases} \dfrac{1}{2t_0}, & \text{for } t \in [-t_0 + \alpha, t_0 + \alpha]; \\ 0, & \text{for } |t - \alpha| > t_0. \end{cases}$$

$$\Longleftrightarrow X(f)\, e^{-j2\pi f \alpha} = \mathbf{sinc}(2ft_0)\, e^{-j2\pi f \alpha}$$
$$= \mathbf{sinc}(2ft_0) \cos 2\pi f \alpha - j\mathbf{sinc}(2ft_0) \sin 2\pi f \alpha.$$

Figure 5.7 Illustrating the time-shift property.

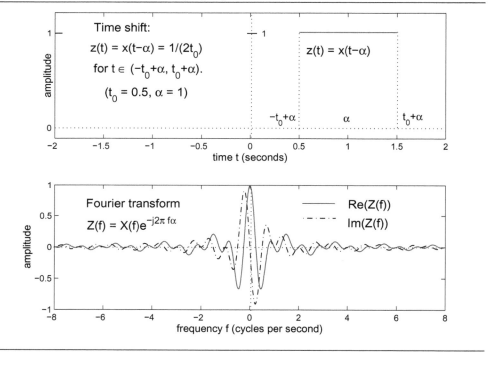

Example 5.6 (Figure 5.8) Given the Fourier transform pair (from Example 5.1)

$$x(t) = \begin{cases} e^{-at}, & \text{for } t \in [0, \infty) \quad (a > 0); \\ 0, & \text{for } t \in (-\infty, 0) \end{cases} \quad \Longleftrightarrow \quad X(f) = \frac{1}{a + j2\pi f},$$

we obtain the pair

$$y(t) = t\,x(t) = \begin{cases} t e^{-at}, & \text{for } t \in [0, \infty) \quad (a > 0); \\ 0, & \text{for } t \in (-\infty, 0) \end{cases} \quad \Longleftrightarrow \quad Y(f) = \frac{1}{(a + j2\pi f)^2}$$

by using the deri vative of the transform property: $-j2\pi \mathcal{F}\{tx(t)\} = X'(f)$, which yields

$$Y(f) = \mathcal{F}\{t\,x(t)\} = \frac{-1}{j2\pi} X'(f) = \frac{-1}{j2\pi}\left[\frac{-j2\pi}{(a + j2\pi f)^2}\right] = \frac{1}{(a + j2\pi f)^2}.$$

Figure 5.8 Illustrating the derivative of the transform property.

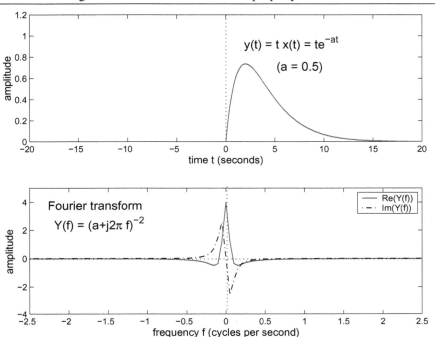

Example 5.7 (Figure 5.9) Given the Fourier transform pair $x(t) \Longleftrightarrow X(f)$, if $\mathcal{F}\{x^{(n)}(t)\}$ exists, it can be found by applying the transform of the derivative property n times, and the following pair is obtained

$$x^{(n)}(t) \Longleftrightarrow (j2\pi f)^n X(f).$$

On the other hand, given $x(t) \Longleftrightarrow X(f)$, we can nd $\mathcal{F}\{t^n x(t)\}$ by applying the derivative of the transform property n times:

$$(-j2\pi)^n t^n x(t) \Longleftrightarrow X^{(n)}(f).$$

Continuing with the result from the last example, we immediately have the transform pair for every $n \geq 1$:

$$z(t) = \begin{cases} t^n e^{-at}, & \text{for } t \in [0, \infty) \quad (a > 0); \\ 0, & \text{for } t \in (-\infty, 0) \end{cases} \Longleftrightarrow Z(f) = \frac{n!}{(a + j2\pi f)^{n+1}}.$$

5.7 Alternate Form of the Fourier Transform

The Fourier transform may be expressed as a function of the angular frequency ω instead of the the rotational frequency f. Noting that because $\omega = 2\pi f$, we have $f = \omega/2\pi$, $df = (1/2\pi)\,d\omega$, and we modify the two de ning integrals by changing the variable as shown below.

$$X(f) = \int_{-\infty}^{\infty} x(t)\,e^{-j2\pi ft}\,dt = \int_{-\infty}^{\infty} x(t)\,e^{-j\omega t}\,dt = \tilde{X}(\omega).$$

Figure 5.9 Illustrating the derivative of the transform property ($n = 2$).

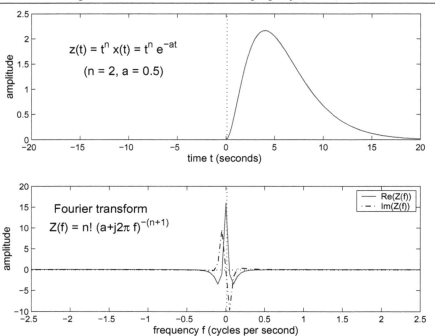

$$x(t) = \int_{-\infty}^{\infty} X(f)\, e^{j2\pi ft}\, df = \frac{1}{2\pi} \int_{-\infty}^{\infty} \tilde{X}(\omega)\, e^{j\omega t}\, d\omega.$$

We have thus obtained the mathematically equivalent definition expressed in the angular frequency ω, namely,

$$\tilde{X}(\omega) = \mathcal{F}\{x(t)\} = \int_{-\infty}^{\infty} x(t)\, e^{-j\omega t}\, dt,$$

$$x(t) = \mathcal{F}^{-1}\{\tilde{X}(\omega)\} = \frac{1}{2\pi} \int_{-\infty}^{\infty} \tilde{X}(\omega)\, e^{j\omega t}\, d\omega.$$

Observe that the mathematically equivalent counterpart of (5.14) is

(5.15) $$\tilde{X}(0) = \int_{-\infty}^{\infty} x(t)\, dt; \quad x(0) = \frac{1}{2\pi} \int_{-\infty}^{\infty} \tilde{X}(\omega)\, d\omega.$$

Using this modified definition, all properties given previously in Table 5.3 can now be re-derived in exactly the same manner. We give the results in Table 5.4.

5.8   Computing the Fourier Transform from Discrete-Time Samples

In this section we return to address several issues raised earlier concerning the sampling theorem, and we show how the theorem can help with the task of constructing and computing the Fourier transform from the discrete-time samples.

Table 5.4 Fourier transform properties (expressed in $\omega = 2\pi f$).

1. Linearity	$\mathcal{F}\{\alpha x(t) + \beta y(t)\} = \alpha \tilde{X}(\omega) + \beta \tilde{Y}(\omega)$
2. Time shift	$\mathcal{F}\{x(t - t_a)\} = \tilde{X}(\omega) e^{-j\omega t_a}$
3. Frequency shift	$\mathcal{F}\{x(t) e^{j2\pi f_a t}\} = \tilde{X}(\omega - 2\pi f_a)$
4. Modulation	$\mathcal{F}\{x(t) \cos 2\pi f_a t\} = \dfrac{1}{2}\left(\tilde{X}(\omega + 2\pi f_a) + \tilde{X}(\omega - 2\pi f_a)\right)$
	$\mathcal{F}\{x(t) \sin 2\pi f_a t\} = \dfrac{j}{2}\left(\tilde{X}(\omega + 2\pi f_a) - \tilde{X}(\omega - 2\pi f_a)\right)$
5. Time scaling	$\mathcal{F}\{x(\alpha t)\} = \dfrac{1}{|\alpha|}\tilde{X}\left(\dfrac{\omega}{\alpha}\right), \alpha \neq 0.$
6. Folding	$\mathcal{F}\{x(-t)\} = \tilde{X}(-\omega)$
7. Transform of a transform	$\mathcal{F}\{\mathcal{F}\{x(t)\}\} = \mathcal{F}\{\tilde{X}(\omega)\} = 2\pi x(-t)$
8. Transform of the derivative	$\mathcal{F}\{x'(t)\} = j\omega \tilde{X}(\omega)$
9. Derivative of the transform	$\tilde{X}'(\omega) = -j\mathcal{F}\{tx(t)\}$

5.8.1 Almost time-limited and band-limited functions

Recall that the sampling theorem is only valid for band-limited signals, and we have expressed the following concerns:

1. To use the interpolating formula given in the theorem, we need an nite number of samples extended from $-\infty$ to ∞.

2. Many signals in the real world are not band limited.

Fortunately, many signals we encounter in the real world are decaying signals the y decay to zero in the long run in both time domain and frequency domain, and they are described to be both *almost time limited* and *almost band limited*. The almost band-limited property allows us to apply the sampling theorem (with acceptable accuracy) when the bandwidth is properly chosen; the almost time-limited property allows us to use a nite interpolating formula (with acceptable accuracy) when the number of terms is properly chosen. The Gaussian function (see Example 5.2) is an obvious example, because its Fourier transform is also a Gaussian function.

In this section we consider a function $x(t)$ which is almost time limited to $[-T/2, T/2]$ and almost band limited to $[-F/2, F/2]$, where T and F are chosen to be suf ciently large so that both $x(t)$ and its Fourier transform $X(f)$ can be deemed essentially zero outside the respective range. Observe that because $\mathcal{F}\{X(f)\} = x(-t)$, so $X(f)$ is also both almost time limited and almost band limited. Therefore, if we can apply the sampling theorem to construct $x(t)$ using a nite number of samples, we can also apply the sampling theorem to construct $X(f)$ using a nite number of samples, and we will see how this plays out below.

It is common in practice to use a sampling rate ve to ten times the minimum Nyquist rate, and we will follow this practice by using the sampling rate $1/\triangle t = 5F = \hat{F}$. Applying the

Sampling Theorem 5.1, we obtain

$$x(t) = \sum_{\ell=-\infty}^{\infty} x(\ell \triangle t) \, \mathbf{sinc}(\hat{F}(t - \ell \triangle t)).$$

Since $x(t)$ is a decaying function and almost zero outside $[-T/2, T/2]$, if we choose a suf - ciently large (but nite) number of samples so that $N \triangle t = 5T = \hat{T}$ and assume $x(t) = 0$ outside $[-\hat{T}/2, \hat{T}/2]$, then we can safely truncate the interpolating formula to N terms:

$$x(t) = \sum_{\ell=-N/2+1}^{N/2} x(\ell \triangle t) \, \mathbf{sinc}(\hat{F}(t - \ell \triangle t)).$$

To obtain the Fourier transform of $x(t)$, recall that we have obtained the following Fourier transform pair (from Example 5.3)

$$p(t) = \mathbf{sinc}(\hat{F}t) \iff P(f) = \begin{cases} 1/\hat{F}, & \text{for } f \in [-\hat{F}/2, \hat{F}/2]; \\ 0, & \text{for } |f| > \hat{F}/2. \end{cases}$$

Since $p(t - \ell \triangle t) \iff P(f) \, e^{-j2\pi f \ell \triangle t}$ by the time-shift property, we have

$$X(f) = \mathcal{F}\{x(t)\} = \sum_{\ell=-N/2+1}^{N/2} x(\ell \triangle t) \, \mathcal{F}\{\mathbf{sinc}(\hat{F}(t - \ell \triangle t))\} \quad \text{(by linearity)}$$

$$= \sum_{\ell=-N/2+1}^{N/2} x(\ell \triangle t) \, \mathcal{F}\{\mathbf{sinc}(\hat{F}t)\} e^{-j2\pi f \ell \triangle t} \quad \text{(by time shift)}$$

$$= \sum_{\ell=-N/2+1}^{N/2} x(\ell \triangle t) P(f) \, e^{-j2\pi f \ell \triangle t}.$$

We may now compute N equally spaced samples of $X(f)$ over the interval $[-\hat{F}/2, \hat{F}/2]$. The sampling interval is thus $\triangle f = \hat{F}/N$, and we compute $X(f_r)$ at $f_r = r\triangle f$ for $-N/2 + 1 \leq r \leq N/2$ using the formula derived above, i.e.,

$$X(f_r) = \sum_{\ell=-N/2+1}^{N/2} x(\ell \triangle t) P(f_r) \, e^{-j2\pi f_r \ell \triangle t}$$

$$= (1/\hat{F}) \sum_{\ell=-N/2+1}^{N/2} x(\ell \triangle t) \, e^{-j2\pi (r\triangle f)\ell \triangle t} \quad (\because P(f_r) = 1/\hat{F}, \ f_r = r\triangle f)$$

$$= \triangle t \sum_{\ell=-N/2+1}^{N/2} x(\ell \triangle t) \, e^{-j2\pi r\ell/N}. \quad (\because \triangle t = 1/\hat{F}, \ \triangle f \triangle t = 1/N)$$

By letting $t_\ell = \ell \triangle t$, $x_\ell = x(t_\ell)$, and $X_r = X(f_r)$, we rewrite the equation above as

$$X_r = N \triangle t \left\{ \frac{1}{N} \sum_{\ell=-N/2+1}^{N/2} x_\ell e^{-j2\pi r\ell/N} \right\}.$$

Letting $\hat{X}_r = X_r/(N\triangle t)$, we obtain the discrete Fourier transform (DFT) of the discrete-time sequence $\{x_\ell\}$:

$$\hat{X}_r = \frac{1}{N} \sum_{\ell=-N/2+1}^{N/2} x_\ell e^{-j2\pi r\ell/N}, \quad \text{for } -N/2+1 \leq r \leq N/2.$$

Since the sequence $\{\hat{X}_r\}$ can be computed using the FFT, and it differs from $\{X_r\}$ only by a constant factor $N\triangle t$, an ef cient method is available to compute the sequence of Fourier transform values $\{X_r\}$ from a sequence of discrete-time samples $\{x_\ell\}$. A graph of $X(f)$ can then be obtained by plotting X_r values versus f_r or index r.

We show next how to construct $X(f)$ (analytically) from its N samples. By the trans-form of the transform property, $\mathcal{F}\{(X(f)\} = x(-t)$, so $X(f)$ is almost band-limited to $[-T/2, T/2]$; with the sampling rate $1/\triangle f = N/\hat{F} = N\triangle t = \hat{T} = 5T$ (ve times the minimum Nyquist rate), we can apply the sampling theorem and obtain

$$X(f) = \sum_{r=-\infty}^{\infty} X(r\triangle f)\, \mathbf{sinc}\big(\hat{T}(f - r\triangle f)\big).$$

Since $X(f)$ is almost time-limited to $[-F/2, F/2]$, assuming $X(f) = 0$ outside $[-\hat{F}/2, \hat{F}/2]$ (recall $\hat{F} = 5F$), we may safely truncate the interpolating formula to N terms:

$$X(f) = \sum_{r=-N/2+1}^{N/2} X(r\triangle f)\, \mathbf{sinc}\big(\hat{T}(f - r\triangle f)\big).$$

5.9 Computing the Fourier Series Coefficients from Discrete-Time Samples

Let us rst recall that if $x(t)$ is a signal time-limited to $[-T/2, T/2]$, we may obtain its periodic extension $x^p(t)$ by repeating $x(t)$ over period T inde n itely. Conversely, given a periodic function $x^p(t)$, we may obtain the time-limited $x(t)$ by restricting $x^p(t)$ to a single period. In either case, $x(t) = x^p(t)$ for $t \in [-T/2, T/2]$, and $x(t) = 0$ otherwise. Assuming that $x^p(t)$ satis es the Dirichlet conditions and it has a Fourier series, we may use the same Fourier series to represent the time-limited $x(t)$ over $[-T/2, T/2]$. This result allows us to relate the Fourier series coef cients for periodic $x^p(t)$ (or time-limited $x(t)$) to the Fourier transform of time-limited $x(t)$ as shown below (previously shown in the middle of Table 5.2):
We also gave an example to demonstrate this connection (Example 5.3).

In this section we shall combine the sampling theorem with the connection above to com-pute the Fourier series coef cients for a periodic function $x^p(t)$ based on its discrete-time samples. We will only need equally spaced samples from a single period of $x^p(t)$, which may be interpreted as samples from a time-limited function $x(t)$, and we assume further that $x(t)$ is almost band-limited (for the sampling theorem to apply.) The latter condition also implies that $x^p(t)$ is almost band-limited, because it is represented by the same Fourier series.

Table 5.5 Connections with time-limited restriction.

Time-limited pair	Fourier transform of $x(t)$	Fourier series connection:
$x(t) \Longleftrightarrow X(f)$ $t \in [-T/2, T/2]$ $f \in (-\infty, \infty)$	$x(t) = \int_{-\infty}^{\infty} X(f) \, e^{j2\pi ft} df,$ $X(f) = \int_{-T/2}^{T/2} x(t) \, e^{-j2\pi ft} dt.$	$x(t) = x^p(t) = \sum_{k=-\infty}^{\infty} C_k e^{j2\pi kt/T},$ for $t \in [-T/2, T/2]$, where coefficient $C_k = \frac{1}{T} X\left(\frac{k}{T}\right),$ for $k = \ldots, -1, 0, 1, \ldots$

5.9.1 Periodic and almost band-limited function

Let $x^p(t)$ denote a periodic and almost band-limited function. As explained above, we define $x(t) = x^p(t)$ over a single period $[-T/2, T/2]$ and $x(t) = 0$ otherwise, and we further assume $x(t)$ to be almost band limited to $[-F/2, F/2]$. Since $x(t)$ is time limited and almost band limited, we may now apply the sampling theorem to compute the discrete Fourier transform of $x(t)$ following the same process developed in the last section with only one adjustment: the sampling interval is now T for time-limited $x(t)$; i.e., we choose the number of samples to satisfy $N \triangle t = T$ (in contrast to choosing $N \triangle t = 5T$ for the almost time-limited function.)

Retracing the steps in the last section, we use the same practical sampling rate $1/\triangle t = 5F = \hat{F}$, apply the sampling theorem, and obtain

$$x(t) = \sum_{\ell=-\infty}^{\infty} x(\ell \triangle t) \, \mathbf{sinc}\big(\hat{F}(t - \ell \triangle t)\big).$$

Since $x(t) = 0$ outside $[-T/2, T/2]$, if N is the total number of samples within $[-T/2, T/2]$, we have $N \triangle t = T$ and there are only N terms in the interpolating formula:

$$x(t) = \sum_{\ell=-N/2+1}^{N/2} x(\ell \triangle t) \, \mathbf{sinc}\big(\hat{F}(t - \ell \triangle t)\big).$$

Following exactly the same derivation in the last section, we obtain the same expressions for $X(f) = \mathcal{F}\{x(t)\}$ and its discrete values $X_r = X(r \triangle f)$, where $\triangle f = 1/(N \triangle t) = 1/T$ (in contrast to $\triangle f = 1/(5T)$ in the last section), namely,

$$X_r = N \triangle t \left\{ \frac{1}{N} \sum_{\ell=-N/2+1}^{N/2} x_\ell e^{-j2\pi r\ell/N} \right\} = T \left\{ \frac{1}{N} \sum_{\ell=-N/2+1}^{N/2} x_\ell e^{-j2\pi r\ell/N} \right\}.$$

Since the Fourier coefficient

$$C_r = \frac{1}{T} X\left(\frac{r}{T}\right) = \frac{1}{T} X(r \triangle f) = \frac{1}{T} X_r,$$

the N Fourier series coefficients can be obtained directly from the discrete Fourier transform (**DFT**) of the N discrete-time samples, i.e.,

$$C_r = \frac{1}{N} \sum_{\ell=-N/2+1}^{N/2} x_\ell e^{-j2\pi r\ell/N} \quad \text{for } r = -N/2+1, \ldots, -1, 0, 1, \ldots, N/2.$$

From the reciprocity relation $\triangle f \triangle t = 1/N$, it can be easily verified that the N coefficients C_r correspond to discrete frequencies $f_r \in [-\hat{F}/2, \hat{F}/2] = [-5F/2, 5F/2]$ as one would have expected.

References

1. A. Ambardar. *Analog and Digital Signal Processing*. Brooks/Cole Publishing Company, Paci c Grove, CA, second edition, 1999.

2. W. L. Briggs and V. E. Hensen. *The DFT: An Owner's Manual for the Discrete Fourier Transform*. The Society for Industrial and Applied Mathematics, Philadelphia, PA, 1995.

3. R. W. Hamming. *Digital Filters*. Prentice-Hall, Inc., Englewood Cliffs, NJ, third edition, 1989.

4. B. Porat. *A Course in Digital Signal Processing*. John Wiley & Sons, Inc., New York, 1997.

5. H. J. Weaver. *Applications of Discrete and Continuous Fourier Analysis*. John Wiley & Sons, Inc., New York, 1983.

Chapter 6

Sampling and Reconstruction of Functions—Part III

In this chapter we build on the material developed in Chapters 3 and 5 when we study the impulse functions and the Fourier transform theorems on convolution, and we show how these mathematical tools interplay in developing the sampling theorem and other digital signal processing tools.

6.1 Impulse Functions and Their Properties

We begin with the familiar rectangular pulse function (with area $= 1$):

$$d_\tau(t) = \begin{cases} \dfrac{1}{2\tau}, & \text{for } t \in [-\tau, \tau]; \\ 0, & \text{for } |t| > \tau > 0. \end{cases}$$

Since the pulse represented by $d_\tau(t)$ gets taller and narrower when the interval of length 2τ gets smaller, it is commonly used to describe a sudden force of large magnitude over very short time intervals. Recall that we have derived the Fourier transform of a rectangular pulse function in Example 5.3, so we immediately obtain

$$\mathcal{F}\{d_\tau(t)\} = \mathbf{sinc}(2f\tau) = \frac{\sin(2\pi f\tau)}{2\pi f\tau}.$$

In the last chapter, the Fourier transform pair consisting of the rectangular pulse and the **sinc** function was shown to play a key role in sampling and recovering a function and its Fourier transform. In this chapter, we base our development on a generalized function de ned as the limit of d_τ as $\tau \to 0$ (see Figure 6.1), i.e.,

$$\delta(t) = \lim_{\tau \to 0} d_\tau(t).$$

The generalized function $\delta(t)$ is usually called the Dirac delta function and it has the following properties:

Figure 6.1 De n ing the Dirac delta function.

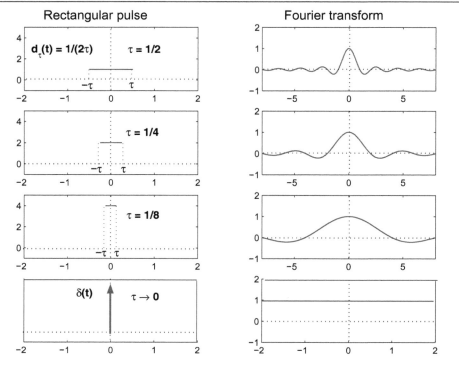

1. When $t \neq 0$,

$$\delta(t) = \lim_{\tau \to 0} d_\tau(t) = 0.$$

2. The impulse of $\delta(t)$ is de n ed by $\int_{-\infty}^{\infty} \delta(t)dt$. Accordingly,

$$\int_{-\infty}^{\infty} \delta(t)\, dt = \int_{-\infty}^{\infty} \lim_{\tau \to 0} d_\tau(t)\, dt = \lim_{\tau \to 0} \int_{-\tau}^{\tau} \frac{1}{2\tau}\, dt = \lim_{\tau \to 0} \frac{\tau - (-\tau)}{2\tau} = 1.$$

Since $\delta(t) = 0$ for $t \neq 0$, it is said to impart a unit impulse at $t = 0$, and $\delta(t)$ is called the unit impulse function. Graphically, $\delta(t)$ is represented by a single spike with unit strength at $t = 0$. The unit strength may be explicitly expressed by putting the label (1) next to the spike. Note that $\delta(t)$ is formally continuous in time, but it is not continuous in amplitude, so it is not an analog signal which refers to signals continuous in both time and amplitude (see Figure 6.2).

3. From the two properties above, we also have (via change of variables)

$$\delta(t - t_a) = 0, \quad t \neq t_a; \qquad \int_{-\infty}^{\infty} \delta(t - t_a)\, dt = 1.$$

Graphically, $\delta(t - t_a)$ is represented by a single spike at $t = t_a$ (see Figure 6.2).

4. The impulse of the product of a signal $x(t)$ and $\delta(t - t_a)$ can be obtained by scaling the

Figure 6.2 Illustrating properties of the unit impulse function.

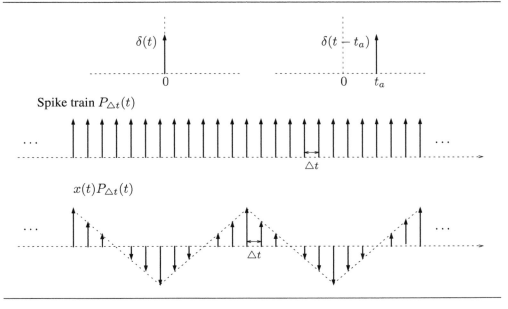

unit impulse imparted at $t = t_a$ by $x(t_a)$:

$$\int_{-\infty}^{\infty} x(t)\,\delta(t - \boldsymbol{t}_a)\,dt = \int_{t_a^-}^{t_a^+} x(t)\,\delta(t - \boldsymbol{t}_a)\,dt$$

$$= x(t_a) \int_{t_a^-}^{t_a^+} \delta(t - \boldsymbol{t}_a)\,dt$$

$$= x(t_a) \int_{-\infty}^{\infty} \delta(t - \boldsymbol{t}_a)\,dt$$

$$= x(t_a).$$

The result expressed as

$$\int_{-\infty}^{\infty} x(t)\,\delta(t - \boldsymbol{t}_a)\,dt = x(t_a)$$

is also called the sifting property of the impulse function; in addition, the following expression of equality

$$x(t)\,\delta(t - t_a) = x(t_a)\,\delta(t - t_a)$$

is commonly inferred from the equality of their impulses.

5. An in nite sequence (or train) of equally spaced impulses may now be de ned as

$$P_{\triangle t}(t) = \sum_{\ell=-\infty}^{\infty} \delta(t - \ell\triangle t).$$

Graphically, $P_{\triangle t}(t)$ is represented by a spike train (see Figure 6.2).

From computing the impulse of the product of a signal $x(t)$ and $P_{\triangle t}$, the expression of equality

$$x(t) \sum_{\ell=-\infty}^{\infty} \delta(t - \ell\triangle t) = \sum_{\ell=-\infty}^{\infty} x(\ell\triangle t)\,\delta(t - \ell\triangle t)$$

is again inferred from the equality of their impulses.

6.2   Generating the Fourier Transform Pairs

With properties of the Fourier transform and the impulse functions available, we can now generate the following Fourier transform pairs:

1. $x(t) = \delta(t) \Longleftrightarrow X(f) = 1$ (Figure 6.3). This pair can be generated in two ways: (a) We make use of the Fourier transform of the rectangular pulse function, and we obtain

$$X(f) = \mathcal{F}\{\delta(t)\} = \mathcal{F}\left\{\lim_{\tau \to 0} d_\tau(t)\right\} = \lim_{\tau \to 0} \mathcal{F}\{d_\tau(t)\} = \lim_{\tau \to 0} \frac{\sin 2\pi f\tau}{2\pi f\tau} = 1.$$

(b) We interpret the integral $\mathcal{F}\{\delta(t)\}$ as the impulse of the product of $g(t) = e^{-j2\pi ft}$ and $\delta(t) = \delta(t - 0)$ so that we can apply the sifting property of the impulse function:

$$\begin{aligned}
X(f) = \mathcal{F}\{\delta(t)\} &= \int_{-\infty}^{\infty} \delta(t)\, e^{-j2\pi ft}\, dt \\
&= \int_{-\infty}^{\infty} g(t)\, \delta(t - 0)\, dt \qquad (\because g(t) = e^{-j2\pi ft}) \\
&= g(0) \qquad \text{(by sifting property of } \delta(t - t_a) \text{ with } t_a = 0) \\
&= 1. \qquad (\because g(0) = e^0 = 1)
\end{aligned}$$

Figure 6.3 Fourier transform pairs involving the impulse function.

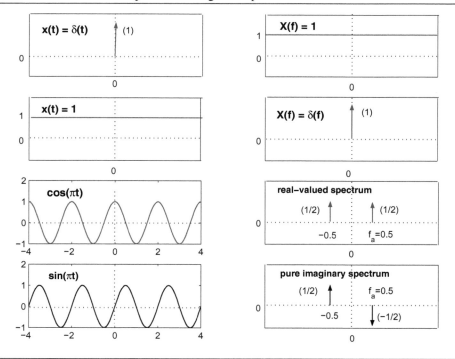

2. $x(t) = 1 \iff X(f) = \delta(f)$ (Figure 6.3). We generate this pair by showing $\mathcal{F}^{-1}\{\delta(f)\} = 1$:

$$
\begin{aligned}
x(t) = \mathcal{F}^{-1}\{\delta(f)\} &= \int_{-\infty}^{\infty} \delta(f)\, e^{j2\pi ft}\, df \\
&= \int_{-\infty}^{\infty} q(f)\, \delta(f-0)\, df \qquad (\because q(f) = e^{j2\pi ft}) \\
&= q(0) \qquad \text{(by sifting property of } \delta(f - f_a) \text{ with } f_a = 0) \\
&= 1. \qquad (\because q(0) = e^0 = 1)
\end{aligned}
$$

3. By applying the Fourier transform frequency-shift property

$$
x(t)\, e^{j2\pi f_a t} \iff X(f - f_a)
$$

to the pair $x(t) = 1 \iff X(f) = \delta(f)$, we immediately obtain

$$
e^{j2\pi f_a t} \iff \delta(f - f_a).
$$

By linearity, we have

$$
e^{j2\pi f_a t} + e^{j2\pi f_b t} \iff \delta(f - f_a) + \delta(f - f_b).
$$

If we let $f_b = -f_a$ and apply Euler s formula, we can express the Fourier transform of $\cos 2\pi f_a t$ and $\sin 2\pi f_a t$ in terms of the shifted impulse functions. The two Fourier transform pairs so obtained are

$$
\cos 2\pi f_a t \iff \frac{\delta(f - f_a) + \delta(f + f_a)}{2},
$$

$$
\sin 2\pi f_a t \iff \frac{\delta(f - f_a) - \delta(f + f_a)}{2j},
$$

and they are shown in Figure 6.3.

6.3   Convolution and Fourier Transform

The convolution of two functions $x(t)$ and $h(t)$ in the time domain is another function of t de ned as

(6.1) $\qquad w(t) = x(t) * h(t) = \int_{-\infty}^{\infty} x(\lambda)\, h(t - \lambda)\, d\lambda = \int_{-\infty}^{\infty} x(t - \lambda)\, h(\lambda)\, d\lambda.$

We note that each particular value of t is treated as a constant (with respect to the variable λ) in the integrand. For every t, $w(t)$ computes the area under the curve of the point-wise product of $x(\lambda)$ and $h(t - \lambda)$. Since the curve of $h(t - \lambda)$ continues to shift along the λ-axis when t takes on each different value, the curve of the product changes with t; hence, the area computed by the convolution integral is a function of t. This process is illustrated graphically in Figure 6.4. (The convolution steps illustrated in this gure will be re-examined and discussed further when we study how to obtain numerical approximation to the convolution result in Chapter 9.) For $x(t)$ and $h(t)$ used in the example given in Figure 6.4, the convolution result $w(t) = x(t) * h(t)$ is shown in Figure 6.5.

Figure 6.4 Illustrating the steps in convolving $x(t)$ with $h(t)$.

Step 1.   Choose stationary function $x(t)$, and change t to λ.

Step 2.   Fold $h(\lambda)$ to obtain the moving function $h(-\lambda)$.

Step 3.   Examples of shifted $h(t - \lambda)$ for $t = -0.5$, $t = 0.5$, and $t = 1.5$:

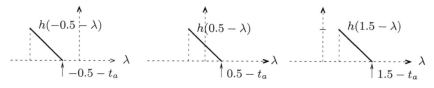

Step 4.   Sample values of $w(t)$ at $t = -0.5$, $t = 0.5$, and $t = 1.5$ are equal to the three shaded areas:

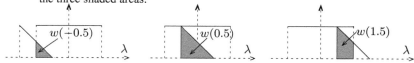

Figure 6.5 The result of continuous convolution $w(t) = x(t) * h(t)$.

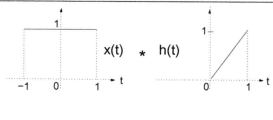

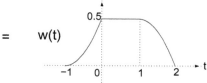

Using this de n ition, the convolution of two functions $X(f)$ and $H(f)$ in the frequency domain is expressed as

$$Z(f) = X(f) * H(f) = \int_{-\infty}^{\infty} X(\lambda) H(f - \lambda) \, d\lambda = \int_{-\infty}^{\infty} X(f - \lambda) H(\lambda) \, d\lambda.$$

The convolution integral together with its Fourier transform plays a crucial role in wide-ranging applications. Given the two Fourier transform pairs

$$x(t) \Longleftrightarrow X(f), \quad h(t) \Longleftrightarrow H(f),$$

it is not immediately clear how the Fourier transform of either the convolution $x(t) * h(t)$ or the product $x(t)h(t)$ can be expressed in terms of $X(f)$ and $H(f)$. To answer this question, we have the following two theorems.

Theorem 6.1 (Convolution theorem) If $\mathcal{F}\{x(t)\} = X(f)$ and $\mathcal{F}\{h(t)\} = H(f)$, then

$$\mathcal{F}\{x(t) * h(t)\} = X(f)H(f).$$

Proof: By de n ition,

$$
\begin{aligned}
\mathcal{F}\{x(t) * h(t)\} &= \int_{-\infty}^{\infty} [x(t) * h(t)] \, e^{-j2\pi ft} \, dt \\
&= \int_{-\infty}^{\infty} \left[\int_{-\infty}^{\infty} x(\lambda) \, h(t - \lambda) \, d\lambda \right] e^{-j2\pi ft} \, dt \\
&= \int_{-\infty}^{\infty} \int_{-\infty}^{\infty} x(\lambda) \, h(t - \lambda) \, e^{-j2\pi f(t-\lambda+\lambda)} \, d\lambda \, dt \\
&= \int_{-\infty}^{\infty} x(\lambda) \left[\int_{-\infty}^{\infty} h(t - \lambda) \, e^{-j2\pi f(t-\lambda)} \, dt \right] e^{-j2\pi f\lambda} \, d\lambda \\
&= \int_{-\infty}^{\infty} x(\lambda) \left[\int_{-\infty}^{\infty} h(s) \, e^{-j2\pi fs} \, ds \right] e^{-j2\pi f\lambda} \, d\lambda \quad (\because s = t - \lambda, \ ds = dt) \\
&= \int_{-\infty}^{\infty} x(\lambda) H(f) \, e^{-j2\pi f\lambda} \, d\lambda \qquad \left(\because H(f) = \int_{-\infty}^{\infty} h(s) \, e^{-j2\pi fs} \, ds \right) \\
&= \left[\int_{-\infty}^{\infty} x(\lambda) \, e^{-j2\pi f\lambda} \, d\lambda \right] H(f) \\
&= X(f)H(f).
\end{aligned}
$$

$\blacksquare$

The following corollary on the convolution of identical functions is an immediate result of Theorem 6.1.

Corollary 6.2 If $w(t) = x(t) * x(t)$, then

$$w(0) = \int_{-\infty}^{\infty} x(\lambda) \, x(-\lambda) \, d\lambda = \int_{-\infty}^{\infty} X(f)^2 \, df.$$

Proof: By de n ition,

$$w(t) = x(t) * x(t) = \int_{-\infty}^{\infty} x(\lambda) \, x(t - \lambda) \, d\lambda.$$

Applying Theorem 6.1 to $w(t)$, we obtain $\mathcal{F}\{w(t)\} = X(f)^2$ and the Fourier integral representation of $w(t)$:

$$w(t) = \mathcal{F}^{-1}\{X(f)^2\} = \int_{-\infty}^{\infty} X(f)^2 e^{j2\pi ft}\, df.$$

At $t = 0$, we have

$$w(0) = \int_{-\infty}^{\infty} X(f)^2\, df.$$

By evaluating $w(t) = x(t) * x(t)$ at $t = 0$, we also have

$$w(0) = \int_{-\infty}^{\infty} x(\lambda)\, x(-\lambda)\, d\lambda.$$

Hence,

$$w(0) = \int_{-\infty}^{\infty} x(\lambda)\, x(-\lambda)\, d\lambda = \int_{-\infty}^{\infty} X(f)^2\, df.$$

$\blacksquare$

Theorem 6.3 (Product theorem) If $\mathcal{F}\{x(t)\} = X(f)$ and $\mathcal{F}\{h(t)\} = H(f)$, then

$$\mathcal{F}\{x(t)\, h(t)\} = X(f) * H(f).$$

Proof: We apply the proof given for Theorem 6.1 to show

$$\mathcal{F}^{-1}\{X(f) * H(f)\} = \int_{-\infty}^{\infty} [X(f) * H(f)]\, e^{j2\pi ft}\, df = x(t)\, h(t).$$

Because the two de n ing integrals for direct and inverse Fourier transforms differ only in the sign of the exponent of the exponential function in the integrand   one uses $e^{-j2\pi ft}$ and the other uses uses $e^{+j2\pi ft}$ i   t is straightforward to accommodate the difference in the proof given for Theorem 6.1, and we omit redundant details here. $\blacksquare$

.

6.4   Periodic Convolution and Fourier Series

Given the Fourier series of two periodic functions (which can also be the protracted version of two time-limited functions)

$$x(t) = \sum_{k=-\infty}^{\infty} X_k e^{j2\pi kt/T}, \quad h(t) = \sum_{k=-\infty}^{\infty} H_k e^{j2\pi kt/T},$$

it will be useful if we can compute the Fourier coef cients of the periodic convolution $y^p(t) = x(t) \otimes h(t)$ and the product $g(t) = x(t)h(t)$ from the available X_k s and H_k s. For these tasks, we he the periodic convolution theorem and the discrete convolution theorem   the former involves continuous convolution of two periodic functions over a single period, and the latter involves the discrete convolution of two sets of Fourier coef cients.

Theorem 6.4 (Periodic convolution) The Fourier series coefficients of the periodic convolution $y^p(t) = x(t) \otimes h(t)$ can be obtained by multiplying the corresponding coefficients from the two individual Fourier series. That is, if

$$y^p(t) = x(t) \otimes h(t) = \frac{1}{T} \int_{-T/2}^{T/2} x(\lambda) \, h(t - \lambda) \, d\lambda,$$

then

$$y^p(t) = \sum_{k=-\infty}^{\infty} Y_k e^{j2\pi kt/T} = \sum_{k=-\infty}^{\infty} (X_k H_k) \, e^{j2\pi kt/T}.$$

Proof: To obtain the Fourier coefficients of $y^p(t) = x(t) \otimes h(t)$, we use the defining formula (3.12) for coefficient Y_k on the convolution integral $y^p(t)$:

$$Y_k = \frac{1}{T} \int_{-T/2}^{T/2} \left\{ \frac{1}{T} \int_{-T/2}^{T/2} x(\lambda) \, h(t - \lambda) \, d\lambda \right\} e^{-j2\pi kt/T} \, dt$$

$$= \frac{1}{T} \int_{-T/2}^{T/2} \left\{ \frac{1}{T} \int_{-T/2}^{T/2} x(\lambda) \, h(t - \lambda) \, e^{-j2\pi k(t-\lambda+\lambda)/T} \, dt \right\} d\lambda$$

$$= \frac{1}{T} \int_{-T/2}^{T/2} x(\lambda) \left\{ \frac{1}{T} \int_{-T/2}^{T/2} h(t - \lambda) \, e^{-j2\pi k(t-\lambda)/T} \, dt \right\} e^{-j2\pi k\lambda/T} \, d\lambda$$

$$= \left\{ \frac{1}{T} \int_{-T/2}^{T/2} x(\lambda) \, e^{-j2\pi k\lambda/T} \, d\lambda \right\} \left\{ \frac{1}{T} \int_{-T/2-\lambda}^{T/2-\lambda} h(\mu) \, e^{-j2\pi k \mu/T} \, d\mu \right\}$$

$$= X_k \, H_k \, .$$

Note that after the change of variable from t to $\mu = t - \lambda$ in the second integral, we obtain the defining formula for the Fourier coefficients H_k of the periodic function $h(t)$, because the interval defined by the limits of integration from $\mu = -T/2 - \lambda$ to $\mu = T/2 - \lambda$ constitutes a single period of $h(t)$. ∎

The next theorem concerns the discrete convolution of two sequences. The convolution of sequence $\{X_k\}$ by sequence $\{H_k\}$ results in another sequence $\{G_k\}$ defined as

$$G_k = \sum_{\ell=-\infty}^{\infty} X_\ell H_{k-\ell}, \quad \text{for all } k \in (-\infty, \infty).$$

The convolution of sequences is commonly denoted by $\{G_k\} = \{X_k\} * \{H_k\}$.

Theorem 6.5 (Discrete convolution) The Fourier series coefficients of the product $g(t) = x(t)h(t)$ can be obtained by convolving the two available sets of coefficients, i.e,

$$g(t) = x(t) \, h(t) = \sum_{k=-\infty}^{\infty} \left(\sum_{\ell=-\infty}^{\infty} X_\ell H_{k-\ell} \right) e^{j2\pi kt/T}.$$

Proof: It is clearer and more convenient in this proof if we use the power series forms of $x(t)$ and $h(t)$ obtained by changing the variable from t to $\theta = 2\pi t/T$ and letting $z = e^{j\theta}$:

$$\tilde{x}(\theta) = \sum_{k=-\infty}^{\infty} X_k z^k, \quad \tilde{h}(\theta) = \sum_{k=-\infty}^{\infty} H_k z^k.$$

The Fourier series of $\tilde{x}(\theta)\tilde{h}(\theta)$ is thus the product of two power series:

$$\tilde{g}(\theta) = \tilde{x}(\theta)\,\tilde{h}(\theta) = \left(\sum_{\ell=-\infty}^{\infty} X_\ell z^\ell\right)\left(\sum_{m=-\infty}^{\infty} H_m z^m\right)$$

$$= \sum_{k=-\infty}^{\infty}\left(\sum_{\ell+m=k} X_\ell H_m\right) z^k$$

$$= \sum_{k=-\infty}^{\infty}\left(\sum_{\ell=-\infty}^{\infty} X_\ell H_{k-\ell}\right) z^k. \qquad (\because \ell+m = k \,\therefore\, m = k - \ell)$$

Since the values of the Fourier coefficients are not affected by the change of variable from t to θ or vice versa, we have thus proved

$$g(t) = x(t)\,h(t) = \sum_{k=-\infty}^{\infty}\left(\sum_{\ell=-\infty}^{\infty} X_\ell H_{k-\ell}\right) e^{j2\pi kt/T}.$$

∎

6.5   Convolution with the Impulse Function

When we convolve a signal $x(t)$ with the unit impulse function $\delta(t)$, we have the following results.

1. The convolution of a signal $x(t)$ and $\delta(t)$ recovers the original signal:

$$x(t) * \delta(t) = \int_{-\infty}^{\infty} x(t-\lambda)\,\delta(\lambda)\,d\lambda$$

$$= \int_{0^-}^{0^+} x(t-\lambda)\,\delta(\lambda)\,d\lambda \quad (\because \delta(\lambda) = 0,\ \lambda \neq 0)$$

$$= x(t) \int_{0^-}^{0^+} \delta(\lambda)\,d\lambda$$

$$= x(t) \int_{-\infty}^{\infty} \delta(\lambda)\,d\lambda$$

$$= x(t). \qquad\qquad \text{(the original signal recoverd)}$$

2. The convolution of $x(t)$ and $\delta(t - t_a)$ produces the shifted signal $x(t - t_a)$:

$$x(t) * \delta(t - t_a) = x(t) * \tilde{\delta}(t) \qquad \qquad \text{(define } \tilde{\delta}(t) = \delta(t - t_a).)$$

$$= \int_{-\infty}^{\infty} x(t - \lambda) \, \tilde{\delta}(\lambda) \, d\lambda$$

$$= \int_{-\infty}^{\infty} x(t - \lambda) \, \delta(\lambda - t_a) \, d\lambda$$

$$= \int_{t_a^-}^{t_a^+} x(t - \lambda) \, \delta(\lambda - t_a) \, d\lambda \qquad (\because \delta(\lambda - t_a) = 0, \ \lambda \neq t_a)$$

$$= x(t - t_a) \int_{t_a^-}^{t_a^+} \delta(\lambda - t_a) \, d\lambda$$

$$= x(t - t_a) \int_{-\infty}^{\infty} \delta(\lambda - t_a) \, d\lambda$$

$$= x(t - t_a). \qquad \qquad \text{(the shifted signal obtained)}$$

We note that the first result is simply the special case with time shift $t_a = 0$.

3. The convolution of $x(t)$ and the impulse train $P_{\triangle t}(t)$ produces a periodic signal:

$$x(t) * P_{\triangle t}(t) = x(t) * \sum_{\ell = -\infty}^{\infty} \delta(t - \ell \triangle t)$$

$$= \sum_{\ell = -\infty}^{\infty} x(t) * \delta(t - \ell \triangle t)$$

$$= \sum_{\ell = -\infty}^{\infty} x(t - \ell \triangle t)$$

$$= z(t).$$

The resulting $z(t)$ is a superposition of copies of $x(t)$, each displaced by multiples of $\triangle t$. Observe that $z(t + \triangle t) = z(t)$. Examples are shown in Figure 6.6.

6.6   Impulse Train as a Generalized Function

In Section 6.1, we introduced the impulse train as an infinite sequence of equally spaced impulses defined by

$$P_{\triangle t}(t) = \sum_{\ell = -\infty}^{\infty} \delta(t - \ell \triangle t).$$

Since the impulse train $P_{\triangle t}(t)$ is defined for all t, it may also be viewed as a function (which is known as the comb function or shah function in the literature). In fact, it may be interpreted as a *periodic* function with period $T = \triangle t$, and it has a Fourier series representation given below in Theorem 6.6, which is known as the Poisson sum formula.

Figure 6.6 The periodic signal resulted from convolving $x(t)$ with an impulse train.

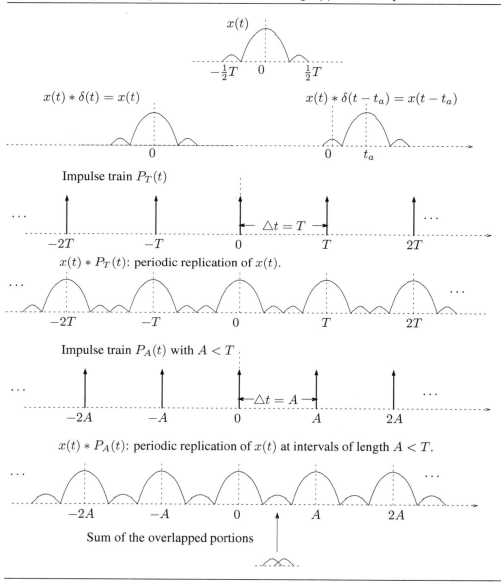

Theorem 6.6 The impulse train in the time domain can be represented by the Poisson sum formula:

$$(6.2) \qquad P_{\triangle t}(t) = \sum_{\ell=-\infty}^{\infty} \delta(t - \ell\triangle t) = \underbrace{\frac{1}{\triangle t} \sum_{k=-\infty}^{\infty} e^{j2\pi kt/\triangle t}}_{\text{Fourier series expansion}}.$$

Since $e^{j\theta} + e^{-j\theta} = 2\cos\theta$, the Poisson sum formula may also be stated as

$$(6.3) \qquad P_{\triangle t}(t) = \sum_{\ell=-\infty}^{\infty} \delta(t - \ell\triangle t) = \underbrace{\frac{1}{\triangle t}\left(1 + 2\sum_{k=1}^{\infty}\cos\frac{2\pi kt}{\triangle t}\right)}_{\text{Fourier series expansion}}.$$

In either form the impulse train is a periodic function representing an infinite sum of sinusoids. Observe that $P_{\triangle t}(t + \triangle t) = P_{\triangle t}(t)$; hence, the sampling interval $\triangle t$ is termed the period of the impulse train. The dual form of this formula in the frequency domain is given by

$$(6.4) \qquad P_{1/\triangle t}(f) = \sum_{\ell=-\infty}^{\infty} \delta\left(f - \frac{\ell}{\triangle t}\right) = \triangle t \sum_{k=-\infty}^{\infty} e^{j2\pi kf\triangle t}.$$

Proof: Since $P_{\triangle t}$ is periodic with period $T = \triangle t$, we express its Fourier series as

$$P_{\triangle t}(t) = \sum_{k=-\infty}^{\infty} C_k e^{j2\pi kt/\triangle t},$$

where the coefficients C_k s are defined as

$$C_k = \frac{1}{\triangle t}\int_{-\triangle t/2}^{\triangle t/2} P_{\triangle t}(t)\, e^{-j2\pi kt/\triangle t}\, dt$$

$$= \frac{1}{\triangle t}\int_{-\triangle t/2}^{\triangle t/2} \left\{\sum_{\ell=-\infty}^{\infty}\delta(t - \ell\triangle t)\right\} e^{-j2\pi kt/\triangle t}\, dt$$

$$= \frac{1}{\triangle t}\int_{-\triangle t/2}^{\triangle t/2} \delta(t)\, e^{-j2\pi kt/\triangle t}\, dt \qquad (\because \delta(t - \ell\triangle t) = 0 \text{ for } \ell \neq 0 \text{ over}$$

$$\text{the interval } [-\triangle t/2, \triangle t/2])$$

$$= \frac{1}{\triangle t}\int_{0-}^{0+} \delta(t)\, e^{-j2\pi kt/\triangle t}\, dt \qquad (\because \delta(t) = 0,\ t \neq 0)$$

$$= \frac{1}{\triangle t}\, e^0 \int_{0-}^{0+} \delta(t)\, dt$$

$$= \frac{1}{\triangle t}\int_{-\infty}^{\infty} \delta(t)\, dt$$

$$= 1/\triangle t.$$

We thus have

$$P_{\triangle t}(t) = \sum_{\ell=-\infty}^{\infty}\delta(t - \ell\triangle t) = \sum_{k=-\infty}^{\infty} C_k e^{j2\pi kt/\triangle t} = \frac{1}{\triangle t}\sum_{k=-\infty}^{\infty} e^{j2\pi kt/\triangle t}.$$

To obtain its dual form in the frequency domain, we substitute $\triangle f = 1/T = 1/\triangle t$ in the corresponding Poisson sum

$$P_{\triangle f}(f) = \sum_{\ell=-\infty}^{\infty} \delta(f - \ell \triangle f) = \frac{1}{\triangle f} \sum_{k=-\infty}^{\infty} e^{j2\pi k f/\triangle f} = \frac{1}{\triangle f} \sum_{k=-\infty}^{\infty} e^{-j2\pi k f/\triangle f}$$

to obtain

$$P_{1/\triangle t}(f) = \sum_{\ell=-\infty}^{\infty} \delta\left(f - \frac{\ell}{\triangle t}\right) = \triangle t \sum_{k=-\infty}^{\infty} e^{j2\pi k f \triangle t} = \triangle t \sum_{k=-\infty}^{\infty} e^{-j2\pi k f \triangle t}.$$

∎

In the next theorem we show that the Fourier transform of an impulse train in t is an impulse train in f, up to a scale factor.

Theorem 6.7 (Fourier transform of the impulse train)

$$\mathcal{F}\{P_{\triangle t}(t)\} = \frac{1}{\triangle t} P_{1/\triangle t}(f) = \underbrace{\frac{1}{\triangle t} \sum_{k=-\infty}^{\infty} \delta\left(f - \frac{k}{\triangle t}\right)}_{\text{Fourier transform of } \{P_{\triangle t}(t)\}}.$$

Proof:

$$\mathcal{F}\{P_{\triangle t}(t)\} = \int_{-\infty}^{\infty} \left[\sum_{\ell=-\infty}^{\infty} \delta(t - \ell \triangle t) \right] e^{-j2\pi ft} \, dt$$

$$= \sum_{\ell=-\infty}^{\infty} \left[\int_{-\infty}^{\infty} \delta(t - \ell \triangle t) \, e^{-j2\pi ft} \, dt \right]$$

$$= \sum_{\ell=-\infty}^{\infty} e^{-j2\pi f \ell \triangle t} \qquad \text{(by sifting property of } \delta(t - t_a) \text{ with } t_a = \ell \triangle t)$$

$$= \sum_{\ell=-\infty}^{\infty} e^{j2\pi f(-\ell) \triangle t}$$

$$= \sum_{k=-\infty}^{\infty} e^{j2\pi f k \triangle t} \qquad (\because -\infty < k = -\ell < \infty)$$

$$= \frac{1}{\triangle t} \sum_{k=-\infty}^{\infty} \delta\left(f - \frac{k}{\triangle t}\right). \qquad \text{(by Theorem 6.6)}$$

∎

The relationship between the impulse train $P_{\triangle t}(t)$ and its Fourier transform is demonstrated for several choices of $\triangle t$ in Figure 6.7.

Mathematically, the Poisson sum formula in Theorem 6.6 is a special case of a more general result in our next theorem, which relates the periodic replication of $x(t)$ to sampled values of its transform $X(f)$ via the Fourier series expansion. Recall that the periodic replication of $x(t)$ with period $T = \triangle t$ is the convolution of $x(t)$ and the shifted impulse train $P_{\triangle t}(t)$, i.e.,

$$z(t) = x(t) * P_{\triangle t}(t) = \sum_{\ell=-\infty}^{\infty} x(t) * \delta(t - \ell \triangle t) = \sum_{\ell=-\infty}^{\infty} x(t - \ell \triangle t).$$

Figure 6.7 The relationship between impulse train and its Fourier transform.

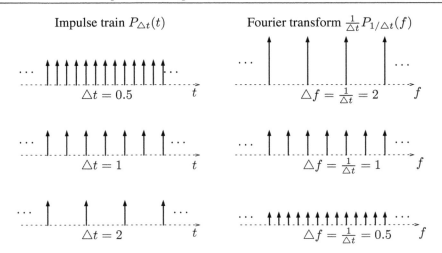

The resulting $z(t)$ is a superposition of copies of $x(t)$, each displaced by multiples of $\triangle t$, and we have $z(t + \triangle t) = z(t)$. Examples were given in Figure 6.6, and several more examples are given in Figure 6.8.

Theorem 6.8 (The generalized Poisson sum) Given the Fourier transform $X(f) = \mathcal{F}\{x(t)\}$, the Poisson sum is the Fourier series expansion of the periodic $x(t) * P_{\triangle t}(t)$, which is given by

$$(6.5) \qquad \underbrace{\sum_{\ell=-\infty}^{\infty} x(t - \ell \triangle t)}_{\text{periodic } x(t) * P_{\triangle t}(t)} = \underbrace{\frac{1}{\triangle t} \sum_{k=-\infty}^{\infty} X\left(\frac{k}{\triangle t}\right) e^{j2\pi k t/\triangle t}}_{\text{Fourier Series expansion of } x(t) * P_{\triangle t}(t)}.$$

Proof: Let $z(t)$ denote the periodic replication of $x(t)$ with period $T = \triangle t$. If we express $z(t)$ by its Fourier series expansion

$$z(t) = \sum_{\ell=-\infty}^{\infty} x(t - \ell \triangle t) = \sum_{k=-\infty}^{\infty} C_k e^{j2\pi(k/\triangle t)t},$$

Figure 6.8 Several more examples of $z(t) = x(t) * P_T(t)$.

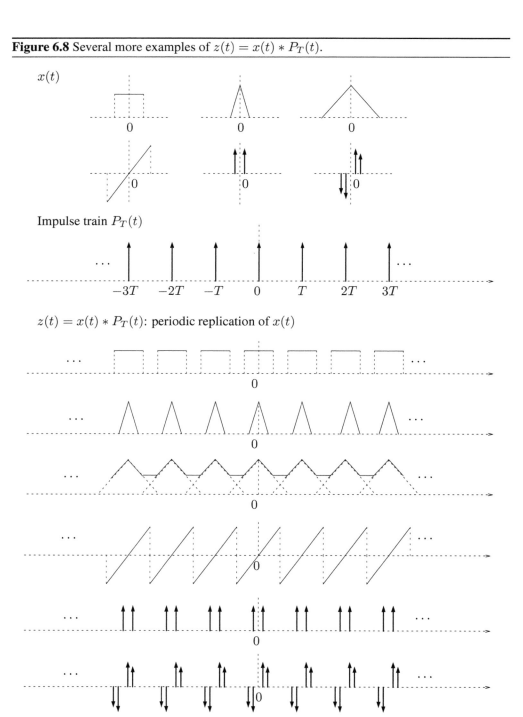

then the coef cients C_k s are de ned as

$$C_k = \frac{1}{\triangle t} \int_0^{\triangle t} z(t) \, e^{-j2\pi(k/\triangle t)t} \, dt$$

$$= \frac{1}{\triangle t} \int_0^{\triangle t} \left[\sum_{\ell=-\infty}^{\infty} x(t - \ell\triangle t) \right] e^{-j2\pi kt/\triangle t} \, dt$$

$$= \frac{1}{\triangle t} \sum_{\ell=-\infty}^{\infty} \left[\int_0^{\triangle t} x(t - \ell\triangle t) \, e^{-j2\pi kt/\triangle t} \, dt \right]$$

$$= \frac{1}{\triangle t} \sum_{\ell=-\infty}^{\infty} \left[\int_{0-\ell\triangle t}^{\triangle t-\ell\triangle t} x(\lambda) \, e^{-j2\pi k(\lambda+\ell\triangle t)/\triangle t} \, d\lambda \right] \qquad (\, \text{let } \lambda = t - \ell\triangle t)$$

$$= \frac{1}{\triangle t} \sum_{\ell=-\infty}^{\infty} \left[\int_{-\ell\triangle t}^{(-\ell+1)\triangle t} x(\lambda) \, e^{-j2\pi k\lambda/\triangle t} \, d\lambda \right] \qquad (\because e^{-j2\pi k\ell\triangle t/\triangle t} = e^{-j2\pi k\ell} = 1)$$

$$= \frac{1}{\triangle t} \int_{-\infty}^{\infty} x(\lambda) \, e^{-j2\pi k\lambda/\triangle t} \, d\lambda \qquad (\text{assume } x(\lambda) \text{ continuous at } \lambda = -\ell\triangle t \text{ for every } \ell)$$

$$= \frac{1}{\triangle t} \int_{-\infty}^{\infty} x(t) \, e^{-j2\pi(k/\triangle t)t} \, dt \qquad (\text{change variable: let } t = \lambda \text{ for clarity})$$

$$= \frac{1}{\triangle t} X\left(\frac{k}{\triangle t}\right). \qquad \left(\because X(f) = \int_{-\infty}^{\infty} x(t) \, e^{-j2\pi ft} \, dt \right)$$

Using C_k in the Fourier series expansion, we obtain

$$z(t) = \sum_{\ell=-\infty}^{\infty} x(t - \ell\triangle t) = \sum_{k=-\infty}^{\infty} C_k e^{j2\pi(k/\triangle t)t} = \frac{1}{\triangle t} \sum_{k=-\infty}^{\infty} X\left(\frac{k}{\triangle t}\right) e^{j2\pi kt/\triangle t}.$$

∎

Corollary 6.9 (Speci c Poisson sums) The two formulas given in this corollary are special cases of Theorem 6.8.

(6.6)
$$\underbrace{\sum_{\ell=-\infty}^{\infty} \delta(t - \ell\triangle t)}_{\text{impulse train } P_{\triangle t}(t)} = \underbrace{\frac{1}{\triangle t} \sum_{k=-\infty}^{\infty} e^{j2\pi kt/\triangle t}}_{\text{Fourier series expansion}},$$

(6.7)
$$\underbrace{\sum_{\ell=-\infty}^{\infty} x(\ell\triangle t)}_{x(t) * P_{\triangle t}(t) \text{ at } t=0} = \underbrace{\frac{1}{\triangle t} \sum_{k=-\infty}^{\infty} X\left(\frac{k}{\triangle t}\right)}_{\text{Fourier series at } t=0}.$$

Proof: To obtain Formula (6.6), we apply Theorem 6.8 to $x(t) = \delta(t)$. Because $X(f) = \mathcal{F}\{\delta(t)\} = 1$, we have $X\left(\frac{k}{\triangle t}\right) = 1$ for all $k \in (-\infty, \infty)$ in the right side of (6.5), and the result is the Poisson sum proved directly in Theorem 6.6.

To obtain Formula (6.7), we evaluate both sides of Formula (6.5) at $t = 0$, and the result follows. ∎

6.7   Impulse Sampling of Continuous-Time Signals

Recall that the product of a signal and the impulse train was introduced in Section 6.1:

$$x(t)P_{\triangle t}(t) = x(t) \sum_{\ell=-\infty}^{\infty} \delta(t - \ell\triangle t) = \sum_{\ell=-\infty}^{\infty} x(\ell\triangle t)\, \delta(t - \ell\triangle t).$$

Since the right-hand side is a weighted linear combination of the shifted delta functions, and the weight (or strength) of $\delta(t - \ell\triangle t)$ is exactly the value of the continuous-time signal $x(t)$ sampled at $t = t_\ell = \ell\triangle t$, we have

$$\int_{\alpha^-}^{\alpha^+} x(t)P_{\triangle t}(t)\, dt = \begin{cases} x(t_\ell), & \text{if } \alpha = t_\ell, \text{ where } t_\ell = \ell\triangle t \text{ for integer } \ell \in (-\infty, \infty); \\ 0, & \text{otherwise.} \end{cases}$$

Therefore, the information conveyed by the weighted impulse train $x_I(t) = x(t)P_{\triangle t}(t)$ is limited to the sequence of sample values $\{x(\ell\triangle t)\}$, and the function

(6.8)
$$x_I(t) = \sum_{\ell=-\infty}^{\infty} x(\ell\triangle t)\, \delta(t - \ell\triangle t)$$

is said to represent a signal ideally sampled by the impulse train. Since a signal cannot be physically sampled by an impulse train, the latter is a mathematical tool we use to model the sampling process, and the function $x_I(t)$ is called the **ideally sampled signal**.

There is also a direct connection between the continuous-time signal $x(t)$ and the ideally sampled signal $x_I(t)$ as shown in the following theorem.

Theorem 6.10 (Approximation via the weighted impulse train)

(6.9) $$x(t) = \lim_{\triangle t \to 0} \sum_{\ell=-\infty}^{\infty} \triangle t\, x(\ell\triangle t)\, \delta(t - \ell\triangle t) \approx \triangle t \underbrace{\sum_{\ell=-\infty}^{\infty} x(\ell\triangle t)\, \delta(t - \ell\triangle t)}_{\text{weighted impulse train } x_I(t)}.$$

Proof: Recall from Section 6.5 that the convolution of a signal $x(t)$ and the unit impulse function $\delta(t)$ recovers the original signal. Accordingly

$$x(t) = x(t) * \delta(t) = \int_{-\infty}^{\infty} x(\lambda)\, \delta(t - \lambda)\, d\lambda$$

$$= \lim_{\triangle\lambda \to 0} \sum_{\ell=-\infty}^{\infty} x(\ell\triangle\lambda)\, \delta(t - \ell\triangle\lambda)\triangle\lambda$$

$$= \lim_{\triangle t \to 0} \sum_{\ell=-\infty}^{\infty} \triangle t\, x(\ell\triangle t)\, \delta(t - \ell\triangle t) \quad \text{(use } \triangle t \text{ to denote } \triangle\lambda\text{)}$$

$$\approx \triangle t \sum_{\ell=-\infty}^{\infty} x(\ell\triangle t)\, \delta(t - \ell\triangle t).$$

∎

6.8 Nyquist Sampling Rate Rediscovered

Recall that we need to physically point sample a band-limited signal $x(t)$ at a rate greater than the minimum Nyquist rate in order to recover its frequency content. For impulse-sampled signal $x_I(t) = x(t)P_{\triangle t}(t)$, we now investigate how to set the sampling rate so that we can recover the Fourier transform of the continuous-time signal $x(t)$. We first find out how the Fourier transform of the impulse-sampled signal $x_I(t) = x(t)P_{\triangle t}(t)$ is connected to the Fourier transform of $x(t)$ in the next theorem.

Theorem 6.11 If $X(f) = \mathcal{F}\{x(t)\}$, then the Fourier transform of the impulse-sampled signal is given by

$$(6.10) \qquad X_I(f) = \mathcal{F}\{x_I(t)\} = \mathcal{F}\{x(t)P_{\triangle t}(t)\} = \underbrace{\frac{1}{\triangle t}\sum_{k=-\infty}^{\infty} X\left(f - \frac{k}{\triangle t}\right)}_{\text{Fourier transform of } x(t)P_{\triangle t}(t)}.$$

Proof: By Product Theorem 6.3, a product in the time domain corresponds to a convolution in the frequency domain; we thus have

$$X_I(f) = \mathcal{F}\{x_I(t)\} = X(f) * \mathcal{F}\{P_{\triangle t}(t)\}$$

$$= X(f) * \left\{\frac{1}{\triangle t}\sum_{k=-\infty}^{\infty} \delta\left(f - \frac{k}{\triangle t}\right)\right\} \qquad \text{(by Theorem 6.7)}$$

$$= \frac{1}{\triangle t}\sum_{k=-\infty}^{\infty} X(f) * \delta\left(f - \frac{k}{\triangle t}\right)$$

$$= \frac{1}{\triangle t}\sum_{k=-\infty}^{\infty} X\left(f - \frac{k}{\triangle t}\right). \qquad \text{(convolution result from Section 6.5)}$$

$\blacksquare$

According to Theorem 6.11, the Fourier transform of the impulse-sampled signal $x_I(t)$ is a superposition of copies of $X(f) = \mathcal{F}\{x(t)\}$, each displaced by multiples of $1/\triangle t$. (See Figures 6.9 and 6.10 presented with Example 6.1 later in this section.) Therefore, the shifted replications of $X(f)$ will not overlap only if the following conditions are met: (1) $X(f)$ is zero outside a band $[-f_{max}, f_{max}]$; (2) the period of replication $1/\triangle t \geq F = 2f_{max}$. Combining these two conditions, we conclude that the Fourier transform $X(f)$ can be recovered in full if $x(t)$ is band limited to $[-f_{max}, f_{max}]$ and the sampling rate $1/\triangle t \geq F = 2f_{max}$, which is, of course, the Nyquist sampling rate rediscovered for the impulse sampling of $x(t)$. (In other words, the frequency contents of band-limited $x(t)$ are preserved by impulse sampling at the Nyquist rate.) This result is formally given below as an corollary of Theorem 6.11.

Corollary 6.12 If a band-limited signal $x(t)$ with bandwidth F is sampled at (or greater than) the Nyquist rate, then

$$(6.11) \qquad X(f) = \mathcal{F}\{x(t)\} = \triangle t\,\mathcal{F}\{x(t)P_{\triangle t}(t)\} = \triangle t\,X_I(f) \text{ for } f \in [-F/2, F/2].$$

Proof: As discussed above, the shifted replications of $X(f)$ on the right-hand side of Equation (6.10) will not overlap if $x(t)$ is band limited to $[-f_{max}, f_{max}] = [-F/2, F/2]$ and the sampling rate $1/\triangle t \geq F$ b oth conditions are met in this corollary. Therefore, by re-stricting the values of f to the   nite band $[-F/2, F/2]$, we simplify the right-hand side of Equation (6.10) to obtain

$$X_I(f) = \mathcal{F}\{x(t)P_{\triangle t}(t)\} = \frac{1}{\triangle t} \sum_{\ell=-\infty}^{\infty} X\left(f - \frac{\ell}{\triangle t}\right)$$

$$= \frac{1}{\triangle t} X(f), \qquad (\because f \in [-F/2, F/2], 1/\triangle t \geq F),$$

which yields

$$X(f) = \triangle t \, X_I(f) \text{ for } f \in [-F/2, F/2].$$

■

Two more related formulas are also given as corollaries of Theorem 6.11.

Corollary 6.13 (The generalized inverse Poisson sum) The Fourier transform of an impulse-sampled signal has two forms, and by equating the two forms one obtains the generalized inverse Poisson summation formula:

(6.12)
$$\underbrace{\sum_{\ell=-\infty}^{\infty} x(\ell\triangle t)\, e^{-j2\pi f\ell\triangle t}}_{\text{Fourier transform of } x(t)P_{\triangle t}(t)} = \underbrace{\frac{1}{\triangle t} \sum_{k=-\infty}^{\infty} X\left(f - \frac{k}{\triangle t}\right)}_{X(f)*\mathcal{F}\{P_{\triangle t}(t)\}}.$$

Proof: We prove this equality by showing that the left side is an alternative formula for the Fourier transform of the impulse-sampled signal $x_I(t) = x(t)P_{\triangle t}(t)$.

$$\mathcal{F}\{x(t)P_{\triangle t}(t)\} = \int_{-\infty}^{\infty} \left[\sum_{\ell=-\infty}^{\infty} x(\ell\triangle t)\, \delta(t - \ell\triangle t)\right] e^{-j2\pi ft} \, dt$$

$$= \sum_{\ell=-\infty}^{\infty} x(\ell\triangle t) \left[\int_{-\infty}^{\infty} \delta(t - \ell\triangle t)\, e^{-j2\pi ft} \, dt\right]$$

$$= \sum_{\ell=-\infty}^{\infty} x(\ell\triangle t)\, e^{-j2\pi f\ell\triangle t}.$$

By Theorem 6.11, we also have

$$\mathcal{F}\{x(t)P_{\triangle t}(t)\} = \frac{1}{\triangle t} \sum_{k=-\infty}^{\infty} X\left(f - \frac{k}{\triangle t}\right).$$

Because the two formulas represent the same result, we have

$$\sum_{\ell=-\infty}^{\infty} x(\ell\triangle t)\, e^{-j2\pi f\ell\triangle t} = \frac{1}{\triangle t} \sum_{k=-\infty}^{\infty} X\left(f - \frac{k}{\triangle t}\right).$$

■

Observe that Formula (6.12) in Corollary 6.13 is the dual of the generalized Poisson sum in Theorem 6.8.

Example 6.1 (Figures 6.9 and 6.10) Recall the Fourier transform pair from Example 5.1:

$$x(t) = \begin{cases} e^{-at}, & \text{for } t \in [0, \infty) \quad (a > 0); \\ 0, & \text{for } t \in (-\infty, 0) \end{cases} \iff X(f) = \frac{1}{a + j2\pi f}.$$

The discrete-time samples of $x(t)$ are de ned by

$$x(\ell \triangle t) = \begin{cases} e^{-a\ell \triangle t}, & \text{for } \ell \geq 0 \quad (a > 0) \\ 0, & \text{otherwise,} \end{cases}$$

and the Fourier transform of the impulse-sampled signal $x_I(t)$ can be obtained in two forms according to Corollary 6.11. Using the summation formula, we obtain

$$X_I(f) = \sum_{\ell=-\infty}^{\infty} x(\ell \triangle t) e^{-j2\pi f \ell \triangle t} = \sum_{\ell=0}^{\infty} e^{-(a\triangle t + j2\pi f \triangle t)\ell} = \frac{1}{1 - e^{-(a+j2\pi f)\triangle t}}.$$

Using the known $X(f) = 1/(a + j2\pi f)$, we obtain

$$X_I(f) = \frac{1}{\triangle t} \sum_{k=-\infty}^{\infty} X\left(f - \frac{k}{\triangle t}\right) = \sum_{k=-\infty}^{\infty} \frac{1}{a\triangle t + j2\pi(f\triangle t - k)}.$$

By equating the two forms of $X_I(f)$, we obtain

$$\boxed{\sum_{k=-\infty}^{\infty} \frac{1}{a\triangle t + j2\pi(f\triangle t - k)} = \frac{1}{1 - e^{-(a+j2\pi f)\triangle t}},}$$

where the right side provides a closed-form expression for the in nite sum on the left side.

The graphs of $x(t)$, $X(f)$, $\{x(\ell \triangle t)\}$, and $X_I(f)$ for $\triangle t = 1$ and $\triangle t = 0.5$ are shown in Figures 6.9 and 6.10. Observe that the central period of $X_I(f)$ deviates from $X(f)$ due to the effect of aliasing. Since $x(t)$ is not band limited, the effect of aliasing cannot be eliminated although it is reduced when the signal is sampled at a higher rate.

Corollary 6.14 (Speci c inverse Poisson sums) Two special cases of (6.12) are

(6.13)
$$\underbrace{\sum_{\ell=-\infty}^{\infty} e^{-j2\pi f \ell \triangle t}}_{(1/\triangle t)P_{1/\triangle t}(f)} = \underbrace{\frac{1}{\triangle t} \sum_{k=-\infty}^{\infty} \delta\left(f - \frac{k}{\triangle t}\right)}_{\text{Fourier transform of } P_{\triangle t}(t)} \quad \text{and}$$

(6.14)
$$\underbrace{\sum_{\ell=-\infty}^{\infty} x(\ell \triangle t)}_{\mathcal{F}\{x(t)P_{\triangle t}(t)\} \text{ at } f=0} = \underbrace{\frac{1}{\triangle t} \sum_{k=-\infty}^{\infty} X\left(\frac{k}{\triangle t}\right)}_{X(f)*\mathcal{F}\{P_{\triangle t}(t)\} \text{ at } f=0}.$$

Proof: To obtain Formula (6.13), we apply Corollary 6.13 to $x(t) = 1$, which results in $x(\ell \triangle t) = 1$ for all $\ell \in (-\infty, \infty)$ on the left side of (6.12); because the Fourier transform $X(f) = \mathcal{F}\{x(t) = 1\} = \delta(f)$, we have $X(f - k/\triangle t) = \delta(f - k/\triangle t)$ on the right side of (6.12), and the result follows. (This result is the dual of (6.6) in Corollary 6.9.)

The result in (6.14) is obtained by setting $f = 0$ on both sides of (6.12). (This result is identical to (6.7) in Corollary 6.9, which was obtained by setting $t = 0$ in the generalized Poisson sum.) ∎

Figure 6.9 Fourier transform of the sequence sampled from $x(t) = e^{-at}$.

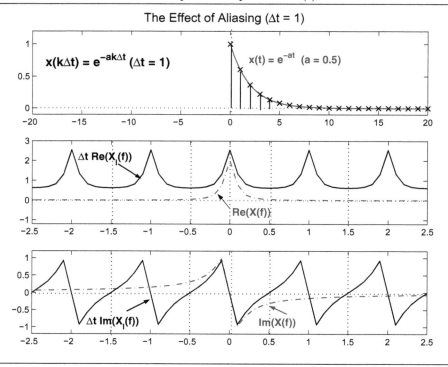

Figure 6.10 Reducing the effect of aliasing by increasing sampling rate.

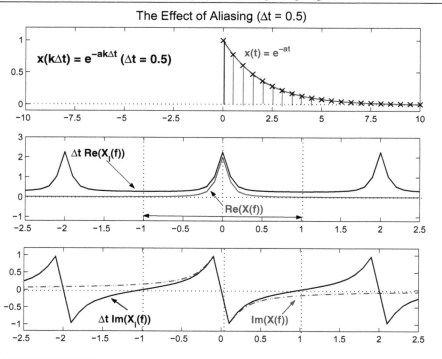

6.9   Sampling Theorem for Band-Limited Signal

The results developed in preceding sections have paved the way for the sampling theorem to re-emerge. We are speci cally interested in applying those results to band-limited signals, for which the sampling theorem is to be developed once again. We speci cally recall formula (6.12) from Corollary 6.13, which is cited below for easy reference.

$$X_I(f) = \mathcal{F}\{x(t)P_{\triangle t}(t)\} = \sum_{\ell=-\infty}^{\infty} x(\ell \triangle t)\, e^{-j2\pi f\ell \triangle t} = \frac{1}{\triangle t} \sum_{\ell=-\infty}^{\infty} X\left(f - \frac{\ell}{\triangle t}\right).$$

Observe that this formula is de ned for all values $f \in (-\infty, \infty)$. Applying this result and the Nyquist sampling rate to a band-limited signal, we have the following lemma.

Lemma 6.15 If a band-limited signal $x(t)$ with bandwidth F is sampled at the Nyquist rate, its Fourier transform my be expressed as a Fourier series whose coef cients are discrete-time samples of $x(t)$ (multiplied by a scale factor). That is,

$$(6.15) \qquad X(f) = \mathcal{F}\{x(t)\} = \frac{1}{F} \sum_{k=-\infty}^{\infty} x\left(-\frac{k}{F}\right) e^{j2\pi kf/F}, \ f \in [-F/2, F/2].$$

Proof: Since the given signal $x(t)$ is band limited with bandwidth F, we have $X(f) = 0$ outside the band $[-F/2, F/2]$. Following Corollary 6.12, we have

$$X(f) = \triangle t\, X_I(f) \quad \text{for } f \in [-F/2, F/2].$$

Following Corollary 6.13, we have

$$X_I(f) = \sum_{\ell=-\infty}^{\infty} x(\ell \triangle t)\, e^{-j2\pi f\ell \triangle t} \quad \text{for } f \in [-\infty, \infty].$$

Combining these two results, we obtain

$$X(f) = \triangle t \sum_{\ell=-\infty}^{\infty} x(\ell \triangle t)\, e^{-j2\pi f\ell \triangle t} \quad \text{for } f \in [-F/2, F/2]$$

$$= \triangle t \sum_{k=-\infty}^{\infty} x(-k\triangle t)\, e^{j2\pi fk\triangle t} \qquad (\because k = -\ell)$$

$$= \frac{1}{F} \sum_{k=-\infty}^{\infty} x\left(-\frac{k}{F}\right) e^{j2\pi kf/F}. \qquad (\because \triangle t = 1/F)$$

∎

Since the result we obtain in Lemma 6.15 is *identical* to an earlier result given by Equations (5.10), we simply repeat the derivation beginning with Equation (5.11), and **Sampling Theorem 5.1** re-emerges! The derivation serves as a direct proof of the sampling theorem.

Alternatively, we can prove the same theorem based on the properties of the impulse train and the convolution theorem. To demonstrate this process, we state the sampling theorem again and provide the second proof below.

Theorem 6.16 (Sampling theorem) If the signal $x(t)$ is known to be band-limited with bandwidth $F = 2f_{max}$, then $x(t)$ can be sampled at the Nyquist rate $1/\triangle t = F$, and we can determine a unique $x(t)$ by interpolating the sequence of samples according to the formula

$$x(t) = \sum_{\ell=-\infty}^{\infty} x(\ell\triangle t) \, \frac{\sin(\pi F(t - \ell\triangle t))}{\pi F(t - \ell\triangle t)}.$$

Proof: Recall from Corollary 6.12 that

$$\mathcal{F}\{x(t)\} = X(f) = \begin{cases} \triangle t \, X_I(f) = \frac{1}{F} X_I(f), & f \in [-F/2, F/2]; \\ 0, & \text{otherwise.} \end{cases}$$

To facilitate further mathematical manipulations, an equivalent definition for $X(f)$ is given by a single formula

(6.16) $$X(f) = X_I(f)Y^{\text{rect}}_{F/2}(f), \quad f \in (-\infty, \infty),$$

where $Y^{\text{rect}}_{F/2}(f)$ is the familiar rectangular pulse function (with area = 1) defined as

(6.17) $$Y^{\text{rect}}_{F/2}(f) = \begin{cases} 1/F, & \text{for } f \in [-F/2, F/2]; \\ 0, & \text{otherwise.} \end{cases}$$

Since $X(f)$ is the product of two functions in the frequency domain according to Equation (6.16), we apply **Convolution Theorem 6.1**, and we obtain

(6.18)
$$\begin{aligned} x(t) = \mathcal{F}^{-1}\{X(f)\} &= \mathcal{F}^{-1}\{X_I(f)Y^{\text{rect}}_{F/2}(f)\} \\ &= \mathcal{F}^{-1}\{X_I(f)\} * \mathcal{F}^{-1}\{Y^{\text{rect}}_{F/2}(f)\} \\ &= x_I(t) * y(t). \end{aligned}$$

To evaluate the convolution on the right-hand side, we substitute

(6.19) $$x_I(t) = x(t)P_{\triangle t}(t) = \sum_{\ell=-\infty}^{\infty} x(\ell\triangle t)\, \delta(t - \ell\triangle t),$$

(6.20) $$y(t) = \mathcal{F}^{-1}\{Y^{\text{rect}}_{F/2}(f)\} = \text{sinc}(Ft), \quad \text{(a result from Example 5.4)},$$

and we obtain

(6.21)
$$\begin{aligned} x(t) = x_I(t) * y(t) &= \int_{-\infty}^{\infty} x_I(t - \lambda)\, y(\lambda)\, d\lambda \\ &= \int_{-\infty}^{\infty} \left[\sum_{\ell=-\infty}^{\infty} x(\ell\triangle t)\, \delta(t - \lambda - \ell\triangle t) \right] \text{sinc}(F\lambda)\, d\lambda \\ &= \sum_{\ell=-\infty}^{\infty} x(\ell\triangle t) \left[\int_{-\infty}^{\infty} \text{sinc}(F\lambda)\, \delta(-\lambda + t - \ell\triangle t)\, d\lambda \right] \\ &= \sum_{\ell=-\infty}^{\infty} x(\ell\triangle t) \left[\int_{-\infty}^{\infty} \text{sinc}(F\lambda)\, \delta(\lambda - (t - \ell\triangle t))\, d\lambda \right] \\ &= \sum_{\ell=-\infty}^{\infty} x(\ell\triangle t)\, \text{sinc}(F(t - \ell\triangle t)) \quad \text{(by sifting property)} \\ &= \sum_{\ell=-\infty}^{\infty} x(\ell\triangle t)\, \frac{\sin(\pi F(t - \ell\triangle t))}{\pi F(t - \ell\triangle t)}. \end{aligned}$$

∎

6.10   Sampling of Band-Pass Signals

A signal $x(t)$ is called a band-pass signal if $X(f) \neq 0$ for $f \in [f_1, f_2] \cup [-f_2, -f_1]$, where $0 < f_1 < f_2$. (Examples of band-pass signals include radar signals and AM and FM radio signals.) On the one hand, we note that because $X(f) = 0$ outside the band $[-f_2, f_2]$, the signal $x(t)$ is band limited to $f_{max} = f_2$, and we can simply treat the band-pass signal as a band-limited signal and sample it at the Nyquist rate $1/\triangle t = 2f_2$.

On the other hand, we have $X(f) = 0$ for $f \in [-f_1, f_1] \subset [-f_2, f_2]$, so the bandwidth of a band-pass signal can be de ned as $2(f_2 - f_1)$ (for the two nonzero bands). If the bandwidth $2(f_2 - f_1)$ is a small fraction of $2f_2$, we would like to know whether a lower sampling rate determined by the bandwidth $2(f_2 - f_1)$ can allow full recovery of $X(f)$ from $X_I(f)$ in Theorem 6.11.

To avoid complication, we consider rst the case when $f_2 = m(f_2 - f_1)$, where m is a positive integer. If we lower the sampling rate to

$$(6.22) \qquad \frac{1}{\triangle t} = 2(f_2 - f_1) = \frac{2f_2}{m},$$

the Fourier transform of the impulse-sampled signal is given by Theorem 6.11 as

$$(6.23) \qquad X_I(f) = 2(f_2 - f_1) \sum_{k=-\infty}^{\infty} X\big(f - 2k(f_2 - f_1)\big).$$

The right-hand side is again a superposition of copies of $X(f) = \mathcal{F}\{x(t)\}$, each displaced by multiples of $2(f_2 - f_1)$. To show that the nonzero portions of the shifted replications of $X(f)$ will not overlap, we need to determine, for every k, the ranges of f over which $X\big(f - 2k(f_2 - f_1)\big) \neq 0$. To simplify the notation, we let $\beta = f_2 - f_1$, $f_2 = m\beta$, and note that $f_1 = f_2 - \beta = (m - 1)\beta$.

1. For the right band, $X(f - 2k\beta) \neq 0$ if $f - 2k\beta \in [f_1, f_2]$; we thus require

$$(m - 1)\beta \leq f - 2k\beta \leq m\beta.$$

Solving for f, we obtain

$$(2k + m - 1)\beta \leq f \leq (2k + m)\beta.$$

Therefore, the kth nonzero right band of width β begins at $(2k + m - 1)\beta$.

2. For the left band, $X(f - 2k\beta) \neq 0$ if $f - 2k\beta \in [-f_2, -f_1]$, we thus require

$$-m\beta \leq f - 2k\beta \leq -(m - 1)\beta.$$

Solving for f, we obtain

$$(2k - m)\beta \leq f \leq (2k - m + 1)\beta.$$

Therefore, the kth nonzero left band of width β begins at $(2k - m)\beta$.

If m is even, then $(2k + m - 1)$ is an odd number for all k, and $(2k - m)$ is an even number for all k, so the nonzero bands cannot overlap; if m is odd, we have even $(2k + m - 1)$ and odd $(2k - m)$ for all k, so the nonzero bands cannot overlap, either.

For the general case, instead of assuming $f_2 = m(f_2 - f_1)$, we can always nd $f_0 \leq f_1$ so that $f_2 = m(f_2 - f_0)$, and $m \geq 1$ is an integer. Our derivation above shows that if we set the sampling rate $1/\triangle t = 2(f_2 - f_0)$, then the portions of $X(f)$ over $[f_0, f_2]$ and $[-f_2, -f_0]$ will not overlap the corresponding portions of the shifted replications of $X(f)$. Since $[f_1, f_2] \subset [f_0, f_2]$, the nonzero bands will not overlap each other. As a result, the nonzero frequency content of a band-pass signal is preserved by the sampling process as desired.

References

1. A. Ambardar. *Analog and Digital Signal Processing*. Brooks/Cole Publishing Company, Paci c Grove, CA, second edition, 1999.

2. W. L. Briggs and V. E. Hensen. *The DFT: An Owner's Manual for the Discrete Fourier Transform*. The Society for Industrial and Applied Mathematics, Philadelphia, PA, 1995.

3. E. O. Brigham. *The Fast Fourier Transform and Its Applications*. Prentice-Hall, Inc., Upper Saddle River, NJ, 1988.

4. B. Porat. *A Course in Digital Signal Processing*. John Wiley & Sons, Inc., New York, 1997.

5. W. H. Press, S. A. Teukolsky, W. T. Vetterling, and B. P. Flannery. *Numerical Recipes in C++: The Art of Scientific Computing*. Cambridge University Press, Cambridge, UK, second edition, 2001.

6. H. J. Weaver. *Applications of Discrete and Continuous Fourier Analysis*. John Wiley & Sons, Inc., New York, 1983.

Chapter 7

The Fourier Transform of a Sequence

In Chapter 6 we were able to treat a discrete-time signal as a formally continuous function by representing the sampled signal as a weighted impulse train

$$(7.1) \qquad x_I(t) = x(t)P_{\triangle t}(t) = \sum_{\ell=-\infty}^{\infty} x(t)\, \delta(t - \ell\triangle t) = \sum_{\ell=-\infty}^{\infty} x(\ell\triangle t)\, \delta(t - \ell\triangle t),$$

and we derived its Fourier transform in two forms (recall Theorem 6.11 and Corollary 6.13):

$$(7.2) \qquad X_I(f) = \mathcal{F}\{x_I(t)\} = \underbrace{\sum_{\ell=-\infty}^{\infty} x(\ell\triangle t)\, e^{-j2\pi f\ell\triangle t}}_{\text{Fourier series of } X_I(f)} = \underbrace{\frac{1}{\triangle t}\sum_{k=-\infty}^{\infty} X\left(f - \frac{k}{\triangle t}\right)}_{\mathcal{F}\{x(t)\}*\mathcal{F}\{P_{\triangle t}(t)\}},$$

where $X(f) = \mathcal{F}\{x(t)\}$. Note that $X_I(f + 1/\triangle t) = X_I(f)$, so $X_I(f)$ is a periodic function with period equal to the sampling rate $\mathbb{R} = 1/\triangle t$. Recall that if $x(t)$ is band-limited with bandwidth $F \leq \mathbb{R}$, then we may extract $X(f) = \mathcal{F}\{x(t)\}$ from the central period of $X_I(f)$; otherwise the shifted replicas of $X(f)$ will overlap, and $X_I(f) \neq X(f)$ for $f \in [-\mathbb{R}/2,\ \mathbb{R}/2]$.

As we indicated in Chapter 6, we used the impulse train $P_{\triangle t}(t)$ (also known as the comb function) as a mathematical tool to model the sampling process, but we cannot physically sample a signal by the impulse train. In order to process the point-sampled signal data in the digital world, we need to formally de ne the Fourier transform on a sequence of discrete-time samples *without* explicitly involving the sampling interval $\triangle t$. Such a de n ition can be derived from relating $x(\ell\triangle t)$ in (7.2) to the Fourier series coef cient of the periodic $X_I(f)$ as shown in the next section.

7.1 Deriving the Fourier Transform of a Sequence

To derive the Fourier transform of a sequence $\{\ldots, x_{-1}, x_0, x_1, \ldots\}$, we denote $x(\ell\triangle t)$ in Formula (7.2) by x_ℓ and we express $X_I(f)$ as a function of the digital frequency $\mathbb{F} = f\triangle t$ (cycles per sample) by a simple change of variable in Formula (7.2):

$$(7.3) \qquad \hat{X}_I(\mathbb{F}) = \sum_{\ell=-\infty}^{\infty} x_\ell\, e^{-j2\pi \ell\mathbb{F}}.$$

Since the period of $X_I(f)$ is equal to the sampling rate $\mathbb{R} = 1/\triangle t$, the period of $\hat{X}_I(\mathbb{F})$ is expected to be $\mathbb{R}\triangle t = 1$ after the change of variable. We can easily verify the period of $\hat{X}_I(\mathbb{F})$ by showing that

$$\hat{X}_I(\mathbb{F} \pm 1) = \sum_{\ell=-\infty}^{\infty} x_\ell \, e^{-j2\pi\ell(\mathbb{F}\pm 1)} = \sum_{\ell=-\infty}^{\infty} x_\ell \, e^{-j2\pi\ell\mathbb{F}} = \hat{X}_I(\mathbb{F}).$$

Hence the central period of $\hat{X}_I(\mathbb{F})$ is $[-1/2, \, 1/2]$. Because $\hat{X}_I(\mathbb{F})$ is a periodic function (with period $\Omega = 1$), we can also interpret the right-hand side of Equation (7.3) as its Fourier series expansion, namely,

$$
\begin{aligned}
\hat{X}_I(\mathbb{F}) &= \sum_{\ell=-\infty}^{\infty} x_\ell \, e^{-j2\pi\ell\mathbb{F}} \\
(7.4) \qquad &= \sum_{\ell=-\infty}^{\infty} x_{-\ell} \, e^{-j2\pi(-\ell)\mathbb{F}} \\
&= \sum_{\ell=-\infty}^{\infty} \hat{c}_\ell \, e^{j2\pi\ell\mathbb{F}/\Omega}, \qquad \text{(rename } x_{-\ell} \text{ as } \hat{c}_\ell, \text{ recall } \Omega = 1)
\end{aligned}
$$

where the Fourier series coef cients $\{\hat{c}_\ell\}$ are de n ed in Dirichlet s theorem 3.1 by

$$(7.5) \qquad \hat{c}_\ell = \frac{1}{\Omega} \int_{-\Omega/2}^{\Omega/2} \hat{X}_I(\mathbb{F}) \, e^{-j2\pi\ell\mathbb{F}/\Omega} d\mathbb{F}, \quad \Omega = 1, \quad \ell \in (-\infty, \, \infty).$$

Because $x_\ell = \hat{c}_{-\ell}$, we obtain

$$(7.6) \qquad x_\ell = \frac{1}{\Omega} \int_{-\Omega/2}^{\Omega/2} \hat{X}_I(\mathbb{F}) \, e^{-j2\pi(-\ell)\mathbb{F}/\Omega} d\mathbb{F} = \int_{-1/2}^{1/2} \hat{X}_I(\mathbb{F}) \, e^{j2\pi\ell\mathbb{F}} d\mathbb{F}.$$

We can now extract the following Fourier transform pair from Formulas (7.3) and (7.6):

$$(7.7) \qquad \boxed{\hat{X}_I(\mathbb{F}) = \underbrace{\sum_{\ell=-\infty}^{\infty} x_\ell \, e^{-j2\pi\ell\mathbb{F}}}_{\text{Fourier series expansion}} \Longleftrightarrow x_\ell = \underbrace{\int_{-1/2}^{1/2} \hat{X}_I(\mathbb{F}) \, e^{j2\pi\ell\mathbb{F}} d\mathbb{F}}_{\text{Fourier series coef cient } \hat{c}_{-\ell}}.}$$

Because the continuous $\hat{X}_I(\mathbb{F})$ is periodic and it is formally the Fourier transform of equis-paced discrete-time samples $\{\ldots, x_{-1}, x_0, x_1, \ldots\}$, it is said that sampling in the time domain leads to periodicity in the frequency domain.

It is also common to express the Fourier transform pair in the angular digital frequency $\theta = 2\pi\mathbb{F}$ (radians per sample), which is obtained by changing the variable in Formula (7.7):

$$(7.8) \qquad \boxed{\tilde{X}_I(\theta) = \sum_{\ell=-\infty}^{\infty} x_\ell \, e^{-j\ell\theta} \Longleftrightarrow x_\ell = \frac{1}{2\pi} \int_{-\pi}^{\pi} \tilde{X}_I(\theta) \, e^{j\ell\theta} d\theta.}$$

Note that we have used θ instead of the previously proposed $\mathbb{W}$ to denote the angular digital frequency in (7.8), and that we have $\tilde{X}(\theta + 2\pi) = \tilde{X}(\theta)$.

Remark 1: By changing the physical frequency variable f to the digital frequency $\mathbb{F} = f\triangle t$, we were able to focus on transforming the data sequence $\{x_\ell\}$ itself and present a derivation which is more direct and clear than using $X_I(f)$. Nevertheless, the mathematically equivalent pair derived from $X_I(f)$ can now be obtained directly from the pair $\hat{X}_I(\mathbb{F}) \Longleftrightarrow \{x_\ell\}$ in (7.7) by changing variable $\mathbb{F}$ back to f th e following result is immediately obtained:

$$(7.9) \quad \boxed{X_I(f) = \sum_{\ell=-\infty}^{\infty} x(\ell\triangle t)e^{-j2\pi \ell f \triangle t} \Longleftrightarrow x(\ell\triangle t) = \triangle t \int_{-1/(2\triangle t)}^{1/(2\triangle t)} X_I(f)e^{j2\pi \ell f \triangle t} df.}$$

Recall that the second form of $X_I(f)$ in (7.2) was obtained in Theorem 6.11. With the relationship (7.9) now established between $X_I(f)$ and $\{x(\ell\triangle t)\}$, we can verify the second form directly from evaluating $x(t) = \mathcal{F}^{-1}\{X(f)\}$ at $t = \ell\triangle t$: we rst express $x(t)$ in the integral form

$$x(t) = \int_{-\infty}^{\infty} X(f)\, e^{j2\pi ft} df,$$

then we evaluate the integral at $t = \ell\triangle t$ for all $\ell \in (-\infty, \infty)$:

$$
\begin{aligned}
x(\ell\triangle t) &= \int_{-\infty}^{\infty} X(f)\, e^{j2\pi f \ell \triangle t}\, df \\
&= \sum_{k=-\infty}^{\infty} \int_{(-k-\frac{1}{2})\mathbb{R}}^{(-k+\frac{1}{2})\mathbb{R}} X(f)\, e^{j2\pi f \ell \triangle t}\, df \\
&= \sum_{k=-\infty}^{\infty} \int_{-\frac{1}{2}\mathbb{R}}^{\frac{1}{2}\mathbb{R}} X(\lambda - k\mathbb{R})\, e^{j2\pi(\lambda - k\mathbb{R})\ell \triangle t}\, d\lambda \qquad (\text{let } \lambda = f + k\mathbb{R}) \\
&= \sum_{k=-\infty}^{\infty} \int_{-1/(2\triangle t)}^{1/(2\triangle t)} X\left(\lambda - \frac{k}{\triangle t}\right) e^{j2\pi \lambda \ell \triangle t}\, e^{-j2\pi k\ell}\, d\lambda \qquad (\text{let } \mathbb{R} = 1/\triangle t) \\
&= \sum_{k=-\infty}^{\infty} \int_{-1/(2\triangle t)}^{(1/(2\triangle t)} X\left(f - \frac{k}{\triangle t}\right) e^{j2\pi f \ell \triangle t}\, df \qquad (\because e^{-j2\pi k\ell} = 1; \ \text{let } f = \lambda) \\
&= \triangle t \int_{-1/(2\triangle t)}^{1/(2\triangle t)} \underbrace{\left[\frac{1}{\triangle t} \sum_{k=-\infty}^{\infty} X\left(f - \frac{k}{\triangle t}\right)\right]}_{\text{extract the integrand}} e^{j2\pi \ell f \triangle t}\, df.
\end{aligned}
$$

$$\because x(\ell\triangle t) = \triangle t \int_{-1/(2\triangle t)}^{1/(2\triangle t)} \overbrace{\Big[\, X_I(f)\, \Big]}^{\text{extract the integrand}} e^{j2\pi \ell f \triangle t}\, df \quad \text{from (7.9) for all } \ell \in (-\infty, \infty),$$

$$\therefore X_I(f) = \frac{1}{\triangle t} \sum_{k=-\infty}^{\infty} X\left(f - \frac{k}{\triangle t}\right).$$

We have thus derived the same result previously given in Theorem 6.11 by taking an entirely different path.

Remark 2: We have used different function names X_I, $\hat{X}_I$, and $\tilde{X}_I$ to denote the three forms of the same function this is necessary, because if we don t change the function name, we can only explicitly *substitute* the variable of $X_I(f)$ by expressing

$$X_I(f) = X_I\left(\frac{\mathbb{F}}{\triangle t}\right) = X_I\left(\frac{\theta}{2\pi\triangle t}\right), \quad \theta = 2\pi\mathbb{F}.$$

Therefore, to be mathematically correct, we must refer to the result expressed in each new variable by a different name; i.e., we de ne

$$\hat{X}_I(\mathbb{F}) = X_I\left(\frac{\mathbb{F}}{\triangle t}\right),$$

and

$$\tilde{X}_I(\theta) = X_I\left(\frac{\theta}{2\pi\triangle t}\right).$$

Of course, the fact that they represent the same function does not change, and we have

$$X_I(f) = \hat{X}_I(\mathbb{F}) = \tilde{X}_I(\theta).$$

Remark 3: We emphasize the mathematical equivalence of the following formulas:

$$X_I(f \pm 1/\triangle t) = \hat{X}_I(\mathbb{F} \pm 1) = \tilde{X}_I(\theta \pm 2\pi),$$

where $\theta = 2\pi\mathbb{F} = 2\pi f\triangle t;$ and

$$x(\ell\triangle t) = \triangle t \int_{-1/(2\triangle t)}^{1/(2\triangle t)} X_I(f)\, e^{j2\pi\ell f\triangle t} df$$

$$= \int_{-1/2}^{1/2} \hat{X}_I(\mathbb{F})\, e^{j2\pi\ell\mathbb{F}} d\mathbb{F}$$

$$= \frac{1}{2\pi} \int_{-\pi}^{\pi} \tilde{X}_I(\theta)\, e^{j\ell\theta} d\theta = x_\ell, \quad \ell \in (-\infty, \infty).$$

Remark 4: We have shown that sampling $x(t)$ in the time domain leads to periodic replication of its Fourier transform $X(f) = \mathcal{F}\{x(t)\}$ in the frequency domain, which was described by the second form of $X_I(f)$ in (7.2):

$$X_I(f) = \underbrace{\frac{1}{\triangle t} \sum_{k=-\infty}^{\infty} X\left(f - \frac{k}{\triangle t}\right)}_{\text{shifted (scaled) replicas of } X(f)}.$$

Hence discrete-time signals have continuous periodic spectra $X_I(f)$, and the sampling space $\triangle t$ in the time domain is the reciprocal of the period of $X_I(f)$ in the frequency domain.

Remark 5: We emphasize that the central period of $X_I(f)$, which covers frequencies $f \in [-\mathbb{R}/2, \mathbb{R}/2]$, is *not* required to be equal to $X(f) = \mathcal{F}\{x(t)\}$ r ecall that the shifted replicas of $X(f)$ in the periodic replication described by (7.2) overlap each other if the the sampling rate $\mathbb{R} = 1/\triangle t$ does not exceed the bandwidth of $x(t)$. (Recall Figures 6.9 and 6.10 as well as our discussion on the Nyquist rate in Section 6.8, Chapter 6.)

Table 7.1 Properties of the Fourier transform $\hat{X}_I(\mathbb{F})$ of a sequence.

1. Linearity	$\mathcal{F}\{\alpha\{x_\ell\} + \beta\{y_\ell\}\} = \alpha\hat{X}_I(\mathbb{F}) + \beta\hat{Y}_I(\mathbb{F})$
2. Time shift	$\mathcal{F}\{\{x_{\ell-m}\}\} = \hat{X}_I(\mathbb{F})\,e^{-j2\pi\mathbb{F}m}$
3. Frequency shift	$\mathcal{F}\{\{x_\ell\,e^{j2\pi\mathbb{F}_\alpha\ell}\}\} = \hat{X}_I(\mathbb{F} - \mathbb{F}_\alpha)$
Special case $\mathbb{F}_\alpha = \frac{1}{2}$	$\mathcal{F}\{\{x_\ell\,(-1)^\ell\}\} = \hat{X}_I\left(\mathbb{F} - \frac{1}{2}\right)$
4. Modulation	$\mathcal{F}\{\{x_\ell\,\cos 2\pi\mathbb{F}_\alpha\ell\}\} = \frac{1}{2}\hat{X}_I(\mathbb{F} + \mathbb{F}_\alpha) + \frac{1}{2}\hat{X}_I(\mathbb{F} - \mathbb{F}_\alpha)$
	$\mathcal{F}\{\{x_\ell\,\sin 2\pi\mathbb{F}_\alpha\ell\}\} = \frac{j}{2}\hat{X}_I(\mathbb{F} + \mathbb{F}_\alpha) - \frac{j}{2}\hat{X}_I(\mathbb{F} - \mathbb{F}_\alpha)$
5. Folding	$\mathcal{F}\{\{x_{-\ell}\}\} = \hat{X}_I(-\mathbb{F})$
6. Derivative of the transform	$\hat{X}_I'(\mathbb{F}) = -j2\pi\mathcal{F}\{\{\ell\,x_\ell\}\}$

7.2 Properties of the Fourier Transform of a Sequence

A subset of the Fourier transform properties (for continuous-time signals) in Table 5.3 is directly adapted for discrete-time signals in Table 7.1, and their derivation from the de n ing formula for $\hat{X}_I(\mathbb{F})$ follows.

1. Linearity

$$\mathcal{F}\{\alpha\{x_\ell\} + \beta\{y_\ell\}\} = \mathcal{F}\{\{\alpha\,x_\ell + \beta\,y_\ell\}\}$$

$$= \sum_{\ell=-\infty}^{\infty} (\alpha\,x_\ell + \beta\,y_\ell)\,e^{-j2\pi\mathbb{F}\ell} \qquad \text{by Formula (7.3)}$$

$$= \alpha \sum_{\ell=-\infty}^{\infty} x_\ell\,e^{-j2\pi\mathbb{F}\ell} + \beta \sum_{\ell=-\infty}^{\infty} y_\ell\,e^{-j2\pi\mathbb{F}\ell}$$

$$= \alpha\,\hat{X}_I(\mathbb{F}) + \beta\,\hat{Y}_I(\mathbb{F}). \qquad \text{by Formula (7.3)}$$

2. The Time-Shift Property

$$\mathcal{F}\{\{x_{\ell-m}\}\} = \sum_{\ell=-\infty}^{\infty} x_{\ell-m}\,e^{-j2\pi\mathbb{F}\ell} \qquad \text{by Formula (7.3)}$$

$$= \sum_{k=-\infty}^{\infty} x_k\,e^{-j2\pi\mathbb{F}(k+m)} \qquad (\text{let } k = \ell - m)$$

$$= \left[\sum_{k=-\infty}^{\infty} x_k\,e^{-j2\pi\mathbb{F}k}\right] e^{-j2\pi\mathbb{F}m}$$

$$= \hat{X}_I(\mathbb{F})\,e^{-j2\pi\mathbb{F}m}. \qquad \text{by Formula (7.3)}$$

3. The Frequency-Shift Property

$$\mathcal{F}\{\{x_\ell\, e^{j2\pi\mathbb{F}_\alpha\ell}\}\} = \sum_{\ell=-\infty}^{\infty} x_\ell\, e^{j2\pi\mathbb{F}_\alpha\ell}\, e^{-j2\pi\mathbb{F}\ell} \quad \text{by Formula (7.3)}$$

$$= \sum_{\ell=-\infty}^{\infty} x_\ell\, e^{-j2\pi(\mathbb{F}-\mathbb{F}_\alpha)\ell}$$

$$= \hat{X}_I\big(\mathbb{F}-\mathbb{F}_\alpha\big). \qquad\qquad \text{by Formula (7.3)}$$

Observe that when $\mathbb{F}_\alpha = \frac{1}{2}$, the frequency shift $e^{j2\pi\mathbb{F}_\alpha\ell} = e^{j\pi\ell} = (-1)^\ell$, and we have $\mathcal{F}\{\{x_\ell\,(-1)^\ell\}\}$ on the left-hand side and $\hat{X}_I\big(\mathbb{F}-\frac{1}{2}\big)$ on the right-hand side for this special case.

4. The Modulation Property

$$\mathcal{F}\{\{x_\ell\,\cos 2\pi\mathbb{F}_\alpha\ell\}\} = \sum_{\ell=-\infty}^{\infty} \big(x_\ell\,\cos 2\pi\mathbb{F}_\alpha\ell\big)\, e^{-j2\pi\mathbb{F}\ell} \quad \text{by Formula (7.3)}$$

$$= \frac{1}{2}\sum_{\ell=-\infty}^{\infty} x_\ell\left[e^{-j2\pi\mathbb{F}_\alpha\ell} + e^{j2\pi\mathbb{F}_\alpha\ell}\right]e^{-j2\pi\mathbb{F}\ell} \text{ (by Euler s formula)}$$

$$= \frac{1}{2}\sum_{\ell=-\infty}^{\infty} x_\ell\left[e^{-j2\pi(\mathbb{F}+\mathbb{F}_\alpha)\ell} + e^{-j2\pi(\mathbb{F}-\mathbb{F}_\alpha)\ell}\right]$$

$$= \frac{1}{2}\sum_{\ell=-\infty}^{\infty} x_\ell\, e^{-j2\pi(\mathbb{F}+\mathbb{F}_\alpha)\ell} + \frac{1}{2}\sum_{\ell=-\infty}^{\infty} x_\ell\, e^{-j2\pi(\mathbb{F}-\mathbb{F}_\alpha)\ell}$$

$$= \frac{1}{2}\hat{X}_I\big(\mathbb{F}+\mathbb{F}_\alpha\big) + \frac{1}{2}\hat{X}_I\big(\mathbb{F}-\mathbb{F}_\alpha\big).$$

Similarly,

$$\mathcal{F}\{\{x_\ell\,\sin 2\pi\mathbb{F}_\alpha\ell\}\} = \frac{j}{2}\hat{X}_I\big(\mathbb{F}+\mathbb{F}_\alpha\big) - \frac{j}{2}\hat{X}_I\big(\mathbb{F}-\mathbb{F}_\alpha\big).$$

5. The Folding Property

$$\mathcal{F}\{\{x_{-\ell}\}\} = \sum_{\ell=-\infty}^{\infty} x_{-\ell}\, e^{-j2\pi\mathbb{F}\ell} \qquad \text{by Formula (7.3)}$$

$$= \sum_{k=-\infty}^{\infty} x_k\, e^{-j2\pi\mathbb{F}(-k)} \quad (\text{let } k=-\ell)$$

$$= \sum_{k=-\infty}^{\infty} x_k\, e^{-j2\pi(-\mathbb{F})k}$$

$$= \hat{X}_I(-\mathbb{F}).$$

6. Derivative of the Transform

$$\hat{X}'_I(\mathbb{F}) = \frac{d}{d\mathbb{F}} \left[\sum_{\ell=-\infty}^{\infty} x_\ell \, e^{-j2\pi\mathbb{F}\ell} \right]$$

$$= \sum_{\ell=-\infty}^{\infty} x_\ell \, \frac{d}{d\mathbb{F}} \, e^{-j2\pi\mathbb{F}\ell}$$

$$= \sum_{\ell=-\infty}^{\infty} x_\ell \, (-j2\pi\ell) \, e^{-j2\pi\mathbb{F}\ell}$$

$$= -j2\pi \sum_{\ell=-\infty}^{\infty} (\ell\, x_\ell) \, e^{-j2\pi\mathbb{F}\ell}$$

$$= -j2\pi \mathcal{F}\{\{\ell\, x_\ell\}\}.$$

In Table 7.2 we list the same properties in terms of the alternate form $\tilde{X}_I(\theta)$ de ned by (7.8), where $\theta = 2\pi\mathbb{F}$.

Table 7.2 Properties of the Fourier transform $\tilde{X}_I(\theta)$ of a sequence ($\theta = 2\pi\mathbb{F}$).

1. Linearity	$\mathcal{F}\{\alpha\{x_\ell\} + \beta\{y_\ell\}\} = \alpha\tilde{X}_I(\theta) + \beta\tilde{Y}_I(\theta)$
2. Time shift	$\mathcal{F}\{\{x_{\ell-m}\}\} = \tilde{X}_I(\theta) \, e^{-jm\theta}$
3. Frequency shift	$\mathcal{F}\{\{x_\ell \, e^{j\ell\theta_\alpha}\}\} = \tilde{X}_I(\theta - \theta_\alpha)$
Special case $\theta_\alpha = \pi$	$\mathcal{F}\{\{x_\ell \, (-1)^\ell\}\} = \tilde{X}_I(\theta - \pi)$
4. Modulation	$\mathcal{F}\{\{x_\ell \, \cos \ell\theta_\alpha\}\} = \frac{1}{2}\tilde{X}_I(\theta + \theta_\alpha) + \frac{1}{2}\tilde{X}_I(\theta - \theta_\alpha)$
	$\mathcal{F}\{\{x_\ell \, \sin \ell\theta_\alpha\}\} = \frac{j}{2}\tilde{X}_I(\theta + \theta_\alpha) - \frac{j}{2}\tilde{X}_I(\theta - \theta_\alpha)$
5. Folding	$\mathcal{F}\{\{x_{-\ell}\}\} = \tilde{X}_I(-\theta)$
6. Derivative of the transform	$\tilde{X}'_I(\theta) = -j\mathcal{F}\{\{\ell\, x_\ell\}\}$

7.3 Generating the Fourier Transform Pairs

7.3.1 The Kronecker delta sequence

The counterpart of the impulse function for discrete-time signals is the Kronecker delta sequence $\{z_\ell = \delta(\ell)\}$, where

$$\delta(\ell) = \begin{cases} 1 & \text{if } \ell = 0, \\ 0 & \text{if } \ell \neq 0. \end{cases}$$

To obtain the Fourier transform of the Kronecker delta sequence, we apply the defining formula to obtain

$$(7.10) \qquad \hat{Z}_I(\mathbb{F}) = \sum_{\ell=-\infty}^{\infty} z_\ell \, e^{-j2\pi\ell\mathbb{F}} = \sum_{\ell=-\infty}^{\infty} \delta(\ell) \, e^{-j2\pi\ell\mathbb{F}} = \delta(0) = 1.$$

We denote a shifted Kronecker delta sequence by $\{z_{\ell-m} = \delta(\ell - m)\}$, where

$$\delta(\ell - m) = \begin{cases} 1 & \text{if } m = \ell, \\ 0 & \text{if } m \neq \ell, \end{cases}$$

and we obtain its Fourier transform using the time-shift property:

$$\{z_{\ell-m} = \delta(\ell - m)\} \Longleftrightarrow \hat{Z}_I(\mathbb{F})e^{-j2\pi\mathbb{F}m} = e^{-j2\pi\mathbb{F}m}.$$

7.3.2   Representing signals by Kronecker delta

In analogy to the impulse train, we may use the *weighted* sum of the shifted Kronecker delta to represent a discrete-time signal:

$$x_\ell = \sum_{m=-\infty}^{\infty} x_m \, \delta(\ell - m), \quad \ell \in (-\infty, \infty).$$

For each specified value of ℓ, the right-hand side has only one nonzero term $x_\ell \, \delta(0)$ corresponding to $m = \ell$.

Example 7.1 We denote the discrete unit step function by $\{z_\ell = u(\ell)\}$, where

$$(7.11) \qquad u(\ell) = \begin{cases} 1 & \text{if } \ell \geq 0, \\ 0 & \text{if } \ell < 0. \end{cases}$$

The definition is satisfied by the sum of the shifted Kronecker delta

$$(7.12) \qquad z_\ell = \sum_{m=0}^{\infty} \delta(\ell - m), \quad \ell \in (-\infty, \infty);$$

or the equally valid

$$(7.13) \qquad z_\ell = \sum_{m=-\infty}^{\ell} \delta(m), \quad \ell \in (-\infty, \infty).$$

Example 7.2 We denote the discrete unit ramp function by $\{z_\ell = r(\ell)\}$, where

$$(7.14) \qquad r(\ell) = \begin{cases} \ell & \text{if } \ell \geq 0, \\ 0 & \text{if } \ell < 0. \end{cases}$$

The definition is satisfied by the weighted sum of Kronecker delta expression

$$(7.15) \qquad z_\ell = \sum_{m=0}^{\infty} \ell \, \delta(\ell - m), \quad \ell \in (-\infty, \infty),$$

or the equally valid

$$(7.16) \qquad z_\ell = \sum_{m=-\infty}^{\ell} \ell\, \delta(m), \quad \ell \in (-\infty, \infty).$$

The discrete unit ramp can also be expressed as a sum of the discrete unit step as

$$(7.17) \qquad z_\ell = \sum_{m=-\infty}^{\ell} u(m), \quad \ell \in (-\infty, \infty).$$

7.3.3 Fourier transform pairs

Example 7.3 (a) (Figure 7.1) We denote the discrete exponential function by $\{z_\ell = \alpha^\ell u(\ell)\}$, where $|\alpha| < 1$, and we obtain its Fourier transform by the de ning formula

$$(7.18) \quad
\begin{aligned}
\hat{Z}_I(\mathbb{F}) &= \sum_{\ell=-\infty}^{\infty} z_\ell\, e^{-j2\pi\ell\mathbb{F}} = \sum_{\ell=-\infty}^{\infty} \alpha^\ell u(\ell)\, e^{-j2\pi\ell\mathbb{F}} \\
&= \sum_{\ell=0}^{\infty} \alpha^\ell\, e^{-j2\pi\ell\mathbb{F}} \qquad (\because u(\ell) = 0,\ \ell < 0) \\
&= \sum_{\ell=0}^{\infty} \left[\alpha\, e^{-j2\pi\mathbb{F}}\right]^\ell \qquad \left(\text{note } \left|\alpha\, e^{-j2\pi\mathbb{F}}\right| < 1\right) \\
&= \frac{1}{1 - \alpha\, e^{-j2\pi\mathbb{F}}}. \qquad \text{(sum of geometric series)}
\end{aligned}$$

(b) (Figure 7.2) We obtain the Fourier transform of $\{y_\ell = \ell\, \alpha^\ell u(\ell)\}$, where $|\alpha| < 1$, by relating it to the result from part **(a)** through the derivative of transform property

$$\boxed{\ \frac{d}{d\mathbb{F}} \hat{Z}_I(\mathbb{F}) = -j2\pi \mathcal{F}\{\{\ell\, z_\ell\}\} \ \text{ or } \ \mathcal{F}\{\{\ell\, z_\ell\}\} = \frac{j}{2\pi} \frac{d}{d\mathbb{F}} \hat{Z}_I(\mathbb{F}).\ }$$

Letting $z_\ell = \alpha^\ell u(\ell)$, we have $y_\ell = \ell\, z_\ell$; hence,

$$
\begin{aligned}
\hat{Y}_I(\mathbb{F}) &= \mathcal{F}\{\{y_\ell\}\} = \mathcal{F}\{\{\ell\, z_\ell\}\} \\
&= \frac{j}{2\pi} \frac{d}{d\mathbb{F}} \hat{Z}_I(\mathbb{F}) \\
&= \frac{j}{2\pi} \frac{d}{d\mathbb{F}} \left[\frac{1}{1 - \alpha\, e^{-j2\pi\mathbb{F}}}\right] \qquad \text{(result from part (a))} \\
&= \frac{j}{2\pi} \frac{-j2\pi\alpha\, e^{-j2\pi\mathbb{F}}}{\left[1 - \alpha\, e^{-j2\pi\mathbb{F}}\right]^2} \\
&= \frac{\alpha\, e^{-j2\pi\mathbb{F}}}{\left[1 - \alpha\, e^{-j2\pi\mathbb{F}}\right]^2}.
\end{aligned}
$$

(c) (Figure 7.3) We next make use of the property of linearity to obtain the Fourier transform

Figure 7.1 Discrete exponential function and its Fourier transform.

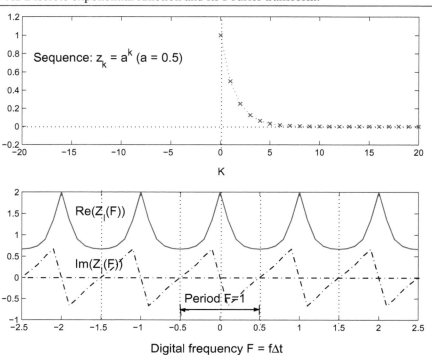

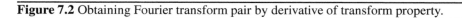

Figure 7.2 Obtaining Fourier transform pair by derivative of transform property.

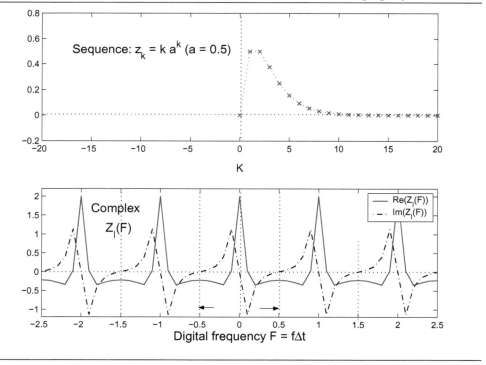

of sequence $\{(\ell + 1)\alpha^\ell u(\ell)\}$ by adding the results from (a) and (b):

$$\mathcal{F}\{\{(\ell + 1)\alpha^\ell u(\ell)\}\} = \mathcal{F}\{\{\ell\,\alpha^\ell u(\ell)\}\} + \mathcal{F}\{\{\alpha^\ell u(\ell)\}\} \quad \text{(by linearity)}$$

$$= \frac{\alpha\,e^{-j2\pi\mathbb{F}}}{\left[1 - \alpha\,e^{-j2\pi\mathbb{F}}\right]^2} + \frac{1}{1 - \alpha\,e^{-j2\pi\mathbb{F}}} \quad \text{(results form (a), (b))}$$

$$= \frac{1}{\left[1 - \alpha\,e^{-j2\pi\mathbb{F}}\right]^2}.$$

Figure 7.3 Obtaining Fourier transform pair by the property of linearity.

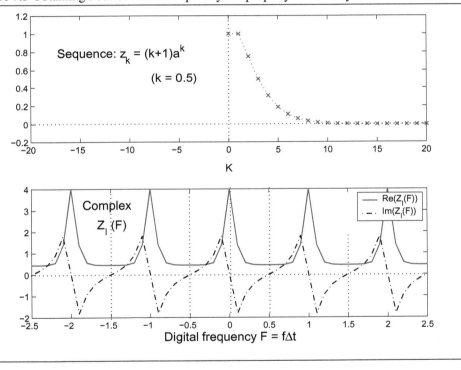

Example 7.4 (Figure 7.4) In this example we evaluate the Fourier transform of a sequence $\{x_\ell\}$ de ned by discrete exponential for $\ell \in (-\infty, \infty)$:

$$x_\ell = \begin{cases} \alpha^\ell & \ell \geq 0 \\ \alpha^{-\ell} & \ell < 0 \end{cases}, \quad \text{where } |\alpha| < 1.$$

Using the technique from Example 7.3(a), we obtain

$$
\begin{aligned}
\mathcal{F}\{\{x_\ell\}\} &= \sum_{\ell=0}^{\infty} \alpha^\ell\, e^{-j2\pi\ell\mathbb{F}} + \sum_{\ell=-\infty}^{-1} \alpha^{-\ell}\, e^{-j2\pi\ell\mathbb{F}} \\
&= \sum_{\ell=0}^{\infty} \left[\alpha\, e^{-j2\pi\mathbb{F}}\right]^\ell + \sum_{k=0}^{\infty} \left[\alpha\, e^{j2\pi\mathbb{F}}\right]^k - 1 \qquad\qquad \text{(let } k=-\ell) \\
&= \frac{1}{1-\alpha\, e^{-j2\pi\mathbb{F}}} + \frac{1}{1-\alpha\, e^{j2\pi\mathbb{F}}} - 1 \qquad \text{(sum the two geometric series)} \\
&= \frac{1-\alpha^2}{\left(1-\alpha\, e^{-j2\pi\mathbb{F}}\right)\left(1-\alpha\, e^{j2\pi\mathbb{F}}\right)} \\
&= \frac{1-\alpha^2}{1-2\alpha\,\cos(2\pi\mathbb{F})+\alpha^2}.
\end{aligned}
$$

Figure 7.4 The Fourier transform of a bilateral exponential function.

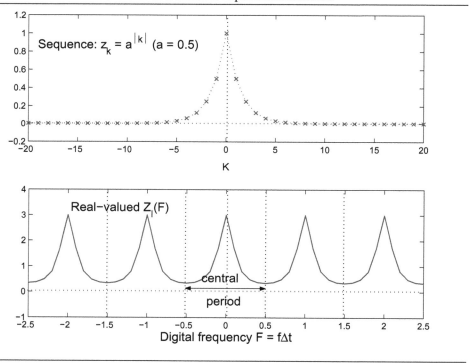

Example 7.5 Recall that given a Fourier transform pair $x(t) \iff X(f)$, we may obtain $X_I(f)$, the periodic Fourier transform of the sample sequence $\{x_\ell\}$, in two forms: (i) $X_I(f)$ may be expressed as a Fourier series with its coefcien ts appropriately de n ed by the data sample; (ii) $X_I(f)$ may be expressed as the sum of the shifted replicas of the known $X(f)$. Therefore, if we can obtain a closed-form expression for $X_I(f)$ in one of the two forms, we would have also found the closed-form expression for the in nite sum in the other form a result which may not be obtained by working with the in nite sum directly. For the pair $x(t) = e^{-at} \iff X(f) = 1/(a + j2\pi f)$, we have shown such results in Example 6.1 (Figures 6.9 and 6.10).

In this example we combine this approach with the use of the derivative of transform property: we will show that for $x(t) = e^{-at}u(t)$, the two forms of $\hat{Y}_I(\theta)$ for $y(t) = t\,x(t)$ can be obtained from the known $X(f)$ and $X_I(f)$. In order to make full use of previously obtained results, we connect relevant examples in a diagram shown in Figure 7.5. The results obtained in those examples are labeled as such and they are displayed in the shaded boxes; the new tasks are those leading to the results displayed in the unshaded boxes.

Figure 7.5 Connecting previously obtained results to new tasks.

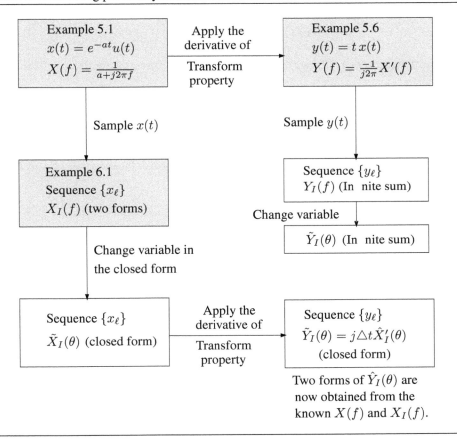

Following the action plan in Figure 7.5, we use the closed-form $X_I(f)$ from Example 6.1 to obtain the closed-form $\hat{X}_I(\theta)$ by changing the variable f to the angular digital frequency θ:

$$\{x_\ell = e^{-a\ell \triangle t}u(\ell)\} \iff \tilde{X}_I(\theta) = \frac{1}{1 - e^{-(a\triangle t + j\theta)}}, \quad \theta = 2\pi f \triangle t.$$

Because $y(t) = t\,x(t)$, we have $y_\ell = (\ell \triangle t)\,x_\ell = \triangle t\,(\ell\,x_\ell)$, and we can use the derivative of the transform property to obtain the closed-form $\tilde{Y}_I(\theta)$:

$$\tilde{Y}_I(\theta) = \triangle t\,\mathcal{F}\{\{\ell\,x_\ell\}\} = j\triangle t\,\tilde{X}_I'(\theta) = j\triangle t\,\frac{d}{d\theta}\left[\frac{1}{1 - e^{-(a\triangle t + j\theta)}}\right] = \frac{\triangle t\,e^{-(a\triangle t + j\theta)}}{\left[1 - e^{-(a\triangle t + j\theta)}\right]^2}.$$

To obtain the other form, we use $Y(f) = \dfrac{1}{(a + j2\pi f)^2}$ from Example 5.6 to express $Y_I(f)$ as an in n ite sum:

$$Y_I(f) = \frac{1}{\triangle t} \sum_{k=-\infty}^{\infty} Y\left(f - \frac{k}{\triangle t}\right) = \sum_{k=-\infty}^{\infty} \frac{\triangle t}{\left[a\triangle t + j2\pi(f\triangle t - k)\right]^2}.$$

Letting $f = \dfrac{\theta}{2\pi\triangle t}$ in $Y_I(f)$, we obtain

$$\tilde{Y}_I(\theta) = \sum_{k=-\infty}^{\infty} \frac{\triangle t}{\left[a\triangle t + j(\theta - 2\pi k)\right]^2}.$$

By equating the two forms of $\tilde{Y}_I(\theta)$, we obtain the closed-form expression for the in nite sum:

$$\boxed{\sum_{k=-\infty}^{\infty} \frac{1}{\left[a\triangle t + j(\theta - 2\pi k)\right]^2} = \frac{e^{-(a\triangle t + j\theta)}}{\left[1 - e^{-(a\triangle t + j\theta)}\right]^2}.}$$

Example 7.6 In this example we shall derive the Fourier transform of the sequence of constant 1, denoted by $\{x_\ell = 1\}$.

Method 1. We apply the de ning formula (7.2) to $x_\ell = x(\ell\triangle t) = 1$, and we obtain

$$\begin{aligned}
X_I(f) &= \sum_{\ell=-\infty}^{\infty} x(\ell\triangle t)\, e^{-j2\pi f\ell\triangle t} \\
&= \sum_{\ell=-\infty}^{\infty} e^{-j2\pi f\ell\triangle t} && (\because x(\ell\triangle t) = 1) \\
&= \sum_{k=-\infty}^{\infty} e^{j2\pi k\ell\triangle t} && (\text{let } k = -\ell) \\
&= \frac{1}{\triangle t} \sum_{k=-\infty}^{\infty} \delta\left(f - \frac{k}{\triangle t}\right) && (\text{by Poisson sum from Theorem 6.6 }) \\
&= \sum_{k=-\infty}^{\infty} \left[\frac{1}{\triangle t}\delta\left(\frac{\mathbb{F} - k}{\triangle t}\right)\right] && (\because f = \mathbb{F}/\triangle t) \\
&= \sum_{k=-\infty}^{\infty} \delta(\mathbb{F} - k). && \left(\because \int_{-\infty}^{\infty}\left[\frac{1}{\alpha}\delta(\frac{\lambda}{\alpha})\right]d\lambda = \int_{-\infty}^{\infty}\delta(\lambda)\,d\lambda\right)
\end{aligned}$$

Accordingly,

$$(7.19) \qquad\qquad \hat{X}_I(\mathbb{F}) = \mathcal{F}\{\{x_\ell = 1\}\} = \sum_{k=-\infty}^{\infty} \delta(\mathbb{F} - k),$$

and $\hat{X}_I(\mathbb{F}) = \delta(\mathbb{F})$ over the principal period $[-\frac{1}{2}, \frac{1}{2}]$. Note that $\hat{X}_I(\mathbb{F})$ is always periodic with unit period.

Method 2. We may use our knowledge about the Dirac delta function to obtain $x_\ell = 1$ by integrating $\delta(\mathbb{F})e^{j2\pi\ell\mathbb{F}}$ over the principal period:

$$x_\ell = \int_{-\frac{1}{2}}^{\frac{1}{2}} \delta(\mathbb{F}) \, e^{j2\pi\ell\mathbb{F}} d\mathbb{F}$$

$$= \int_{-\infty}^{\infty} g(\mathbb{F}) \, \delta(\mathbb{F}-0) \, d\mathbb{F}, \quad \text{where } g(\mathbb{F}) = e^{j2\pi\ell\mathbb{F}} \quad (\text{recall } \delta(\mathbb{F}) = 0, \ \mathbb{F} \neq 0)$$

$$= g(0) \qquad\qquad (\text{by sifting property of } \delta(\mathbb{F} - \mathbb{F}_a) \text{ with } \mathbb{F}_a = 0)$$

$$= 1. \qquad\qquad (\because g(0) = e^0 = 1)$$

Hence $\delta(\mathbb{F})$ plays the role of $\hat{X}_I(\mathbb{F})$ for $\mathbb{F} \in [-\frac{1}{2}, \frac{1}{2}]$ in Formula (7.7). The periodic extension of $\delta(\mathbb{F})$ yields $\hat{X}_I(\mathbb{F})$ with unit period:

$$\hat{X}_I(\mathbb{F}) = \sum_{k=-\infty}^{\infty} \delta(\mathbb{F} - k).$$

To obtain the alternative form $\tilde{X}_I(\theta)$, observe that we must obtain $x_\ell = 1$ from $\tilde{X}_I(\theta)$ using Formula (7.8), namely,

$$x_\ell = \frac{1}{2\pi} \int_{-\pi}^{\pi} \tilde{X}_I(\theta) \, e^{j\ell\theta} d\theta = 1.$$

By recognizing that

$$\frac{1}{2\pi} \int_{-\pi}^{\pi} \left[2\pi \, \delta(\theta) \right] e^{j\ell\theta} d\theta = \int_{-\pi}^{\pi} \delta(\theta) \, e^{j\ell\theta} d\theta = 1,$$

we determine that

$$\tilde{X}_I(\theta) = 2\pi \, \delta(\theta), \quad \theta \in [-\pi, \pi].$$

Since $\tilde{X}_I(\theta)$ is always periodic with period 2π, we have

(7.20) $$\tilde{X}_I(\theta) = \mathcal{F}\{\{x_\ell = 1\}\} = 2\pi \sum_{k=-\infty}^{\infty} \delta(\theta - 2k\pi).$$

Method 3. Since we derived previously the continuous Fourier transform pair $x(t) = 1 \Longleftrightarrow X(f) = \delta(f)$, we can obtain the second form of $X_I(f)$ directly from the given $X(f)$ according to Formula (7.2):

$$X_I(f) = \frac{1}{\Delta t} \sum_{k=-\infty}^{\infty} X\left(f - \frac{k}{\Delta t} \right)$$

$$= \frac{1}{\Delta t} \sum_{k=-\infty}^{\infty} \delta\left(f - \frac{k}{\Delta t} \right) \qquad (\because X(f) = \delta(f))$$

$$= \sum_{k=-\infty}^{\infty} \delta(\mathbb{F} - k). \qquad (\text{by steps identical to Method 1})$$

Remarks: Comparing the results from this example with those of the last three examples, we see that when $\hat{X}_I(\mathbb{F})$ involves an impulse train, the infinite sum is reduced to a single impulse within the principal period $[-\frac{1}{2}, \frac{1}{2}]$. In such case, it is common practice to express $\hat{X}_I(\mathbb{F})$ for the principal period only. Therefore, when the following expression is used,

$$\{x_\ell = 1\} \iff \hat{X}_I(\mathbb{F}) = \delta(\mathbb{F}),$$

it is understood that the values of $\mathbb{F}$ are restricted to the range $[-\frac{1}{2}, \frac{1}{2}]$. Confusion occurs when this practice is adopted without quali cation, because this simpli ed expression is *neither* periodic *nor* valid for $\mathbb{F}$ outside the principal period, while $\hat{X}_I(\mathbb{F})$ was supposed to represent a periodic function valid for all $\mathbb{F} \in (-\infty, \infty)$. Such discrepancy in the de n ition of $\hat{X}_I(\mathbb{F})$ is often tolerated, probably because we usually only need to study or plot one period of a periodic function.

Example 7.7 Using the modulation properties on the pair $\{y_\ell = 1\} \iff \delta(\mathbb{F})$, $\mathbb{F} \in [-\frac{1}{2}, \frac{1}{2}]$, from last example, we obtain

$$(7.21) \qquad \mathcal{F}\{\{y_\ell \cos 2\pi \mathbb{F}_a \ell\}\} = \mathcal{F}\{\{\cos 2\pi \mathbb{F}_a \ell\}\} = \frac{1}{2}\delta(\mathbb{F} + \mathbb{F}_a) + \frac{1}{2}\delta(\mathbb{F} - \mathbb{F}_a),$$

$$(7.22) \qquad \mathcal{F}\{\{y_\ell \sin 2\pi \mathbb{F}_b \ell\}\} = \mathcal{F}\{\{\sin 2\pi \mathbb{F}_b \ell\}\} = \frac{j}{2}\delta(\mathbb{F} + \mathbb{F}_b) - \frac{j}{2}\delta(\mathbb{F} - \mathbb{F}_b).$$

Recall from Section 1.9.1 that a discrete-time sinusoid has the general form

$$x_\ell = D_\alpha \cos(2\pi \mathbb{F}_\alpha \ell - \phi_\alpha), \quad \ell = 0, 1, 2, \dots$$

To obtain its formal Fourier transform, we apply trigonometric identity to express

$$x_\ell = D_\alpha \cos \phi_\alpha \cos(2\pi \mathbb{F}_\alpha \ell) + D_\alpha \sin \phi_\alpha \sin(2\pi \mathbb{F}_\alpha \ell),$$

and, by linearity, we superpose the transform results from (7.21) and (7.22) to obtain

$$\hat{X}_I(\mathbb{F}) = \frac{1}{2}D_\alpha \cos \phi_\alpha \left[\delta(\mathbb{F} + \mathbb{F}_\alpha) + \delta(\mathbb{F} - \mathbb{F}_\alpha)\right] + \frac{j}{2}D_\alpha \sin \phi_\alpha \left[\delta(\mathbb{F} + \mathbb{F}_\alpha) - \delta(\mathbb{F} - \mathbb{F}_\alpha)\right]$$

$$= \frac{1}{2}D_\alpha e^{j\phi_\alpha} \delta(\mathbb{F} + \mathbb{F}_\alpha) + \frac{1}{2}D_\alpha e^{-j\phi_\alpha} \delta(\mathbb{F} - \mathbb{F}_\alpha) \qquad (\because \cos \theta \pm j \sin \theta = e^{\pm j\theta})$$

$$= \pi D_\alpha e^{j\phi_\alpha} \delta(\theta + \theta_\alpha) + \pi D_\alpha e^{-j\phi_\alpha} \delta(\theta - \theta_\alpha). \qquad (\because \delta(\mathbb{F}) = 2\pi\delta(\theta) \text{ from (7.20)})$$

7.4   Duality in Connection with the Fourier Series

Observe that the relationship between the pair $\hat{X}_I(\mathbb{F}) \iff \{x_\ell\}$ mirrors that of a periodic function $y^p(t)$ with period T and its Fourier series coef cients $\{C_k\}$:

$$(7.23) \qquad y^p(t) = \sum_{k=-\infty}^{\infty} C_k e^{j2\pi kt/T} \iff C_k = \frac{1}{T}\int_{-T/2}^{T/2} y^p(t) e^{-j2\pi kt/T} dt.$$

Recall that if we extract one period of $y^p(t)$ to de n e the time-limited function

$$y_1(t) = \begin{cases} y^p(t), & t \in [-T/2, T/2] \\ 0, & \text{otherwise} \end{cases},$$

then $y^p(t)$ is formally the periodic extension of $y_1(t)$, and it can be expressed as the convolution of $y_1(t)$ and the impulse train $P_T(t)$ (for examples, see Figures 6.6 and 6.8 in Chapter 6):

$$(7.24) \qquad y^p(t) = \overbrace{y_1(t) * P_T(T)}^{\text{convolution}} = \underbrace{y_1(t) * \sum_{\ell=-\infty}^{\infty} \delta(t - \ell T)}_{\text{periodic extension of } y_1(t)} = \underbrace{\sum_{\ell=-\infty}^{\infty} y_1(t - \ell T)}_{\text{shifted replicas of } y_1(t)}.$$

Assuming that the time-limited $y_1(t)$ has Fourier transform

$$(7.25) \qquad Y_1(f) = \int_{-\infty}^{\infty} y_1(t)\, e^{-j2\pi ft}\, dt = \int_{-T/2}^{T/2} y^p(t)\, e^{-j2\pi ft}\, dt,$$

the Fourier series coefficients C_k of $y^p(t) = y_1(t) * P_T(t)$ are equally spaced samples of $Y_1(f)$ scaled by the factor $1/T$, i.e.,

$$C_k = \frac{1}{T} Y_1\left(\frac{k}{T}\right), \quad k \in (-\infty, \infty).$$

Hence the sequence $\{C_k\}$ represents the impulse sampled $Y_1(f)$ scaled by the factor $1/T$:

$$\frac{1}{T} Y_1(f)\, P_{1/T}(f) = \sum_{k=-\infty}^{\infty} \frac{1}{T} Y_1\left(\frac{k}{T}\right) \delta\left(f - \frac{k}{T}\right) = \sum_{k=-\infty}^{\infty} C_k\, \delta\left(f - \frac{k}{T}\right),$$

and we see that sampling $Y_1(f) = \mathcal{F}\{y_1(t)\}$ in the frequency domain results in the periodic extension of $y_1(t)$ to $y^p(t)$ in the time domain.

Since the sampling space $\triangle f = f_{k+1} - f_k = \frac{1}{T}$ is the reciprocal of the period of $y^p(t)$, the Fourier series pair $y^p(t) \Longleftrightarrow \{C_k\}$ is said to be the dual of the Fourier transform pair $\hat{X}_I(\mathbb{F}) \Longleftrightarrow \{x_\ell\}$. In both cases, periodic extension in one domain is the consequence of sampling in the other domain. Observe that the Fourier series pair $y^p(t) \Longleftrightarrow \{C_k\}$ represents the result expected from Convolution Theorem 6.1:

$$(7.26) \qquad \underbrace{y_1(t) * P_T(t)}_{\text{periodic } y^p(t)} \Longleftrightarrow \underbrace{\mathcal{F}\{y_1(t)\} \cdot \mathcal{F}\{P_T(t)\}}_{\text{impulse sampled } Y_1(f)},$$

where the Fourier transform of the impulse train is given by Theorem 6.7:

$$(7.27) \qquad \mathcal{F}\{P_T(t)\} = \frac{1}{T} P_{1/T}(f) = \frac{1}{T} \sum_{k=-\infty}^{\infty} \delta\left(f - \frac{k}{T}\right).$$

7.4.1 Periodic convolution and discrete convolution

In this section we obtain the duals of Theorem 6.4 (Periodic Convolution) and Theorem 6.5 (Discrete Convolution) for discrete-time signals.

Theorem 7.1 (Frequency-domain periodic convolution) The convolution of periodic $\hat{X}_I(\mathbb{F})$ and $\hat{H}_I(\mathbb{F})$ in the frequency domain corresponds to the multiplication of sequences $\{x_\ell\}$ and $\{h_\ell\}$ in the time domain. That is,

$$\hat{G}_I(\mathbb{F}) = \mathcal{F}\{\{x_\ell\, h_\ell\}\} = \hat{X}_I(\mathbb{F}) \otimes \hat{H}_I(\mathbb{F}),$$

where

$$\hat{X}_I(\mathbb{F}) \otimes \hat{H}_I(\mathbb{F}) = \int_{-1/2}^{1/2} \hat{X}_I(\lambda)\, \hat{H}_I(\mathbb{F} - \lambda)\, d\lambda.$$

Proof: Recall from Equation (7.4) we may express $\hat{X}_I(\mathbb{F})$ and $\hat{H}_I(\mathbb{F})$ by their Fourier series expansions, namely,

$$\hat{X}_I(\mathbb{F}) = \sum_{\ell=-\infty}^{\infty} \hat{c}_\ell \, e^{j2\pi\ell\mathbb{F}/\Omega}, \qquad \text{where } \hat{c}_\ell = x_{-\ell}, \, \Omega = 1;$$

$$\hat{H}_I(\mathbb{F}) = \sum_{\ell=-\infty}^{\infty} \hat{d}_\ell \, e^{j2\pi\ell\mathbb{F}/\Omega}, \qquad \text{where } \hat{d}_\ell = h_{-\ell}, \, \Omega = 1.$$

By the Periodic Convolution Theorem 6.4 for the Fourier series, we immediately have the Fourier series expansion of $\hat{X}_I(\mathbb{F}) \otimes \hat{H}_I(\mathbb{F})$:

$$\hat{X}_I(\mathbb{F}) \otimes \hat{H}_I(\mathbb{F}) = \sum_{\ell=-\infty}^{\infty} \left(\hat{c}_\ell \, \hat{d}_\ell \right) e^{j2\pi\ell\mathbb{F}/\Omega}, \quad \text{where } \Omega = 1 \quad \text{(by Theorem 6.4)}$$

$$= \sum_{\ell=-\infty}^{\infty} \left(x_{-\ell} \, h_{-\ell} \right) e^{j2\pi\ell\mathbb{F}}$$

$$= \sum_{k=-\infty}^{\infty} \left(x_k \, h_k \right) e^{-j2\pi k\mathbb{F}} \qquad (\text{let } k = -\ell)$$

$$= \mathcal{F}\{\{x_k \, h_k\}\} \qquad\qquad \text{by Formula (7.3)}$$

$$= \hat{G}_I(\mathbb{F}). \qquad\qquad\qquad\qquad\qquad\blacksquare$$

We show next that the multiplication of the continuous Fourier transforms in the frequency domain corresponds to the convolution of the discrete-time signals in the time domain. Recall that the discrete linear convolution of two sequences was de n ed in Section 6.4 as

$$g_k = \sum_{\ell=-\infty}^{\infty} x_\ell \, h_{k-\ell}, \quad \text{for all } k \in (-\infty, \infty).$$

We also express the discrete convolution as $\{g_k\} = \{x_k\} * \{h_k\}$.

Theorem 7.2 (Time-Domain Discrete Convolution) The Fourier transform of the discrete convolution of the two sequences $\{x_\ell\}$ and $\{h_\ell\}$ is the product of $\hat{X}_I(\mathbb{F})$ and $\hat{H}_I(\mathbb{F})$. That is,

$$\hat{X}_I(\mathbb{F}) \, \hat{H}_I(\mathbb{F}) = \sum_{k=-\infty}^{\infty} \left(\sum_{\ell=-\infty}^{\infty} x_\ell \, h_{k-\ell} \right) e^{-j2\pi k\mathbb{F}}.$$

Proof: We again use Equation (7.4) to express $\hat{X}_I(\mathbb{F})$ and $\hat{H}_I(\mathbb{F})$ by their Fourier series expansions,

$$\hat{X}_I(\mathbb{F}) = \sum_{\ell=-\infty}^{\infty} \hat{c}_\ell \, e^{j2\pi\ell\mathbb{F}/\Omega}, \qquad \text{where } \hat{c}_\ell = x_{-\ell}, \text{ and } \Omega = 1;$$

$$\hat{H}_I(\mathbb{F}) = \sum_{\ell=-\infty}^{\infty} \hat{d}_\ell \, e^{j2\pi\ell\mathbb{F}/\Omega}, \qquad \text{where } \hat{d}_\ell = h_{-\ell}, \text{ and } \Omega = 1.$$

By Theorem 6.5, we immediately obtain the Fourier series expansion for the product of two Fourier series as

$$
\hat{X}_I(\mathbb{F})\,\hat{H}_I(\mathbb{F}) = \sum_{k=-\infty}^{\infty}\left(\sum_{\ell=-\infty}^{\infty}\hat{c}_\ell\,\hat{d}_{k-\ell}\right)e^{j2\pi k\mathbb{F}/\Omega}, \quad \text{where } \Omega = 1 \quad \text{(by Theorem 6.5)}
$$

$$
= \sum_{k=-\infty}^{\infty}\left(\sum_{\ell=-\infty}^{\infty}x_{-\ell}\,h_{-k+\ell}\right)e^{j2\pi k\mathbb{F}}
$$

$$
= \sum_{k=-\infty}^{\infty}\left(\sum_{m=-\infty}^{\infty}x_m\,h_{-k-m}\right)e^{j2\pi k\mathbb{F}} \quad (\text{let } m = -\ell)
$$

$$
= \sum_{r=-\infty}^{\infty}\left(\sum_{m=-\infty}^{\infty}x_m\,h_{r-m}\right)e^{-j2\pi r\mathbb{F}} \quad (\text{let } r = -k)
$$

$$
= \sum_{k=-\infty}^{\infty}\left(\sum_{\ell=-\infty}^{\infty}x_\ell\,h_{k-\ell}\right)e^{-j2\pi k\mathbb{F}}. \quad (\text{relabel } r = k,\ m = \ell)
$$

$\blacksquare$

7.5 The Fourier Transform of a Periodic Sequence

In Section 7.4 we have derived the Fourier transform of the envelope signal $y^p(t)$ of the periodic sequence $\{y_\ell\}$, namely,

$$
(7.28) \quad \text{Periodic } y^p(t) = y_1(t) * P_T(t) \iff \mathcal{F}\{y_1(t)\}\cdot\mathcal{F}\{P_T(t)\} = Y_1(f)\cdot\frac{1}{T}P_{1/T}(f),
$$

where

$$
(7.29) \quad Y_1(f)\cdot\frac{1}{T}P_{1/T}(f) = \sum_{k=-\infty}^{\infty}\overbrace{\frac{1}{T}Y_1\!\left(\frac{k}{T}\right)}^{C_k}\delta\!\left(f-\frac{k}{T}\right).
$$

Fourier transform of $y^p(t)$

Hence an impulse train weighted by the Fourier series coef cients shall be adopted as the formal Fourier transform of a periodic signal, which is simply another useful mathematical expression for the same frequency contents represented by the Fourier series coef cients. Recall that $C_k \to 0$ as $k \to \pm\infty$, so the impulse train weighted by decaying C_k s will not be periodic. (If the periodic signal is band limited, we shall have only a nite number of nonzero C_k s.)

We show next that it is indeed justi ab le to treat $\mathcal{F}\{\text{periodic } y^p(t)\}$ g iven by (7.29) as the formal Fourier transform of $y^p(t)$, because we get back the original signal by invoking the inverse Fourier transform as usual:

$$\mathcal{F}^{-1}\{\mathcal{F}\{\text{periodic } y^p(t)\}\} = \int_{-\infty}^{\infty} \mathcal{F}\{\text{periodic } y^p(t)\} \, e^{j2\pi ft} df$$

$$= \int_{-\infty}^{\infty} \left[\sum_{k=-\infty}^{\infty} C_k \delta\left(f - \frac{k}{T}\right) \right] e^{j2\pi ft} df$$

$$= \sum_{k=-\infty}^{\infty} \int_{-\infty}^{\infty} C_k \delta\left(f - \frac{k}{T}\right) e^{j2\pi ft} df$$

$$= \sum_{k=-\infty}^{\infty} C_k \int_{-\infty}^{\infty} \delta\left(f - \frac{k}{T}\right) e^{j2\pi ft} df$$

$$= \sum_{k=-\infty}^{\infty} C_k \, e^{j2\pi kt/T}$$

$$= y^p(t). \qquad (\because \text{Fourier series of } y^p(t) \text{ has been obtained})$$

Assuming that N equally spaced samples are taken from each period of $y^p(t)$, we can now express the periodic sequence $\{y_\ell\}$ as the shifted replicas of the impulse sampled $y_1(t)$, which is time limited to a single period of $y^p(t)$. That is,

(7.30)
$$\overbrace{\text{Periodic Sequence } y_I^p(t)}^{y^p(t) P_{\triangle t}(t):\ N \text{ samples/period}} = \overbrace{\left(\underbrace{y_1(t) \, P_{\triangle t}(t)}_{N \text{ samples}} \right) * P_{N\triangle t}(t)}^{\text{shifted replicas of } N \text{ samples}},$$

where $N\triangle t = T$, which is the time duration of $y_1(t)$ as well as the period of the extended $y^p(t) = y_1(t) * P_{N\triangle t}(t)$. For the corresponding sequence $\{y_\ell\}$, we have $y_{\ell+N} = y_\ell$ due to the shifted replicas of the N samples (or weighted impulses.) Using Convolution Theorem 6.1 on Equation (7.30), we obtain

$$\mathcal{F}\{\text{periodic sequence } y_I^p(t)\}$$

$$= \mathcal{F}\{\left(y_1(t) \cdot P_{\triangle t}(t)\right) * \{P_T(t)\}, \quad \text{where } T = N\triangle t,$$

$$= \mathcal{F}\{y_1(t) \cdot P_{\triangle t}(t)\} \cdot \mathcal{F}\{P_T(t)\} \quad \text{by Convolution Theorem 6.1}$$

$$= \frac{1}{T} \underbrace{\left[\frac{1}{\triangle t} \sum_{k=-\infty}^{\infty} Y_1\left(f - \frac{k}{\triangle t}\right) \right]}_{\text{periodic } \mathcal{Y}_I(f):\ \text{period } \mathcal{R} = 1/\triangle t} \underbrace{\sum_{k=-\infty}^{\infty} \delta\left(f - \frac{k}{T}\right)}_{P_{1/T}(f)}$$

(7.31)
$$= \frac{1}{T} \sum_{k=-\infty}^{\infty} \underbrace{\left[\frac{1}{\triangle t} \sum_{k=-\infty}^{\infty} Y_1\left(f - \frac{k}{\triangle t}\right) \right]}_{\text{periodic } \mathcal{Y}_I(f)} \delta\left(f - \frac{k}{T}\right)$$

$$= \frac{1}{T} \sum_{k=-\infty}^{\infty} \mathcal{Y}_I\left(\frac{k}{T}\right) \delta\left(f - \frac{k}{T}\right)$$

$$= \underbrace{\frac{1}{\triangle t} \sum_{k=-\infty}^{\infty} \frac{1}{N} \mathcal{Y}_I\left(\frac{k}{T}\right) \delta\left(f - \frac{k}{T}\right)}_{\text{periodic sequence: } N \text{ samples per period } \mathbb{R}}, \quad \frac{1}{T} = \frac{1}{N\triangle t} = \frac{\mathbb{R}}{N}.$$

Observe that the continuous periodic function $\mathcal{Y}_I(f)$ is impulse sampled at intervals of $\mathbb{R}/N = 1/T$. Since the sampling rate $\mathbb{R} = 1/\triangle t$ is the period of $\mathcal{Y}_I(f)$, there are exactly N equally spaced samples over each period of $\mathcal{Y}_I(f)$. If the periodic signal $y^p(t)$ is not band limited, the N samples taken over one period of $\frac{1}{N}\mathcal{Y}_I(f)$ will not be identical to the corresponding N samples of $\frac{1}{N\triangle t}Y_1(f) = \frac{1}{T}Y_1(f)$ due to the effect of aliasing.

On the other hand, if the periodic signal $y^p(t) = y^p(t+T)$ is band limited, then there are only a finite number of nonzero Fourier series coefficients:

$$(7.32) \qquad \text{band-limited periodic } y^p(t) = \sum_{k=-n}^{n} C_k\, e^{j2\pi kt/T}.$$

Using the Fourier transform pair $e^{j2\pi f_a t} \Longleftrightarrow \delta(f - f_a)$ from Chapter 6, we obtain

$$(7.33) \quad \mathcal{F}\{\text{band-limited periodic } y^p(t)\} = \sum_{k=-n}^{n} C_k\, \mathcal{F}\{e^{j2\pi kt/T}\} = \sum_{k=-n}^{n} C_k\, \delta\!\left(f - \frac{k}{T}\right).$$

The Fourier transform of the impulse-sampled band-limited periodic function $y_I^p(t)$ can now be expressed as

$$\mathcal{F}\{\text{band-limited periodic sequence } y_I^p(t)\}$$

$$= \mathcal{F}\{y^p(t) \cdot P_{\triangle t}(t)\}, \qquad \text{where sampling interval } \triangle t = \frac{T}{N}$$

$$= \mathcal{F}\{y^p(t)\} * \mathcal{F}\{P_{\triangle t}(t)\} \qquad\qquad \text{by Convolution Theorem}$$

$$= \left[\sum_{k=-n}^{n} C_k\, \delta\!\left(f - \frac{k}{T}\right)\right] * \frac{1}{\triangle t} P_{1/\triangle t}(f)$$

$$= \left[\sum_{k=-n}^{n} C_k\, \delta\!\left(f - \frac{k}{T}\right)\right] * \left[\frac{1}{\triangle t} \sum_{m=-\infty}^{\infty} \delta\!\left(f - \frac{m}{\triangle t}\right)\right]$$

$$(7.34)$$

$$= \frac{1}{\triangle t}\left[\sum_{k=-n}^{n} C_k\, \delta\!\left(f - \frac{k}{T}\right)\right] * \sum_{m=-\infty}^{\infty} \delta\!\left(f - \frac{m}{\triangle t}\right)$$

$$= \frac{1}{\triangle t}\left[\sum_{k=-n}^{n} C_k\, \delta\!\left(f - k\triangle f\right)\right] * \sum_{m=-\infty}^{\infty} \delta\!\left(f - mN\triangle f\right), \;\; \triangle f = \frac{1}{T} = \frac{1}{N\triangle t}$$

$$= \frac{1}{\triangle t}\underbrace{\left[\sum_{k=-n}^{n} C_k\, \delta\!\left(f - k\frac{\mathbb{R}}{N}\right)\right]}_{\text{Replicate } N \text{ weighted impulses spaced by } \triangle f = \mathbb{R}/N} * \sum_{m=-\infty}^{\infty} \delta\!\left(f - m\mathbb{R}\right), \;\; \mathbb{R} = \frac{1}{\triangle t} = N\triangle f.$$

Note that the $N = 2n+1$ Fourier coefficients C_k s become the strengths of the N impulses spaced by $\triangle f = 1/T$ over the frequency range $[-n\triangle f, n\triangle f]$. We also identify one period of the Fourier transform as $\mathbb{R} = N\triangle f = (2n+1)\triangle f$, which begins with $(-n + 0.5)\triangle f$ and ends with $(n + 0.5)\triangle f$. Since $f_{max} = n/T$, $N = 2n+1$, the choice of sampling interval $\triangle t = T/N < 1/\{2f_{max}\}$, and the Nyquist condition is satisfied. Hence the periodic replication of the N impulses (spaced by $\triangle f$) over the distance $\mathbb{R} = N\triangle f$ does not cause overlap, and no aliasing effect will result from sampling a band-limited periodic signal at or above Nyquist rate.

7.6   The DFT Interpretation

In this section we show that equispaced sample values of the Fourier transform of an in nite sequence $\{x_\ell\}$ can be interpreted as the **Discrete Fourier Transform (DFT)** of a formally n ite sequence $\{u_k\}$. We begin with the de n ing formula (7.2):

$$X_I(f) = \mathcal{F}\{x_I(t)\} = \sum_{\ell=-\infty}^{\infty} x(\ell\triangle t)\, e^{-j2\pi f\ell\triangle t}.$$

Suppose that we choose the DFT length to be $N = 2n+1$. Since N samples are taken over each interval of duration $\mathbb{T} = N\triangle t$, we can express the sample point $\ell\triangle t = (k + mN)\triangle t$, where $-n \le k \le n$, $-\infty < m < \infty$, which allows us to express the summation involving $x(\ell\triangle t)$ as a double sum:

$$
\begin{aligned}
X_I(f) &= \sum_{\ell=-\infty}^{\infty} x(\ell\triangle t)\, e^{-j2\pi f\ell\triangle t} \\
(7.35)\qquad &= \sum_{m=-\infty}^{\infty} \sum_{k=-n}^{n} x\big((k+mN)\triangle t\big)\, e^{-j2\pi f(k+mN)\triangle t} \\
&= \sum_{k=-n}^{n} \sum_{m=-\infty}^{\infty} x\big((k+mN)\triangle t\big)\, e^{-j2\pi f(k+mN)\triangle t}.
\end{aligned}
$$

If we now evaluate $\frac{1}{N}X_I(f)$ at $f = f_r = r/\mathbb{T} = r/(N\triangle t)$, we obtain

$$
\begin{aligned}
\underbrace{\frac{1}{N}X_I\left(\frac{r}{N\triangle t}\right)}_{U_r \text{ in } (7.37)} &= \frac{1}{N} \sum_{k=-n}^{n} \sum_{m=-\infty}^{\infty} x\big((k+mN)\triangle t\big)\, e^{-j2\pi r(k+mN)/N} \\
&= \frac{1}{N} \sum_{k=-n}^{n} \sum_{m=-\infty}^{\infty} x\big((k+mN)\triangle t\big)\, e^{-j2\pi rk/N} \\
(7.36)\qquad &= \frac{1}{N} \sum_{k=-n}^{n} \left[\sum_{m=-\infty}^{\infty} x\big((k+mN)\triangle t\big) \right] e^{-j2\pi rk/N} \\
&= \frac{1}{N} \sum_{k=-n}^{n} u_k\, \omega_N^{-rk},
\end{aligned}
$$

$$\text{where}\quad u_k = \sum_{m=-\infty}^{\infty} x\big((k+mN)\triangle t\big),\quad \omega_N = e^{j2\pi/N}.$$

Observe that the familiar DFT formula emerges from the right side of (7.36).

Since $X_I(f)$ de ned by (7.2) is periodic with period determined by the sampling rate $\mathbb{R} = 1/\triangle t$, we obtain N equally spaced samples of $X_I(f)$ over one period $\mathbb{R}$ when they are spaced by

$$\triangle f = \frac{\mathbb{R}}{N} = \frac{1}{N\triangle t} = \frac{1}{\mathbb{T}}.$$

If we let

$$U_r = \frac{1}{N}X_I\big(r\triangle f\big) = \frac{1}{N}X_I\left(\frac{r}{N\triangle t}\right),\quad \text{where } N = 2n+1,\ -n \le r \le n,$$

then $U_{r+N} = U_r$; that is, the sequence $\{U_r\}$ is periodic with period N, and we can obtain U_r by evaluating the DFT de ned by (7.36); i.e., we compute

$$(7.37) \qquad U_r = \frac{1}{N} \sum_{k=-n}^{n} u_k \omega_N^{-rk}, \quad \text{where } N = 2n + 1, \ -n \leq r \leq n.$$

Remark 1: Observe that because

$$u_k = \sum_{m=-\infty}^{\infty} x\big((k + mN)\Delta t\big) \approx \sum_{m=-n}^{n} x\big((k + mN)\Delta t\big),$$

in practice the values of U_r can be approximated using the DFT formula (7.37) on $\{u_k\}$ computed from *finite* sample sequence $\{x_\ell\}$.

Remark 2: The DFT relationship

$$U_r = \frac{1}{N} X_I\left(\frac{r}{N\Delta t}\right) = \frac{1}{N} \sum_{k=0}^{N-1} u_k \omega_N^{-rk}, \quad 0 \leq r \leq N - 1,$$

holds with the N-sample sequence $\{u_k\}$ de ned by

$$u_k = \sum_{m=-\infty}^{\infty} x\big((k + mN)\Delta t\big), \quad 0 \leq k \leq N - 1.$$

Example 7.8 Recall the following result from Example 7.3(a) and Figure 7.1:

$$\{z_\ell = \alpha^\ell u(\ell)\} \iff \hat{Z}_I(\mathbb{F}) = \frac{1}{1 - \alpha e^{-j2\pi\mathbb{F}}}, \quad \text{where } |\alpha| < 1.$$

Letting $\mathbb{F}_r = f_r \Delta t = \dfrac{r}{N}$, we obtain

$$U_r = \frac{1}{N} \hat{Z}_I\left(\frac{r}{N}\right) = \frac{1}{N(1 - \alpha e^{-j2\pi r/N})} = \frac{1}{N(1 - \alpha \omega_N^{-r})}, \quad 0 \leq r \leq N - 1.$$

We show next that if we de n e

$$u_k = \sum_{m=-\infty}^{\infty} z_{k+mN} = \sum_{m=0}^{\infty} \alpha^{k+mN} = \alpha^k \sum_{m=0}^{\infty} (\alpha^N)^m = \frac{\alpha^k}{1 - \alpha^N},$$

then we obtain the same value of U_r by performing DFT on $\{u_k\}$. That is, we compute

$$U_r = \frac{1}{N} \sum_{k=0}^{N-1} u_k \omega_N^{-rk} = \frac{1}{N} \sum_{k=0}^{N-1} \left[\frac{\alpha^k}{1 - \alpha^N}\right] \omega_N^{-rk}, \qquad 0 \leq r \leq N - 1$$

$$= \frac{1}{N(1 - \alpha^N)} \sum_{k=0}^{N-1} (\alpha \omega_N^{-r})^k$$

$$= \frac{1}{N(1 - \alpha^N)} \left[\frac{1 - \alpha^N \omega_N^{-rN}}{1 - \alpha \omega_N^{-r}}\right]$$

$$= \frac{1}{N(1 - \alpha \omega_N^{-r})}. \qquad\qquad (\because \omega_N^{-rN} = 1)$$

We have thus veri ed that

$$Z_I\left(\frac{r}{N\triangle t}\right) = \hat{Z}_I\left(\frac{r}{N}\right) = \sum_{k=0}^{N-1}\left[\sum_{-\infty}^{\infty} z_{k+mN}\right]\omega_N^{-rk}.$$

Note that if we apply the DFT directly to the truncated N-sample sequence $\{z_k = \alpha^k\}$, $0 \le k \le N-1$, then

$$Z_r = \frac{1}{N}\sum_{k=0}^{N-1} z_k\,\omega_N^{-rk} = \frac{1}{N}\sum_{k=0}^{N-1}\alpha^k\,\omega_N^{-rk}, \qquad 0 \le r \le N-1$$

$$= \frac{1}{N}\left[\frac{1 - \alpha^N\omega_N^{-rN}}{1 - \alpha\,\omega_N^{-r}}\right]$$

$$= \frac{1 - \alpha^N}{N\left(1 - \alpha\,\omega_N^{-r}\right)}$$

$$\ne \frac{1}{N\left(1 - \alpha\,\omega_N^{-r}\right)} = \frac{1}{N}Z_I\left(\frac{r}{N\triangle t}\right).$$

Therefore, using nite N samples we can only expect $Z_r \approx \frac{1}{N}Z_I(r/\mathbb{T})$ for $-n \le r \le n$, $n = (N-1)/2$, $\mathbb{T} = N\triangle t$.

7.6.1   The interpreted DFT and the Fourier transform

We can also relate the DFT of formally n ite sequence $\{u_k\}$ to the Fourier transform of the continuous-time signal $x(t)$ through the second form of $X_I(f)$ given by (7.2), namely,

$$X_I(f) = \mathcal{F}\{x_I(t)\} = \frac{1}{\triangle t}\sum_{k=-\infty}^{\infty} X\left(f - \frac{k}{\triangle t}\right),$$

where $X(f) = \mathcal{F}\{x(t)\}$. Accordingly, we can express U_r by values of $X(f)$:

$$U_r = \frac{1}{N}X_I\left(\frac{r}{N\triangle t}\right) = \frac{1}{N\triangle t}\sum_{k=-\infty}^{\infty} X\left(\frac{r}{N\triangle t} - \frac{k}{\triangle t}\right)$$

(7.38)

$$= \frac{1}{\mathbb{T}}\sum_{k=-\infty}^{\infty} X\left((r - kN)\frac{1}{\mathbb{T}}\right) \quad (\because \mathbb{T} = N\triangle t)$$

$$= \triangle f\sum_{k=-\infty}^{\infty} X\left((r - kN)\triangle f\right) \quad \left(\because \triangle f = \frac{1}{\mathbb{T}} = \frac{\mathbb{R}}{N}\right)$$

$$= \triangle f\sum_{m=-\infty}^{\infty} X\left((r + mN)\triangle f\right). \quad (\text{let } m = -k)$$

Hence the equality of the two forms of $X_I(f)$ given by (7.2) allows us to relate the sampled $x(t)$ to the samples of its transform $X(f) = \mathcal{F}\{x(t)\}$ through the DFT:

(7.39) $$U_r = \frac{1}{N}\sum_{k=-n}^{n} u_k\,\omega_N^{-rk}, \quad \text{for } N = 2n+1, \; -n \le r \le n,$$

where

$$u_k = \sum_{m=-\infty}^{\infty}\overbrace{x\big((k+mN)\triangle t\big)}^{\text{samples of } x(t)}; \; U_r = \triangle f\sum_{m=-\infty}^{\infty}\overbrace{X\big((r+mN)\triangle f\big)}^{\text{samples of } X(f)}, \; \triangle f = \frac{1}{N\triangle t}.$$

Note that $u_{k+N} = u_k$ and $U_{r+N} = U_r$.

Example 7.9 The following Fourier transform pair was obtained in Example 6.1 (see Figures 6.9 and 6.10):

$$\{x_\ell = e^{-a\ell\Delta t}u(\ell)\} \iff X_I(f) = \sum_{k=-\infty}^{\infty} \frac{1}{a\Delta t + j2\pi(f\Delta t - k)}.$$

Recall that the sequence $\{x_\ell\}$ was sampled from $x(t) = e^{-at}u(t)$, and we show the complex-valued $X(f) = \dfrac{1}{a + j2\pi f}$ in Example 5.1 (see Figure 5.2). In this example we verify that

$$U_r = \frac{1}{N}X_I\left(\frac{r}{N\Delta t}\right) = \Delta f \sum_{m=-\infty}^{\infty} X((r + mN)\Delta f).$$

The DFT U_r was defined by

$$U_r = \frac{1}{N}X_I\left(\frac{r}{N\Delta t}\right) = \frac{1}{N}\sum_{k=-\infty}^{\infty} \frac{1}{a\Delta t + j2\pi(r - kN)/N}$$

$$= \frac{1}{N\Delta t}\sum_{k=-\infty}^{\infty} \frac{\Delta t}{a\Delta t + j2\pi(r - kN)/N}$$

$$= \Delta f \sum_{k=-\infty}^{\infty} \frac{1}{a + j2\pi(r - kN)\Delta f} \qquad \left(\because \Delta f = \frac{1}{\mathbb{T}} = \frac{1}{N\Delta t}\right)$$

$$= \Delta f \sum_{m=-\infty}^{\infty} \frac{1}{a + j2\pi(r + mN)\Delta f}.$$

The same result can also be obtained from summing the values of $X(f) = \dfrac{1}{a + j2\pi f}$:

$$U_r = \Delta f \sum_{m=-\infty}^{\infty} X((r + mN)\Delta f) = \Delta f \sum_{m=-\infty}^{\infty} \frac{1}{a + j2\pi(r + mN)\Delta f}.$$

7.6.2 Time-limited case

If $x(t) = 0$ for $t < -\mathbb{T}/2$ or $t > \mathbb{T}/2$, then for $\Delta t = \mathbb{T}/N = \mathbb{T}/(2n+1)$, we have $x(\ell\Delta t) = 0$ for $\ell < -n$ or $\ell > n$, and the inner sum defining the DFT input u_k is reduced to one single term:

$$u_k = \sum_{m=-\infty}^{\infty} x((k + mN)\Delta t) = x_k, \quad -n \le k \le n, \quad n = (N - 1)/2.$$

We can thus reinterpret (7.39) as the DFT of the N-sample sequence $\{x_k\}$:

(7.40)
$$\boxed{U_r = \frac{1}{N} \sum_{k=-n}^{n} x_k \, \omega_N^{-rk}, \quad -n \le r \le n, \quad N = 2n + 1,}$$

where

(7.41) $$U_r = \frac{1}{N}X_I\left(\frac{r}{N\Delta t}\right) = \Delta f \sum_{m=-\infty}^{\infty} X\left((r + mN)\Delta f\right), \quad \Delta f = \frac{1}{N\Delta t} = \frac{1}{\mathbb{T}}.$$

When a time-limited function $x(t)$ is not band limited, the Nyquist condition will not be satisfied by any choice of $\triangle t$, and we must account for the aliased frequencies in this case. We show next how the Fourier series coefficients of $x^p(t)$ the periodic extension of $x(t)$ are aliased into the DFT of the sampled sequence $\{x_\ell\}$.

Recall that if we interpret the time-limited $x(t)$ as one period of its periodic extension function $x^p(t)$ (with period $\mathbb{T} = N\triangle t$), then for $t \in [-\mathbb{T}/2, \mathbb{T}/2]$ we can express $x(t)$ by the Fourier series expansion of $x^p(t)$:

$$x(t) = x^p(t) = \sum_{k=-\infty}^{\infty} C_k e^{j2\pi kt/\mathbb{T}}, \quad C_k = \frac{1}{\mathbb{T}} X\left(\frac{k}{\mathbb{T}}\right),$$

where $X(f) = \mathcal{F}\{x(t)\}$. We can now rewrite U_r in (7.41) as

$$U_r = \frac{1}{N} X_I\left(\frac{r}{N\triangle t}\right) = \frac{1}{\mathbb{T}} \sum_{m=-\infty}^{\infty} X\left(\frac{r+mN}{\mathbb{T}}\right) = \sum_{m=-\infty}^{\infty} C_{r+mN},$$

and we have once again proved the relationship between the DFT $\{U_r\}$ and the Fourier series coefficients $\{C_k\}$:

(7.42)
$$\boxed{U_r = \sum_{m=-\infty}^{\infty} C_{r+mN}.}$$

Recall that we proved the same result from a different perspective in Section 3.11. The Fourier transform and the Fourier series expansion of a time-limited rectangular pulse function were given in Example 5.3 (Figures 5.4 and 5.5).

7.6.3   Band-limited case

If $x(t)$ is band-limited with bandwidth $F = 2f_{\max}$ and we have chosen the sampling rate $\mathbb{R} = 1/\triangle t > F$, then the Nyquist condition is satisfied, and from Corollary 6.12, we have

$$X_I(f) = \frac{1}{\triangle t} \sum_{k=-\infty}^{\infty} X\left(f - \frac{k}{\triangle t}\right) = \frac{1}{\triangle t} X(f), \quad \text{for } f \in [-\mathbb{R}/2, \mathbb{R}/2].$$

Note that $[-F/2, F/2] \subset [-\mathbb{R}/2, \mathbb{R}/2]$. Because $f_r = \dfrac{r}{N\triangle t} \in [-\mathbb{R}/2, \mathbb{R}/2]$ for $|r| \leq \frac{N-1}{2}$, we can express U_r by a single sample value of $X(f)$ in the right side:

$$U_r = \frac{1}{N} X_I\left(\frac{r}{N\triangle t}\right) = \frac{1}{N\triangle t} X\left(\frac{r}{N\triangle t}\right), \quad -n \leq r \leq n, \quad N = 2n+1.$$

Letting $\mathbb{T} = N\triangle t$, we can now obtain the sample values of $X(f) = \mathcal{F}\{x(t)\}$ through the DFT on $\{u_k\}$:

(7.43)
$$\frac{1}{\mathbb{T}} X\left(\frac{r}{\mathbb{T}}\right) = U_r = \frac{1}{N} \sum_{k=-n}^{n} u_k \omega_N^{-rk}, \quad -n \leq r \leq n, \quad N = 2n+1,$$

where, as before,

$$u_k = \sum_{m=-\infty}^{\infty} x\big((k+mN)\triangle t\big).$$

A band-limited example was given in Example 5.4 (Figure 5.6).

7.6.4 Periodic and band-limited case

If the periodic signal $x^p(t) = x^p(t + T)$ is band-limited, then it has a finite Fourier series expansion:

$$x^p(t) = \sum_{r=-n}^{n} C_r e^{j2\pi rt/T}.$$

When $N = 2n + 1$ samples of $x^p(t)$ are taken over one period $T = N\triangle t$, we obtain

$$x_\ell = x^p(\ell\triangle t) = \sum_{r=-n}^{n} C_r \omega_N^{r\ell}, \text{ where } \omega_N \equiv e^{j2\pi/N}, \ \ell = 0, 1, \ldots, N-1,$$

which is identical to the system of equations (in complex exponential modes) described by Equation (2.5) in Chapter 2, and the latter leads to the DFT formula given by Equation (2.7), by its alternate form (2.8) we obtain the N Fourier series coefficients, i.e.,

$$(7.44) \qquad C_r = X_r = \frac{1}{N} \sum_{\ell=-n}^{n} x_\ell \omega_N^{-r\ell}, \text{ for } -n \le r \le n, \ N = 2n + 1.$$

Recall also that in Example 4.5 (Table 4.1, Figure 4.2) in Chapter 4, we computed DFT coefficients from samples taken from one period (and three periods) of a band-limited periodic function

$$y(t) = 4.5 \cos(1.2\pi t) + 7.2 \cos(1.8\pi t),$$

and we show how to identify the Fourier series coefficients (expressed as complex-valued Y_r or real-valued A_r and B_r) from the computed DFT coefficients displayed in Table 4.1.

According to Formulas (7.33) and (7.34), the C_rs are the strengths of the $N = 2n + 1$ impulses which define $\mathcal{F}\{x^p(t)\}$, and the periodic replicas (with no overlap) of these N impulses (scaled by $1/\triangle t$) define $\mathcal{F}\{x_I^p(t)\}$, the Fourier transform of the periodic sequence formed by taking N samples over one period (or multiple full periods) of duration T. Hence this is the case (and the only case) in which we can use DFT formulas on N samples to recover a signal s true frequency content. The DFT formulas for computing the N Fourier series coefficients were given in Chapter 2 and Chapter 4; there we also addressed the related issues including the sampling rate, sampling period, sample size, and alternate forms of the DFT. After we derived the DFT and IDFT formulas in Chapter 4, we further explained the possible frequency distortion by leakage and the effects of zero padding in Sections 4.5 and 4.6. We shall revisit some of these issues from a different perspective after we discuss the windowing of a sequence for DFT computation in the next chapter.

References

1. A. Ambardar. *Analog and Digital Signal Processing*. Brooks/Cole Publishing Company, Pacific Grove, CA, second edition, 1999.

2. W. L. Briggs and V. E. Hensen. *The DFT: An Owner's Manual for the Discrete Fourier Transform*. The Society for Industrial and Applied Mathematics, Philadelphia, PA, 1995.

3. E. O. Brigham. *The Fast Fourier Transform and Its Applications*. Prentice-Hall, Inc., Upper Saddle River, NJ, 1988.

4. B. Porat. *A Course in Digital Signal Processing*. John Wiley & Sons, Inc., New York, 1997.

5. W. H. Press, S. A. Teukolsky, W. T. Vetterling, and B. P. Flannery. *Numerical Recipes in C++: The Art of Scientific Computing*. Cambridge University Press, Cambridge, UK, second edition, 2001.

6. H. J. Weaver. *Applications of Discrete and Continuous Fourier Analysis*. John Wiley & Sons, Inc., New York, 1983.

Chapter 8

The Discrete Fourier Transform of a Windowed Sequence

In Chapter 7 we established that after a signal $x(t)$ is sampled, we can only hope to compute the values of $X_I(f) = \mathcal{F}\{x(t)P_{\triangle t}(t)\}$, which is the Fourier transform of the sampled sequence. As discussed initially in Chapter 6 and more than once in Chapter 7, whether the central period of $X_I(f)$ agrees with or closely approximates $X(f) = \mathcal{F}\{x(t)\}$ is determined by the chosen sampling rate $\mathbb{R} = 1/\triangle t$, which cannot be changed after the signal has been sampled. When they don t agree with each other, the Fourier transform of the sequence $X_I(f)$ is said to contain *aliased* frequencies. While we were concerned with the mathematical relationship between the sample values of $X_I(f)$ and the sample values of the signal $x(t)$ in Chapter 7, in this chapter we are concerned with computing the numerical values of $X_I(f)$ from a nite sequence of N samples, assuming that we have some knowledge about the duration or periodicity of the signal $x(t)$ so that we can decide on the sample size N.

8.1 A Rectangular Window of Infinite Width

To set the stage, we begin with the simplest case involving time-limited signals, because the optimal sample size N can be determined by the chosen sampling rate $\mathbb{R} = 1/\triangle t$ and the n ite duration $\mathbb{T}$ of the signal $x(t)$, i.e., $N = \mathbb{T}/\triangle t$. If N is odd, the DFT of the N-sample sequence $\{x_k\}$ was given by Equation (7.40), which computes

$$(8.1) \qquad U_r = \frac{1}{N}X_I\left(\frac{r}{N\triangle t}\right) = \frac{1}{N}\sum_{k=-n}^{n} x_k\,\omega_N^{-rk}, \quad -n \leq r \leq n, \ \ N = 2n+1.$$

If N is even, the corresponding DFT computes

$$(8.2) \qquad U_r = \frac{1}{N}X_I\left(\frac{r}{N\triangle t}\right) = \frac{1}{N}\sum_{k=-n}^{n+1} x_k\,\omega_N^{-rk}, \quad -n \leq r \leq n+1, \ \ N = 2n+2.$$

Recall that when N is odd, the N samples do not reach either end of the interval $[-\mathbb{T}/2, \mathbb{T}/2]$ (see Figure 2.9), whereas when N is even, we must include a sample at one end of the interval $[-\mathbb{T}/2, \mathbb{T}/2]$ (see Figure 2.11). Because a window function for either case can be easily

239

modi ed to take care of the other case, we shall assume $N = 2n+1$ in the remainder of this chapter.

To unify our treatment of all windows, let us represent the sampled time-limited signal by the impulse train $x_I(t) = x(t)P_{\triangle t}(t)$. Here we interpret $x_I(t)$ as containing N impulses with potentially nonzero strengths as well as an nite number of impulses with zero strengths. To arti cially perform a null windowing operation, we multiply $x_I(t)$ by a rectangular window of *infinite width*, which is simply the constant function $w_\infty(t) = 1$ for all $t \in (-\infty, \infty)$. Recall that $W_\infty(f) = \mathcal{F}\{w_\infty(t)\} = \delta(f)$; hence, by invoking Product Theorem 6.3 we may express the Fourier transform of the windowed sequence $\mathcal{F}\{x_I(t) \cdot w_\infty(t)\}$ by

(8.3) $$\mathcal{F}\{x_I(t) \cdot w_\infty(t)\} = X_I(f) * W_\infty(f) = X_I(f) * \delta(f) = X_I(f).$$

As expected, the null windowing operation does not alter $X_I(f)$, which is a periodic function with period $\mathbb{R} = 1/\triangle t$. The N equally spaced samples taken from one period of $X_I(f)$ span one period of the sampled sequence $[X_I(f) * W_\infty(f)] \cdot P_{\mathbb{R}/N}(f)$, and their values are obtained by DFT formula (8.1). Note that the samples of $X_I(f)$ are spaced by $\triangle f = \mathbb{R}/N = 1/(N\triangle t) = 1/\mathbb{T}$.

Observe that the periodic sequence $\frac{1}{\mathbb{T}}[X_I(f) * W_\infty(t)] \cdot P_{1/\mathbb{T}}(f)$ is the Fourier transform of $[x_I(t) \cdot w_\infty(t)] * P_{\mathbb{T}}(t)$, and the latter is a periodic sequence resulting from replicating the N samples of $x(t)$ over intervals equal to its duration $T = N\triangle t$. Accordingly, the periodic sequence in the time domain repeats the original N samples with no overlap, the periodic sequence in the frequency domain takes N samples from each period of $X_I(f)$, and the two sequences form a Fourier transform pair:

(8.4) $$\overbrace{[x_I(t) \cdot w_\infty(t)]}^{\text{identical to } x_I(t)} * P_{\mathbb{T}}(t) \underset{\text{periodic sequence in time domain}}{\Longleftrightarrow} \frac{1}{\mathbb{T}}\overbrace{[X_I(f) * W_\infty(f)]}^{\text{identical to } X_I(f)} \cdot P_{1/\mathbb{T}}(f).$$

periodic sequence in frequency domain

Note that in obtaining the transform pair we invoke Convolution Theorem 6.1, and we make use of the Fourier transform of the impulse train given by Theorem 6.7:

(8.5) $$P_{\mathbb{T}}(t) \iff \frac{1}{\mathbb{T}} P_{1/\mathbb{T}}(f), \quad \text{where } \mathbb{T} = N\triangle t.$$

We conclude this section with the following remarks:

Remark 1. Sampling $X_I(f) * W_\infty(f)$ in the frequency domain results in periodic extension of the windowed sequence $x_I(t) \cdot w_\infty(t)$ in the time domain.

Remark 2. The DFT and the IDFT directly relate the two periodic sequences. Using the DFT we obtain

(8.6) $$U_r = \triangle t \left[\frac{1}{\mathbb{T}}X_I\left(\frac{r}{\mathbb{T}}\right)\right] = \frac{1}{N}X_I\left(\frac{r}{\mathbb{T}}\right) = \frac{1}{N}\sum_{k=-n}^{n} x_k\, \omega_N^{-rk}, \quad -n \le r \le n;$$

using the IDFT we obtain

(8.7) $$x_k = \sum_{r=-n}^{n} U_r\, \omega_N^{rk}, \quad -n \le k \le n, \quad N = 2n+1.$$

8.2 A Rectangular Window of Appropriate Finite Width

In this section we consider windowing a periodic sequence obtained by sampling a periodic signal $y^p(t) = x(t) * P_T(t)$, where $x(t)$ is time limited and has finite duration T. We assume that a sampling rate $\mathbb{R} = 1/\triangle t$ has been chosen, and that with $N = T/\triangle t$ we are able to obtain N samples over each period T. As before, we assume $N = 2n+1$. The periodic sequence $y_I^p(t)$ may be viewed as the periodic extension of the N-sample sequence defined by the impulse train $x_I(t) = x(t)P_{\triangle t}(t)$; i.e., $y_I^p(t) = x_I(t) * P_T(t)$, where $T = N\triangle t$. To extract N samples over a finite duration T from the infinite impulse train $y_I^p(t) = x_I(t) * P_T(t)$, we multiply $y_I^p(t)$ by a rectangular window function of width T, which is defined by

$$(8.8) \qquad w_{\text{rect}}(t) = \begin{cases} 1, & \text{for } t \in (-T/2, T/2); \\ 0, & \text{for } |t| \geq T/2. \end{cases}$$

We recall from Equations (7.30) and (7.31) in Section 7.5 that the Fourier transform of an infinite periodic sequence is given by

$$\mathcal{F}\{y_I^p(t)\} = \mathcal{F}\{x_I(t) * P_T(t)\}$$

$$= \mathcal{F}\{x(t)P_{\triangle t}(t)\} \cdot \mathcal{F}\{P_T(t)\}$$

$$(8.9) \qquad = \frac{1}{T} \underbrace{\left[\frac{1}{\triangle t} \sum_{k=-\infty}^{\infty} X\left(f - \frac{k}{\triangle t}\right) \right]}_{\text{periodic } X_I(f), \text{ period } \mathbb{R} = 1/\triangle t} \cdot P_{1/T}(f)$$

$$= \underbrace{\frac{1}{T} \sum_{r=-\infty}^{\infty} X_I\left(\frac{r}{T}\right) \delta\left(f - \frac{r}{T}\right).}_{\text{Fourier transform of } y_I^p(t)}$$

We may now determine the Fourier transform of the truncated N-sample sequence $\mathcal{F}\{y_I^p(t) \cdot w_{rect}(t)\}$ by invoking Product Theorem 6.3:

$$\mathcal{F}\{\underbrace{y_I^p(t) \cdot w_{rect}(t)}_{\text{one period}}\} = \mathcal{F}\{y_I^p(t)\} * \mathcal{F}\{w_{rect}(t)\}$$

$$(8.10) \qquad = \underbrace{\left[\frac{1}{T} \sum_{r=-\infty}^{\infty} X_I\left(\frac{r}{T}\right) \delta\left(f - \frac{r}{T}\right) \right]}_{\text{transform of periodic sequence } y_I^p(t)} * T\left[\frac{\sin(\pi T f)}{\pi T f} \right]$$

$$= \underbrace{\sum_{r=-\infty}^{\infty} X_I\left(\frac{r}{T}\right) \frac{\sin \pi T \left(f - \frac{r}{T}\right)}{\pi T \left(f - \frac{r}{T}\right)}.}_{\text{transform of truncated } N \text{ samples}}$$

Once again, sampling the periodic transform $\mathcal{F}\{y_I^p(t) \cdot w_{rect}(t)\}$ results in the periodic extension of the truncated sequence $y^p(t) \cdot w_{rect}(t)$, and we obtain the transform pair

$$(8.11) \qquad \underbrace{[y_I^p(t) \cdot w_{\text{rect}}(t)] * P_T(t)}_{\text{periodic sequence in time domain}} \Longleftrightarrow \underbrace{\frac{1}{T} \left[\sum_{r=-\infty}^{\infty} X_I\left(\frac{r}{T}\right) \frac{\sin \pi T \left(f - \frac{r}{T}\right)}{\pi T \left(f - \frac{r}{T}\right)} \right] \cdot P_{1/T}(f)}_{\text{periodic sequence in frequency domain}}$$

We show next that the sampled transform returns the values of $X_I(f)$ at $f = \frac{r}{T}$ for all $r \in (-\infty, \infty)$:

$$\overbrace{\frac{1}{T}\left[\sum_{r=-\infty}^{\infty} X_I\left(\frac{r}{T}\right)\frac{\sin \pi T\left(f - \frac{r}{T}\right)}{\pi T\left(f - \frac{r}{T}\right)}\right]\cdot P_{1/T}(f)}^{\mathcal{F}\{y_I^p(t)\cdot w_{\text{rect}}(t)\}}$$

$$= \frac{1}{T}\sum_{m=-\infty}^{\infty}\left[\sum_{r=-\infty}^{\infty} X_I\left(\frac{r}{T}\right)\frac{\sin \pi T\left(f - \frac{r}{T}\right)}{\pi T\left(f - \frac{r}{T}\right)}\right]\delta\left(f - \frac{m}{T}\right)$$

$$= \frac{1}{T}\sum_{r=-\infty}^{\infty} X_I\left(\frac{r}{T}\right)\left[\sum_{m=-\infty}^{\infty}\frac{\sin \pi T\left(f - \frac{r}{T}\right)}{\pi T\left(f - \frac{r}{T}\right)}\delta\left(f - \frac{m}{T}\right)\right]$$

(8.12)

$$= \frac{1}{T}\sum_{r=-\infty}^{\infty} X_I\left(\frac{r}{T}\right)\left[\sum_{m=-\infty}^{\infty}\frac{\sin \pi T\left(\frac{m}{T} - \frac{r}{T}\right)}{\pi T\left(\frac{m}{T} - \frac{r}{T}\right)}\delta\left(f - \frac{m}{T}\right)\right]$$

$$\underbrace{= \frac{1}{T}\sum_{r=-\infty}^{\infty} X_I\left(\frac{r}{T}\right)\delta\left(f - \frac{r}{T}\right), \quad T = N\triangle t,}_{\text{same as Fourier transform of } y_I^p(t)}$$

$$= \frac{1}{\triangle t}\sum_{r=-\infty}^{\infty}\frac{1}{N}X_I\left(\frac{r}{T}\right)\delta\left(f - \frac{r}{T}\right).$$

In deriving the result above, we have made use of the fact that

(8.13) $$\frac{\sin \pi T\left(\frac{m}{T} - \frac{r}{T}\right)}{\pi T\left(\frac{m}{T} - \frac{r}{T}\right)} = \begin{cases} 0, & \text{if } m \neq r; \\ 1, & \text{if } m = r. \quad \text{(de ned by the limit of } \sin 0/0) \end{cases}$$

Note that these are the same results obtained when evaluating $\mathbf{sinc}(x) = \sin \pi x/(\pi x)$ at zero or nonzero integers.

When DFT is applied to the truncated one period (N-sample) sequence $y_I^p(t)\cdot w_{\text{rect}}(t) = x(t)P_{\triangle t}(t)$, we know from Equation (8.6) that the DFT produces the periodic sequence

(8.14) $$U_r = \frac{1}{N}X_I\left(\frac{r}{T}\right) = \frac{1}{N}\sum_{k=-n}^{n} x_k \omega_N^{-rk}, \quad -n \leq r \leq n, \quad N = 2n + 1.$$

As before, the IDFT transforms the U_r s back to the x_k s. Therefore, the DFT and the IDFT directly relate the sampled periodic sequence to its transform provided that the width of the truncating window $w_{\text{rect}}(t)$ is appropriately chosen to be the period T of the envelope signal.

The DFT results are equally valid if they are applied to $M > N$ samples over multiple full periods $\mathbb{T} = \kappa T$, where κ is an integer. (This case was studied in detail in Section 4.3; see also Example 4.5, Table 4.1, and Figure 4.2 in Chapter 4.) This means that the width of the window function may be chosen so that M samples span the duration $M\triangle t = \kappa T = \mathbb{T}$.

To incorporate this case in our derivation above, we can simply allow the time-limited $x(t)$ (which we use to generate the envelope periodic function $y^p(t)$) to represents κ periods of $z^p(t)$ of period $T_o = T/\kappa$, where κ is an integer. Clearly, the envelope functions $y^p(t) = z^p(t)$; hence, the sampled sequence $y_I^p(t) = z_I^p(t)$. Because we are talking about two identical periodic sequences, the Fourier transform results we derived in terms of $X_I(f)$ in Equation (8.9) remain valid for the sequence $\{z_I^p\}$.

8.3 Frequency Distortion by Improper Truncation

While we assume, as before, that the signal $y^p(t)$ is periodic with period T, and that it has been sampled at intervals of $\triangle t$ with $T = N\triangle t$, we now consider modifying the width of the rectangular window: we shall assume that the periodic sequence $y_I^p(t) = [\,x(t)P_{\triangle t}(t)\,] {*} P_T(t) = x_I(t) * P_T(t)$ is truncated by a rectangular window of width αT, where α is *not* an integer. Hence we have either $\alpha T < T$ or $\alpha T > T$, and the window function is de ned by

$$(8.15) \qquad w_{\text{rect}}(t) = \begin{cases} 1, & \text{for } t \in (-\alpha T/2, \alpha T/2); \\ 0, & \text{for } |t| \ge \alpha T/2, \end{cases} \qquad \text{where } \alpha \text{ is not an integer.}$$

Without loss of generality we assume that $\alpha < 1$, and that there are L samples in the windowed sequence; i.e., the truncated sequence spans the duration $L\triangle t = \alpha T$. Consequently, when the DFT is applied to the truncated L samples de ned by the product $y_I^p(t) {\cdot} w_{\text{rect}}(t)$, the results are samples of the transform given by

$$
\begin{aligned}
\mathcal{F}\left\{y_I^p(t){\cdot}w_{\text{rect}}(t)\right\} &= \mathcal{F}\left\{y_I^p(t)\right\} * \mathcal{F}\left\{w_{\text{rect}}(t)\right\} \\
&= \mathcal{F}\left\{x_I(t) * P_T(t)\right\} * \mathcal{F}\left\{w_{\text{rect}}(t)\right\} \\
&= \left[\frac{1}{T}X_I(f){\cdot}P_{1/T}(f)\right] * \alpha T\left[\frac{\sin(\pi\alpha Tf)}{\pi\alpha Tf}\right] \\
&= \underbrace{\left[\frac{1}{T}\sum_{r=-\infty}^{\infty}X_I\left(\frac{r}{T}\right)\delta\left(f - \frac{r}{T}\right)\right]}_{\text{transform of periodic sequence } y_I^p(t)} * \alpha T\left[\frac{\sin(\pi\alpha Tf)}{\pi\alpha Tf}\right] \\
&= \underbrace{\sum_{r=-\infty}^{\infty}\alpha X_I\left(\frac{r}{T}\right)\frac{\sin\pi\alpha T\left(f - \frac{r}{T}\right)}{\pi\alpha T\left(f - \frac{r}{T}\right)}}_{\text{transform of truncated } L \text{ samples}}, \qquad \alpha < 1.
\end{aligned}
$$

(8.16)

If we sample the Fourier transform by the impulse train $P_{1/\alpha T}(t)$, we obtain the transform pair:

$$\overbrace{\left[y_I^p(t) \cdot w_{\text{rect}}(t) \right] * P_{\alpha T}(t)}^{\text{periodic extension of } z_I(t)}$$

$z_I(t)$: L samples

(8.17)

$$\Longleftrightarrow \quad \frac{1}{\alpha T} \overbrace{\underbrace{\left[\sum_{r=-\infty}^{\infty} \alpha X_I\left(\frac{r}{T}\right) \frac{\sin \pi \alpha T \left(f - \frac{r}{T}\right)}{\pi \alpha T \left(f - \frac{r}{T}\right)} \right]}_{Z_I(f): \text{ transform of } z_I(t)} \cdot P_{1/\alpha T}(f)}^{\text{sampling } Z_I(f) \text{ at intervals of } \triangle f = 1/(\alpha T)}$$

$$= \frac{1}{\alpha T} \sum_{m=-\infty}^{\infty} \underbrace{\left[\sum_{r=-\infty}^{\infty} \alpha X_I\left(\frac{r}{T}\right) \frac{\sin \pi \alpha T \left(f - \frac{r}{T}\right)}{\pi \alpha T \left(f - \frac{r}{T}\right)} \right]}_{Z_I(f)} \delta\left(f - \frac{m}{\alpha T}\right)$$

$$= \frac{1}{\alpha T} \sum_{m=-\infty}^{\infty} \underbrace{\left[\sum_{r=-\infty}^{\infty} \alpha X_I\left(\frac{r}{T}\right) \frac{\sin \pi \alpha T \left(\frac{m}{\alpha T} - \frac{r}{T}\right)}{\pi \alpha T \left(\frac{m}{\alpha T} - \frac{r}{T}\right)} \right]}_{\text{value of } Z_I(f) \text{ at } f = m/(\alpha T)} \delta\left(f - \frac{m}{\alpha T}\right)$$

$$= \frac{1}{\triangle t} \sum_{m=-\infty}^{\infty} \frac{1}{L} Z_I\left(\frac{m}{L\triangle t}\right) \delta\left(f - \frac{m}{L\triangle t}\right), \quad \because \alpha T = L\triangle t.$$

What we have obtained above is the Fourier transform of the periodic sequence defined by the convolution product $z_I(t) * P_{\alpha T}(t)$. Because $z_I(t) * P_{\alpha T}(t)$ does not represent the original periodic sequence $y_I^p(t)$, we expect discrepancies in the frequency domain as well. In particular, we note the following:

- $\sin \pi \alpha T \left(\frac{m}{\alpha T} - \frac{r}{T}\right) = \sin \pi(m - \alpha \cdot r) \neq 0$ for all integer $r \neq 0$. ($\because 0 < \alpha < 1$)

- the frequency content has been distorted:

  (8.18) $$Z_I\left(\frac{m}{L\triangle t}\right) = \sum_{r=-\infty}^{\infty} \alpha X_I\left(\frac{r}{N\triangle t}\right) \frac{\sin \pi(m - \alpha \cdot r)}{\pi(m - \alpha \cdot r)} \neq X_I\left(\frac{m}{L\triangle t}\right)$$

- using the DFT on L samples we can only obtain the distorted transform values:

  (8.19) $$U_r = \frac{1}{L} Z_I\left(\frac{m}{L\triangle t}\right) = \frac{1}{L} \sum_{k=-\tau}^{\tau} x_k \omega_N^{-rk}, \quad -\tau \leq r \leq \tau, \quad L = 2\tau + 1$$

- the periodic extension of the L signal samples in the time domain does not represent the original periodic sequence because

  (8.20) $$y_I^p = x_I(t) * P_T(t) \neq z_I(t) * P_{\alpha T}(t).$$

8.4   Windowing a General Nonperiodic Sequence

Recall that in Section 8.1 we performed the windowing of a finite sequence $x_I(t)$ by $w_\infty(t)$ for DFT computation. To perform the task on an infinite sequence $x_I(t)$, we only need to replace the transform pair $w_\infty(t) \Longleftrightarrow \delta(f)$ u sed in Section 8.1 with a rectangular window of nite width $\mathbb{T}$:

$$(8.21) \qquad w_{\text{rect}}(t) = \begin{cases} 1, & \text{for } t \in (-\mathbb{T}/2, \mathbb{T}/2); \\ 0, & \text{for } |t| \geq \mathbb{T}/2, \end{cases}$$

and its transform:

$$(8.22) \qquad W(f) = \mathcal{F}\{w_{\text{rect}}(t)\} = \mathbb{T}\left[\frac{\sin \pi \mathbb{T} f}{\pi \mathbb{T} f}\right].$$

Assuming that $\mathbb{T} = N\triangle t$, where $\triangle t$ denotes the predetermined sampling interval, the Fourier transform of the truncated N-sample sequence becomes

$$(8.23) \qquad U_I(f) = \mathcal{F}\{\underbrace{x_I(t) \cdot w_{\text{rect}}(t)}_{N \text{ samples}}\} = X_I(f) * W(f) = \underbrace{X_I(f) * \mathbb{T}\left[\frac{\sin \pi \mathbb{T} f}{\pi \mathbb{T} f}\right]}_{\text{ripples are added to } X_I(f)}.$$

Therefore, the DFT of the N samples computes the values of $U_I(f)$, and Equation (8.1) may be used on $U_I(f)$ in exactly the same manner:

$$(8.24) \qquad U_r = \frac{1}{N}U_I\left(\frac{r}{\mathbb{T}}\right) = \frac{1}{N}\sum_{k=-n}^{n} x_k\, \omega_N^{-rk}, \quad -n \leq r \leq n, \quad N = 2n+1.$$

As has been done in every case, using the IDFT we obtain

$$(8.25) \qquad x_k = \sum_{r=-n}^{n} U_r\, \omega_N^{rk}, \quad -n \leq k \leq n, \quad N = 2n+1.$$

As always, we have $U_{r+N} = U_r$ and $x_{k+N} = x_k$; hence, the DFT and IDFT results represent two periodic sequences which form a Fourier transform pair (regardless of the fact that the N samples were truncated from a nonperiodic signal):

$$(8.26) \qquad \overbrace{[\, x_I(t) \cdot w_{\text{rect}}(t)\,] * P_{\mathbb{T}}(t)}^{\text{periodic extension of } N \text{ samples}} \Longleftrightarrow \overbrace{\frac{1}{\mathbb{T}}\, U_I(f) \cdot P_{1/\mathbb{T}}(f)}^{\text{sampling } U_I(f)}, \quad \text{where } \mathbb{T} = N\triangle t,$$

$$= \frac{1}{\triangle t}\sum_{r=-\infty}^{\infty} \frac{1}{N}U_I\left(\frac{r}{\mathbb{T}}\right)\delta\left(f - \frac{r}{\mathbb{T}}\right).$$

Since a nonperiodic signal of in nite duration must be truncated before we can apply the DFT, the resulting frequency distortion cannot be completely eliminated. To improve the accuracy and resolving power of the DFT, we need to study further the roles played by various window functions in a quantitative manner. It turns out that a tapered window can truncate and modify the sampled signal values at the same time; the latter role is particularly important when jump discontinuities are caused by abrupt truncation. We shall examine the properties of windows in the next section.

8.5 Frequency-Domain Properties of Windows

In order to compare different windows without bias, they are assumed to have the same length $\mathbb{T} = N\triangle t$, where $\triangle t$ denotes the sampling interval, and $N = 2n+1$ represents the number of samples in the windowed signal sequence. The actual length of each window can be set according to the application and the resolution required. In this section we shall characterize each window based on the properties of its Fourier transform $W(f)$, because the severity of frequency distortion, i.e., the extent of smearing and leakage, caused by the convolution of $W(f)$ and the signal s Fourier transform can be alleviated if $W(f)$ has good properties.

8.5.1   The rectangular window

Recall the Fourier transform of a rectangular window given by

$$(8.27) \qquad W_{\text{rect}}(f) = \mathcal{F}\{w_{\text{rect}}(t)\} = \mathbb{T}\left[\frac{\sin(\pi\mathbb{T}f)}{\pi\mathbb{T}f}\right] = \mathbb{T}\left[\frac{\sin(\pi\lambda)}{\pi\lambda}\right] = \mathbb{T}\,\textbf{sinc}(\lambda),$$

where $\lambda = \mathbb{T}f = (N\triangle t)f$. Since the length $\mathbb{T}$ is x ed for all windows in this section (and it is canceled out when convolving with the transform of the signal sequence), it is suf cient to consider the normalized transform

$$\widetilde{W}_{\text{rect}}(\lambda) = \textbf{sinc}(\lambda) = \frac{\sin(\pi\lambda)}{\pi\lambda},$$

whose magnitude is plotted in Figure 8.1, where we identify the mainlobe as the central peak between $\lambda = -1$ and $\lambda = 1$ th e nearest two zeros of $\widetilde{W}_{\text{rect}}(\lambda)$ surrounding the origin. The cooresponding two zeros of $W_{\text{rect}}(f)$ are $f_{-1} = -1/\mathbb{T}$ and $f_1 = 1/\mathbb{T}$. Note that the maximum height of the mainlobe in the normalized transform $\widetilde{W}_{\text{rect}}(\lambda)$ is now one. In Figure 8.1 we also identify the sidelobes to be those between two adjacent zeros of $\widetilde{W}_{\text{rect}}(\lambda)$, namely the peaks and valleys between $\lambda = k$ and $\lambda = k+1$, and they correspond to $f_k = k/\mathbb{T}$ and $f_{k+1} = (k+1)/\mathbb{T}$ the neighboring zeros of $W_{\text{rect}}(f)$.

To illustrate the disproportionally small sidelobes at high frequencies, we plot the magnitude of $\widetilde{W}_{\text{rect}}(\lambda)$ using logarithm representations in Figure 8.1. Customarily, the logarithm of the normalized magnitude spectrum

$$\log_{10}|\widetilde{W}_{\text{rect}}(\lambda)|$$

is further scaled by 20 and expressed as

$$(8.28) \qquad\qquad 20\cdot\log_{10}|\widetilde{W}_{\text{rect}}(\lambda)| \quad \text{d ecibel units   or dB.}$$

Since the maximum height has been normalized to one, we have $0 \leq |\widetilde{W}_{\text{rect}}(\lambda)| \leq 1$, which leads to zero or negative decibel values. For example, we obtain

$$\begin{aligned} 0 \text{ dB} &= 20\cdot\log_{10} 1 && \text{at } \lambda = 0; \\ -20 \text{ dB} &= 20\cdot\log_{10} 0.1 && \text{for a 10-fold reduction in magnitude;} \\ -40 \text{ dB} &= 20\cdot\log_{10} 0.01 && \text{for a 100-fold reduction in magnitude;} \end{aligned}$$

and so on.

To measure how $\widetilde{W}_{\text{rect}}(\lambda)$ deviates from the ideal unit impulse $\delta(\lambda)$, we need to quantify the width of the mainlobe as well as the relative magnitudes of the sidelobes. The narrower the mainlobe and the lower the sidelobes (compared with the mainlobe), the closer the transform $\widetilde{W}_{\text{rect}}(\lambda)$ approximates $\delta(\lambda)$. The following quantities are commonly used to describe the spectral characteristics of a rectangular window:

Figure 8.1 The rectangular window and its magnitude spectrum.

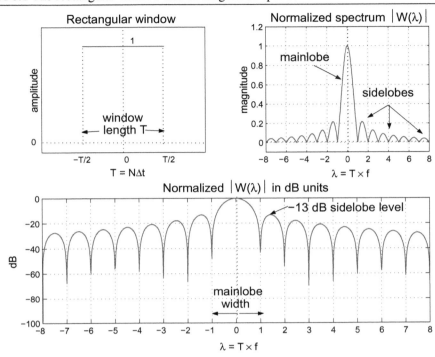

1. The 3-dB bandwidth of the mainlobe in the graph of $20 \cdot \log_{10} |\widetilde{W}_{\text{rect}}(\lambda)|$: This quantity measures the width of the mainlobe on either side of the origin when its height is reduced to $\frac{1}{\sqrt{2}} \approx 0.707$, because

$$-3 \text{ dB} \approx 20 \cdot \log_{10} 0.707.$$

By solving $\widetilde{W}_{\text{rect}}(\lambda) = \sin(\pi\lambda)/(\pi\lambda) = 0.707$ for unknown λ, one obtains the 3 dB bandwidth $\lambda_{3db} \approx 0.443$ or $f_{3db} = \lambda/\mathbb{T} \approx 0.443/\mathbb{T}$, where $\mathbb{T} = N\triangle t$.

2. The highest sidelobe level in the graph of $20\log_{10} |\widetilde{W}_{\text{rect}}(\lambda)|$: For the rectangular window, its spectrum in Figure 8.1 shows that the sidelobe level reaches as high as -13 dB at $\lambda = 1.5$ (or $f = 1.5/\mathbb{T}$):

(8.29) $$20 \cdot \log_{10} |\widetilde{W}_{\text{rect}}(1.5)| \approx -13 \text{ dB}.$$

8.5.2 The triangular window

The triangular window is also known as the Bartlett window. A triangular window of length $\mathbb{T}$ and unit height is de ned by

(8.30) $$w_{\text{tri}}(t) = \begin{cases} 1 - 2|t|/\mathbb{T}, & \text{for } t \in (-\mathbb{T}/2, \mathbb{T}/2); \\ 0, & \text{for } |t| \geq \mathbb{T}/2. \end{cases}$$

Note that a triangular window can be obtained by convolving two identical rectangular windows of half length. If we require the triangular window to have certain height, the convolution product can be scaled accordingly. We can thus express a triangular window of length $\mathbb{T}$ and unit height as

$$(8.31) \qquad w_{\text{tri}}(t) = \frac{2}{\mathbb{T}}\left[\hat{w}_{\text{rect}}(t) * \hat{w}_{\text{rect}}(t)\right],$$

where the rectangular window $\hat{w}_{\text{rect}}(t)$ and its Fourier transform $\hat{W}_{\text{rect}}(f)$ are obtained by using $\mathbb{T}/2$ t o replace every occurrence of length $\mathbb{T}$ in Equations (8.8) and (8.27), i.e.,

$$(8.32) \qquad \hat{w}_{\text{rect}}(t) = \begin{cases} 1, & \text{for } t \in (-\mathbb{T}/4, \mathbb{T}/4); \\ 0, & \text{for } |t| \geq T/4. \end{cases} \iff \hat{W}_{\text{rect}}(f) = \frac{\mathbb{T}}{2}\left[\frac{\sin(\frac{1}{2}\pi\mathbb{T}f)}{\frac{1}{2}\pi\mathbb{T}f}\right].$$

By invoking Convolution Theorem 6.1, we obtain the Fourier transform of the triangular window as the square of $\hat{W}_{\text{rect}}(f)$:

$$(8.33) \qquad W_{\text{tri}}(f) = \mathcal{F}\{w_{\text{tri}}(t)\} = \frac{2}{\mathbb{T}}\left[\hat{W}_{\text{rect}}(f)\cdot\hat{W}_{\text{rect}}(f)\right] = \frac{\mathbb{T}}{2}\left[\frac{\sin\left(\frac{1}{2}\pi\mathbb{T}f\right)}{\frac{1}{2}\pi\mathbb{T}f}\right]^2.$$

Before we quantify the properties of $W_{\text{tri}}(f)$, we normalize it so its mainlobe has unit height, and we obtain

$$(8.34) \qquad \widetilde{W}_{\text{tri}}(\lambda) = \left[\frac{\sin\left(\frac{1}{2}\pi\lambda\right)}{\frac{1}{2}\pi\lambda}\right]^2 = \mathbf{sinc}^2\left(\tfrac{1}{2}\lambda\right), \text{ where } \lambda = \mathbb{T}f = (N\triangle t)f.$$

In Figure 8.2 we compare the magnitude spectrum $|\widetilde{W}_{\text{tri}}(\lambda)|$ of the triangular window with that of the rectangular window. While the triangular window shows lower sidelobe levels, its mainlobe is wider than that of the rectangular window. Quantitatively, we obtain

$$(8.35) \qquad 20\cdot\log_{10}|\widetilde{W}_{\text{tri}}(0.64)| \approx -3 \text{ dB}; \quad 20\cdot\log_{10}|\widetilde{W}_{\text{tri}}(3.0)| \approx -27 \text{ dB}.$$

Hence, the triangular window is characterized by the 3-dB bandwidth $\lambda_{3\text{db}} \approx 0.64$ (or $f_{3\text{db}} \approx 0.64/\mathbb{T}$) and the highest sidelobe level of -27 dB. Observe further that $|\widetilde{W}_{\text{tri}}(\lambda)|$ has zeros at $\lambda = \pm 2, \pm 4, \cdots$, and that the peak of its r st sidelobe reaches -27 dB at $\lambda = 3.0$.

8.5.3   The von Hann window

A von Hann window of length $\mathbb{T}$ and unit height is de ned by

$$(8.36) \qquad w_{\text{hann}}(t) = \begin{cases} 0.5 + 0.5\cos\frac{2\pi t}{\mathbb{T}}, & \text{for } t \in (-\mathbb{T}/2, \mathbb{T}/2); \\ 0, & \text{for } |t| \geq \mathbb{T}/2. \end{cases}$$

Figure 8.2 The triangular window and its magnitude spectrum.

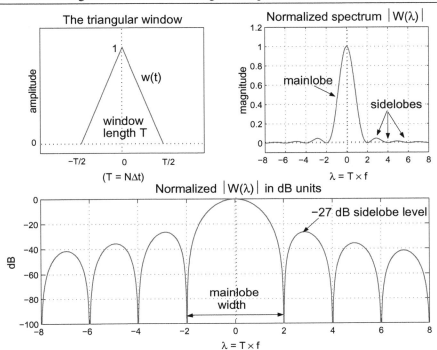

Its Fourier transform is given by

$$W_{\text{hann}}(f) = \mathcal{F}\{w_{\text{hann}}(t)\}$$

$$= \int_{-\infty}^{\infty} \left[0.5 + 0.5\cos\frac{2\pi t}{\mathbb{T}}\right] e^{-j2\pi ft}\, dt$$

$$= 0.25 \int_{-\mathbb{T}/2}^{\mathbb{T}/2} \left[2e^{-j2\pi ft} + e^{-j2\pi(f-1/\mathbb{T})t} + e^{-j2\pi(f+1/\mathbb{T})t}\right] dt$$

(8.37)

$$= 0.5\mathbb{T} \cdot \frac{\sin \pi\mathbb{T}f}{\pi\mathbb{T}f} + 0.25\mathbb{T} \cdot \frac{\sin(\pi\mathbb{T}f - \pi)}{\pi\mathbb{T}f - \pi} + 0.25\mathbb{T} \cdot \frac{\sin(\pi\mathbb{T}f + \pi)}{\pi\mathbb{T}f + \pi}$$

$$= \mathbb{T} \cdot \frac{\sin(\pi\mathbb{T}f)}{\pi} \left[\frac{0.5}{\mathbb{T}f} + \frac{0.25}{1 - \mathbb{T}f} - \frac{0.25}{1 + \mathbb{T}f}\right]$$

$$= \mathbb{T} \frac{\sin(\pi\mathbb{T}f)}{\pi\mathbb{T}f} \left[\frac{0.5}{1 - \mathbb{T}^2 f^2}\right].$$

We denote the normalized $\frac{2}{\mathbb{T}} W_{\text{hann}}(f)$ by

(8.38) $$\widetilde{W}_{\text{hann}}(\lambda) = \frac{\sin(\pi\lambda)}{\pi\lambda(1 - \lambda^2)} = \frac{\text{sinc}(\lambda)}{1 - \lambda^2}, \quad \text{where } \lambda = \mathbb{T}f = (N\triangle t)f.$$

Quantitatively, we obtain

(8.39) $$20 \cdot \log_{10}|\widetilde{W}_{\text{hann}}(0.721)| \approx -3 \text{ dB}; \quad 20 \cdot \log_{10}|\widetilde{W}_{\text{hann}}(2.5)| \approx -32 \text{ dB}.$$

Therefore, the von Hann window s 3-dB bandwidth is $\lambda_{3db} \approx 0.721$ (or $f_{3db} \approx 0.721/\mathbb{T}$), and its highest sidelobe level is -32 dB reached at $\lambda = 2.5$ or $f = 2.5/\mathbb{T}$ as shown in Figure 8.3.

Observe further that the zeros of $\widetilde{W}_{\text{hann}}(\lambda)$ occur at $\lambda = \pm k$ for integer $k \geq 2$. For $\lambda = \pm 1$, because $\widetilde{W}_{\text{hann}}(\pm 1) = 0/0$ is in an indeterminate form, we establish $\widetilde{W}_{\text{hann}}(\lambda) = 1/2$ in the limit as $\lambda \to \pm 1$ by applying L Hospital s rule:

$$\lim_{\lambda \to \pm 1} \widetilde{W}_{\text{hann}}(\lambda) = \lim_{\lambda \to \pm 1} \frac{\sin(\pi\lambda)}{\pi\lambda(1 - \lambda^2)} = \lim_{\lambda \to \pm 1} \frac{\cos(\pi\lambda)}{1 - 3\lambda^2} = \frac{1}{2}.$$

Hence the height of the mainlobe has dropped 50% at $\lambda = 1$. Because $20 \log_{10} \frac{1}{2} = -6$ dB, we have obtained the 6-dB bandwidth $\lambda_{6db} = 1$ for the von Hann window.

Figure 8.3 The von Hann window and its magnitude spectrum.

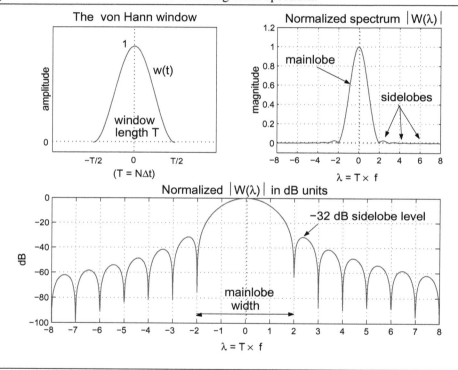

8.5.4 The Hamming window

A Hamming window of length $\mathbb{T}$ and unit height is de ned by

(8.40)
$$w_{\text{ham}}(t) = \begin{cases} 0.575 + 0.425 \cos \frac{2\pi t}{\mathbb{T}}, & \text{for } t \in (-\mathbb{T}/2, \mathbb{T}/2); \\ 0, & \text{for } |t| \geq \mathbb{T}/2. \end{cases}$$

Its Fourier transform follows directly from our derivation of $W(f)$ for the closely related von Hann window:

$$
\begin{aligned}
W_{\text{ham}}(f) &= \mathcal{F}\{w_{\text{ham}}(t)\} \\
&= \int_{-\infty}^{\infty} \left[0.575 + 0.425 \cos \frac{2\pi t}{\mathbb{T}} \right] e^{-j2\pi ft}\, dt \\
&= 0.575\mathbb{T} \cdot \frac{\sin \pi \mathbb{T} f}{\pi \mathbb{T} f} + 0.2125\mathbb{T} \cdot \frac{\sin(\pi \mathbb{T} f - \pi)}{\pi \mathbb{T} f - \pi} + 0.2125\mathbb{T} \cdot \frac{\sin(\pi \mathbb{T} f + \pi)}{\pi \mathbb{T} f + \pi} \\
&= \mathbb{T} \cdot \frac{\sin(\pi \mathbb{T} f)}{\pi} \left[\frac{0.575}{\mathbb{T} f} + \frac{0.2125}{1 - \mathbb{T} f} - \frac{0.2125}{1 + \mathbb{T} f} \right] \\
&= \mathbb{T} \cdot \frac{\sin(\pi \mathbb{T} f)}{\pi \mathbb{T} f} \left[\frac{0.575 - 0.15\mathbb{T}^2 f^2}{1 - \mathbb{T}^2 f^2} \right].
\end{aligned}
$$

(8.41)

Again we normalize $W_{\text{ham}}(f)$ so that its height is reduced to unity, and the result is denoted by

(8.42) $\qquad \widetilde{W}_{\text{ham}}(\lambda) = \frac{1}{0.575} \frac{\sin(\pi\lambda)}{\pi\lambda} \left[\frac{0.575 - 0.15\lambda^2}{1 - \lambda^2} \right]$, where $\lambda = \mathbb{T} f = (N\triangle t)f$.

Quantitatively, we obtain

(8.43) $\qquad 20 \cdot \log_{10}|\widetilde{W}_{\text{ham}}(0.608)| \approx -3\text{ dB}; \qquad 20 \cdot \log_{10}|\widetilde{W}_{\text{ham}}(3.5)| \approx -35\text{ dB}.$

Therefore, as shown in Figure 8.4, the Hamming window s 3-dB bandwidth is $\lambda_{3db} \approx 0.608$ or $f_{3db} \approx 0.608/\mathbb{T}$, and its highest sidelobe level is -35 dB, which is the peak of the second sidelobe reached at $\lambda = 3.5$ or $f = 3.5/\mathbb{T}$.

8.5.5 The Blackman window

A Blackman window of length $\mathbb{T}$ and unit height is de ned by

(8.44) $\qquad w_{\text{bkm}}(t) = \begin{cases} 0.42 + 0.5 \cos \frac{2\pi t}{\mathbb{T}} + 0.08 \cos \frac{4\pi t}{\mathbb{T}}, & \text{for } t \in (-\mathbb{T}/2, \mathbb{T}/2); \\ 0, & \text{for } |t| \geq \mathbb{T}/2. \end{cases}$

Its Fourier transform follows directly from our derivation of $W_{\text{ham}}(f)$ for the Hamming window in the last section:

$$
\begin{aligned}
W_{\text{bkm}}(f) &= \mathcal{F}\{w_{\text{bkm}}(t)\} \\
&= \int_{-\infty}^{\infty} \left[0.42 + 0.5 \cos \frac{2\pi t}{\mathbb{T}} + 0.08 \cos \frac{4\pi t}{\mathbb{T}} \right] e^{-j2\pi ft}\, dt \\
&= \mathbb{T} \cdot \frac{\sin(\pi \mathbb{T} f)}{\pi} \left[\frac{0.42}{\mathbb{T} f} + \frac{0.25}{1 - \mathbb{T} f} - \frac{0.25}{1 + \mathbb{T} f} - \frac{0.04}{2 - \mathbb{T} f} + \frac{0.04}{2 + \mathbb{T} f} \right] \\
&= \mathbb{T} \cdot \frac{\sin(\pi \mathbb{T} f)}{\pi \mathbb{T} f} \left[\frac{1.68 - 0.18\,\mathbb{T}^2 f^2}{(1 - \mathbb{T}^2 f^2)(4 - \mathbb{T}^2 f^2)} \right].
\end{aligned}
$$

(8.45)

The function $W_{\text{bkm}}(f)$ is then normalized to have unit height at the origin. We denote the result by

(8.46) $\qquad \widetilde{W}_{\text{bkm}}(\lambda) = \frac{1}{0.42} \frac{\sin(\pi\lambda)}{\pi\lambda} \left[\frac{1.68 - 0.18\lambda^2}{(1 - \lambda^2)(4 - \lambda^2)} \right]$, where $\lambda = \mathbb{T} f = (N\triangle t)f$.

Figure 8.4 The Hamming window and its magnitude spectrum.

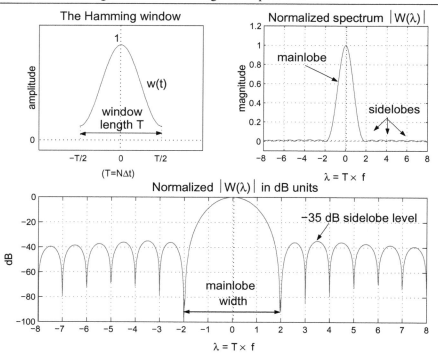

We now have

(8.47) $20 \cdot \log_{10} |\widetilde{W}_{\text{bkm}}(0.82)| \approx -3$ dB; $20 \cdot \log_{10} |\widetilde{W}_{\text{bkm}}(3.5)| \approx -58$ dB.

Accordingly, as shown in Figure 8.5, the Blackman window s 3-dB bandwidth is $\lambda_{3db} \approx 0.82$ or $f_{3db} \approx 0.82/\mathbb{T}$, and its highest sidelobe level is -58 dB, which is the peak of the rst sidelobe reached at $\lambda = 3.5$ or $f = 3.5/\mathbb{T}$.

8.6   Applications of the Windowed DFT

For easy reference in this section, we summarize the performance characteristics of the ve windows discussed in the last section in Table 8.1.

To illustrate how windows can impact various DFT-based applications, we shall use different windows in the task of determining the sinusoidal components of an unknown signal $x(t)$ from a sequence of its samples. Since the signal underlying the given samples is unknown, we may encounter one of the scenarios discussed below.

8.6.1   Several scenarios

The signal $x(t)$ is periodic and band-limited, but we do not know its period. Let T denote the signal s unknown period, and let $x(t)$ be represented by a nite Fourier series:

(8.48) $x(t) = \sum_{k=-n}^{n} C_k e^{j2\pi kt/T}.$

Figure 8.5 The Blackman window and its magnitude spectrum.

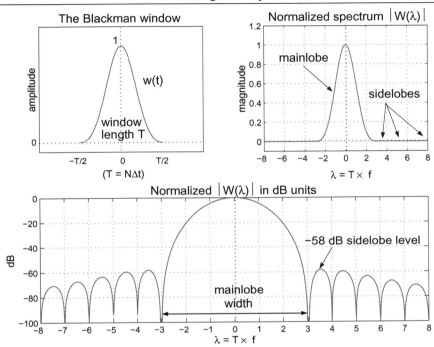

Table 8.1 Spectral characteristics of ve windows ($\lambda = \mathbb{T}f = (N\triangle t)f$).

Window of length $\mathbb{T}$: $w(t)$	Fourier transform $W(f) = \mathcal{F}\{w(t)\}$	Mainlobe 3-dB width	Sidelobe peak level	Zeros of Fourier transform $\widetilde{W}(\lambda)$
Rectangular	$\mathbb{T}\cdot\widetilde{W}_{\text{rect}}(\lambda)$	$\lambda_{3db} = 0.443$	-13 dB	$\lambda = \pm 1, \pm 2, \pm 3, \cdots$
Triangular	$0.5\mathbb{T}\cdot\widetilde{W}_{\text{tri}}(\lambda)$	$\lambda_{3db} = 0.640$	-27 dB	$\lambda = \pm 2, \pm 4, \pm 6, \cdots$
von Hann	$0.5\mathbb{T}\cdot\widetilde{W}_{\text{hann}}(\lambda)$	$\lambda_{3db} = 0.721$	-32 dB	$\lambda = \pm 2, \pm 3, \pm 4, \cdots$
Hamming	$0.575\mathbb{T}\cdot\widetilde{W}_{\text{ham}}(\lambda)$	$\lambda_{3db} = 0.608$	-35 dB	$\lambda = \pm 2, \pm 3, \pm 4, \cdots$
Blackman	$0.42\mathbb{T}\cdot\widetilde{W}_{\text{bkm}}(\lambda)$	$\lambda_{3db} = 0.820$	-58 dB	$\lambda = \pm 3, \pm 4, \pm 5, \cdots$

Our objective is to determine the numerical value of $f_k = k/T$ for each sinusoidal component with $C_k \neq 0$.

Because we do not know the period T, we cannot choose the length $\mathbb{T}$ of the rectangular window to be one or multiple full periods; we must assume that the sampled interval $\mathbb{T} = N\triangle t \neq mT$ for any integer $m \geq 1$. Since the effect of aliasing is caused by the sampling rate $\mathbb{R} = 1/\triangle t$ alone, we may assume, without loss of generality, that the signal has been sampled above the Nyquist rate in this study, which means that we have

$$(8.49) \qquad\qquad 2f_{\max} = 2f_n = \frac{2n}{T} < \frac{1}{\triangle t}.$$

Recall that

$$(8.50) \qquad X(f) = \mathcal{F}\{x(t)\} = \sum_{k=-n}^{n} C_k \cdot \delta\left(f - \frac{k}{T}\right), \quad \text{where } |f| \leq \frac{n}{T} < \frac{1}{2\triangle t}.$$

and it follows that

$$(8.51) \qquad X_I(f) = \mathcal{F}\{x(t) \cdot P_{\triangle t}(t)\} = \frac{1}{\triangle t} \cdot \sum_{r-\infty}^{\infty} X\left(f - \frac{r}{\triangle t}\right),$$

which represents the shifted replications of the entire $X(f)$ over intervals of length $1/\triangle t$. Because $X(f)$ consists of $2n+1$ impulses weighted by C_k for $-n \leq k \leq n$, and they spread over the distance of $2f_{\max} < 1/\triangle t$, there is no overlap when the entire group of $2n+1$ impulses are replicated over intervals of length $1/\triangle t$ to form the periodic $X_I(f)$. Hence $X_I(f)$ is a periodic sequence, with $2n+1$ weighted impulses per period. Again, the period of $X_I(f)$ is the sampling rate $\mathbb{R} = 1/\triangle t$, and we have $X_I(f) = \frac{1}{\triangle t} \cdot X(f)$ in the central period, i.e.,

$$(8.52) \qquad \begin{aligned} X_I(f) &= \frac{1}{\triangle t} \cdot X(f) = \frac{1}{\triangle t} \sum_{k=-n}^{n} C_k \cdot \delta\left(f - \frac{k}{T}\right) \quad \text{for } |f| \leq \frac{1}{2\triangle t}, \\ &= \frac{1}{\triangle t} \sum_{k=-n}^{n} C_k \cdot \delta\left(f - \frac{\alpha \cdot k}{\mathbb{T}}\right) \quad \text{if } \mathbb{T} = \alpha \cdot T. \end{aligned}$$

Observe that when $\mathbb{T} = \alpha \cdot T$ and α is not an integer, $\alpha \cdot k$ is not an integer, either.

Now, suppose all C_k s are equal to zero except for $C_{\pm 3} \neq 0$, $C_{\pm 4} \neq 0$, and $C_{\pm n} \neq 0$. How could we detect them and obtain the numerical values of $f_3 = 3/T$, $f_4 = 4/T$, and $f_n = n/T$? Remember that we know *neither* the value of k *nor* the value of T, so what we shall try to obtain are the numerical values of

$$f_{3\alpha} = \frac{3\alpha}{\mathbb{T}}, \quad f_{4\alpha} = \frac{4\alpha}{\mathbb{T}}, \quad \text{and } f_{n\alpha} = \frac{n \cdot \alpha}{\mathbb{T}},$$

where $\mathbb{T} = \alpha T$, with both α and T representing values unknown to us. For example, if it happens that $\mathbb{T} = 2.2\,T$, then our task is to detect the presence of

$$f_{3\alpha} = \frac{6.6}{\mathbb{T}}, \quad f_{4\alpha} = \frac{8.8}{\mathbb{T}}, \quad \text{and } f_{n\alpha} = \frac{2.2n}{\mathbb{T}}$$

from processing the N samples collected over the duration $\mathbb{T} = N\triangle t$.

To proceed with such a task of frequency detection, we assume, at rst, that N samples have been truncated from the sampled signal by a rectangular window of length $\mathbb{T} = N\triangle t = \alpha T$, and that $\mathbb{T} > T$. Recall that

$$\mathcal{F}\{\overbrace{x_I(t) \cdot w_{\text{rect}}(t)}^{\text{truncated sequence}}\} = X_I(f) * W_{\text{rect}}(f)$$

$$= \left[\frac{1}{\triangle t} \sum_{k=-n}^{n} C_k \cdot \delta\left(f - \frac{\alpha \cdot k}{\mathbb{T}}\right)\right] * W_{\text{rect}}(f)$$

(8.53)
$$= \frac{1}{\triangle t} \sum_{k=-n}^{n} C_k \cdot W_{\text{rect}}\left(f - \frac{\alpha \cdot k}{\mathbb{T}}\right)$$

$$= \frac{\mathbb{T}}{\triangle t} \sum_{k=-n}^{n} C_k \cdot \left[\frac{\sin \pi \mathbb{T}\left(f - \frac{\alpha \cdot k}{\mathbb{T}}\right)}{\pi \mathbb{T}\left(f - \frac{\alpha \cdot k}{\mathbb{T}}\right)}\right]$$

$$= N \sum_{k=-n}^{n} C_k \cdot \mathbf{sinc}\left(\mathbb{T}f - \alpha \cdot k\right). \qquad (\because \mathbb{T} = N\triangle t)$$

For convenience we associate the factor $1/N$ with the left-hand side, i.e., we de ne

(8.54)
$$U_I(f) = \frac{1}{N}\mathcal{F}\{x_I(t) \cdot w_{\text{rect}}(t)\} = \sum_{k=-n}^{n} C_k \cdot \mathbf{sinc}\left(\mathbb{T}f - \alpha \cdot k\right).$$

Because $X_I(f)$ repeats the $2n+1$ weighted impulses in each period, we obtain $U_I(f) = \frac{1}{N}X_I(f) * W_{\text{rect}}(f)$ by replicating and summing up $\mathbf{sinc}(\mathbb{T}f - \alpha \cdot k)$, multiplied by C_k, at locations $f = \alpha \cdot k/\mathbb{T}$. Since all C_k s are zero except for $C_{\pm 3}$, $C_{\pm 4}$ and $C_{\pm n}$, we obtain the one-sided spectrum of $U_I(f)$, as shown in Figure 8.6, for $\alpha = 2.2$ and $n = 11$. Observe that $U_I(f)$ has local maxima at $f = 6.6/\mathbb{T}$, $8.8/\mathbb{T}$, and $2.2\,n/\mathbb{T}$. Hence, by detecting the local maxima of $U_I(f)$ we also determine the numerical values of f (at which each local maximum occurs), and they represent the frequencies of the sinusoidal components present in the sampled signal.

Observe also that $U_I(f)$ is nonzero within all mainlobes and sidelobes present over the entire spectrum. Since the mainlobe of the Fourier transform of a rectangular window of length $\mathbb{T}$ covers a subinterval of $2/\mathbb{T}$ centered at the location of each impulse in $X_I(f)$, we see that it is desirable to choose $\mathbb{T}$ long enough so that any neighboring impulses in $X_I(f)$ are separated by more than $2/\mathbb{T}$, otherwise the two adjacent mainlobes overlap and will *smear out* (and merge) the closely spaced local maxima. Using our example, when $\mathbb{T} = 2.2 \cdot T$, the distance between $3/T = 6.6/\mathbb{T}$ and $4/T = 8.8/\mathbb{T}$ is $2.2/\mathbb{T}$, and there is no overlap of mainlobes as shown in Figure 8.7; if $\mathbb{T} = 1.5 \cdot T$, then the distance between $3/T = 4.5/\mathbb{T}$ and $4/T = 6/\mathbb{T}$ is $1.5/\mathbb{T}$, and the two adjacent mainlobes will overlap as shown in Figures 8.8 and 8.9. It is clear that when a signi cant portion of the two mainlobes overlap, we won t be able to distinguish two closely spaced local maxima if they are of similar strengths. Recall that the mainlobe width varies with both the length and type of the window. Therefore, if we replace the rectangular window (which has the narrowest mainlobe) with a data-weighting window (with increased mainlobe width), the reduction of frequency resolution is expected unless we increase the length of the data-weighting window to compensate for that.

Assuming that the local maxima are distinguishable in $U_I(f)$, we still have to address how to obtain them. Recall that the DFT computes the sample values of $U_I(f)$ at $f_k = k/\mathbb{T}$ for integer $-n \leq k \leq n$: for example, as shown in Figure 8.10, the local maximum at $f = 6.6/\mathbb{T}$ falls between f_6 and f_7; the local maximum at $f = 8.8/\mathbb{T}$ falls between f_8 and f_9.

Figure 8.6 The one-sided spectrum of $U_I(f) = \frac{1}{N}\mathcal{F}\{x_I(t) \cdot w_{\text{rect}}(t)\}$.

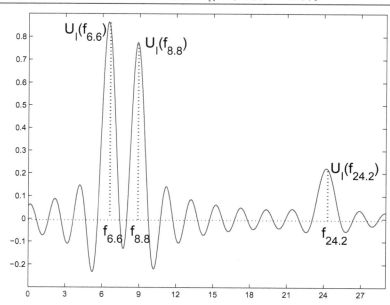

Consequently, if we do not proceed with further search of local maximum between f_6 and f_7 (by chirp Fourier transform to be introduced in Chapter 9), either f_6 or f_7 would be our best estimate of $f = 6.6/\mathbb{T}$, and in either case the error is bounded by $\triangle f = 1/\mathbb{T}$. It is clear that the longer the sampled duration $\mathbb{T}$ or the larger the number of samples N (because $\mathbb{T} = N\triangle t$), the smaller the error in the estimated frequency detected by the DFT.

A data-weighting window can play an important role in frequency detection when we need to resolve large differences in signal amplitudes. Since the rectangular window has high side-lobe levels, a weak local maxima of $U_I(f)$ may not rise above the frequency leakage from the sidelobes of a strong component. For example, the peak sidelobe level of the rectangular window is -13 dB, which represents the reduction in magnitude from 100% to 22%; therefore, a weak local maxima with magnitude being lower than 22% of the strong component is not distinguishable within the sidelobes of the latter. Note that when $X_I(f)$ is a weighted impulse train, the sidelobes represent additional frequency contents (which were not present in the original signal), and the higher the sidelobe levels the more likely a weak local maximum is masked. The sidelobe leakage can be reduced if a suitable data-weighting window presented in last section is used, and the Fourier transform of the weighted or so-called windowed sequence is given by

(8.55)
$$U_I(f) = \frac{1}{N}\mathcal{F}\{\overbrace{x_I(t) \cdot w_{\text{name}}(t)}^{\text{windowed sequence}}\} = \frac{1}{N}X_I(f) * W_{\text{name}}(f)$$
$$= \sum_{k=-n}^{n} C_k \cdot \frac{1}{\mathbb{T}}W_{\text{name}}\left(f - \frac{\alpha \cdot k}{\mathbb{T}}\right).$$

In Figure 8.10 we show the sidelobe leakage caused by a rectangular window. In Figures 8.11 we show the Fourier transform $U_I(f) = \frac{1}{N}\mathcal{F}\{z_I(t) \cdot w_{\text{name}}(t)\}$ using four different windows of

Figure 8.7 Non-overlapped mainlobes and separate local maxima.

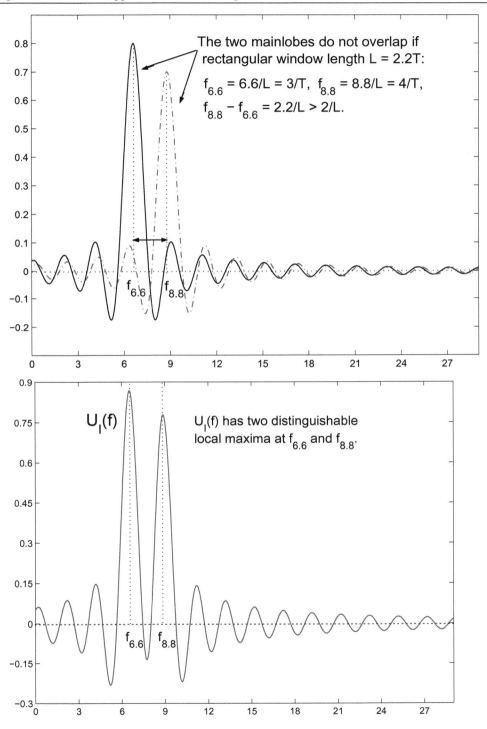

Figure 8.8 The merging of local maxima due to overlapped mainlobes.

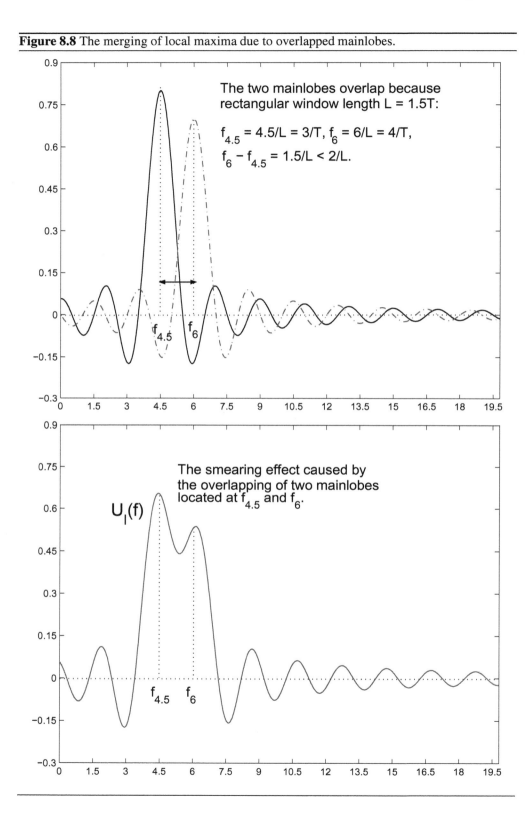

Figure 8.9 A local maximum is smeared out by overlapped mainlobes.

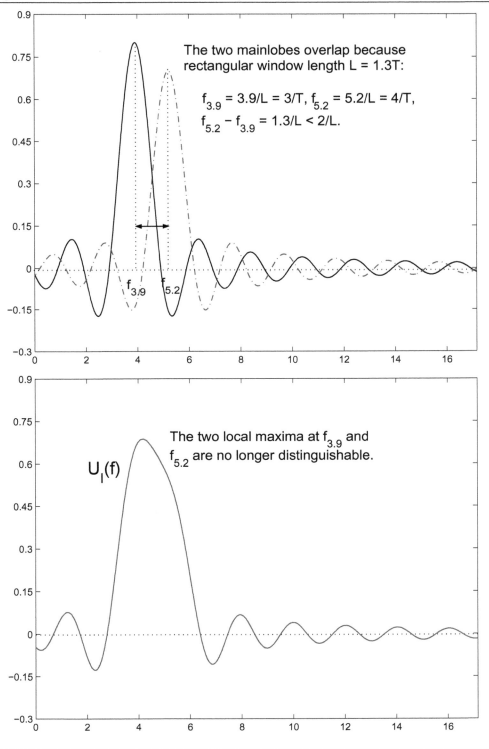

The two mainlobes overlap because rectangular window length L = 1.3T:

$f_{3.9} = 3.9/L = 3/T$, $f_{5.2} = 5.2/L = 4/T$,

$f_{5.2} - f_{3.9} = 1.3/L < 2/L$.

The two local maxima at $f_{3.9}$ and $f_{5.2}$ are no longer distinguishable.

$U_l(f)$

Figure 8.10 Values of $U_I(f_k)$ obtainable by the DFT, where $f_k = k/\mathbb{T}$ ($\mathbb{T} = 2.2T$).

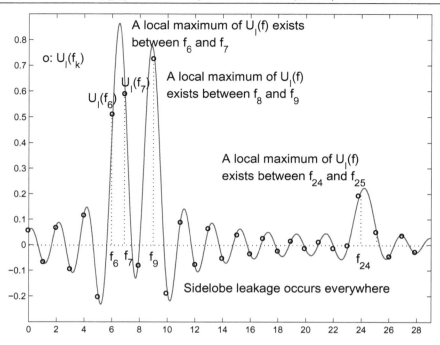

length $\mathbb{T}$, where $z_I(t)$ are sampled from $z(t)$ given by

$$z(t) = 1.2\cos\frac{6\pi t}{T} + 1.8\cos\frac{8\pi t}{T} + 0.2\cos\frac{22\pi t}{T},$$

and the window length $\mathbb{T} = 2.2T$. In Figure 8.11 we illustrate that a weak local maximum masked by the high sidelobes of a rectangular window can be unmasked when the rectangular window is replaced by a triangular window, a von Hann window, or a Blackman window (of the same length). Observe that while the Blackman window has the lowest sidelobe level, it has the widest mainlobe and the smearing effect caused by that is also evident in Figure 8.11.

In Figures 8.12, 8.13, 8.14, and 8.15, we plot the one-sided discrete spectrum of $z_I(t) \cdot w_{\text{name}}(t)$. In each case the spectrum we show consists of the computed DFT coefficients $\{Z_0, Z_1, \ldots, Z_{29}\}$, which come from a set of $N = 58$ DFT coefficients $\{Z_{-28} \ldots, Z_0, \ldots, Z_{29}\}$ we have obtained by performing the DFT on 58 samples of the windowed sequence $z_I(t) \cdot w_{\text{name}}(t)$. The samples of $z(t)$ are truncated and weighted by each window defined on the interval $(-\mathbb{T}/2, \mathbb{T}/2] = (-1/2, 1/2]$. The graphs of the one-sided Fourier transform $U_I(f)$ are also drawn in dotted lines in Figures 8.12, 8.13, 8.14, and 8.15. We demonstrate that in each case the numerical values of the DFT coefficients are samples of $U_I(f) = \frac{1}{N}\mathcal{F}\{z_I(t) \cdot w_{\text{name}}(t)\}$ taken at $f = k/\mathbb{T}$ for integer k. Recall that the window length $\mathbb{T} = 2.2T$; hence, the sampled sequence does not span an integer number of periods of $z(t)$, and the local maxima may occur between the computed DFT coefficients as shown in these figures.

Figure 8.11 Fourier transforms of $z_I(t)$ weighted by four different windows.

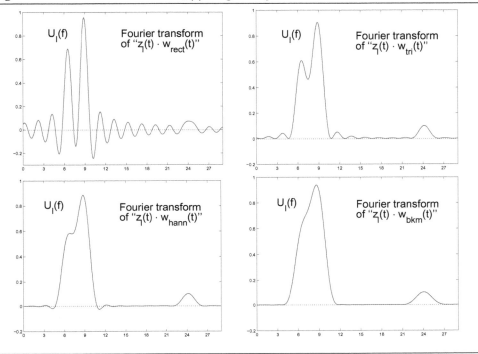

Figure 8.12 The computed DFT of $z_I(t)$ truncated by a rectangular window.

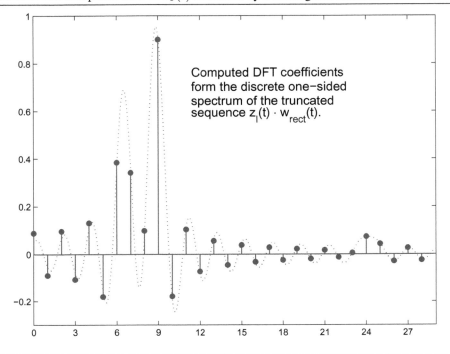

Figure 8.13 The computed DFT of $z_I(t)$ weighted by a triangular window.

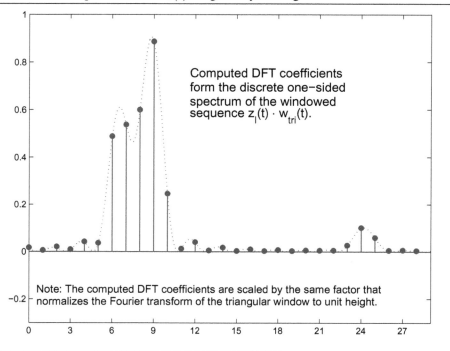

Computed DFT coefficients form the discrete one-sided spectrum of the windowed sequence $z_I(t) \cdot w_{tri}(t)$.

Note: The computed DFT coefficients are scaled by the same factor that normalizes the Fourier transform of the triangular window to unit height.

Figure 8.14 The computed DFT of $z_I(t)$ weighted by a von Hann window.

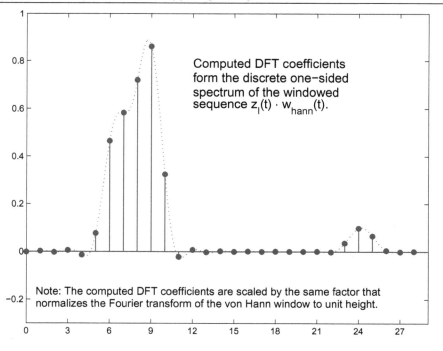

Computed DFT coefficients form the discrete one-sided spectrum of the windowed sequence $z_I(t) \cdot w_{hann}(t)$.

Note: The computed DFT coefficients are scaled by the same factor that normalizes the Fourier transform of the von Hann window to unit height.

Figure 8.15 The computed DFT of $z_I(t)$ weighted by a Blackman window.

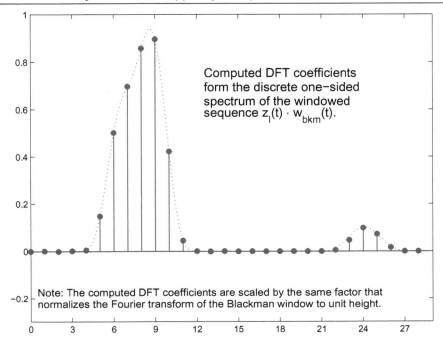

Computed DFT coefficients form the discrete one-sided spectrum of the windowed sequence $z_I(t) \cdot w_{bkm}(t)$.

Note: The computed DFT coefficients are scaled by the same factor that normalizes the Fourier transform of the Blackman window to unit height.

8.6.2   Selecting the length of DFT in practice

From our discussion above, the length of the window $\mathbb{T} = N\triangle t$ also de nes the DFT of length N. This would be the case if computing the N samples of $U_I(f)$ spaced by $1/\mathbb{T}$ serves our purpose. However, in practice, this may not be the nal step, and we may want to change the length of DFT for a number of reasons. However, let us rst say that changing the DFT length to suit a particular FFT implementation is not our major concern here, because, as will be shown in Part II of this book, FFT algorithms for arbitrary composite N or prime N are all available, so the use of FFT imposes no restriction on the length of DFT. The need to change the DFT length may arise under the following circumstances:

1. We recall the usage of zero padding from Section 4.6 in Chapter 4: when the DFT N-sample spectrum is too sparse for us to visualize a continuous analog spectrum $U_I(f)$, one may wish to decrease the spectral spacing $\triangle f$ on the frequency grid. Recall that $\triangle f = 1/(N\triangle t)$; hence, $\triangle f$ can be reduced if we enlarge N by adding zeros to the signal samples.

    The method and effects of zero padding the truncated signal (before the DFT) were studied in detail in Section 4.6.1 in Chapter 4 (with examples illustrated in Figures 4.7, 4.8, and 4.9 and Table 4.3), and we show the same effects of zero padding the windowed (truncated and modi ed) signal sequence in Figure 8.16, where the plot on the left shows the 29 DFT coef cients of the 58-sample sequence $z_I(t) \cdot w_{tr}(t)$ before zero padding, and the plot on the right shows the effect after the sequence is doubled by zero padding. Observe that by zero padding we do not change $U_I(f)$; what we obtain are additional sample values of the same $U_I(f)$, and the additional data points are obtained by interpolating between the original N sample values.

Figure 8.16 The effects of zero padding a windowed sequence.

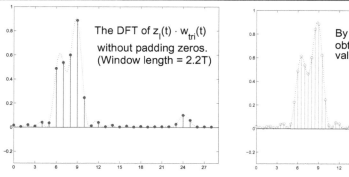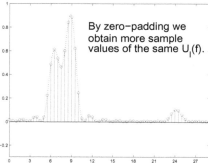

2. In case we can re-sample the signal, we may want to experiment the DFT on longer sample sequences for better frequency resolution, because the mainlobe width of any chosen window is inversely proportional to the duration of the signal.

   In Figure 8.17 we compare the two different $U_I(f) = \frac{1}{N}X_I(f) * W_{tri}(f)$ obtained from using triangular windows of lengths $\mathbb{T} = 2.2T$ and $\mathbb{T} = 4.4T$ for the same example in Figures 8.13. Recall that we compute the DFT coef cients of $z_I(t){\cdot}w_{tri}(t)$ using window length $\mathbb{T}=2.2T$, and they are shown to be the sample values of the corresponding $U_I(f)$ at $f = k/\mathbb{T}$ in Figure 8.13. Consequently, when we double the length of the triangular window, we not only halve the spacing $\triangle f$ on the frequency grid, the computed DFT coef cients are actually samples of a different $U_I(f)$, and we illustrate the improved DFT results in Figure 8.18.

3. In case we can re-sample the signal at different sampling rates, we may want to exper-iment DFT on longer sample sequence obtained over the same duration $\mathbb{T}$ at increased sampling rate (or decreased sampling interval $\triangle t$) until there is no change in the spec-trum. This is one experimental way to make certain that the sampling rate is high enough to eliminate the aliasing effect. In Figure 8.19 we compare the DFT results of $z_I(t)$ weighted by a triangular window of length $\mathbb{T}=4.4T$ before and after the sampling rate is increased to satisfy the Nyquist condition. Observe that because the sampling duration $\mathbb{T}=4.4T$ does not change, we have doubled the number of samples by doubling the sam-pling rate; while the spacing $\triangle f = 1/\mathbb{T}$ on the frequency grid remains unchanged, the range of frequencies we can detect is extended because $f_{max} = 1/(2\triangle t)$ doubles when the sampling interval $\triangle t$ is halved.

4. Recall that the error associated with the frequency measurement using the N-sample sequence is bounded by the spacing $\triangle f = 1/(N\triangle t)$. While this error bound may be reduced to $\triangle f/N = 1/(N^2\triangle t)$ using a N^2-sample sequence, it may not be ef cient (and it is not necessary) to compute the DFT of the whole N^2 sample sequence. It turns out that by using the chirp Fourier transform (to be covered in Chapter 9, Section 9.3), instead of computing N^2 samples, we can compute only N samples of $U_I(f)$ over a chosen segment of length $1/(N\triangle t)$, assuming that the particular segment is known to contain the local maximum of interest to us. By searching the N samples over $1/(N\triangle t)$,

Figure 8.17 Improving $U_I(f) = \frac{1}{N}\mathcal{F}\{z_I(t)\cdot w_{\text{tri}}(t)\}$ by changing window length.

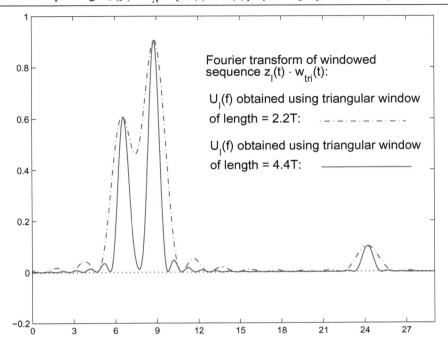

Figure 8.18 The computed DFT of $z_I(t)\cdot w_{\text{tri}}(f)$ after doubling the window length.

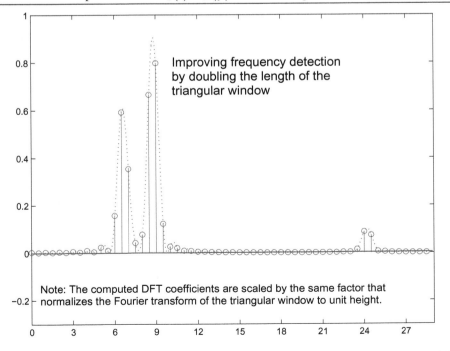

Figure 8.19 Improving frequency detection by doubling the sampling rate.

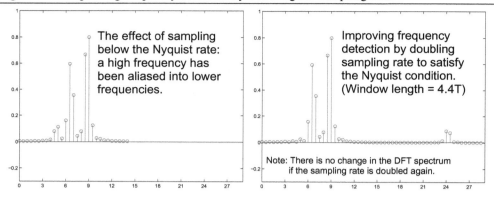

we obtain a more accurate local maximum, with error bound reduced to $1/(N^2 \triangle t)$ as desired.

References

1. A. Ambardar. *Analog and Digital Signal Processing*. Brooks/Cole Publishing Company, Paci c Grove, CA, second edition, 1999.

2. E. O. Brigham. *The Fast Fourier Transform and Its Applications*. Prentice-Hall, Inc., Upper Saddle River, NJ, 1988.

3. T. Butz. *Fourier Transformation for Pedestrians*. Springer-Verlag, Berlin, 2006.

4. R. W. Hamming. *Digital Filters*. Prentice-Hall, Inc., Englewood Cliffs, NJ, third edition, 1989.

5. B. Porat. *A Course in Digital Signal Processing*. John Wiley & Sons, Inc., New York, 1997.

Chapter 9

Discrete Convolution and the DFT

In this chapter we shall introduce rst the linear convolution of two nite sequences, which, on the one hand, allows us to approximate continuous convolution using sampled function values, and on the other hand, is algebraically equivalent to the multiplication of two polynomials. Because we are familiar with the operations involved in multiplying polynomials of arbitrary degrees, the algebraic de nitio n of linear convolution is easy to understand and verify by constructing examples in this context. However, for diverse applications in signal processing and system analysis, it is essential that we interpret the de nition as the discrete counterpart of the continuous convolution. For computational ef ciency we must also learn how to turn the linear convolution into a DFT-based periodic convolution, because the discrete cyclic convolution theorem (to be given in this chapter) tells us that a periodic convolution can be implemented by the DFT, and we already know that DFT can be ef ciently computed by the fast Fourier transform (FFT) algorithms. The periodic convolution is also useful in computing the chirp Fourier transform, which computes a partial DFT in the neighborhood of a particular frequency in order to measure it to greater accuracy a topic to be covered in Section 9.3 in this chapter.

9.1   Linear Discrete Convolution

Because of its close connection to the continuous convolution, the *linear* discrete convolution is also referred to as the *regular* or *conventional* discrete convolution in signal processing literature.

9.1.1   Linear convolution of two finite sequences

We shall begin with a review of the continuous convolution of two nite-duration continuous-time signals. In Figure 9.1 we illustrate again the convolution of two signals $g(t)$ and $h(t)$ as de ned by Equation (6.1) in Section 6.3 of Chapter 6, which we restate here for easy reference:

$$(9.1) \qquad u(t) = g(t) * h(t) \stackrel{\text{def}}{=} \int_{-\infty}^{\infty} g(\lambda) \cdot h(t - \lambda) \, d\lambda, \quad t \in \Re.$$

We indicate in Chapter 6 that we will re-examine and discuss further the convolution steps illustrated in Figure 9.1 when we study how to obtain numerical approximation to the convolution result in this chapter; we now explain the illustrated process step by step below.

Figure 9.1 The steps in performing continuous convolution $u(t) = g(t) * h(t)$.

Step 1.   Choose stationary function $g(t)$, and change t to λ.

Step 2.   Fold $h(\lambda)$ to obtain the moving function $h(-\lambda)$.

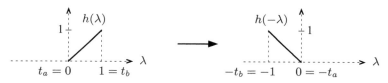

Step 3.   Examples of shifted $h(t - \lambda)$ for $t = -0.5$, $t = 0.5$, and $t = 1.5$.

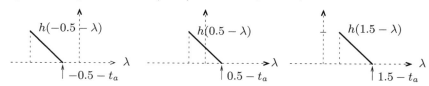

Step 4.   Sample values of $u(t)$ at $t = -0.5$, $t = 0.5$, and $t = 1.5$ are equal to the three shaded areas:

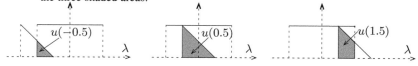

Step 1. Choose one function to be stationary. In Figure 9.1 we let $g(t)$ be the stationary function. By renaming the variable t as λ we obtain $g(\lambda)$ in the integrand.

Step 2. We first obtain $h(\lambda)$ by renaming the variable t as λ, then we obtain the moving function $h(-\lambda)$ by folding $h(\lambda)$ with respect to the ordinate. Observe that the folded function $h(-\lambda)$ is nonzero for $-t_b \leq \lambda \leq -t_a = 0$ so that $0 = t_a \leq -\lambda \leq t_b$ as per the original definition of $h(t)$. (These results remain valid if $t_a \neq 0$.)

Step 3. For each value of t, we obtain shifted $h(t - \lambda)$ by moving $h(-\lambda)$ so that its origin $\lambda = 0$ is positioned at t, which dictates that the right end of $h(-\lambda)$ is positioned at $t - t_a$. (There is no restriction on the value of t_a.)

Step 4. For each value of t, the convolution result $u(t)$ can be obtained analytically (if possible) or by numerical integration of (9.1) over the finite interval where $g(\lambda)$ and $h(t - \lambda)$ overlap. In either case, the value of $u(t)$ represents the area under the curve of $g(\lambda) \cdot h(t - \lambda)$ over the finite interval in which the two overlap.

The continuous convolution result $u(t) = g(t) * h(t)$ is nonzero for $-1 \leq t \leq 2$, which is shown in Figure 9.2.

Figure 9.2 The result of continuous convolution $u(t) = g(t) * h(t)$.

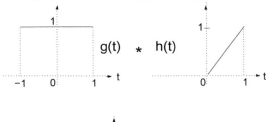

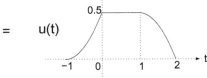

With the steps of continuous convolution laid out above, we can now introduce the linear convolution of two sampled signals $\{g_0, g_1, \cdots, g_{N-1}\}$ and $\{h_0, h_1, \cdots h_{N-1}\}$ in a straightforward manner. The following steps are illustrated in Figure 9.3 for $N = 5$:

Step 1. Choose one sequence to be stationary. In Figure 9.3 we choose $\{g_0, g_1, \cdots, g_{N-1}\}$ to be the stationary sequence.

Step 2. We obtain the moving sequence $\{h_{N-1}, h_{N-2}, \cdots, h_0\}$ by reversing the elements in the given sequence.

Step 3. Beginning with $\ell = 0$, we obtain the linear convolution result $u_0 = g_0 \cdot h_0$ by overlapping the right-most element h_0 of the reversed sequence with the first element g_0 of the stationary sequence.

For each value $0 < \ell \leq N - 1$, we move h_0 to overlap with g_ℓ, then compute the convolution result u_ℓ by summing the pairwise products of all overlapped elements. Note

Figure 9.3 The steps in performing linear discrete convolution $\{u_\ell\} = \{g_\ell\} * \{h_\ell\}$.

Step 1.   Choose $\{g_\ell\} = \{g_0, g_1, g_2, g_3, g_4\}$ as the stationary sequence.

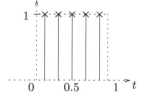

Step 2.   Reverse $\{h_0, h_1, h_2, h_3, h_4\}$ to obtain $\{h_4, h_3, h_2, h_1, h_0\} = \{0.9, 0.7, 0.5, 0.3, 0.1\}$.

Step 3.   Computing u_0:

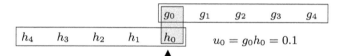

$$u_0 = g_0 h_0 = 0.1$$

Step 4.   Computing $u_1, u_2, \ldots, u_8$:

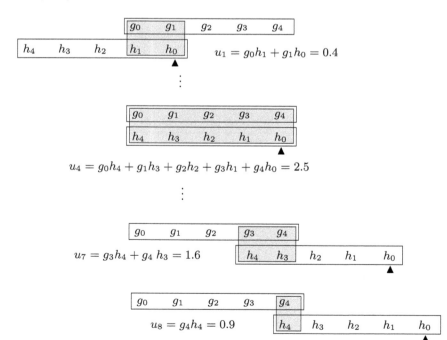

$$u_1 = g_0 h_1 + g_1 h_0 = 0.4$$

$$u_4 = g_0 h_4 + g_1 h_3 + g_2 h_2 + g_3 h_1 + g_4 h_0 = 2.5$$

$$u_7 = g_3 h_4 + g_4 h_3 = 1.6$$

$$u_8 = g_4 h_4 = 0.9$$

that the product $u_\ell \triangle t$, where $\triangle t$ denotes the spacing between samples, approximates the area under the curve of $g(\lambda) \cdot h(t - \lambda)$ over the same interval by composite midpoint rule. Hence, by linear discrete convolution, we compute one value of the corresponding continuous convolution each time.

Step 4. For each value $N - 1 < \ell \le 2N - 2$, we continue to move the reversed sequence one position to the right, and compute u_ℓ by summing the pairwise products of all overlapped elements. (As explained in Step 3, each product $u_\ell \triangle t$ approximates one value of the continuous convolution.)

The discrete convolution result $\{u_\ell\} = \{g_\ell\} * \{h_\ell\}$ is a sequence of length $2N - 1$, which is shown in Figure 9.4. In the same gure we also compare the sequence $\{u_k \triangle t\}$ directly with the continuous convolution $u(t) = g(t) * h(t)$ when $\triangle t = 0.2$, and we show in Figure 9.5 the improvement in the discrete approximation when $\triangle t$ is reduced to 0.1 and 0.05.

Figure 9.4 The result of discrete convolution $\{u_k\} = \{g_k\} * \{h_k\}$.

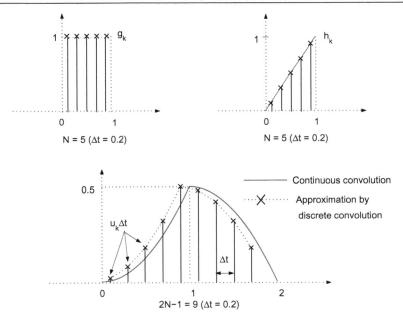

We comment further on the nature of linear discrete convolution below.

- The algebraic equivalence of linear discrete convolution and the multiplication of two polynomials can be easily veri ed. If we let

$$G_4(x) = g_0 + g_1 x + g_2 x^2 + g_3 x^3 + g_4 x^4 = 1 + x + x^2 + x^3 + x^4,$$
$$H_4(x) = h_0 + h_1 x + h_2 x^2 + h_3 x^3 + h_4 x^4 = 1 + 2x + 3x^2 + 4x^3 + 5x^4,$$

then we obtain their product as

$$U_8(x) = (1 + x + x^2 + x^3 + x^4) \times (1 + 2x + 3x^2 + 4x^3 + 5x^4)$$
$$= 1 + 3x + 6x^2 + 10x^3 + 15x^4 + 14x^5 + 12x^6 + 9x^7 + 5x^8$$
$$= u_0 + u_1 x + u_2 x^2 + u_3 x^3 + u_4 x^4 + u_5 x^5 + u_6 x^6 + u_7 x^7 + u_8 x^8,$$

Figure 9.5 The results of discrete convolution $\{u_k\} = \{g_k\} * \{h_k\}$.

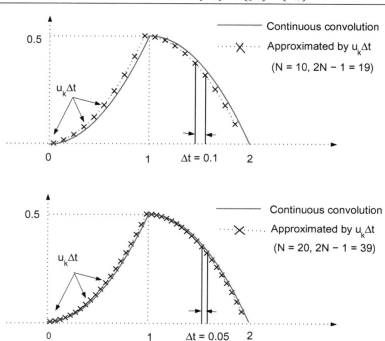

which is a polynomial of degree eight with its coef cients

$$\{u_0,\ u_1,\ u_2,\ u_3,\ u_4,\ u_5,\ u_6,\ u_7\ u_8\} = \{1,\ 3,\ 6,\ 10,\ 15,\ 14,\ 12,\ 9,\ 5\}.$$

By following the steps illustrated in Figure 9.3, we obtain the same result by performing linear convolution on the two coef cient sequences $\{g_0,\ g_1,\ g_2,\ g_3,\ g_4\}$ and $\{h_0,\ h_1,\ h_2,\ h_3,\ h_4\}$, i.e,

$$\{u_0,\ u_1,\ \ldots\ u_8\} = \{1,1,1,1,1\} * \{1,2,3,4,5\} = \{1,3,6,10,15,14,12,9,5\}.$$

- Although we illustrate the linear convolution using two sequences of the same length, the steps for convolving two sequences of different lengths are exactly the same. (In the latter case, it is common to choose the longer sequence as the stationary sequence, because in signal processing when a signal is sampled in real-time and processed by a digital lter, the in nitely long real-time input is convolved with the nite moving sequence which represents the lter. Obviously we *cannot reverse* a general sequence of *infinite length*.)

- If the two sequences are of lengths L_1 and L_2, the result of linear convolution is a sequence of length L_1+L_2-1, which is simply the result of passing the reversed sequence through the stationery sequence until the two separate.

As expected, we can verify this result easily from the equivalent polynomial multiplication. Since a polynomial of degree $N-1$ has N coef cien ts, it is represented by a coef cient sequence of length N. Consequently, if we multiply two polynomials of degree L_1-1 and L_2-1, the resulting product is a polynomial of degree L_1+L_2-2, which has exactly L_1+L_2-1 coef cients .

- For two sequences of lengths L_1 and L_2, the linear convolution involves $L_1 \times L_2$ multiplications. For two sequences of the same length N, the arithmetic cost is proportional to N^2, which can be reduced to $\alpha N \log_2 N$ *after* we learn in Section 9.2.2 how to convert the linear convolution to a periodic convolution, and compute the latter by the DFT (via the FFT.)

  For two sequences of different lengths L_1 and L_2, we will learn in Section 9.2.2 that they must be both zero-padded to the same length $N = L_1 + L_2 - 1$ *before* they can be turned into a periodic convolution and computed via the FFT, which again incurs $\alpha N \log_2 N$ multiplications.

9.1.2 Sectioning a long sequence for linear convolution

As noted at the end of Section 9.1.1, the linear convolution of two sequences of lengths L_1 and L_2 requires $\alpha N \log_2 N$ multiplications (via the FFT), where $N = L_1 + L_2 - 1$. Consequently, when L_1 is very large, the required computer time and/or storage may still be too costly to compute all N results in a single convolution. In such case, the long stationary sequence can be sectioned into segments of length $S_1 \ll L_1$, and each segment of length S_1 is convolved with the moving segment of length $L_2 \le S_1$ at a cost of $\alpha M \log_2 M$ with $M = S_1 + L_2 - 1$. Assuming that $L_1/S_1 = K$, there are K linear convolutions to be performed, and the total cost is given by $\alpha K M \log_2 M$. To illustrate this process, an example with sequence lengths $L_1 = 10$, $L_2 = 5$, and segment length $S_1 = 5$ is given in Figure 9.6. By summing the results from two (short) convolutions we get back the result of the original (long) convolution.

9.2 Periodic Discrete Convolution

The periodic discrete convolution plays a key role in making the FFT (which implements the DFT) ubiquitous for diverse applications in signal processing and system analysis, because the conventional or linear convolution used in these applications can be implemented by a properly formulated periodic convolution, and the latter can be implemented by the DFT (via the FFT) according to the discrete cyclic convolution theorems to be covered in this section. The periodic convolution is also useful in the development of the chirp Fourier transform algorithm (to be covered in Section 9.3) as well as the fast Fourier transform algorithm for arbitrary prime N (to be covered in Chapter 14.)

9.2.1 Definition based on two periodic sequences

To de ne the periodic convolution, we assume that $\{P_\ell\}$ and $\{Q_\ell\}$ are two periodic sequences and they have the same period N. Note that the convolution is to be performed for one period only. However, the result is different from the linear convolution of two truncated nite periods, because the two sequences overlap all the way due to their periodic nature. We illustrate the convolution process in Figure 9.7, and the steps involved are explained further below.

Figure 9.6 Performing linear convolution $\{u_k\} = \{g_k\} * \{h_k\}$ in two sections.

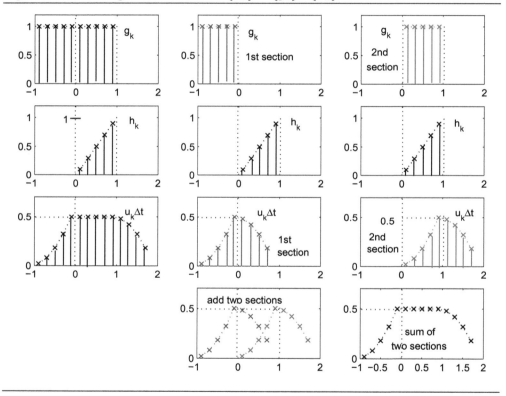

Step 1. Choose one periodic sequence to be stationary. In Figure 9.7 we let one period of $\{P_\ell\}$ be the stationary sequence of length N.

Step 2. We obtain the moving sequence by reversing the elements within each period of the other sequence $\{Q_\ell\}$.

Step 3. We then align the stationary sequence and the reversed moving sequence so that Q_0 overlaps P_0 as shown in Figure 9.7, and we compute u_0, the rst convolution result, as the sum of N pairwise products of the overlapped elements identi ed for this step in Figure 9.7.

Step 4. To continue, we shift the periodic moving sequence one position to the right, and we compute the second convolution result as the sum of N pairwise products of the overlapped elements identi ed for this step in Figure 9.7; we do the same to obtain the remaining $N-2$ convolution results. At this time, the head of our moving sequence has passed all N elements within one period of the stationary sequence $\{P_\ell\}$, and we have completed the periodic convolution de ned on these two sequences.

Observe that the periodic convolution of two sequences of period N is, by de n ition, a nite sequence of length N given by the following equation:

$$(9.2) \qquad U_k = \sum_{\ell=0}^{N-1} P_\ell \cdot Q_{k-\ell}, \quad \text{for } k = 0, 1, \cdots, N-1,$$

where $P_\ell = P_{\ell \pm N}$ and $Q_{k-\ell} = Q_{k-\ell \pm N}$ are satis ed due to their periodicity, which ensures that $U_k = U_{k \pm N}$. Hence, continuing the convolution process beyond one period would simply result in a periodic extension of the rst N results.

9.2.2 Converting linear to periodic convolution

In anticipation of the bene t from using the FFT to compute the discrete convolution, we show next how to implement a linear convolution by a periodic convolution. This turns out to be a simple procedure: suppose we are given two sequences $\{g_\ell\}$ and $\{h_\ell\}$ of lengths L_1 and L_2, the conversion from linear to periodic convolution involves the following steps:

Step 1. Zero-pad both sequences to length $L_1 + L_2 - 1$.

Step 2. Obtain two periodic sequences by extending the two zero-padded sequences periodically. Hence they both have period $N = L_1 + L_2 - 1$.

Step 3. Perform periodic convolution to obtain N results as explained in Section 9.2.1.

We illustrate these steps in Figure 9.8. Since the $N = L_1 + L_2 - 1$ results in Figure 9.8 are identical to those obtained from linear convolution of the original two sequences in Figure 9.3, this process provides an alternative way to compute the same results. It is this alternative process which would bring us the bene t of FFT.

9.2.3 Defining the equivalent cyclic convolution

When we re-examine the periodic convolution process illustrated in Figure 9.7, it is clear that all elements involved come from a single period of $\{P_\ell\}$ and $\{Q_\ell\}$. Therefore, it should

Figure 9.7 The steps in performing periodic discrete convolution.

Step 1.   Choose onr period of $\{P_\ell\}$ as the stationary sequence.

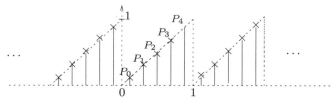

Step 2.   Reverse the periodic sequence $\{Q_\ell\}$, which is identical to $\{P_\ell\}$ for this example.

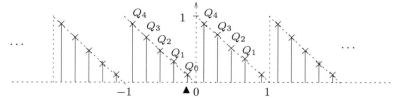

Step 3.   Align Q_0 of the moving sequence with P_0 and compute U_0:

				P_0	P_1	P_2	P_3	P_4		
$\cdots$	Q_4	Q_3	Q_2	Q_1	Q_0	Q_4	Q_3	Q_2	Q_1	$\cdots$ $\cdots$

$$U_0 = P_0 Q_0 + P_1 Q_4 + P_2 Q_3 + P_3 Q_2 + P_4 Q_1.$$

Step 4.   Compute U_1, U_2, U_3, and U_4:

				P_0	P_1	P_2	P_3	P_4		
$\cdots$	Q_4	Q_3	Q_2	Q_1	Q_0	Q_4	Q_3	Q_2	Q_1	$\cdots$

$$U_1 = P_0 Q_1 + P_1 Q_0 + P_2 Q_4 + P_3 Q_3 + P_4 Q_2.$$

				P_0	P_1	P_2	P_3	P_4		
$\cdots$	Q_4	Q_3	Q_2	Q_1	Q_0	Q_4	Q_3	Q_2	Q_1

$$U_2 = P_0 Q_2 + P_1 Q_1 + P_2 Q_0 + P_3 Q_4 + P_4 Q_3.$$

			P_0	P_1	P_2	P_3	P_4		
$\cdots$	$\cdots$	Q_4	Q_3	Q_2	Q_1	Q_0	Q_4	Q_3	Q_2

$$U_3 = P_0 Q_3 + P_1 Q_2 + P_2 Q_1 + P_3 Q_0 + P_4 Q_4.$$

			P_0	P_1	P_2	P_3	P_4		
$\cdots$	$\cdots$	$\cdots$	Q_4	Q_3	Q_2	Q_1	Q_0	Q_4	Q_3

$$U_4 = P_0 Q_4 + P_1 Q_3 + P_2 Q_2 + P_3 Q_1 + P_4 Q_0.$$

Note: $\{U_0, U_1, U_2, U_3 . U_4\}$ form one period of the output sequence.

Figure 9.8 Converting linear to periodic discrete convolution.

Step 1. Zero-pad $\{g_0, g_1, g_2, g_3, g_4\}$ to obtain $\{g_0, g_1, g_2, g_3., g_4, 0, 0, 0, 0\}$,
which forms one period of the stationary periodic sequence:

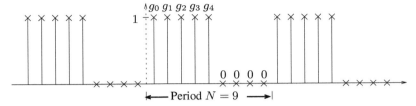

Step 2. Zero-pad $\{h_0, h_1, h_2, h_3, h_4\}$ to obtain $\{h_0, h_1, h_2, h_3, h_4, 0, 0, 0, 0\}$,
then reverse the zero-padded sequence to obtain $\{0, 0, 0, 0, h_4, h_3, h_2, h_1, h_0\}$,
which forms one period of the moving periodic sequence:

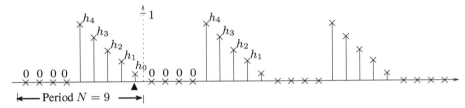

Step 3. Align h_0 of the moving sequence with g_0, and compute $u_0 = g_0 h_0 + 0 + \cdots + 0$:

g_0	g_1	g_2	g_3	g_4	0	0	0	0	

| 0 | 0 | 0 | 0 | h_4 | h_3 | h_2 | h_1 | h_0 | 0 | 0 | 0 | 0 | h_4 | h_3 | h_2 | h_1 | $\cdots$ |

$$u_0 = g_0 h_0 + 0 + \cdots + 0 = g_0 h_0.$$

Step 4. Compute $u_1, u_2, \cdots, u_8$:

g_0	g_1	g_2	g_3	g_4	0	0	0	0

$\cdots$ 0 0 0 0 h_4 h_3 h_2 | h_1 h_0 0 0 0 0 h_4 h_3 h_2 | h_1 $\cdots$

$$u_1 = g_0 h_1 + g_1 h_0 + 0 + \cdots + 0 = g_0 h_1 + g_1 h_0.$$

g_0	g_1	g_2	g_3	g_4	0	0	0	0

$\cdots$ $\cdots$ 0 0 0 0 h_4 h_3 | h_2 h_1 h_0 0 0 0 0 h_4 h_3 | h_2 h_1 $\cdots$

$$u_2 = g_0 h_2 + g_1 h_1 + g_2 h_0 + 0 + \cdots + 0 = g_0 h_2 + g_1 h_1 + g_2 h_0.$$

$$\vdots$$

g_0	g_1	g_2	g_3	g_4	0	0	0	0

$\cdots$ $\cdots$ $\cdots$ $\cdots$ $\cdots$ | 0 0 0 0 h_4 h_3 h_2 h_1 h_0 | 0 0 0 0

$$u_8 = 0 + \ldots + 0 + g_4 h_4 + 0 + \cdots + 0 = g_4 h_4.$$

Note: $\{u_0, u_1, u_2, u_3, u_4, u_5, u_6, u_7., u_8\}$ forms one period of the output sequence.

not be surprising that we can compute the N periodic convolution results without explicitly constructing the periodic sequences. The so-called *cyclic* or *circular* convolution is such a scheme, which computes identical results using only N elements from each periodic sequence. So, we can again work with two finite sequences of length N, and the steps are given below.

Step 1. Choose one period of $\{P_\ell\}$ as the stationary sequence. Arrange the N elements *clockwise* on a ring as shown in Figure 9.9.

Step 2. Arrange the N elements of $\{Q_\ell\}$ *counterclockwise* in an inner ring, which forms the reversed moving sequence. Note that when overlapping the elements on the two rings initially, the first element in the sequence $\{Q_\ell\}$, i.e., Q_0, should overlap P_0, the first element of the stationary sequence $\{P_\ell\}$.

Step 3. Compute the first convolution result by summing the N pairwise products involving all elements on the two rings configured in Step 2; compute the second convolution result in the same manner after turning the inner ring (which houses the moving sequence) *clockwise* one position, and so on. After we obtain all N periodic convolution results, the inner ring has been turned clockwise exactly $N-1$ times. Since the inner ring returns to its initial position after it is turned N times, it is explicitly clear that the same N results are obtained if one repeats the cycle.

The steps of cyclic convolution are illustrated in Figure 9.9 and are equivalent to the steps of periodic convolution shown in Figure 9.7 in Section 9.2.1.

9.2.4 The cyclic convolution in matrix form

It is also useful to express the cyclic convolution of two length-N sequences $\{P_\ell\}$ and $\{Q_\ell\}$ as the product of an $N \times N$ circulant matrix A and an $N \times 1$ vector V, which is illustrated below for $N=5$:

$$(9.3) \qquad Z = \{P_\ell\} \odot \{Q_\ell\} = \begin{bmatrix} Q_0 & Q_4 & Q_3 & Q_2 & Q_1 \\ Q_1 & Q_0 & Q_4 & Q_3 & Q_2 \\ Q_2 & Q_1 & Q_0 & Q_4 & Q_3 \\ Q_3 & Q_2 & Q_1 & Q_0 & Q_4 \\ Q_4 & Q_3 & Q_2 & Q_1 & Q_0 \end{bmatrix} \begin{bmatrix} P_0 \\ P_1 \\ P_2 \\ P_3 \\ P_4 \end{bmatrix}.$$

Observe that the second row of matrix A is obtained by cyclic-shifting its first row one position to the right, and, in general, we cyclic-right-shift the kth row to obtain the $(k+1)$st row until we have the N-by-N circulant matrix. Furthermore, the components of the vector $Z = \{P_\ell\} \odot \{Q_\ell\} = AV$ can be expressed by the algebraic equation:

$$(9.4) \qquad Z_k = \sum_{\ell=0}^{N-1} P_\ell \cdot Q_{(k-\ell) \bmod N}, \quad 0 \le k \le N-1,$$

which denotes the inner product of the kth row of matrix A and the vector V and serves as another mathematical definition for the cyclic convolution. Note that the evaluation of the index $(k-\ell) \bmod N$ is done according to the following rule: for any integer $m = k - \ell$, let

$$(9.5) \qquad R = \text{remainder of } \frac{|m|}{N},$$

Figure 9.9 De n ing the equivalent cyclic convolution.

Recall Step 3 of periodic convolution from Figure 9.7:

Align Q_0 of the moving sequence with P_0 and compute U_0:

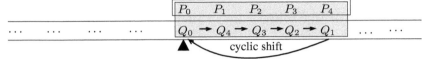

The equivalent cyclic convolution:

$$U_0 = P_0Q_0 + P_1Q_4 + P_2Q_3 + P_3Q_2 + P_4Q_1.$$

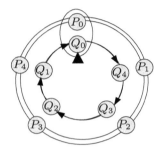

$U_1 = P_0Q_1 + P_1Q_0 + P_2Q_4 + P_3Q_3 + P_4Q_2.$ $U_2 = P_0Q_2 + P_1Q_1 + P_2Q_0 + P_3Q_4 + P_4Q_3.$

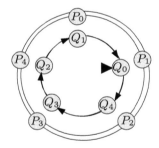   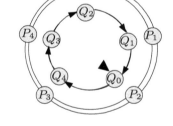

$U_3 = P_0Q_3 + P_1Q_2 + P_2Q_1 + P_3Q_0 + P_4Q_4.$ $U_4 = P_0Q_4 + P_1Q_3 + P_2Q_2 + P_3Q_1 + P_4Q_0.$

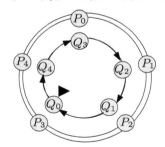   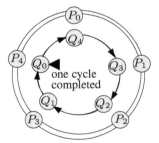

Note: $\{U_0, U_1, U_2, U_3. U_4\}$ forms one cycle of the output sequence.

then the unique value $r = m \bmod N$ is given by

(9.6)
$$r = \begin{cases} R & \text{if } m \geq 0; \\ N - R & \text{if } m < 0 \text{ and } R \neq 0; \\ 0 & \text{if } m < 0 \text{ and } R = 0. \end{cases}$$

This is, of course, the de n ition of the residue r of m modulo N i n linear algebra. Since r takes on integer values from the set $\{0, 1, \cdots, N-1\}$, it is straightforward to verify that Equation (9.4) involves exactly the kth row from the circulant matrix A in Equation (9.3). For the matrix-vector product de ned by (9.3), we have $N = 5$, $0 \leq k, \ell \leq 4$, and $-4 \leq m = k - \ell \leq 4$. Hence, the remainder $R = |m|/5$ in (9.5) takes on integer values from 0 to 4, and the corresponding residue $r = m \bmod 5$ is given by (9.6); with that we can uniquely identify $Q_r = Q_{m \bmod N}$. In particular, for $m = k - \ell = -1, -2, -3, -4$, we have $R = 1, 2, 3, 4$ and $r = N - R = 4, 3, 2, 1$, so we can uniquely identify $Q_4 = Q_{-1 \bmod 5}$, $Q_3 = Q_{-2 \bmod 5}$, $Q_2 = Q_{-3 \bmod 5}$, and $Q_1 = Q_{-4 \bmod 5}$, which are all we need when evaluating Z_k using the cyclic convolution formula given by (9.4).

Remark: The cyclic convolution de ned by (9.4) is equivalent to the periodic convolution de ned by (9.2) if $\{P_\ell\}$ and $\{Q_\ell\}$ are interpreted as one period of a periodic sequence.

9.2.5   Converting linear to cyclic convolution

We have illustrated in Figure 9.8 and explained in Section 9.2.2 the steps involved in converting linear to periodic convolution. Hence all we need to do now is to implement the periodic convolution by cyclic convolution. Assuming that we are given two sequences $\{g_\ell\}$ and $\{h_\ell\}$ of lengths L_1 and L_2, the conversion from linear to cyclic convolution involves the following steps:

Step 1. Zero-pad both sequences to length $L_1 + L_2 - 1$.

Step 2. Arrange the stationary sequence and the other sequence on the outer and inner rings as explained in Section 9.2.3.

Step 3. Perform cyclic convolution on the two rings as explained in Section 9.2.3.

Corresponding to the steps of periodic convolution illustrated in Figure 9.8, the steps of cyclic convolution are shown in Figure 9.10.

9.2.6   Two cyclic convolution theorems

Recall that we have used extensively the continuous convolution theorem and product theorem which generate the following Fourier transform pairs

$$x(t) * g(t) \Longleftrightarrow X(f) \cdot G(f); \qquad x(t) \cdot g(t) \Longleftrightarrow X(f) * G(f),$$

where $X(f) = \mathcal{F}\{x(t)\}$ and $G(f) = \mathcal{F}\{g(t)\}$. Note that the   rst pair relates time-domain convolution to frequency-domain multiplication, and it allows us to express $x(t) * g(t) = \mathcal{F}^{-1}\{X(f) \cdot G(f)\}$. A similar relationship exists between the periodic convolution of two sequences and the point-wise product of their respective discrete Fourier transforms, which is proved in the discrete convolution theorem given below.

Figure 9.10 Converting linear to cyclic convolution.

Recall Step 3 of converting linear to periodic convolution from Figure 9.8:

Align h_0 of the moving sequence with g_0 and compute $u_0 = g_0 h_0 + 0 + \cdots + 0$:

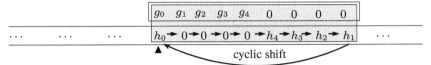

The equivalent cyclic convolution:

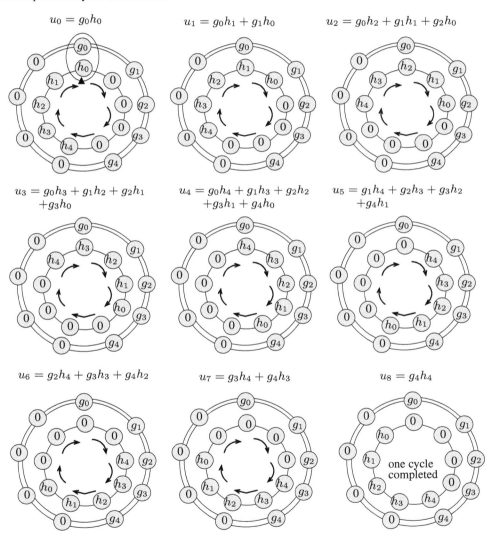

$$u_0 = g_0 h_0$$

$$u_1 = g_0 h_1 + g_1 h_0$$

$$u_2 = g_0 h_2 + g_1 h_1 + g_2 h_0$$

$$u_3 = g_0 h_3 + g_1 h_2 + g_2 h_1 + g_3 h_0$$

$$u_4 = g_0 h_4 + g_1 h_3 + g_2 h_2 + g_3 h_1 + g_4 h_0$$

$$u_5 = g_1 h_4 + g_2 h_3 + g_3 h_2 + g_4 h_1$$

$$u_6 = g_2 h_4 + g_3 h_3 + g_4 h_2$$

$$u_7 = g_3 h_4 + g_4 h_3$$

$$u_8 = g_4 h_4$$

Note: $\{u_0, u_1, u_2, u_3, u_4, u_5, u_6, u_7, u_8\}$ forms one cycle of output sequence.

Theorem 9.1 (Time-Domain Cyclic Convolution Theorem) Let the cyclic convolution of sequences $\{x_\ell\}$ and $\{g_\ell\}$ of length (or period) N be denoted by $\{x_\ell\} \odot \{g_\ell\}$. If the discrete Fourier transforms of the two sequences are given by $\{X_r\} = \mathrm{DFT}\big[\{x_\ell\}\big]$ and $\{G_r\} = \mathrm{DFT}\big[\{g_\ell\}\big]$, then

$$(9.7) \qquad \{x_\ell\} \odot \{g_\ell\} = N \cdot \mathrm{IDFT}\big[\{X_r \cdot G_r\}\big].$$

Proof: By Equation (9.2) (which is equivalent to (9.4) for periodic sequences), we obtain $\{u_k\} = \{x_\ell\} \odot \{g_\ell\}$ with its elements given by

$$(9.8) \qquad u_k = \sum_{\ell=0}^{N-1} x_\ell \cdot g_{k-\ell}, \quad \text{for } k = 0, 1, \cdots, N-1,$$

where $g_{k-\ell} = g_{k-\ell+N}$ due to the assumed periodicity. Assuming that the DFT coefficients $\{X_r\}$ and $\{G_r\}$ are computed by Formula (2.7), we use the corresponding IDFT formula (2.6) to express

$$x_\ell = \sum_{r=0}^{N-1} X_r\, \omega_N^{\ell r}, \quad g_{k-\ell} = \sum_{r=0}^{N-1} G_r\, \omega_N^{(k-\ell)r}, \quad \text{where } \omega_N = e^{j2\pi/N},$$

and we rewrite (9.8) as

$$
\begin{aligned}
(9.9)\qquad u_k &= \sum_{\ell=0}^{N-1} \overbrace{\left[\sum_{m=0}^{N-1} X_m\, \omega_N^{\ell m}\right]}^{x_\ell} \times \overbrace{\left[\sum_{r=0}^{N-1} G_r\, \omega_N^{(k-\ell)r}\right]}^{g_{k-\ell}} \\
&= \sum_{\ell=0}^{N-1} \left[\sum_{r=0}^{N-1} G_r\, \omega_N^{kr}\, \omega_N^{-\ell r}\right] \times \left[\sum_{m=0}^{N-1} X_m\, \omega_N^{\ell m}\right] \\
&= \sum_{\ell=0}^{N-1} \sum_{r=0}^{N-1} \left[G_r\, \omega_N^{kr} \sum_{m=0}^{N-1} X_m\, \omega_N^{\ell(m-r)}\right] \\
&= \sum_{r=0}^{N-1} G_r\, \omega_N^{kr} \left[\sum_{m=0}^{N-1}\sum_{\ell=0}^{N-1} X_m\, \omega_N^{\ell(m-r)}\right] \\
&= \sum_{r=0}^{N-1} G_r\, \omega_N^{kr} \left[\sum_{m=0}^{N-1} X_m \sum_{\ell=0}^{N-1} \omega_N^{\ell(m-r)}\right] \\
&= N \sum_{r=0}^{N-1} \{X_r \cdot G_r\}\, \omega_N^{kr}.
\end{aligned}
$$

Note that in the last step we have used the orthogonality property proved in Chapter 4 on page 112:

$$\sum_{\ell=0}^{N-1} \omega^{\ell(m-r)} = \begin{cases} 0 & \text{if } m \neq r, \\ N & \text{if } m = r. \end{cases}$$

■

Corresponding to the other Fourier transform pair $X(f) * G(f) = \mathcal{F}\{x(t) \cdot g(t)\}$ in the continuous convolution theorem, we have the discrete Fourier transform pair given by the next theorem.

Theorem 9.2 (Frequency-domain cyclic convolution theorem) Let $\{X_r\}$ and $\{G_r\}$ denote two DFT sample sequences of length (or period) N. If $\{x_\ell\} = \text{IDFT}\big[\{X_r\}\big]$ and $\{g_\ell\} = \text{IDFT}\big[\{G_r\}\big]$, then

$$(9.10) \qquad \{X_r\} \odot \{G_r\} = \text{DFT}\big[\{x_\ell \cdot g_\ell\}\big].$$

Proof: By Equation (9.2) we obtain $\{U_k\} = \{X_r\} \odot \{G_r\}$ with its elements given by

$$(9.11) \qquad U_k = \sum_{r=0}^{N-1} X_r \cdot G_{k-r}, \quad \text{for } k = 0, 1, \cdots, N-1,$$

where $G_{k-r} = G_{k-r+N}$ due to the assumed periodicity. Using the DFT formula (2.7), we express

$$X_r = \frac{1}{N} \sum_{\ell=0}^{N-1} x_\ell \, \omega_N^{-r\ell}, \quad G_{k-r} = \frac{1}{N} \sum_{\ell=0}^{N-1} g_\ell \, \omega_N^{-(k-r)\ell}, \quad \text{where } \omega_N = e^{j2\pi/N},$$

and we rewrite (9.11) as

$$(9.12) \qquad
\begin{aligned}
U_k &= \sum_{r=0}^{N-1} \overbrace{\left[\frac{1}{N} \sum_{m=0}^{N-1} x_m \, \omega_N^{-rm} \right]}^{X_r} \times \overbrace{\left[\frac{1}{N} \sum_{\ell=0}^{N-1} g_\ell \, \omega_N^{-(k-r)\ell} \right]}^{G_{k-r}} \\
&= \frac{1}{N^2} \sum_{r=0}^{N-1} \left[\sum_{\ell=0}^{N-1} g_\ell \omega_N^{-k\ell} \, \omega_N^{r\ell} \right] \times \left[\sum_{m=0}^{N-1} x_m \, \omega_N^{-rm} \right] \\
&= \frac{1}{N^2} \sum_{\ell=0}^{N-1} g_\ell \, \omega_N^{-k\ell} \left[\sum_{m=0}^{N-1} x_m \sum_{r=0}^{N-1} \omega_N^{-r(m-\ell)} \right] \\
&= \frac{1}{N} \sum_{\ell=0}^{N-1} \{x_\ell \cdot g_\ell\} \, \omega_N^{-k\ell}.
\end{aligned}$$

Hence, we have proved

$$\{U_k\} = \{X_r\} \odot \{G_r\} = \text{DFT}\big[\{x_\ell \cdot g_\ell\}\big].$$

∎

These two discrete convolution theorems establish that the cyclic convolution of two sequences of length N (in *either* time *or* frequency domain) can be computed via a combination of DFT and IDFT, which incurs arithmetic cost proportional to $N \log_2 N$ when FFT and IFFT are used to compute the DFT and IDFT. Therefore, all fast convolution algorithms involve FFT/IFFT, and they make FFT/IFFT ubiquitous in diverse application areas.

9.2.7 Implementing sectioned linear convolution

Recall our discussion on sectioning a long sequence for linear convolution in Section 9.1.2. As stated there, all we need to do is to convert the linear convolution de ned on each section of length S_1 to a periodic convolution, and we compute the latter via DFT/IDFT as described in Theorems 9.1 and 9.2. Observe that the DFT of the moving sequence only needs to be computed once.

9.3   The Chirp Fourier Transform

We are now in a position to introduce the chirp Fourier transform and explain why it is needed. For its ef cien t computation, we show how to turn the transform into the convolution of two appropriately de ned sequences. However, let us point out that the equivalent discrete convolution de ned on the original data (without extension and restructuring) is *neither* linear *nor* periodic per se   it is, instead, a *partial* linear convolution, so we must learn another way to turn it into a cyclic convolution so that we can use the FFT algorithms for its computation.

9.3.1   The scenario

To set the stage, let us assume that we are given a sequence $\{\tilde{x}_\ell\}$ consisting of $N=64$ samples taken from a time-limited signal $x(t)$ over its duration $\mathbb{T}$. Assuming that $\mathbb{T}=N\triangle t$, the DFT of the N-sample sequence $\{\tilde{x}_\ell\}$ computes $\{\widetilde{U}_r\}$ according to Equation (8.2), namely,

$$(9.13) \qquad \widetilde{U}_r = \frac{1}{N} X_I\left(\frac{r}{N\triangle t}\right) = \frac{1}{N}\sum_{\ell=-n}^{n+1}\tilde{x}_\ell\,\omega_N^{-r\ell}, \quad -n\le r\le n+1, \quad N=2n+2.$$

Recall that the N samples of $X_I(f)$ are spaced by $\triangle f = 1/(N\triangle t) = 1/\mathbb{T}$. If we want more details of $X_I(f)$ between two particular frequencies at f_m and $f_{m+1}=f_m+\triangle f$, we would need to obtain more sample values of $X_I(f)$ between f_m and f_{m+1}. Suppose $K-1$ more values of $X_I(f)$ are needed between f_m and f_{m+1}; with the desired spacing now reduced to $\triangle f/K$, we can accomplish that by two different approaches:

1. **Full range interpolation by zero padding.** Recall that we discussed the effects of zero padding in Section 4.6 in Chapter 4. In particular, we explained that when the N-sample DFT spectrum is too sparse for us to visualize a continuous spectrum $X_I(f)$, we may decrease the spacing $\triangle f = 1/(N\triangle t)$ on the entire frequency grid if we enlarge N by adding zeros (see Figures 4.8 and 4.9). That is, for $K=N=64$, we extend the 64-sample sequence $\{\tilde{x}_\ell\}$ to length $L=K\times N=4096$ by appending 2016 zeros at each end, and we perform the DFT on the resulting 4096-sample sequence

$$\{\tilde{x}_\ell\} = \{0,0,\ldots,0,\tilde{x}_{-31}\ldots,\tilde{x}_{-1},\tilde{x}_0,\tilde{x}_1,\ldots,\tilde{x}_{32},0,0,\ldots,0\}$$

   to obtain

$$(9.14) \qquad \hat{U}_r = \frac{1}{L}\sum_{\ell=-2047}^{2048}\tilde{x}_\ell\omega_L^{-r\ell}, \quad -2047\le r\le 2048, \quad L=K\times N=4096.$$

   This approach gives us $\{\hat{U}_{-2047},\cdots,\hat{U}_{-1},\hat{U}_0,\hat{U}_1,\cdots,\tilde{U}_{2048}\}$ a   total of $L=4096$ samples of $\frac{1}{L}X_I(f)$, while we only need $K=64$ samples in a particular subinterval and for that we pay the high cost of computing the DFT of length $L=4096$. This approach is evidently too costly unless K is very small relative to N.

2. **Partial DFT by chirp Fourier transform.** Using this approach we will compute K samples of $\frac{1}{L}X_I(f)$ inside a single subinterval of length $\triangle f = 1/(N\triangle t)$. For convenience in dealing with an arbitrary subinterval in our analysis, recall the alternate form of the DFT given by (4.9), which is restated below:

$$(9.15) \qquad U_r = \frac{1}{N}\sum_{\ell=0}^{N-1}x_\ell\omega_N^{-r\ell}, \quad 0\le r\le N-1, \quad N=2n+2, \quad \text{where}$$

the reordered samples in $\{x_\ell\}$, $0 \leq \ell \leq N - 1$, are de n ed by

$$x_\ell = \begin{cases} \tilde{x}_\ell & \text{for } 0 \leq \ell \leq n + 1; \\ \tilde{x}_{\ell-N} & \text{for } n + 2 \leq \ell \leq 2n + 1; \end{cases}$$

and the DFT samples in $\{\tilde{U}_r\}$, $-n \leq r \leq n + 1$, can be recovered from the reordered samples in $\{U_r\}$ de ned by (9.15) using the relationship (4.14) established in Section 4.2:

$$U_r = \begin{cases} \tilde{U}_r & \text{for } 0 \leq r \leq n + 1; \\ \tilde{U}_{r-N} & \text{for } n + 2 \leq r \leq 2n + 1. \end{cases}$$

Assuming that the arbitrary subinterval is between $f_m = m\triangle f$ and $f_{m+1} = (m+1)\triangle f$, we now compute the K samples by performing a partial DFT on the given N-sample sequence $\{x_\ell\}$. That is, we compute U_r according to Equation (9.15) for $r = mK + \lambda$, $0 \leq \lambda \leq K - 1$, without zero-padding $\{x_\ell\}$:

$$(9.16) \quad U_{mK+\lambda} = \frac{1}{L} \sum_{\ell=0}^{N-1} x_\ell \omega_L^{-(mK+\lambda)\ell}, \quad 0 \leq \lambda \leq K-1 = 63, \quad L = K \times N = 4096.$$

Note that only $K = 64$ elements $\{U_{mK}, U_{mK+1}, \cdots, U_{mK+63}\}$ are computed by this formula, and that its right-hand side shows only the $N = 64$ nonzero terms with $\omega_L = e^{j2\pi/L}$ unchanged.

9.3.2 The equivalent partial linear convolution

Since the chirp Fourier transform is a partial DFT, it cannot be computed by the FFT. For its ef cient computation, we shall rst turn the partial DFT into a partial linear convolution, which can then be converted to a cyclic convolution, because the latter can be computed via DFT and IDFT.

We begin by rewriting the partial DFT from (9.16) as

$$(9.17) \quad \begin{aligned} U_{mK+\lambda} &= \frac{1}{L} \sum_{\ell=0}^{N-1} \left[x_\ell \omega_L^{-mK\ell} \right] \omega_L^{-\lambda\ell}, \quad \text{for } \lambda = 0, 1, \cdots, K - 1, \\ &= \frac{1}{L} \sum_{\ell=0}^{N-1} \left[x_\ell \omega_N^{-m\ell} \right] \omega_L^{0.5[(\lambda-\ell)^2 - \lambda^2 - \ell^2]} \\ &= \frac{\omega_L^{-0.5\lambda^2}}{L} \sum_{\ell=0}^{N-1} \left[x_\ell \omega_N^{-m\ell} \omega_L^{-0.5\ell^2} \right] \omega_L^{0.5(\lambda-\ell)^2} \\ &= \frac{\omega_L^{-0.5\lambda^2}}{L} \sum_{\ell=0}^{N-1} g_\ell \cdot h_{\lambda-\ell}, \end{aligned}$$

where we de ne

$$g_\ell = x_\ell \omega_N^{-m\ell} \omega_L^{-0.5\ell^2} \quad \text{and} \quad h_n = \omega_L^{0.5n^2}, \quad \text{so that} \quad h_{\lambda-\ell} = \omega_L^{0.5(\lambda-\ell)^2}.$$

Observe that $\{y_0, y_1, \cdots, y_{K-1}\}$ computed by

$$(9.18) \quad y_\lambda = \sum_{\ell=0}^{N-1} g_\ell \cdot h_{\lambda-\ell}, \quad \text{for } \lambda = 0, 1, \cdots, K - 1,$$

are the middle K elements (beginning with the Nth) obtained from the linear convolution of the length-N sequence

$$\{g_0, g_1, \cdots, g_{N-1}\}$$

and the length-$(N+K-1)$ sequence

$$\{f_0, f_1, \cdots, f_N, f_{N+1}, \cdots, f_{N+K-2}\} = \{h_{-N+1}, h_{-N+2}, \cdots, h_0, h_1, \cdots, h_{K-1}\}.$$

Note that we have explicitly stored the data $\{h_{-N+1}, \cdots, h_{-1}, h_0, h_1, \cdots, h_{K-1}\}$ in the array f in the specied order, so that f_0 refers to the rst element in the sequence, and f_k refers to the $(k+1)$st element in the sequence as before.

To be specic, for $N=5$ and $K=4$, we have the stationary sequence

$$\{g_0, g_1, g_2, g_3, g_4\},$$

which is of length $N=5$, and the moving sequence

$$\{f_0, f_1, f_2, f_3, f_4, f_5, f_6, f_7\} = \{h_{-4}, h_{-3}, h_{-2}, h_{-1}, h_0, h_1, h_2, h_3\},$$

which is of length $N+K-1=8$, and their partial linear convolution is illustrated in Figure 9.11. Hence, the $K=4$ elements in sequence $\{y_0, y_1, y_2, y_3\}$ are only partial results of the linear convolution, and $U_{mK+\lambda} = y_\lambda \cdot \omega_L^{-0.5\lambda^2}/L$ for $\lambda = 0, 1, 2, 3$.

Remark: Observe that $h_n = h_{-n}$ and h_n is dened for all integer n, but $h_{n+N} \neq h_n$. Therefore, the formula given by (9.18) represents *neither* a full *nor* a partial periodic convolution.

9.3.3   The equivalent partial cyclic convolution

From the partial linear convolution example with $N=5$ and $K=4$, it is not dicult to see that $\{y_0, y_1, y_2, y_3\}$ can be computed by the cyclic convolution of the sequences $\{g_0, g_1, g_2, g_3, g_4, 0, 0, 0\}$ and $\{h_0, h_1, h_2, h_3, h_{-4}, h_{-3}, h_{-2}, h_{-1}\}$ (See Figure 9.12). Both sequences are of length $N+K-1=8$, and the partial cyclic convolution process can be conveniently displayed in its matrix form:

$$(9.19) \quad \begin{bmatrix} y_0 \\ y_1 \\ y_2 \\ y_3 \\ \times \\ \times \\ \times \\ \times \end{bmatrix} = \begin{bmatrix} h_0 & h_{-1} & h_{-2} & h_{-3} & h_{-4} & h_3 & h_2 & h_1 \\ h_1 & h_0 & h_{-1} & h_{-2} & h_{-3} & h_{-4} & h_3 & h_2 \\ h_2 & h_1 & h_0 & h_{-1} & h_{-2} & h_{-3} & h_{-4} & h_3 \\ h_3 & h_2 & h_1 & h_0 & h_{-1} & h_{-2} & h_{-3} & h_{-4} \\ \times & \times & \times & \times & \times & \times & \times & \times \\ \times & \times & \times & \times & \times & \times & \times & \times \\ \times & \times & \times & \times & \times & \times & \times & \times \\ \times & \times & \times & \times & \times & \times & \times & \times \end{bmatrix} \begin{bmatrix} g_0 \\ g_1 \\ g_2 \\ g_3 \\ g_4 \\ 0 \\ 0 \\ 0 \end{bmatrix}.$$

Since a partial cyclic convolution cannot be implemented by FFT/IFFT, we shall compute a full cyclic convolution of length $M = N+K-1 = 8$ via FFT/IFFT at a cost proportional to $M \log_2 M$, and take the rst $K=4$ results from the computed $\{y_0, y_1, \cdots, y_{M-1}\}$. After y_λ s are available, we can compute $U_{mK+\lambda} = y_\lambda \cdot \omega_L^{-0.5\lambda^2}/L$ for $\lambda = 0, 1, \cdots, K-1$.

Remarks: To avoid unnecessary complications, we have assumed that the N signal samples are taken from a time-limited signal $x(t)$ over its duration $\mathbb{T}$. It is, of course, very likely that the N samples are truncated from an initely long $x_I(t) = x(t) \cdot P_{\triangle t}(t)$ by either a

Figure 9.11 Interpreting chirp Fourier transform as a partial linear convolution.

Stationary sequence: $\{g_0, g_1, g_2, g_3, g_4\}$.

| g_0 | g_1 | g_2 | g_3 | g_4 |

Moving sequence: $\{f_0, f_1, \cdots \cdots, f_7\} = \{h_{-4}, h_{-3}, h_{-2}, h_{-1}, h_0, h_1, h_2, h_3\}$.

| | h_{-4} | h_{-3} | h_{-2} | h_{-1} | h_0 | h_1 | h_2 | h_3 |
| array $\boldsymbol{f}$: | f_0 | f_1 | f_2 | f_3 | f_4 | f_5 | f_6 | f_7 |

Recall Step 4 of the linear discrete convolution from Figure 9.3: a partial convolution computes $\{u_4, u_5, u_6, u_7\} = \{y_0, y_1, y_2, y_3\}$ in this step.

			g_0	g_1	g_2	g_3	g_4	
Shifted reversed array $\boldsymbol{f}$:	h_3	h_2	h_1	h_0	h_{-1}	h_{-2}	h_{-3}	h_{-4}
	f_7	f_6	f_5	f_4	f_3	f_2	f_1	f_0

$$u_4 = g_0 f_4 + g_1 f_3 + g_2 f_2 + g_3 f_1 + g_4 f_0$$
$$= g_0 h_0 + g_1 h_{-1} + g_2 h_{-2} + g_3 h_{-3} + g_4 h_{-4} = y_0.$$

		g_0	g_1	g_2	g_3	g_4		
	h_3	h_2	h_1	h_0	h_{-1}	h_{-2}	h_{-3}	h_{-4}
	f_7	f_6	f_5	f_4	f_3	f_2	f_1	f_0

$$u_5 = g_0 f_5 + g_1 f_4 + g_2 f_3 + g_3 f_2 + g_4 f_1$$
$$= g_0 h_1 + g_1 h_0 + g_2 h_{-1} + g_3 h_{-2} + g_4 h_{-3} = y_1.$$

	g_0	g_1	g_2	g_3	g_4		
h_3	h_2	h_1	h_0	h_{-1}	h_{-2}	h_{-3}	h_{-4}
f_7	f_6	f_5	f_4	f_3	f_2	f_1	f_0

$$u_6 = g_0 f_6 + g_1 f_5 + g_2 f_4 + g_3 f_3 + g_4 f_2$$
$$= g_0 h_2 + g_1 h_1 + g_2 h_0 + g_3 h_{-1} + g_4 h_{-2} = y_2.$$

g_0	g_1	g_2	g_3	g_4			
h_3	h_2	h_1	h_0	h_{-1}	h_{-2}	h_{-3}	h_{-4}
f_7	f_6	f_5	f_4	f_3	f_2	f_1	f_0

$$u_7 = g_0 f_7 + g_1 f_6 + g_2 f_5 + g_3 f_4 + g_4 f_3$$
$$= g_0 h_3 + g_1 h_2 + g_2 h_1 + g_3 h_0 + g_4 h_{-1} = y_3.$$

Note: The chirp Fourier transform results $\{y_0, y_1, y_2, y_3\}$ have now been computed by the equivalent partial linear discrete convolution of $\{g_k\}$ and $\{f_k\}$.

Figure 9.12 Interpreting chirp Fourier transform as a partial cyclic convolution.

Cyclic convolution of stationary sequence

$$\{g_0, g_1, g_2, g_3, g_4, 0, 0, 0\}$$

and moving sequence

$$\{h_0, h_1, h_2, h_3, h_{-4}, h_{-3}, h_{-2}, h_{-1}\}:$$

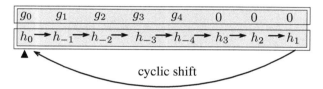

The chirp Fourier transform results $\{y_0, y_1, y_2, y_3\}$ are the first four results of a cyclic convolution:

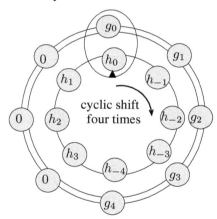

$$y_0 = g_0 h_0 + g_1 h_{-1} + g_2 h_{-2} + g_3 h_{-3} + g_4 h_{-4} + 0 + 0 + 0.$$

$$y_1 = g_0 h_1 + g_1 h_0 + g_2 h_{-1} + g_3 h_{-2} + g_4 h_{-3} + 0 + 0 + 0.$$

$$y_2 = g_0 h_2 + g_1 h_1 + g_2 h_0 + g_3 h_{-1} + g_4 h_{-2} + 0 + 0 + 0.$$

$$y_3 = g_0 h_3 + g_1 h_2 + g_2 h_1 + g_3 h_0 + g_4 h_{-1} + 0 + 0 + 0.$$

rectangular window or one of the data weighing windows $w(t)$ of length $\mathbb{T} = N\triangle t$. Using the DFT on the *windowed* N-sample sequence, we obtain U_r as the samples of $\frac{1}{N}U_I(f) = \frac{1}{N}\{X_I(f) * W(f)\}$, where $W(f) = \mathcal{F}\{w(t)\}$. Therefore, the use of windows changes only the function beneath the DFT samples, but it does not change how we obtain more values of that function, be it $U_I(f)$ or $X_I(f)$. Therefore, the chirp Fourier transform can be applied in the same manner regardless how the N samples are obtained or whether they are modi ed by windows.

References

1. A. Ambardar. *Analog and Digital Signal Processing*. Brooks/Cole Publishing Company, Paci c Grove, CA, second edition, 1999.

2. R. N. Bracewell. *The Fourier Transform and Its Applications*. McGraw-Hill Higher Education, A Division of The McGraw-Hill Companies, San Francisco, CA, third edition, 2000.

3. W. L. Briggs and V. E. Hensen. *The DFT: An Owner's Manual for the Discrete Fourier Transform*. The Society for Industrial and Applied Mathematics, Philadelphia, PA, 1995.

4. E. O. Brigham. *The Fast Fourier Transform and Its Applications*. Prentice-Hall, Inc., Upper Saddle River, NJ, 1988.

5. B. Porat. *A Course in Digital Signal Processing*. John Wiley & Sons, Inc., New York, 1997.

6. W. H. Press, S. A. Teukolsky, W. T. Vetterling, and B. P. Flannery. *Numerical Recipes in C++: The Art of Scientific Computing*. Cambridge University Press, Cambridge, UK, second edition, 2001.

7. C. F. Van Loan. *Computational Frameworks for the Fast Fourier Transform*. The Society for Industrial and Applied Mathematics, Philadelphia, PA, 1992.

Chapter 10

Applications of the DFT in Digital Filtering and Filters

In this chapter we study the roles the DFT (and IDFT) plays in signal filtering operations defined by the linear convolution of two sequences: the sequence $\{x_\ell\}$ of length L_1 represents the sampled input signal, which can be as long as it needs to be; the sequence $\{h_\ell\}$ of length L_2 represents the digital filter, and L_2 is assumed to be much shorter than L_1. Since we have already covered the various mathematical tools and computational algorithms needed for digital filtering and filters to be discussed in this chapter, we shall begin by connecting digital filtering to the various topics we presented in the first part of this book.

10.1   The Background

From what we have learned in Chapter 9, the linear convolution of the length-L_1 sequence $\{x_\ell\}$ with the length-L_2 filter $\{h_\ell\}$ produces the output sequence $\{y_\ell\}$ of length L_1+L_2-1, which we previously expressed as

(10.1) $$\{y_\ell\} = \{x_\ell\}*\{h_\ell\}.$$

The DFT-based fast algorithm for computing the convolution result $\{y_\ell\}$ was presented in Section 9.2.5. For cases when the length of the convolution is too long for a single DFT, we show how to section a long sequence for linear convolution in Section 9.1.2, and we show how to implement the sectioned linear convolution via DFT/IDFT in Section 9.2.7. Therefore, if we are given the filter in the form of a finite sequence $\{h_\ell\}$, we already know how to compute the output $\{y_\ell\}$ via DFT/IDFT, and it goes without saying that the DFT/IDFT are computed by the FFT/IFFT. (The FFT algorithms are covered in Part II of this book.)

After the time-domain output sequence $\{y_\ell\}$ is computed, we can examine its frequency content by computing the DFT of the sequence $\{y_\ell\}$ (possibly windowed) as we have discussed thoroughly in Chapter 8 recall, in particular, that the computed DFT coefficients approximate the sample values of the Fourier transform of the filtered sequence $\{y_\ell\}$; hence, a filter designer can always check whether the filter has the desired effect or it needs further tweaking. (More on filter construction later.) Recall also that according to the Time-Domain Discrete Convolution Theorem 7.2 proved in Chapter 7, we can relate the Fourier transform of the output to the

Fourier transforms of the input and the filter, namely,

$$(10.2) \qquad \mathcal{F}\{\{y_\ell\}\} = \mathcal{F}\{\{x_\ell\} * \{h_\ell\}\} = \mathcal{F}\{\{x_\ell\}\} \cdot \mathcal{F}\{\{h_\ell\}\};$$

hence, we can investigate further how the Fourier transform of the input signal is modified by the filter by forming the analytical $\hat{H}_I(\mathbb{F})$, the Fourier transform of the digital filter, using Formula (7.7) from Chapter 7 on the finite sequence $\{h_\ell\}$.

It turns out that the DFT and IDFT can also be used to construct the digital filter $\{h_\ell\}$ if we know how we want to modify the frequency content of the input signal e.g., the elimination or attenuation of specific frequencies is among many other commonly desired effects. Before we proceed with the filter construction, it is useful to bring the mathematical concepts involved in analog filtering into the picture; the analog filters are used to alter the Fourier transform $X(f)$ of the input signal $x(t)$ so that the Fourier transform $Y(f)$ of the output signal $y(t)$ has the desired amplitude and phase characteristics, and this is accomplished in the time-domain by convolving the time-domain signal $x(t)$ with the filter function $h(t)$. The analog filters are physical devices (e.g., circuits formed by resistors, capacitors, and inductances) which implement the linear convolution of $h(t)$ and the continuous-time signal $x(t)$. By invoking the Convolution Theorem 6.1, we obtain the Fourier transform pair:

$$(10.3) \qquad y(t) = x(t) * h(t) \Longleftrightarrow Y(f) = X(f) \cdot H(f).$$

Since the desired $Y(f)$ is the product of $X(f)$ and $H(f)$, the effects desired on $Y(f)$ lead to the proposed $H(f)$, which can then serve as the frequency-domain specification of the filter. For arbitrarily shaped $H(f)$, the IDFT of appropriately sampled $\{H_\ell\}$ (possibly windowed) gives us the digital filter $\{h_\ell\}$ (which may be further windowed), which convolves with the sampled input signal $\{x_\ell\}$ to produce $\{y_\ell\}$ the sequence which approximates the sampled values of $y(t)$.

10.2 Application-Oriented Terminology

Since digital filtering has been practiced in a broad range of disciplines for a long time, the same mathematical ideas and terms were called different names when they were used in different applications. In this section we shall introduce some commonly used terminology from the application viewpoint. At the same time, we note that the mathematical terms and the relationship the new terminology represents are those we have established and used in a consistent manner in the first nine chapters of this book.

1. $h(t)$: impulse response of the analog filter: This term comes from interpreting the convolution integral

$$\delta(t) * h(t) = \int_{-\infty}^{\infty} \delta(\lambda) h(t - \lambda) \, d\lambda = h(t)$$

as the filtering of an impulse function $\delta(t)$ by $h(t)$ with the output, or *response*, being $h(t)$ itself; this term also represents a physical concept, because the analog filter is a physical device which implements the linear convolution of $h(t)$ and a continuous-time input signal $x(t)$, including $x(t) = \delta(t)$.

Other names for $h(t)$ include

- the lter
- the lter impulse (response) function
- the time-response function of the lter
- the lter (time-domain) impulse response
- the system (time-domain) impulse response

2. $y(t) = x(t) * h(t)$: a linear-system convolution : The phrase refers to a system characterized by the continuous-time output $y(t)$, which results from convolving the system input $x(t)$ with the system impulse response $h(t)$. The system is linear because

(10.4) $$z(t) = \Big(\alpha x_1(t) + \beta x_2(t)\Big) * h(t) = \alpha x_1(t) * h(t) + \beta x_2(t) * h(t).$$

3. $H(f)$: frequency response of an analog lter (or a linear system) : Recall that, in mathematical terms, $H(f)$ is simply the Fourier transform of the impulse response $h(t)$. Other names for $H(f)$ include

- the frequency-response function of the lter
- the lter frequency-domain response
- the system frequency-domain response
- the frequency speci cation of the lter
- the analog frequency function
- the transfer factor
- the transfer function
- the system function

If $H(f)$ is complex-valued, we can express

(10.5) $$H(f) = |H(f)|e^{j\phi(f)},$$

where the magnitude spectrum $|H(f)|$ is called the magni tude response (function) of the lter, and the phase spectrum $\phi(f)$ is called the phas e response (function) of the lter .

4. R eal-valued $H(f)$: a zero-phase-shift lter : From $H(f) = |H(f)|e^{j\phi(f)}$ we see that the phase spectrum $\phi(f) = 0$ if $H(f)$ is real-valued. Since the Fourier transform of an even function is purely real, the impulse response $h(t)$ of the zero-phase-shift lters is an even function. (Note that the phase $\phi(f) = 0$ also results in $H(f) = |H(f)|$; hence, $H(f)$ is also an even function.)

5. F inite-Impulse Response (FIR) lters : This term refers to the discrete linear convolution formula involving sample values (or their approximations) of the impulse response $h(t)$. The sample sequence $\{h_\ell\}$ must be nite (as implied by the name of the lter), and it is called the impulse response of the digital lter.

The FIR lters are also called nonrecursive lters.

6. $H_I(f)$: transfer function of the FIR filter : The Fourier transform of the finite filter sequence $\{h_\ell\}$ is called the transfer function of the digital filter.

Recall that the Fourier transform of a sequence has two forms and they were examined in detail in Chapter 7. When $H_I(f)$ is referred to as the transfer function of the finite sequence $\{h_\ell\}$, it is understood that $H_I(f)$ is expressed in terms of $\{h_\ell\}$ instead of the sum of the shifted replicas of $H(f)$.

When dealing with the finite L_2-sample sequence $\{h_\ell\}$, it is also convenient to use $\hat{H}_I(\mathbb{F})$ or $\tilde{H}(\theta)$, where the digital frequency $\mathbb{F} = f\triangle t$, and the digital angular frequency $\theta = 2\pi\mathbb{F}$. Recall that in Chapter 7 both $\hat{H}_I(\mathbb{F})$ and $\tilde{H}_I(\theta)$ were shown to be equivalent representation of $H(f)$, with $-0.5 \leq \mathbb{F} \leq 0.5$, $-\pi \leq \theta \leq \pi$, and $-f_{max} \leq f \leq f_{max}$. The Nyquist frequency $f_{max} = 1/(2\triangle t)$ corresponds to $\mathbb{F} = 0.5$ and $\theta = \pi$.

7. Ideal low-pass, high-pass, and band-pass FIR filters : These terms refer to those digital filter s with frequency specification $\hat{H}(\mathbb{F})$ defined by the ideal waveforms given below.

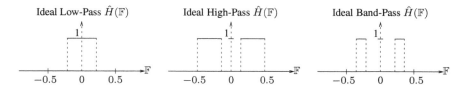

In each case, the name of the filter reflects where the frequency-response $\hat{H}(\mathbb{F})$ is nonzero, because

$$\hat{X}(\mathbb{F})\cdot\hat{H}(\mathbb{F}) = \begin{cases} \hat{X}(\mathbb{F}) & \text{if } \hat{H}(\mathbb{F}) = 1; \quad (f \in \text{passing band}) \\ 0 & \text{if } \hat{H}(\mathbb{F}) = 0. \quad (f \in \text{stop band}) \end{cases}$$

8. Ideal band-stop and notch FIR filters : By reversing the passing band and stop band of an ideal band-pass filter, we obtain a band-stop filter:

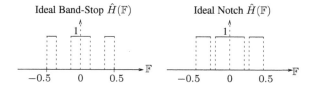

A notch filter is a band-stop filter with very narrow stop band, which is useful in removing an interfering signal of certain frequency e.g., a notch filter can be designed to eliminate the 60 Hz sinusoidal interference from an ECG signal. (The presence of 60 Hz (or 50 Hz) is caused by the electrical power line.)

10.3   Revisit Gibbs Phenomenon from the Filtering Viewpoint

In Chapter 3, Section 3.10.4, we studied the Gibbs phenomenon exhibited by the truncated N-term Fourier series of a periodic function with jump discontinuities (see Figures 3.1, 3.2, 3.6, 3.7, and 3.11.) In Section 3.10.8 we showed that the truncation of a Fourier series can be understood as the result of applying an N-point rectangular frequency-domain window (see Figure 3.17) to the Fourier coef cients of

$$(10.6) \qquad g^p(t) = \sum_{k=-\infty}^{\infty} C_k e^{j2\pi kt/T}.$$

By interpreting the N-point spectral window as the N Fourier series coef cients of a periodic time-domain function $w^p(t)$ (expressed as the Dirichlet kernel in Section 3.10.5), the Gibbs phenomenon was shown to be the result of the periodic convolution of $g^p(t)$ and $w^p(t)$ in the time domain.

To view the Gibbs effect from the ltering viewpoint, we shall interpret the N-point se-quence $\{1, 1 \cdots, 1\}$ as the N equally spaced sample values taken from the frequency response of an ideal low-pass analog lter:

$$(10.7) \qquad H(f) = \begin{cases} 1 & \text{if } -f_c \leq f \leq f_c \\ 0 & \text{otherwise} \end{cases},$$

where f_c is the desired cutoff frequency, and we must have $2f_c = N/T$ to re ect the length of the N-point rectangular spectral window.

In Chapter 7, Section 7.5, we show that the Fourier transform of the periodic $g^p(t)$ can be formally de ned as an impulse train weighted by the Fourier series coef cients, namely,

$$(10.8) \qquad G(f) = \mathcal{F}\{g^p(t)\} = \sum_{k=-\infty}^{\infty} C_k \delta\left(f - \frac{k}{T}\right).$$

Hence, assuming that $N = 2n+1$, the Fourier transform $Y(f)$ of the lter output $y(t)$ can be expressed as

$$\begin{aligned} Y(f) = G(f)H(f) &= \sum_{k=-\infty}^{\infty} C_k \left[H\left(\frac{k}{T}\right) \delta\left(f - \frac{k}{T}\right) \right] \\ &= \sum_{k=-n}^{n} C_k \delta\left(f - \frac{k}{T}\right). \end{aligned}$$

(10.9)

Since the impulse response $h(t) = \mathcal{F}^{-1}\{H(f)\}$, and for the low-pass $H(f)$ proposed above we have obtained the analytical $h(t)$ in Example 5.4 (Figure 5.6),

$$h(t) = \frac{\sin(2\pi f_c t)}{\pi t} = 2f_c \, \mathbf{sinc}(2f_c t),$$

we can therefore express the output of the lter as

$$y^p(t) = g^p(t) * h(t).$$

(As we pointed out before, the function $\mathbf{sinc}(x) = \sin(\pi x)/(\pi x)$ is not periodic.) Hence the Gibbs effect exhibited by the periodic output $y^p(t)$ is caused by the ltering of the periodic input $g^p(t)$ by the impulse response $h(t)$ of an ideal low-pass lter.

10.4   Experimenting with Digital Filtering and Filter Design

In this section we shall demonstrate the use of DFT and IDFT in digital ltering and lter design by constructing a low-pass FIR lter. Let us assume that the signal data to be ltered has frequency components either lower that 30 Hz or higher than 50 Hz, and that a front-end analog anti-aliasing lter bandlimits the signal to 80 Hz. Suppose that we need a low-pass digital lter to eliminate the frequency components above 30 Hz. From the information we have about the signal to be ltered, an ideal low-pass lte r may be de ned with a cutoff frequency $f_c = 40$ Hz:

$$(10.10) \qquad H(f) = \begin{cases} 1 & \text{if } -40 \le f \le 40; \\ 0 & \text{otherwise.} \end{cases}$$

Since we assume that the signal to be ltered is band-limited to $f_{\max} = 80$ Hz, we must choose the sampling rate $\mathbb{R} = 1/\triangle t > 2f_{\max}$ to satisfy the Nyquist condition. Let us set $\triangle t = 0.005$, which allows $f_{\max}$ up to $\mathbb{R}/2 = 100$ Hz; hence, we only need to sample $H(f)$ to cover the range from -100 Hz to 100 Hz as shown below.

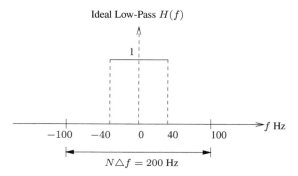

To obtain the FIR lter represented by the nite N-sample impulse response $\{h_\ell\}$, we need to choose the length N. Since the lter length N is a parameter of the design, which can be easily changed if the digital lter $\{h_\ell\}$ turns out to be too long for the ltering operation to be performed ef ciently, we may begin with a suf ciently large N to meet the design speci catio n $H(f)$. If we decide later to shorten the lter sequence $\{h_\ell\}$, it can be truncated using an appropriately chosen data-weighting window (as discussed in Chapter 8) provided that the lter s frequency response $H_I(f)$ still approximates $H(f)$ to our satisfaction.

To demonstrate this process using the proposed low-pass lter, we shall try $N = 256$. With $\triangle t = 0.005$ sec, we have $N\triangle t = 256 \times 0.005 = 1.28$ sec, and $H(f)$ will be sampled at $\triangle f = \mathbb{R}/N = 200/256$ Hz for the IDFT computation. We can now experiment with the digital ltering and lter design as outlined in the following steps:

Step 1. De n e the 256 samples of $H(f)$ by the sequence

$$\{H_{-127}, \cdots, H_{-51}, H_{-50}, \cdots, H_{-1}, H_0, H_1, \cdots, H_{50}, H_{51}, \cdots, H_{128}\}$$
$$= \{\underbrace{0, 0, \ldots, 0}_{76 \text{ zeros}}, \tfrac{1}{2}, \underbrace{1, 1, \ldots, 1, \ldots, 1, 1}_{101 \text{ ones}}, \tfrac{1}{2}, \underbrace{0, 0 \ldots, 0}_{77 \text{ zeros}}\}.$$

Recall that because $N = 256$ is an even number, the sequence $\{H_r\}$ begins at $-100 + \triangle f$ Hz and ends at 100 Hz. Note that, following what we have discussed in detail in Section 4.6.2 in Chapter 4, we have split the one at $f = 51\triangle t = 39.84$ Hz and use one-half to replace the zero at $f = -39.84$ Hz to maintain symmetry in the zero-padded sampled $H(f)$, and this results in setting $H_{\pm 51} = \frac{1}{2}$ (as shown in the defining sequence and also in Figure 10.1).

Figure 10.1 Sampling $H(f)$ to obtain impulse response of a FIR filter.

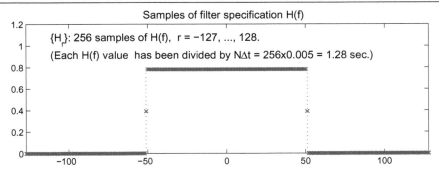

Samples of filter specification H(f)

{H$_r$}: 256 samples of H(f), r = −127, ..., 128.
(Each H(f) value has been divided by NΔt = 256x0.005 = 1.28 sec.)

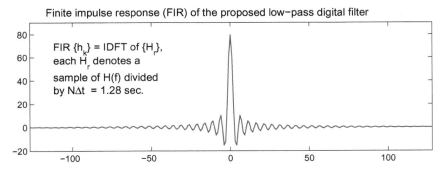

Finite impulse response (FIR) of the proposed low−pass digital filter

FIR {h$_k$} = IDFT of {H$_r$},
each H$_r$ denotes a
sample of H(f) divided
by NΔt = 1.28 sec.

Step 2. For the chosen length $N = 256$, compute the N-sample filter sequence $\{h_\ell\}$ by the IDFT:

$$(10.11) \qquad h_\ell = \sum_{r=-N/2+1}^{N/2} \tilde{H}_r e^{j2\pi r\ell/N}, \quad \text{for } -N/2+1 \le \ell \le N/2,$$

where $\tilde{H}_r = \frac{1}{N\triangle t}H_r$, because ideally $\mathrm{DFT}(\{h_\ell\}) = \frac{1}{N}\{\frac{1}{\triangle t}H_r\}$ when H_r denotes the samples of the Fourier transform $H(f) = \mathcal{F}\{h(t)\}$ evaluated at $f = \frac{r}{N\triangle t}$. (The exact relationships were studied in Chapter 7, Section 7.6.) For $N = 256$ and $\triangle t = 0.005$ seconds, we have $N\triangle t = 1.28$ seconds here. The 256-sample sequence $\{\tilde{H}_r\}$ and the computed 256-sample $\{h_\ell\}$ are shown together in Figure 10.1.

Step 3. Generate a test signal $x(t)$:

$$(10.12) \quad x(t) = 3 - 4\cos(2\pi t) + 3\cos(12\pi t) - 1.5\cos(104\pi t) + 1.2\cos(150\pi t).$$

Step 4. Sample the test signal at intervals of $\triangle t = 0.005$ seconds for the duration of 3 seconds to obtain the M-sample sequence $\{x_\ell\}$, where $M = 600$. The 600-sample sequence $\{x_\ell\}$ and its magnitude spectrum (computed by the DFT) are shown in Figure 10.2.

Figure 10.2 Sampled noisy signal $x(t)$ and its magnitude spectrum.

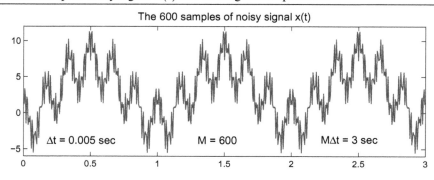

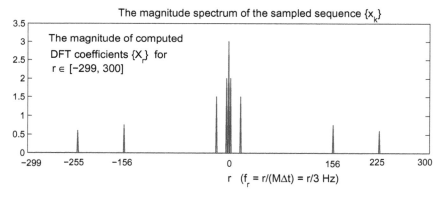

Step 5. Compute the linear convolution of the M-sample sequence $\{x_\ell\}$ and the N-sample lter sequence $\{h_\ell\}$.

Recall from Chapter 9 that we need to convolve $\{x_0, x_1, \ldots, x_{599}\}$ with

$$\{f_0, f_1, \ldots, f_{255}\} = \{h_{-127}, \ldots, h_{-1}, h_0, h_1, \ldots, h_{128}\}.$$

As discussed in Section 9.2.5 in Chapter 9, the linear convolution of $\{x_\ell\}$ and $\{f_\ell\}$ should be converted to a cyclic convolution of two zero-padded sequences with length (of both) equal to $M+N-1$, because the latter can then be computed by two DFTs and one IDFT (via FFTs and IFFT). The result shown in Figure 10.3 is obtained by calling the function `conv(f,x)` available from MATLAB, because it has implemented the fast algorithm for the linear convolution of two sequences of arbitrary lengths.

Figure 10.3 Discrete linear convolution of $\{x_\ell\}$ and FIR lter $\{h_\ell\}$.

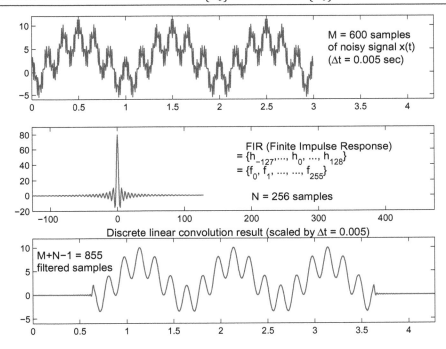

Recall from Chapter 9 that the discrete linear convolution of $\{x_\ell\}$ and $\{h_\ell\}$ (multiplied by $\triangle t$) approximates the continuous linear convolution of the nite-duration noisy signal $x(t)$ and the analog lter $h(t)$, so the end effects in the graph shown in Figure 10.3 are due to the nite length of $x(t)$. However, we can also use the same example to demonstrate how to lter a knowingly periodic (and hence in nite) sequence.

For this particular example, the signal $x(t)$ is periodic and we have sampled it for three full periods, so the entire M-sample sequence $\{x_\ell\}$ can be interpreted as one period of an in n itely long periodic sequence. Because the linear convolution of an in nite periodic function with a nite $h(t)$ can be de ned by equivalent periodic convolution, the same is true for the discrete linear convolution when one sequence is periodic (and hence in n ite). Therefore, in this and other similar cases, we can eliminate the end effects by performing periodic convolution of the noisy sequence $\{x_\ell\}$ and the zero-padded $\{h_\ell\}$. Both sequences have now $M = 600$ samples, and following the Time-Domain Cyclic Convolution Theorem 9.1, we use two DFT and one IDFT to compute $\{x_\ell\} \odot \{f_\ell\}$, where $\{f_\ell\}$ represents the zero-padded $\{h_\ell\}$ as shown in Figure 10.4. The convolution results shown in Figure 10.4 have been multiplied by $\triangle t = 0.005$ so they represent samples of ltered signal $y(t)$.

Figure 10.4 Discrete periodic convolution of $\{x_\ell\}$ and FIR lter $\{h_\ell\}$.

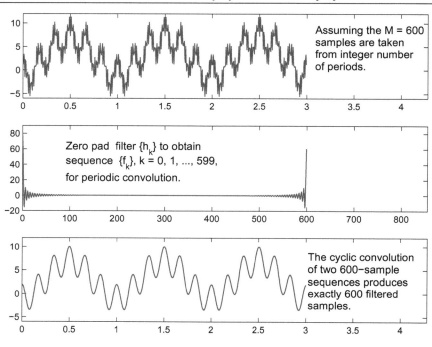

Step 6. Compute the DFT of the M-sample output $\{y_\ell\} = \triangle t\,(\{x_\ell\} \odot \{f_\ell\})$ to evaluate the effect of the lter. That is, we compute

$$(10.13) \qquad Y_r = \frac{1}{M} \sum_{\ell=0}^{M-1} y_\ell e^{-j2\pi r\ell/M}, \quad \text{for } \ell = 0, 1, \ldots, M-1.$$

The ltered signal samples $\{y_\ell\}$ and the magnitude of the computed DFT coef cients of the ltered sample sequence are shown in Figure 10.5.

Step 7. If the digital lter $\{h_\ell\}$ needs to be truncated, one may consider using the data-weighting windows presented in Chapter 8. For each truncated (and possibly modi ed) sequence $\{h_\ell\}$, compute and plot its DFT results, and compare that with the design spec- i catio n $H(f)$. When we are satis ed with the truncated $\{h_\ell\}$, we may repeat Steps 3—6 to determine the effect of the lter on the test signal. (Steps 3–6 can also be repeated for different test signals.)

Note that the entire process (as described by steps 1—7) can be repeated if we need to ac- commodate any of the following changes in the lter design: (i) the bandwidth of the expected

Figure 10.5 Computed DFT coef cients of the ltered sample sequence.

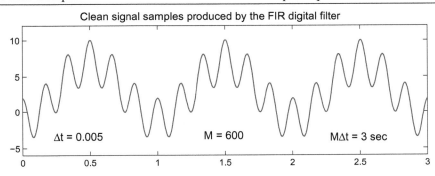

Clean signal samples produced by the FIR digital filter

$\Delta t = 0.005$ M = 600   MΔt = 3 sec

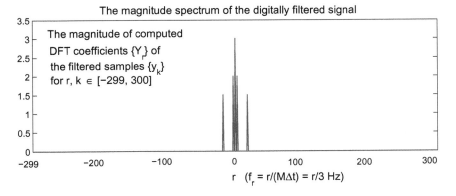

The magnitude spectrum of the digitally filtered signal

The magnitude of computed
DFT coefficients $\{Y_r\}$ of
the filtered samples $\{y_k\}$
for r, k $\in$ [−299, 300]

r $(f_r = r/(M\Delta t) = r/3$ Hz)

signal data is changed; (ii) the length N of the filter is changed; (iii) the desired frequency response $H(f)$ of the filter is changed, including the use of different $H(f)$ for high-pass, band-pass, or notch filter.

Our general discussion of the experimental procedure above demonstrates how the DFT and IDFT (hence the FFT and IFFT) can be used in digital filtering and filter construction, and provides background for readers to pursue specialized filter applications.

References

1. A. Ambardar. *Analog and Digital Signal Processing*. Brooks/Cole Publishing Company, Pacific Grove, CA, second edition, 1999.

2. R. N. Bracewell. *The Fourier Transform and Its Applications*. McGraw-Hill Higher Education, A Division of The McGraw-Hill Companies, San Francisco, CA, third edition, 2000.

3. W. L. Briggs and V. E. Hensen. *The DFT: An Owner's Manual for the Discrete Fourier Transform*. The Society for Industrial and Applied Mathematics, Philadelphia, PA, 1995.

4. E. O. Brigham. *The Fast Fourier Transform and Its Applications*. Prentice-Hall, Inc., Upper Saddle River, NJ, 1988.

5. R. W. Hamming. *Digital Filters*. Prentice-Hall, Inc., Englewood Cliffs, NJ, third edition, 1989.

6. B. Porat. *A Course in Digital Signal Processing*. John Wiley & Sons, Inc., New York, 1997.

Part II

Fast Algorithms

Chapter 11

Index Mapping and Mixed-Radix FFTs

As we indicate in Part I of this book, it is well known that the fast Fourier transform (FFT) algorithm can be used to speed up the computation of the DFT if its length N is a power of two. The family of FFT algorithms specifically designed to handle composite $N = 2^n$ are called the radix-2 FFTs. Although the radix-2 FFT is the most widely known and most widely available, it is, in fact, a special case of the arbitrary factor (mixed-radix) algorithm proposed by Cooley and Tukey in [16]. Since the radix-2 FFT fits the divide-and-conquer paradigm, it is usually presented from that perspective, and the general mixed-radix FFTs are easily left out because the divide-and-conquer paradigm does not apply when N is the product of mixed factors.

In this chapter we explore ways to organize the mixed-radix DFT computation facilitated by index mapping via multidimensional arrays. This approach would allow us to study a large number of mixed-radix FFT algorithms in a systematic manner, including the radix-2 special case, and it also paves the way for the more specialized prime factor FFT algorithms covered in Chapter 13.

11.1   Algebraic DFT versus FFT-Computed DFT

Given N equally spaced discrete-times samples $\{x_0, x_1, \ldots, x_{N-1}\}$, the algebraic DFT formula used throughout Part I of this book was given by

$$(11.1) \qquad X_r = \frac{1}{N} \sum_{\ell=0}^{N-1} x_\ell \, \omega_N^{r\ell}, \quad \omega_N \equiv e^{-j2\pi/N}, \quad \text{for } r = 0, 1, \ldots, N-1,$$

which we derive in Chapter 4 by passing a finite Fourier series as an interpolating trigonometric polynomial through the N given samples; the associated IDFT formula allows us to recover the time-domain sequence:

$$(11.2) \qquad x_\ell = \sum_{r=0}^{N-1} X_r \, \omega_N^{-r\ell}, \quad \text{for } \ell = 0, 1, \ldots, N-1.$$

While the FFTs are simply fast algorithms developed to compute the DFT, we need to be aware of the inconsistency concerning the placement of the scaling factor $1/N$ in the DFT formula, because such inconsistency also exists in the mathematical software which implements the FFT algorithms. It turns out that, in the FFT literature, it is common *not* to include division by N in the computed DFT, because the development of fast algorithms can be quite involved, and it is cumbersome to carry the scaling factor $1/N$ all the way through while its omission has no effect on the resulting algorithms.

Following the convention in the FFT literature, all FFT algorithms derived in Part II of this book compute the DFT without the scaling factor $1/N$; i.e., for algorithmic development the computed DFT is de ned by

$$(11.3) \qquad X_r = \sum_{\ell=0}^{N-1} x_\ell\, \omega_N^{r\ell}, \quad \omega_N \equiv e^{-j2\pi/N}, \quad \text{for } r = 0, 1, \ldots, N-1.$$

Since the IDFT must return the time-domain sequence $\{x_\ell\}$, the computed IDFT is de ned by

$$(11.4) \qquad x_\ell = \frac{1}{N} \sum_{r=0}^{N-1} X_r\, \omega_N^{-r\ell}, \quad \text{for } \ell = 0, 1, \ldots, N-1.$$

Observe that the Inverse FFT (IFFT) algorithms for computing the IDFT can be obtained from the FFT algorithms by rede ning $\omega_N \equiv e^{+j2\pi/N}$ after including division by N. Therefore, the IFFT algorithm is immediately available for every FFT algorithm developed.

11.2   The Role of Index Mapping

Index mapping plays a fundamental role in all mixed-radix algorithms   ef cient algorithms are developed by pairing up different index mapping schemes: one is deployed on the input data sequence, and the other one is deployed on the DFT output. We consider   rst the general case when the factors of N are arbitrary integers. To establish the notation and make explicit the index manipulation required, we begin with a composite N with three factors. The following notations will be used consistently throughout this chapter:

- Given three integers N_0, N_1, and N_2, we shall use $\boldsymbol{A}[N_0, N_1, N_2]$ to declare a 3-D array of dimensions N_0-by-N_1-by-N_2; each element of $\boldsymbol{A}$ will be uniquely identi ed as $A[n_0, n_1, n_2]$, where each dimensional index n_k can take on values between 0 and $N_k - 1$.

- For $N = N_0 \times N_1 \times N_2$, the 1-D input array $\boldsymbol{x}$ of length N is to be identi ed with (or mapped to) the 3-D array $\boldsymbol{A}[N_0, N_1, N_2]$. Assuming (without loss of generality) that x_ℓ s are stored in (or mapped to) $\boldsymbol{A}$ using column-major scheme, we shall use $A[n_0, n_1, n_2]$ to denote x_ℓ for $\ell = n_2 N_1 N_0 + n_1 N_0 + n_0$, where $0 \le n_0 \le N_0 - 1$, $0 \le n_1 \le N_1 - 1$, and $0 \le n_2 \le N_2 - 1$.

- With the dimensions of $\boldsymbol{A}[N_0, N_1, N_2]$ xed , we now require the 1-D output array $\boldsymbol{X}$ (of length N) to be identi ed with (or mapped to) a 3-D array $\boldsymbol{B}[N_2, N_1, N_0]$. Note that because we reverse the dimensions of the output array, we now use $B[\hat{n}_2, \hat{n}_1, \hat{n}_0]$ to denote X_r for $r = \hat{n}_0 N_1 N_2 + \hat{n}_1 N_2 + \hat{n}_2$ according the column-major scheme, where $0 \le \hat{n}_2 \le N_2 - 1$, $0 \le \hat{n}_1 \le N_1 - 1$, and $0 \le \hat{n}_0 \le N_0 - 1$. Observe that the numerical range covered by dimensional index $\hat{n}_k$ is consistent with that covered by n_k.

To develop the three-factor mixed-radix FFT, we shall deploy the index mapping schemes for x_ℓ and X_r to *decouple* the length-N DFT into multiple short DFT or DFT-like transforms of lengths N_0, N_1 and N_2. The index mapping scheme is first deployed on the input data to be transformed by the DFT formula defined by (11.3) (without division by N):

$$(11.5) \qquad X_r = \sum_{\ell=0}^{N-1} x_\ell \, \omega_N^{r\ell}, \quad \omega_N \equiv e^{-j2\pi/N}, \quad \text{for } r = 0, 1, \dots, N - 1;$$

that is, we use $A[n_0, n_1, n_2]$ to denote x_ℓ for $\ell = n_2 N_1 N_0 + n_1 N_0 + n_0$, and we obtain the mathematically equivalent DFT formula involving array A and its indices:

> **for** $r := 0$ **to** $N - 1$ **do**
> $\qquad X_r := \sum_{n_0=0}^{N_0-1} \sum_{n_1=0}^{N_1-1} \sum_{n_2=0}^{N_2-1} A[n_0, n_1, n_2] \, \omega_N^{r(n_2 N_1 N_0 + n_1 N_0 + n_0)}.$
> **end for**

Next the output index mapping $r = \hat{n}_0 N_1 N_2 + \hat{n}_1 N_2 + \hat{n}_2$ is deployed so we can obtain a mathematically equivalent formula involving also array B and its indices:

> **for** $\hat{n}_0 := 0$ **to** $N_0 - 1$ **do**
> $\quad$ **for** $\hat{n}_1 := 0$ **to** $N_1 - 1$ **do**
> $\qquad$ **for** $\hat{n}_2 := 0$ **to** $N_2 - 1$ **do**
> $\qquad\quad r := \hat{n}_0 N_1 N_2 + \hat{n}_1 N_2 + \hat{n}_2$
> $\qquad\quad B[\hat{n}_2, \hat{n}_1, \hat{n}_0] := \sum_{n_0=0}^{N_0-1} \sum_{n_1=0}^{N_1-1} \left(\sum_{n_2=0}^{N_2-1} A[n_0, n_1, n_2] \, \omega_N^{r(n_2 N_1 N_0)} \right) \omega_N^{r(n_1 N_0)} \omega_N^{r \, n_0}$
> $\qquad\quad X_r := B[\hat{n}_2, \hat{n}_1, \hat{n}_0]$
> $\qquad$ **end for**
> $\quad$ **end for**
> **end for**

The decoupling process begins with the innermost partial sum the expression in brackets and the entire process is covered in the following three subsections.

11.2.1 The decoupling process—Stage I

In preparation for the decoupling process to occur, we make explicit the computation of every innermost partial sum the bracketed expression in the summation formula via double for-loops indexed by n_0 and n_1, and each computed partial-sum is saved in a 2-D temporary working array $V[N_0, N_1]$ as shown below:

for $\hat{n}_0 := 0$ **to** $N_0 - 1$ **do**
    for $\hat{n}_1 := 0$ **to** $N_1 - 1$ **do**
        for $\hat{n}_2 := 0$ **to** $N_2 - 1$ **do**
            $r := \hat{n}_0 N_1 N_2 + \hat{n}_1 N_2 + \hat{n}_2$
            for $n_0 := 0$ **to** $N_0 - 1$ **do**
                for $n_1 := 0$ **to** $N_1 - 1$ **do**
                    $V[n_0, n_1] := \sum_{n_2=0}^{N_2-1} A[n_0, n_1, n_2] \omega_N^{r(n_2 N_1 N_0)}$
                end for
            end for
            $B[\hat{n}_2, \hat{n}_1, \hat{n}_0] := \sum_{n_0=0}^{N_0-1} \sum_{n_1=0}^{N_1-1} V[n_0, n_1] \, \omega_N^{r(n_1 N_0)} \omega_N^{r n_0}$
            $X_r := B[\hat{n}_2, \hat{n}_1, \hat{n}_0]$
        end for
    end for
end for

On the surface, the value of each partial sum $V[n_0, n_1]$ appears to depend on all three (output array) indices $\hat{n}_0$, $\hat{n}_1$, $\hat{n}_2$ due to the occurrence of $r = \hat{n}_0 N_1 N_2 + \hat{n}_1 N_2 + \hat{n}_2$ in the exponent of $\omega_N^{r(n_2 N_1 N_0)}$. If such dependency cannot be reduced, we must compute the entire V array for all different values of $\hat{n}_0$, $\hat{n}_1$, and $\hat{n}_2$ as we do here — inside the triple for-loops indexed by $\hat{n}_0$, $\hat{n}_1$, and $\hat{n}_2$, and no decoupling can occur. Fortunately, this does not have to be the case — we show next how to compute $V[n_0, n_1]$ *independent of* both $\hat{n}_0$ and $\hat{n}_1$ by deploying the output index mapping scheme on the exponent itself, and we obtain the following result after simple expansion:

$$
\begin{aligned}
r(n_2 N_1 N_0) &= (\hat{n}_0 N_1 N_2 + \hat{n}_1 N_2 + \hat{n}_2)(n_2 N_1 N_0) \\
&= (\hat{n}_0 N_1 N_2 + \hat{n}_1 N_2)(n_2 N_1 N_0) + \hat{n}_2(n_2 N_1 N_0) \\
&= \hat{n}_0 n_2 N_1 N_2 N_1 N_0 + \hat{n}_1 n_2 N_2 N_1 N_0 + \hat{n}_2(n_2 N_1 N_0) \\
&= N(\hat{n}_0 n_2 N_1 + \hat{n}_1 n_2) + \hat{n}_2(n_2 N_1 N_0) \qquad (\because N = N_0 N_1 N_2) \\
&= N\lambda + \hat{n}_2 n_2 N_1 N_0, \quad \text{where } \lambda = \hat{n}_0 n_2 N_1 + \hat{n}_1 n_2.
\end{aligned}
$$

Because $N = N_0 N_1 N_2$, $\omega_N^N = 1$, and $\omega_N^{N_1 N_0} = \omega_{N_2}$, we can further simplify the term

$$
\omega_N^{r(n_2 N_1 N_0)} = \omega_N^{N\lambda} \omega_N^{\hat{n}_2 n_2 N_1 N_0} = \omega_{N_2}^{\hat{n}_2 n_2},
$$

which is now independent of $\hat{n}_0$ and $\hat{n}_1$ and so is the content of $V[N_0, N_1]$ that — means we only need to update array V as $\hat{n}_2$ varies from 0 to $N_2 - 1$. This also means that we compute V only N_2 times (instead of $N = N_0 N_1 N_2$ times). Since we can save all N_2 copies of $V[N_0, N_1]$ in a 3-D array $A_1[N_0, N_1, N_2]$, we can compute $A_1[n_0, n_1, \hat{n}_2]$ in an *independent* triple for-loop indexed by n_0, n_1, and $\hat{n}_2$ as shown below. (Observe that the occurrence of $V[n_0, n_1]$ in the summation formula has now been replaced by $A_1[n_0, n_1, \hat{n}_2]$.)

$$
\begin{aligned}
&\textbf{for } n_0 := 0 \textbf{ to } N_0 - 1 \textbf{ do} \\
&\quad \textbf{for } n_1 := 0 \textbf{ to } N_1 - 1 \textbf{ do} \\
&\qquad \textbf{for } \hat{n}_2 := 0 \textbf{ to } N_2 - 1 \textbf{ do} \\
&\qquad\quad A_1[n_0, n_1, \hat{n}_2] := \sum_{n_2=0}^{N_2-1} A[n_0, n_1, n_2]\, \omega_{N_2}^{\hat{n}_2 n_2} \\
&\qquad \textbf{end for} \\
&\quad \textbf{end for} \\
&\textbf{end for} \\
&\textbf{for } \hat{n}_0 := 0 \textbf{ to } N_0 - 1 \textbf{ do} \\
&\quad \textbf{for } \hat{n}_1 := 0 \textbf{ to } N_1 - 1 \textbf{ do} \\
&\qquad \textbf{for } \hat{n}_2 := 0 \textbf{ to } N_2 - 1 \textbf{ do} \\
&\qquad\quad r := \hat{n}_0 N_1 N_2 + \hat{n}_1 N_2 + \hat{n}_2 \\
&\qquad\quad B[\hat{n}_2, \hat{n}_1, \hat{n}_0] := \sum_{n_0=0}^{N_0-1} \left(\sum_{n_1=0}^{N_1-1} A_1[n_0, n_1, \hat{n}_2]\, \omega_N^{r(n_1 N_0)} \right) \omega_N^{r n_0} \\
&\qquad\quad X_r := B[\hat{n}_2, \hat{n}_1, \hat{n}_0] \\
&\qquad \textbf{end for} \\
&\quad \textbf{end for} \\
&\textbf{end for}
\end{aligned}
$$

Note that a short DFT of length N_2 is de n ed by the for-loop indexed by $\hat{n}_2$:

$$
\begin{aligned}
&\textbf{for } \hat{n}_2 := 0 \textbf{ to } N_2 - 1 \textbf{ do} \\
&\quad A_1[n_0, n_1, \hat{n}_2] = \sum_{n_2=0}^{N_2-1} A[n_0, n_1, n_2]\, \omega_{N_2}^{\hat{n}_2 n_2} \\
&\textbf{end for}
\end{aligned}
$$

Since each short DFT (of length N_2) is computed by the inner loop of the triple for-loop indexed by n_0, n_1, and $\hat{n}_2$, we need to compute, in total, $N_0 N_1$ (or N/N_2) such short transforms during the rst stage of the decoupled computation.

11.2.2 The decoupling process—Stage II

To prepare for the next stage of decoupling, we shall do the same with the currently innermost partial sum the expression in brackets in the remainder of the summation formula. This time we may use a 1-D temporary working array $V[N_0]$ to save the N_0 partial sums explicitly computed by a newly added for-loop indexed by n_0:

for $n_0 := 0$ **to** $N_0 - 1$ **do**
  for $n_1 := 0$ **to** $N_1 - 1$ **do**
    for $\hat{n}_2 := 0$ **to** $N_2 - 1$ **do**
      $A_1[n_0, n_1, \hat{n}_2] = \sum_{n_2=0}^{N_2-1} A[n_0, n_1, n_2] \, \omega_{N_2}^{\hat{n}_2 n_2}$
    end for
  end for
end for
for $\hat{n}_0 := 0$ **to** $N_0 - 1$ **do**
  for $\hat{n}_1 := 0$ **to** $N_1 - 1$ **do**
    for $\hat{n}_2 := 0$ **to** $N_2 - 1$ **do**
      $r := \hat{n}_0 N_1 N_2 + \hat{n}_1 N_2 + \hat{n}_2$
      for $n_0 := 0$ **to** $N_0 - 1$ **do**
        $V[n_0] := \sum_{n_1=0}^{N_1-1} A_1[n_0, n_1, \hat{n}_2] \, \omega_N^{r(n_1 N_0)}$
      end for
      $B[\hat{n}_2, \hat{n}_1, \hat{n}_0] := \sum_{n_0=0}^{N_0-1} V[n_0] \, \omega_N^{r n_0}$
      $X_r := B[\hat{n}_2, \hat{n}_1, \hat{n}_0]$
    end for
  end for
end for

To decouple the computation of array $V[N_0]$, we again analyze the dependence of the term $\omega_N^{r(n_1 N_0)}$ on the output indices $\hat{n}_k$ by expanding the exponent itself using the index mapping $r = \hat{n}_0 N_1 N_2 + \hat{n}_1 N_2 + \hat{n}_2$ with the following result:

$$
\begin{aligned}
r(n_1 N_0) &= (\hat{n}_0 N_1 N_2 + \hat{n}_1 N_2 + \hat{n}_2)(n_1 N_0) \\
&= \hat{n}_0 n_1 N_0 N_1 N_2 + (\hat{n}_1 N_2 + \hat{n}_2)(n_1 N_0) \\
&= N(\hat{n}_0 n_1) + \hat{n}_1 n_1 N_0 N_2 + \hat{n}_2 (n_1 N_0).
\end{aligned}
$$

Because $N = N_0 N_1 N_2$, $\omega_N^N = 1$, and $\omega_N^{N_0 N_2} = \omega_{N_1}$, we can further simplify the term

$$
\omega_N^{r(n_1 N_0)} = \omega_N^{N(\hat{n}_0 n_1)} \omega_N^{\hat{n}_1 n_1 N_0 N_2} \omega_N^{\hat{n}_2 (n_1 N_0)} = \omega_{N_1}^{\hat{n}_1 n_1} \omega_N^{\hat{n}_2 (n_1 N_0)},
$$

which depends on $\hat{n}_1$ and $\hat{n}_2$, but it is *independent of* $\hat{n}_0$. Therefore, we can again compute the $N_1 \times N_2$ copies of the 1-D array $V[N_0]$ independently, provided we save all of them in a 3-D array $A_2[N_0, N_1, N_2]$ for use in the summation formula. We incorporate the changes below.

$$
\begin{aligned}
&\textbf{for } n_0 := 0 \textbf{ to } N_0 - 1 \textbf{ do} \\
&\quad \textbf{for } n_1 := 0 \textbf{ to } N_1 - 1 \textbf{ do} \\
&\quad\quad \textbf{for } \hat{n}_2 := 0 \textbf{ to } N_2 - 1 \textbf{ do} \\
&\quad\quad\quad A_1[n_0, n_1, \hat{n}_2] = \sum_{n_2=0}^{N_2-1} A[n_0, n_1, n_2]\, \omega_{N_2}^{\hat{n}_2 n_2} \\
&\quad\quad \textbf{end for} \\
&\quad \textbf{end for} \\
&\textbf{end for} \\
&\textbf{for } n_0 := 0 \textbf{ to } N_0 - 1 \textbf{ do} \\
&\quad \textbf{for } \hat{n}_2 := 0 \textbf{ to } N_2 - 1 \textbf{ do} \\
&\quad\quad \textbf{for } \hat{n}_1 := 0 \textbf{ to } N_1 - 1 \textbf{ do} \\
&\quad\quad\quad A_2[n_0, \hat{n}_1, \hat{n}_2] = \sum_{n_1=0}^{N_1-1} A_1[n_0, n_1, \hat{n}_2]\, \omega_N^{\hat{n}_2(n_1 N_0)} \omega_{N_1}^{\hat{n}_1 n_1} \\
&\quad\quad \textbf{end for} \\
&\quad \textbf{end for} \\
&\textbf{end for} \\
&\textbf{for } \hat{n}_0 := 0 \textbf{ to } N_0 - 1 \textbf{ do} \\
&\quad \textbf{for } \hat{n}_1 := 0 \textbf{ to } N_1 - 1 \textbf{ do} \\
&\quad\quad \textbf{for } \hat{n}_2 := 0 \textbf{ to } N_2 - 1 \textbf{ do} \\
&\quad\quad\quad r := \hat{n}_0 N_1 N_2 + \hat{n}_1 N_2 + \hat{n}_2 \\
&\quad\quad\quad B[\hat{n}_2, \hat{n}_1, \hat{n}_0] := \sum_{n_0=0}^{N_0-1} A_2[n_0, \hat{n}_1, \hat{n}_2]\, \omega_N^{r n_0} \\
&\quad\quad\quad X_r := B[\hat{n}_2, \hat{n}_1, \hat{n}_0] \\
&\quad\quad \textbf{end for} \\
&\quad \textbf{end for} \\
&\textbf{end for}
\end{aligned}
$$

Note that a short DFT-like transform of length N_1 is de n ed by the for-loop indexed by $\hat{n}_1$:

$$
\begin{aligned}
&\textbf{for } \hat{n}_1 := 0 \textbf{ to } N_1 - 1 \textbf{ do} \\
&\quad A_2[n_0, \hat{n}_1, \hat{n}_2] = \sum_{n_1=0}^{N_1-1} \left\{ A_1[n_0, n_1, \hat{n}_2]\, \omega_N^{\hat{n}_2(n_1 N_0)} \right\} \omega_{N_1}^{\hat{n}_1 n_1} \\
&\textbf{end for}
\end{aligned}
$$

Since each short DFT (of length N_1) is computed by the inner loop of the triple for-loop indexed by n_0, $\hat{n}_2$, and $\hat{n}_1$, we need to compute, in total, $N_0 N_2$ (or N/N_1) such short transforms during the second stage of the decoupled computation.

11.2.3 The decoupling process—Stage III

To complete the three-factor mixed-radix FFT, we treat the last part of the summation formula in a formal manner with index expansion

$$
r n_0 = (\hat{n}_0 N_1 N_2 + \hat{n}_1 N_2 + \hat{n}_2) n_0 = \hat{n}_0 n_0 N_1 N_2 + (\hat{n}_1 N_2 + \hat{n}_2) n_0
$$

followed by dependence analysis of the term

$$
\omega_N^{r n_0} = \omega_N^{\hat{n}_0 n_0 N_1 N_2} \omega_N^{(\hat{n}_1 N_2 + \hat{n}_2) n_0} = \omega_{N_0}^{\hat{n}_0 n_0} \omega_N^{(\hat{n}_1 N 2 + \hat{n}_2) n_0},
$$

which depends on all three output array indices; hence, we must compute the nal sum inside a triple for-loop indexed by $\hat{n}_0$, $\hat{n}_1$, and $\hat{n}_2$. The complete three-factor mixed-radix FFT algorithm is listed below. Observe that we have saved the N values of nal sum into

$A_3[\hat{n}_0, \hat{n}_1, \hat{n}_2]$ to   be consistent with the usage of $\boldsymbol{A}_1$ and $\boldsymbol{A}_2$. Because $A_3[\hat{n}_0, \hat{n}_1, \hat{n}_2]$ contains the n al sum previously assigned to $B[\hat{n}_2, \hat{n}_1, \hat{n}_0]$, we can directly assign the former value to X_r array $\boldsymbol{B}$ is no longer needed.   Observe that the same index mapping $r = \hat{n}_0 N_1 N_2 + \hat{n}_1 N_2 + \hat{n}_2$ identi es the output element X_r with $A_3[\hat{n}_0, \hat{n}_1, \hat{n}_2]$, meaning that X_r is stored in (or mapped to) the 3-D array $\boldsymbol{A}_3[N_0, N_1, N_2]$ using row-major scheme. Since $\boldsymbol{A}_3[N_0, N_1, N_2]$ is the transpose of $\boldsymbol{B}[N_2, N_1, N_0]$, this is a fully consistent result, and it shows once more that it is the index mapping scheme that counts!

$$
\begin{aligned}
&\underline{\textbf{for}}\ n_0 := 0\ \underline{\textbf{to}}\ N_0 - 1\ \underline{\textbf{do}} \\
&\quad \underline{\textbf{for}}\ n_1 := 0\ \underline{\textbf{to}}\ N_1 - 1\ \underline{\textbf{do}} \\
&\quad\quad \underline{\textbf{for}}\ \hat{n}_2 := 0\ \underline{\textbf{to}}\ N_2 - 1\ \underline{\textbf{do}} \\
&\quad\quad\quad A_1[n_0, n_1, \hat{n}_2] = \sum_{n_2=0}^{N_2-1} A[n_0, n_1, n_2]\, \omega_{N_2}^{\hat{n}_2 n_2} \\
&\quad\quad \underline{\textbf{end for}} \\
&\quad \underline{\textbf{end for}} \\
&\underline{\textbf{end for}} \\
&\underline{\textbf{for}}\ n_0 := 0\ \underline{\textbf{to}}\ N_0 - 1\ \underline{\textbf{do}} \\
&\quad \underline{\textbf{for}}\ \hat{n}_2 := 0\ \underline{\textbf{to}}\ N_2 - 1\ \underline{\textbf{do}} \\
&\quad\quad \underline{\textbf{for}}\ \hat{n}_1 := 0\ \underline{\textbf{to}}\ N_1 - 1\ \underline{\textbf{do}} \\
&\quad\quad\quad A_2[n_0, \hat{n}_1, \hat{n}_2] = \sum_{n_1=0}^{N_1-1} A_1[n_0, n_1, \hat{n}_2]\, \omega_{N}^{\hat{n}_2(n_1 N_0)} \omega_{N_1}^{\hat{n}_1 n_1} \\
&\quad\quad \underline{\textbf{end for}} \\
&\quad \underline{\textbf{end for}} \\
&\underline{\textbf{end for}} \\
&\underline{\textbf{for}}\ \hat{n}_1 := 0\ \underline{\textbf{to}}\ N_1 - 1\ \underline{\textbf{do}} \\
&\quad \underline{\textbf{for}}\ \hat{n}_2 := 0\ \underline{\textbf{to}}\ N_2 - 1\ \underline{\textbf{do}} \\
&\quad\quad \underline{\textbf{for}}\ \hat{n}_0 := 0\ \underline{\textbf{to}}\ N_0 - 1\ \underline{\textbf{do}} \\
&\quad\quad\quad A_3[\hat{n}_0, \hat{n}_1, \hat{n}_2] := \sum_{n_0=0}^{N_0-1} A_2[n_0, \hat{n}_1, \hat{n}_2]\, \omega_{N}^{(\hat{n}_1 N_2 + \hat{n}_2) n_0} \omega_{N_0}^{\hat{n}_0 n_0} \\
&\quad\quad\quad r := \hat{n}_0 N_1 N_2 + \hat{n}_1 N_2 + \hat{n}_2 \\
&\quad\quad\quad X_r := A_3[\hat{n}_0, \hat{n}_1, \hat{n}_2] \\
&\quad\quad \underline{\textbf{end for}} \\
&\quad \underline{\textbf{end for}} \\
&\underline{\textbf{end for}}
\end{aligned}
$$

The short DFT-like transform computed in this stage is of length N_0 and it is de ned by the for-loop indexed by $\hat{n}_0$:

$$
\begin{aligned}
&\underline{\textbf{for}}\ \hat{n}_0 := 0\ \underline{\textbf{to}}\ N_0 - 1\ \underline{\textbf{do}} \\
&\quad A_3[\hat{n}_0, \hat{n}_1, \hat{n}_2] := \sum_{n_0=0}^{N_0-1} \left\{ A_2[n_0, \hat{n}_1, \hat{n}_2]\, \omega_{N}^{(\hat{n}_1 N_2 + \hat{n}_2) n_0} \right\} \omega_{N_0}^{\hat{n}_0 n_0} \\
&\underline{\textbf{end for}}
\end{aligned}
$$

Since each short transform (of length N_0) is computed by inner loop of the triple for-loop indexed by $\hat{n}_0$, $\hat{n}_1$, and $\hat{n}_2$, we need to compute, in total, $N_1 N_2$ (or N/N_0) such short transforms in this n al stage of the decoupled computation.

We have developed the three-factor algorithm step by step, and it is clear that the pairing of the column- and row-index mapping schemes are the driving force behind the scene. Of course, there are many other input and output index mapping schemes, which can *pair up* to achieve similar cost reduction, and the corresponding mixed-radix algorithms can now be developed without any dif culty following the steps we have systematically laid out in this section.

11.3 The Recursive Equation Approach

The term recursive equation refers to the single mathematical equation which de nes how $A_K[\ldots, n_{q-1}, \hat{n}_q, \ldots, \hat{n}_{\nu-1}]$ can be obtained from previously computed $A_{K-1}[\ldots, n_q, \hat{n}_{q+1}, \ldots, \hat{n}_{\nu-1}]$ in a ν-factor mixed-radix FFT algorithm. In this section we shall derive the recursive equation for an arbitrary ν-factor FFT, so its meaning and its usefulness as a mathematical tool in algorithm representation and extension (to arbitrary number of factors) will become clear as we progress through this section. With the index mapping background covered and a three-factor FFT algorithm fully developed in the last section, we can now present a compact (and fully explainable) version of the mixed-radix algorithm for composite $N = N_0 \times N_1 \times N_2$, which forms the basis of the so-called recursive equation approach in the FFT literature:

Step 0. Map x_ℓ to $A[n_0, n_1, n_2]$ for $\ell = n_2 N_1 N_0 + n_1 N_0 + n_0$.

Step 1. Compute $A_1[n_0, n_1, \hat{n}_2] = \sum_{n_2=0}^{N_2-1} A[n_0, n_1, n_2] \omega_{N_2}^{\hat{n}_2 n_2}$.

Step 2. Compute $A_2[n_0, \hat{n}_1, \hat{n}_2] = \sum_{n_1=0}^{N_1-1} \left\{ A_1[n_0, n_1, \hat{n}_2] \omega_N^{\hat{n}_2(n_1 N_0)} \right\} \omega_{N_1}^{\hat{n}_1 n_1}$.

Step 3. Compute $A_3[\hat{n}_0, \hat{n}_1, \hat{n}_2] = \sum_{n_0=0}^{N_0-1} \left\{ A_2[n_0, \hat{n}_1, \hat{n}_2] \omega_N^{(\hat{n}_1 N_2 + \hat{n}_2) n_0} \right\} \omega_{N_0}^{\hat{n}_0 n_0}$.

Step 4. Map $A_3[\hat{n}_0, \hat{n}_1, \hat{n}_2]$ to X_r for $r = \hat{n}_0 N_1 N_2 + \hat{n}_1 N_2 + \hat{n}_2$.

It is understood that the statement compute $A_1[n_0, n_1, \hat{n}_2]$ in Step 1 means compute $A_1[n_0, n_1, \hat{n}_2]$ for all values of $n_0, n_1, \hat{n}_2$. (Recall that it was done by a triple for-loop indexed by n_0, n_1 and $\hat{n}_2$ in the full algorithm presented in the last section.) Such understanding is assumed for all steps performing similar computation.

11.3.1 Counting short DFT or DFT-like transforms

In Step 1, each short DFT computes $A_1[n_0, n_1, \hat{n}_2]$ for $0 \leq \hat{n}_2 \leq N_2 - 1$, and there are $N/N_2 = N_0 N_1$ of them. In Step 2, each short DFT-like transform computes $A_2[n_0, \hat{n}_1, \hat{n}_2]$ for $0 \leq \hat{n}_1 \leq N_1 - 1$, and there are $N/N_1 = N_0 N_2$ of them. In Step 3, each short DFT-like transform computes $A_3[\hat{n}_0, \hat{n}_1, \hat{n}_2]$ for $0 \leq \hat{n}_0 \leq N_0 - 1$, and there are $N/N_0 = N_1 N_2$ of them.

11.3.2 The recursive equation for arbitrary composite N

In this section we show how a single recursive equation, combined with the index mapping schemes, may be used to represent the mixed-radix ν-factor FFT for arbitrary ν. Such a recursive equation needs to be *extracted* from the ν equations which de ne the short transforms for individual stages of the decoupled computation. Of course one can repeat the analysis and simpli catio n we have performed in developing the three-factor FFT to obtain any ν-factor FFT, or we can simply do it one more time for a suf ciently large number of factors and determine the pattern to follow for any given ν (without repeating the analysis every time). We nd $\nu = 5$ (i.e., $N = N_0 \times N_1 \times N_2 \times N_3 \times N_4$) would be suf ciently large for this purpose, and we use the resulting ve-factor FFT to explain an easy-to-follow road map for arbitrary ν:

Step 0. Map x_ℓ to $A[n_0, n_1, n_2, n_3, n_4]$ via the column-major scheme:

$$\ell = n_4 N_3 N_2 N_1 N_0 + n_3 N_2 N_1 N_0 + n_2 N_1 N_0 + n_1 N_0 + n_0.$$

Step 1. Compute the first set of N/N_4 short transforms defined by

$$A_1[n_0, n_1, n_2, n_3, \hat{n}_4] = \sum_{n_4=0}^{N_4-1} A[n_0, n_1, n_2, n_3, n_4]\, \omega_{N_4}^{\hat{n}_4 n_4}.$$

Step 2. Compute the second set of N/N_3 short transforms defined by

$$A_2[n_0, n_1, n_2, \hat{n}_3, \hat{n}_4] = \sum_{n_3=0}^{N_3-1} \left\{ A_1[n_0, n_1, n_2, n_3, \hat{n}_4]\, \omega_N^{\hat{n}_4 (n_3 N_2 N_1 N_0)} \right\} \omega_{N_3}^{\hat{n}_3 n_3}.$$

Observe that the exponent of the so-called twiddle factor ω_N (inside the braces) is the product of $\hat{n}_4$ and $(n_3 N_2 N_1 N_0)$ — the former comes from the last term of the output index map for r as so labeled in Equation (11.7) which defines the output index splitting in the paragraph following Step 6, and the latter represents the second term of the input index map for ℓ as clearly labeled in Equation (11.6) which defines the input index splitting in the paragraph following Step 6.

Step 3. Compute the third set of N/N_2 short transforms defined by

$$A_3[n_0, n_1, \hat{n}_2, \hat{n}_3, \hat{n}_4] = \sum_{n_2=0}^{N_2-1} \left\{ A_2[n_0, n_1, n_2, \hat{n}_3, \hat{n}_4]\, \omega_N^{(\hat{n}_3 N_4 + \hat{n}_4) n_2 N_1 N_0} \right\} \omega_{N_2}^{\hat{n}_2 n_2}.$$

Observe that the exponent of the twiddle factor ω_N is now the product of $(\hat{n}_3 N_4 + \hat{n}_4)$ and $n_2 N_1 N_0$ — the former involves the last two terms of the output index map for r as labeled in Equation (11.7), and the latter represents the third term of the input index map for ℓ as labeled in Equation (11.6).

Step 4. Compute the fourth set of N/N_1 short transforms defined by

$$A_4[n_0, \hat{n}_1, \hat{n}_2, \hat{n}_3, \hat{n}_4] = \sum_{n_1=0}^{N_1-1} \left\{ A_3[n_0, n_1, \hat{n}_2, \hat{n}_3, \hat{n}_4]\, \omega_N^{(\hat{n}_2 N_3 N_4 + \hat{n}_3 N_4 + \hat{n}_4) n_1 N_0} \right\} \omega_{N_1}^{\hat{n}_1 n_1}.$$

Observe again that the exponent of the twiddle factor ω_N is the product of two terms — the first term now involves the last three terms of the output index map for r as labeled in Equation (11.7), and the second represents the fourth term of the input index map for ℓ as labeled in Equation (11.6).

Step 5. Compute the fifth set of short N/N_0 transforms defined by

$$A_5[\hat{n}_0, \hat{n}_1, \hat{n}_2, \hat{n}_3, \hat{n}_4]$$
$$= \sum_{n_0=0}^{N_0-1} \left\{ A_4[n_0, \hat{n}_1, \hat{n}_2, \hat{n}_3, \hat{n}_4]\, \omega_N^{(\hat{n}_1 N_2 N_3 N_4 + \hat{n}_2 N_3 N_4 + \hat{n}_3 N_4 + \hat{n}_4) n_0} \right\} \omega_{N_0}^{\hat{n}_0 n_0}.$$

For the final step, the exponent of the twiddle factor ω_N is again the product of two terms as one would now expect: the first term involves the four terms (from the second to the last) of the output index map for r as labeled in Equation (11.7), and the second represents the very last term of the input index map for ℓ as labeled in Equation (11.6).

Step 6. Map $A_5[\hat{n}_0, \hat{n}_1, \hat{n}_2, \hat{n}_3, \hat{n}_4]$ to X_r via the row-major scheme:

$$r = \hat{n}_0 N_1 N_2 N_3 N_4 + \hat{n}_1 N_2 N_3 N_4 + \hat{n}_2 N_3 N_4 + \hat{n}_3 N_4 + \hat{n}_4.$$

Accordingly, the exponent of the twiddle factor in every step of the ve-factor mixed-radix FFT can be fully speci ed by splitting the two index maps as shown below:

$$(11.6) \qquad \ell = n_4 N_3 N_2 N_1 N_0 + \overbrace{n_3 N_2 N_1 N_0}^{\text{Step (2)}} + \overbrace{n_2 N_1 N_0}^{\text{Step (3)}} + \overbrace{n_1 N_0}^{\text{Step (4)}} + \overbrace{n_0}^{(5)}.$$

$$(11.7) \qquad r = \overbrace{\hat{n}_0 N_1 N_2 N_3 N_4 + \hat{n}_1 N_2 N_3 N_4 + \overbrace{\hat{n}_2 N_3 N_4 + \overbrace{\hat{n}_3 N_4 + \underbrace{\hat{n}_4}_{(2)}}^{\text{Step (4)}}}^{\text{Step (5)}}}.$$

Step (3)

Since the index mapping equations in the forms of (11.6) and (11.7) are readily available for composite N with an arbitrary number of factors, and these two equations alone prescribe the exponent of the twiddle factor in every stage of the decoupled computation, we can fully express all of the short transforms which collectively represent the mixed-radix FFT we are seeking. The algorithm can now be abbreviated to one single recursive equation for the Kth step, namely,

(11.8)

$$A_K[\ldots, n_{q-1}, \hat{n}_q, \ldots, \hat{n}_{\nu-1}] = \sum_{n_q=0}^{N_q-1} \left\{ A_{K-1}[\ldots, n_q, \hat{n}_{q+1}, \ldots, \hat{n}_{\nu-1}] \, \omega_N^{\hat{\lambda}_K \lambda_K} \right\} \omega_{N_q}^{\hat{n}_q n_q},$$

where $K = \nu - q = 1, 2, \ldots, \nu$ for $q = \nu - 1, \nu - 2, \ldots, 0$; $\hat{\lambda}_K$ and λ_K are the two groups of terms identi ed by Step (K) from the two index-map splitting equations for the ν-factor composite $N = N_0 \times N_1 \times \cdots \times N_{\nu-1}$ as explained above. It is understood that $A_0[n_0, \ldots, n_{\nu-1}] \equiv A[n_0, \ldots, n_{\nu-1}]$ and $\hat{\lambda}_1 = \lambda_1 = 0$ when the recursive equation is rst applied to generate $A_1[n_0, \ldots, \hat{n}_{\nu-1}]$ in Step (1).

Remark: The ν-factor FFT developed in this section is referred to as the *Decimation-In-Time* (DIT) FFT, which re ects how the indices ℓ o f the time-domain input elements x_ℓ are split according to Equation (11.6).

11.3.3 Specialization to the radix-2 DIT FFT for $N = 2^\nu$

Since the ν-factor mixed-radix FFT was derived without restricting the values of the factors, it is expected that it would have included radix-2 FFT as a special case for $N = 2^\nu$. This is indeed the case, which can be shown clearly for $\nu = 5$. The key is to explicitly express equations (11.6) and (11.7) as

$$(11.9) \qquad \ell = i_4 2^4 + \overbrace{i_3 2^3}^{(2)} + \overbrace{i_2 2^2}^{(3)} + \overbrace{i_1 2}^{(4)} + \overbrace{i_0}^{(5)}.$$

$$(11.10) \qquad r = \overbrace{\tau_0 2^4 + \tau_1 2^3 + \overbrace{\tau_2 2^2 + \tau_3 2 + \underbrace{\tau_4}_{(2)}}^{(4)}}^{(5)}.$$

Note that we have changed the labels of the dimensional indices to match the binary address—based notation. To relate the binary address in the 1-D array to the multidimensional array used in this chapter, we express $\ell = i_4 i_3 i_2 i_1 i_0$, and $x_\ell = a[\ell] = a[i_4 i_3 i_2 i_1 i_0] = A[i_0, i_1, i_2, i_3, i_4]$, assuming that the 1-D array a, which contains naturally ordered input $\{x_\ell\}$, is stored in the ν-D array A by column-major scheme; for the output we express $r = \tau_0 \tau_1 \tau_2 \tau_3 \tau_4$, and $X_r = b[r] = b[\tau_0 \tau_1 \tau_2 \tau_3 \tau_4] = B[\tau_4, \tau_3, \tau_2, \tau_1, \tau_0]$, assuming that the 1-D array b, which contains naturally ordered output $\{X_r\}$, is stored in the ν-D array B (of dimensions reversed from A) by column-major scheme.

To reveal the radix-2 FFT, observe that when $N = 2^\nu$, each short transform of length $N_q = 2$ represents a Cooley—Tukey butterfly computation. With $\omega_{N_q} = \omega_N^{N/2} = -1$, we obtain the five steps of a radix-2 DIT FFT for $N = 2^5 = 32$:

Step 1. Compute $N/2$ short transforms (or butterflies) defined by

$$A_1[i_0, i_1, i_2, i_3, \tau_4] = A[i_0, i_1, i_2, i_3, 0] + (-1)^{\tau_4} A[i_0, i_1, i_2, i_3, 1],$$

or (using binary addresses with 1-D arrays)

$$a^{(1)}[\tau_4 i_3 i_2 i_1 i_0] = a[0 i_3 i_2 i_1 i_0] + (-1)^{\tau_4} a[1 i_3 i_2 i_1 i_0].$$

Step 2. Compute $N/2$ short transforms (or butterflies) defined by

$$A_2[i_0, i_1, i_2, \tau_3, \tau_4] = A_1[i_0, i_1, i_2, 0, \tau_4] + (-1)^{\tau_3} \omega_N^{2^3 \tau_4} A[i_0, i_1, i_2, 1, \tau_4],$$

or (using binary addresses with 1-D arrays)

$$a^{(2)}[\tau_4 \tau_3 i_2 i_1 i_0] = a^{(1)}[\tau_4 0 i_2 i_1 i_0] + (-1)^{\tau_3} \omega_N^{8\tau_4} a^{(1)}[\tau_4 1 i_2 i_1 i_0].$$

Observe that the twiddle factor associated with the first term is $\omega_N^{\tau_4 i_3 2^3} = 1$ because $i_3 = 0$ in its exponent, and $\omega_N^{\tau_4 i_3 2^3} = \omega_N^{8\tau_4}$ in the second term because $i_3 = 1$ in its exponent.

Step 3. Compute $N/2$ short transforms (or butterflies) defined by

$$A_3[i_0, i_1, \tau_2, \tau_3, \tau_4] = A_2[i_0, i_1, 0, \tau_3, \tau_4] + (-1)^{\tau_2} \omega_N^{2^2 (\tau_3 2 + \tau_4)} A_2[i_0, i_1, 1, \tau_3, \tau_4],$$

or (using binary addresses with 1-D arrays)

$$a^{(3)}[\tau_4 \tau_3 \tau_2 i_1 i_0] = a^{(2)}[\tau_4 \tau_3 0 i_1 i_0] + (-1)^{\tau_2} \omega_N^{4(\tau_3 2 + \tau_4)} a^{(2)}[\tau_4 \tau_3 1 i_1 i_0].$$

Observe that the twiddle factor associated with the first term always $\omega_N^0 = 1$ because its exponent contains $i_2 = 0$, and the values of $\omega_N^4, \omega_N^8, \omega_N^{12}$ are assigned to the second term depending on the actual value of the exponent $i_2 2^2 (\tau_3 2 + \tau_4)$, which is simplified to $4(\tau_3 2 + \tau_4)$ because $i_2 = 1$.

Step 4. Compute $N/2$ short transforms (or butterflies) defined by

$$A_4[i_0, \tau_1, \tau_2, \tau_3, \tau_4] = A_3[i_0, 0, \tau_2, \tau_3, \tau_4] + (-1)^{\tau_1} \omega_N^{2(\tau_2 2^2 + \tau_3 2 + \tau_4)} A_3[i_0, 1, \tau_2, \tau_3, \tau_4],$$

or (using binary addresses with 1-D arrays)

$$a^{(4)}[\tau_4 \tau_3 \tau_2 \tau_1 i_0] = a^{(3)}[\tau_4 \tau_3 \tau_2 0 i_0] + (-1)^{\tau_1} \omega_N^{2(\tau_2 2^2 + \tau_3 2 + \tau_4)} a^{(3)}[\tau_4 \tau_3 \tau_2 1 i_0].$$

The twiddle factors involved in this step are $\omega_N^0, \omega_N^2, \omega_N^4, \ldots, \omega_N^{14}$.

Step 5. Compute $N/2$ short transforms (or butterflies) defined by

$$A_5[\tau_0, \tau_1, \tau_2, \tau_3, \tau_4] = A_4[0, \tau_1, \tau_2, \tau_3, \tau_4] + (-1)^{\tau_0} \omega_N^{\tau_3 2^3 + \tau_2 2^2 + \tau_3 2 + \tau_4} A_4[1, \tau_1, \tau_2, \tau_3, \tau_4],$$

or (using binary addresses with 1-D arrays)

$$a^{(5)}[\tau_4 \tau_3 \tau_2 \tau_1 \tau_0] = a^{(4)}[\tau_4 \tau_3 \tau_2 \tau_1 0] + (-1)^{\tau_0} \omega_N^{\tau_1 2^3 + \tau_2 2^2 + \tau_3 2 + \tau_4} a^{(4)}[\tau_4 \tau_3 \tau_2 \tau_1 1].$$

The $N/2$ twiddle factors involved in this last step are $\omega_N^0, \omega_N^1, \omega_N^2, \ldots, \omega_N^{15}$.

We have shown that Steps 1 through 5 of the five-factor mixed radix FFT now prescribe the in-place radix-2 DIT FFT. For $N = 32 = 2^5$, the five stages of decoupled computation correspond to the five stages of butterfly computation. Since $B[\tau_4, \tau_3, \tau_2, \tau_1, \tau_0] = A_5[\tau_0, \tau_1, \tau_2, \tau_3, \tau_4]$, $X_r = b[\tau_0 \tau_1 \tau_2 \tau_3 \tau_4] = a^{(5)}[\tau_4 \tau_3 \tau_2 \tau_1 \tau_0]$, and X_r is available from $\boldsymbol{a}^{(5)}$ in bit-reversed order. For specific details on binary address-based radix-2 DIT FFTs, interested readers are referred to [13].

11.4 Other Forms by Alternate Index Splitting

In this section we demonstrate how our systematic approach can be applied at once to obtain the *Decimation-In-Frequency* (DIF) form of the mixed-radix FFT, which will also lead us to the radix-2 DIF FFT when $N = 2^\nu$. We begin with the same example for $N = N_0 \times N_1 \times N_2$ from Section 11.2. This time we first deploy the index mapping scheme on the output $\{X_r\}$ computed by the DFT formula

$$\textbf{for } r := 0 \textbf{ to } N - 1 \textbf{ do}$$
$$X_r := \sum_{\ell=0}^{N-1} x_\ell \, \omega_N^{r\ell}$$
$$\textbf{end for}$$

With $X_r = B[\hat{n}_2, \hat{n}_1, \hat{n}_0]$ for $r = \hat{n}_0 N_1 N_2 + \hat{n}_1 N_2 + \hat{n}_2$, we obtain the mathematically equivalent DFT formula:

$$\textbf{for } \hat{n}_0 := 0 \textbf{ to } N_0 - 1 \textbf{ do}$$
$$\quad \textbf{for } \hat{n}_1 := 0 \textbf{ to } N_1 - 1 \textbf{ do}$$
$$\quad\quad \textbf{for } \hat{n}_2 := 0 \textbf{ to } N_2 - 1 \textbf{ do}$$
$$\quad\quad\quad B[\hat{n}_2, \hat{n}_1, \hat{n}_0] := \sum_{\ell=0}^{N-1} x_\ell \, \omega_N^{\hat{n}_2 \ell} \omega_N^{(\hat{n}_1 N_2)\ell} \omega_N^{(\hat{n}_0 N_1 N_2)\ell}$$
$$\quad\quad \textbf{end for}$$
$$\quad \textbf{end for}$$
$$\textbf{end for}$$

Observe that we have split the exponent of $\omega_N^{r\ell}$ b y splitting the given mapping formula for index r, i .e.,

(11.11)
$$
\begin{aligned}
r\ell &= (\hat{n}_0 N_1 N_2 + \hat{n}_1 N_2 + \hat{n}_2)\ell \\
&= \hat{n}_2 \ell + (\hat{n}_1 N_2)\ell + (\hat{n}_0 N_1 N_2)\ell.
\end{aligned}
$$

By repeating the decoupling processes performed in Sections 11.2.1, 11.2.2, and 11.2.3, we obtain alternate sets of recursive equations:

Step 0. Map x_ℓ to $A[n_0, n_1, n_2]$ for $\ell = n_2 N_1 N_0 + n_1 N_0 + n_0$.

Step 1. Compute $A_1[n_0, n_1, \hat{n}_2] = \sum_{n_2=0}^{N_2-1} \left\{ A[n_0, n_1, n_2] \, \omega_N^{\hat{n}_2(n_1 N_0 + n_0)} \right\} \omega_{N_2}^{\hat{n}_2 n_2}$.

Step 2. Compute $A_2[n_0, \hat{n}_1, \hat{n}_2] = \sum_{n_1=0}^{N_1-1} \left\{ A_1[n_0, n_1, \hat{n}_2] \, \omega_N^{\hat{n}_1(n_0 N_2)} \right\} \omega_{N_1}^{\hat{n}_1 n_1}$.

Step 3. Compute $A_3[\hat{n}_0, \hat{n}_1, \hat{n}_2] = \sum_{n_0=0}^{N_0-1} A_2[n_0, \hat{n}_1, \hat{n}_2] \, \omega_{N_0}^{\hat{n}_0 n_0}$.

Step 4. Map $A_3[\hat{n}_0, \hat{n}_1, \hat{n}_2]$ to X_r for $r = \hat{n}_0 N_1 N_2 + \hat{n}_1 N_2 + \hat{n}_2$.

11.4.1   The recursive equation for arbitrary composite N

We now apply the systematic approach in Section 11.3.2 to obtain the alternately split e xponent of the twiddle factor in every step of a   ve-factor mixed-radix FFT:

(11.12)
$$
\ell = n_4 N_3 N_2 N_1 N_0 + n_3 N_2 N_1 N_0 + \overbrace{n_2 N_1 N_0 + n_1 N_0 + \underbrace{n_0}_{(4)}}^{\text{Step (2)}}.
$$

with the overbrace spanning Step (1) and the underbrace n_0 labeled Step (3).

(11.13)
$$
r = \hat{n}_0 N_1 N_2 N_3 N_4 + \overbrace{\hat{n}_1 N_2 N_3 N_4}^{\text{Step (4)}} + \overbrace{\hat{n}_2 N_3 N_4}^{\text{Step (3)}} + \overbrace{\hat{n}_3 N_4}^{\text{Step (2)}} + \overbrace{\hat{n}_4}^{(1)}.
$$

Since the index mapping equations in the forms of (11.12) and (11.13) are readily available for composite N with an arbitrary number of factors, and these two equations alone prescribe the exponent of the twiddle factor in every stage of the decoupled computation, we can fully express all of the short transforms which collectively represent the mixed-radix FFT we are seeking. The algorithm can now be abbreviated to one single recursive equation for the Kth step, namely,

(11.14)
$$
A_K[\ldots, n_{q-1}, \hat{n}_q, \ldots, \hat{n}_{\nu-1}] = \sum_{n_q=0}^{N_q-1} \left\{ A_{K-1}[\ldots, n_q, \hat{n}_{q+1}, \ldots, \hat{n}_{\nu-1}] \, \omega_N^{\hat{\lambda}_K \lambda_K} \right\} \omega_{N_q}^{\hat{n}_q n_q},
$$

where $K = \nu - q = 1, 2, \ldots, \nu$ for $q = \nu - 1, \nu - 2, \ldots, 0$; $\hat{\lambda}_K$ and λ_K are the two groups of terms identi ed by Step K f rom the two index mapping equations for the ν-factor composite $N = N_0 \times N_1 \times \cdots \times N_{\nu-1}$ as explained above.

Remark: The ν-factor FFT developed in this section is referred to as the *Decimation-In-Frequency* (DIF) FFT, which re ects how the indices r o f the frequency-domain output elements X_r are splitted according to Equation (11.13).

11.4.2   Specialization to the radix-2 DIF FFT for $N = 2^\nu$

To adapt the ν-factor mixed-radix FFT for $N = 2^\nu$, the procedure is analogous to that followed in Section 11.3.3. The key is to explicitly express equations (11.12) and (11.13) as

(11.15)
$$\ell = i_4 2^4 + i_3 2^3 + i_2 2^2 + i_1 2 + i_0 ;$$

(11.16)
$$r = \tau_0 2^4 + \tau_1 2^3 + \tau_2 2^2 + \tau_3 2 + \tau_4 .$$

As already explained in Section 11.3.3, we use binary address—based notation to express $\ell = i_4 i_3 i_2 i_1 i_0$, and $x_\ell = a[\ell] = a[i_4 i_3 i_2 i_1 i_0] = A[i_0, i_1, i_2, i_3, i_4]$, assuming that the 1-D array a, which contains naturally ordered input $\{x_\ell\}$, is stored in the ν-D array A by column-major scheme; for the output we express $r = \tau_0 \tau_1 \tau_2 \tau_3 \tau_4$, and $X_r = b[r] = b[\tau_0 \tau_1 \tau_2 \tau_3 \tau_4] = B[\tau_4, \tau_3, \tau_2, \tau_1, \tau_0]$, assuming that the 1-D array b, which contains naturally ordered output $\{X_r\}$, is stored in the ν-D array B (of dimensions reversed from A) by column-major scheme.

To reveal the radix-2 DIF FFT, observe that when $N = 2^\nu$, each short transform of length $N_q = 2$ represents a Gentleman—Sande butterfly computation. With $\omega_{N_q} = \omega_N^{N/2} = -1$, we obtain the five steps for a (five-factor) radix-2 FFT as shown below.

Step 1. Compute $N/2$ short transforms (or butterflies) defined by

$$A_1[i_0, i_1, i_2, i_3, \tau_4] = \left\{ A[i_0, i_1, i_2, i_3, 0] + (-1)^{\tau_4} A[i_0, i_1, i_2, i_3, 1] \right\} \omega_N^{\tau_4(i_3 2^3 + i_2 2^2 + i_1 2 + i_0)},$$

or (using binary addresses with 1-D arrays)

$$a^{(1)}[\tau_4 i_3 i_2 i_1 i_0] = \left\{ a[0 i_3 i_2 i_1 i_0] + (-1)^{\tau_4} a[1 i_3 i_2 i_1 i_0] \right\} \omega_N^{\tau_4(i_3 2^3 + i_2 2^2 + i_1 2 + i_0)}.$$

The first step deploys $N/2 = 16$ twiddle factors: $\omega_N^0, \omega_N^1, \omega_N^2, \ldots, \omega_N^{15}$.

Step 2. Compute $N/2$ short transforms (or butterflies) defined by

$$A_2[i_0, i_1, i_2, \tau_3, \tau_4] = \left\{ A_1[i_0, i_1, i_2, 0, \tau_4] + (-1)^{\tau_3} A[i_0, i_1, i_2, 1, \tau_4] \right\} \omega_N^{\tau_3 2(i_2 2^2 + i_1 2 + i_0)},$$

or (using binary addresses with 1-D arrays)

$$a^{(2)}[\tau_4 \tau_3 i_2 i_1 i_0] = \left\{ a^{(1)}[\tau_4 0 i_2 i_1 i_0] + (-1)^{\tau_3} a^{(1)}[\tau_4 1 i_2 i_1 i_0] \right\} \omega_N^{2(i_2 2^2 + i_1 2 + i_0)\tau_3}.$$

The second step deploys $N/2^2 = 8$ twiddle factors: $\omega_N^0, \omega_N^2, \omega_N^4, \ldots, \omega_N^{14}$.

Step 3. Compute $N/2$ short transforms (or butterflies) defined by

$$A_3[i_0, i_1, \tau_2, \tau_3, \tau_4] = \left\{ A_2[i_0, i_1, 0, \tau_3, \tau_4] + (-1)^{\tau_2} A_2[i_0, i_1, 1, \tau_3, \tau_4] \right\} \omega_N^{\tau_2 2^2(i_1 2 + i_0)},$$

or (using binary addresses with 1-D arrays)

$$a^{(3)}[\tau_4\tau_3\tau_2 i_1 i_0] = \left\{ a^{(2)}[\tau_4\tau_3 0 i_1 i_0] + (-1)^{\tau_2} a^{(2)}[\tau_4\tau_3 1 i_1 i_0] \right\} \omega_N^{4(i_1 2 + i_0)\tau_2}.$$

The third step deploys $N/2^3 = 4$ twiddle factors: $\omega_N^0, \omega_N^4, \omega_N^8,$ and ω_N^{12}

Step 4. Compute $N/2$ short transforms (or butterflies) defined by

$$A_4[i_0, \tau_1, \tau_2, \tau_3, \tau_4] = \left\{ A_3[i_0, 0, \tau_2, \tau_3, \tau_4] + (-1)^{\tau_1} A_3[i_0, 1, \tau_2, \tau_3, \tau_4] \right\} \omega_N^{\tau_1 2^3 i_0},$$

or (using binary addresses with 1-D arrays)

$$a^{(4)}[\tau_4\tau_3\tau_2\tau_1 i_0] = \left\{ a^{(3)}[\tau_4\tau_3\tau_2 0 i_0] + (-1)^{\tau_1} a^{(3)}[\tau_4\tau_3\tau_2 1 i_0] \right\} \omega_N^{8 i_0 \tau_1}.$$

The $N/2^4 = 2$ twiddle factors involved in this step are ω_N^0 and ω_N^8.

Step 5. Compute $N/2$ short transforms (or butterflies) defined by

$$A_5[\tau_0, \tau_1, \tau_2, \tau_3, \tau_4] = A_4[0, \tau_1, \tau_2, \tau_3, \tau_4] + (-1)^{\tau_0} A_4[1, \tau_1, \tau_2, \tau_3, \tau_4],$$

or (using binary addresses with 1-D arrays)

$$a^{(5)}[\tau_4\tau_3\tau_2\tau_1\tau_0] = a^{(4)}[\tau_4\tau_3\tau_2\tau_1 0] + (-1)^{\tau_0} a^{(4)}[\tau_4\tau_3\tau_2\tau_1 1].$$

We have shown that Steps 1 through 5 of the five-factor mixed-radix FFT now prescribe the in-place radix-2 DIF FFT. For $N = 32 = 2^5$, the five stages of decoupled computation correspond to the five stages of butterfly computation. Since $B[\tau_4, \tau_3, \tau_2, \tau_1, \tau_0] = A_5[\tau_0, \tau_1, \tau_2, \tau_3, \tau_4]$, $X_r = b[\tau_0\tau_1\tau_2\tau_3\tau_4] = a^{(5)}[\tau_4\tau_3\tau_2\tau_1\tau_0]$, and X_r is available from $a^{(5)}$ in bit-reversed order. For specific details on binary address—based radix-2 DIF FFT, interested readers are referred to [13].

Chapter 12

Kronecker Product Factorization and FFTs

In this chapter we make explicit the connection between the ν-factor mixed-radix FFT algorithms and the Kronecker product factorization of the DFT matrix. This process results in a *sparse* matrix formulation of the mixed-radix FFT algorithm. Although multidimensional arrays do not appear in the nal equation, they remain to be instrumental in the development, and the index mapping schemes continue to play an essential role this is not surprising because the decoupling processes are enabled by the direct manipulation of the indices. Initially the following two de nitions are needed in our treatment of the two-factor mixed-radix FFT; other properties and rules for Kronecker products will be introduced as we progress. Readers are assumed to be familiar with the content of Chapter 11.

Definition 12.1 Let A be a p-by-q matrix

$$(12.1) \qquad \begin{bmatrix} a_{1,1} & a_{1,2} & \cdots & a_{1,q} \\ a_{2,1} & a_{2,2} & \cdots & a_{2,q} \\ \vdots & \vdots & & \vdots \\ a_{p,1} & a_{p,2} & \cdots & a_{p,q} \end{bmatrix}.$$

Then the **vec** operator stacks the columns of matrix A on top of one another to form a vector

u:

$$(12.2) \qquad u = \text{vec} \begin{bmatrix} a_{1,1} & a_{1,2} & \cdots & a_{1,q} \\ \vdots & \vdots & & \vdots \\ a_{p,1} & a_{p,2} & \cdots & a_{p,q} \end{bmatrix} = \begin{bmatrix} a_{1,1} \\ \vdots \\ a_{p,1} \\ a_{1,2} \\ \vdots \\ a_{p,2} \\ \vdots \\ a_{1,q} \\ \vdots \\ a_{p,q} \end{bmatrix}.$$

The Kronecker product, also known as a *direct product*, or a *tensor product*, is de ned for two matrices of arbitrary dimensions.

Definition 12.2 Let A be a p-by-q matrix and B be an m-by-n matrix. Then the Kronecker product of A and B is de ned as the $p \cdot m$-by-$q \cdot n$ matrix

$$(12.3) \qquad A \otimes B = \begin{bmatrix} a_{1,1}B & a_{1,2}B & \cdots & a_{1,q}B \\ a_{2,1}B & a_{2,2}B & \cdots & a_{2,q}B \\ \vdots & \vdots & & \vdots \\ a_{p,1}B & a_{p,2}B & \cdots & a_{p,q}B \end{bmatrix}.$$

12.1   Reformulating the Two-Factor Mixed-Radix FFT

Recall that the de n ition of a length-N DFT (excluding division by N)

$$(12.4) \qquad X_r = \sum_{\ell=0}^{N-1} x_\ell\, \omega_N^{r\ell} = \sum_{\ell=0}^{N-1} \omega_N^{r\ell}\, x_\ell, \qquad \omega_N \equiv e^{-j2\pi/N}, \qquad \text{for } r = 0, 1, \ldots, N-1,$$

expresses a matrix-vector product $X = \Omega_N x$, where $x = [x_0, x_1, \ldots, x_{N-1}]^T$, $X = [X_0, X_1, \ldots, X_{N-1}]^T$, and Ω_N denotes the N-by-N DFT matrix de ned by $\Omega_N[r, \ell] = \omega_N^{r\ell}$ for $0 \le r, \ell \le N-1$. In this section we shall derive the Kronecker product factorization of the DFT matrix Ω_N for $N = N_0 \times N_1$ directly from the two-factor mixed-radix FFT. When a concrete example is needed to clarify the construction of various sparse matrices, we shall use $N = 12$ with $N_0 = 3$ and $N_1 = 4$. To obtain the two-factor FFT in multidimensional formulation, we use $\nu = 2$ and $N = N_0 \times N_1$ in the recursive equations set up for arbitrary ν-factor composite N in Section 11.3 of Chapter 11.

Step 0. Map x_ℓ to $A[n_0, n_1]$ for $\ell = n_1 N_0 + n_0$.

Step 1. Compute $N/N_1 = N_0$ short DFT transforms of length N_1:

$$\textbf{for } n_0 := 0 \textbf{ to } N_0 - 1 \textbf{ do}$$
$$\quad \textbf{for } \hat{n}_1 := 0 \textbf{ to } N_1 - 1 \textbf{ do}$$
$$\quad\quad A_1[n_0, \hat{n}_1] = \sum_{n_1=0}^{N_1-1} A[n_0, n_1]\, \omega_{N_1}^{\hat{n}_1 n_1}$$
$$\quad \textbf{end for}$$
$$\textbf{end for}$$

Step 2. Compute $N/N_0 = N_1$ short DFT-like transforms of length N_0:

$$\textbf{for } \hat{n}_1 := 0 \textbf{ to } N_1 - 1 \textbf{ do}$$
$$\quad \textbf{for } \hat{n}_0 := 0 \textbf{ to } N_0 - 1 \textbf{ do}$$
$$\quad\quad A_2[\hat{n}_0, \hat{n}_1] = \sum_{n_0=0}^{N_0-1} \left\{ A_1[n_0, \hat{n}_1]\, \omega_N^{\hat{n}_1 n_0} \right\} \omega_{N_0}^{\hat{n}_0 n_0}$$
$$\quad \textbf{end for}$$
$$\textbf{end for}$$

Step 3. Map $A_2[\hat{n}_0, \hat{n}_1]$ to X_r for $r = \hat{n}_0 N_1 + \hat{n}_1$.

Our task is to *rework* the two-dimensional formulation of the two-factor mixed-radix FFT *into* a sequence of matrix-vector products, which begins with the multiplication of the input vector $\boldsymbol{x}$ by an N-by-N sparse matrix $\boldsymbol{F}_N$. Using $N = N_0 \times N_1 = 3 \times 4$, we show below how to construct the sparse matrix $\boldsymbol{F}_N$ for Step 1.

Now it becomes useful to display the mapping in Step 0 in matrix form: the N_0-by-N_1 matrix $\boldsymbol{A}$ (for $N_0 = 3$ and $N_1 = 4$) and its contents are given below. Note that we have used either a_{n_0,n_1} or $A[n_0, n_1]$ to address the individual elements of matrix $\boldsymbol{A}$, and we show $a_{n_0,n_1} = x_\ell$ according to the column-major scheme $\ell = n_1 N_0 + n_0 = n_0 + 3n_1$:

$$(12.5) \qquad \boldsymbol{A} = \begin{bmatrix} a_{0,0} & a_{0,1} & a_{0,2} & a_{0,3} \\ a_{1,0} & a_{1,1} & a_{1,2} & a_{1,3} \\ a_{2,0} & a_{2,1} & a_{2,2} & a_{2,3} \end{bmatrix} = \begin{bmatrix} x_0 & x_3 & x_6 & x_9 \\ x_1 & x_4 & x_7 & x_{10} \\ x_2 & x_5 & x_8 & x_{11} \end{bmatrix}.$$

In Step 1, each of the three short DFTs transforms a row of $N_1 = 4$ elements from matrix $\boldsymbol{A}$; hence, each transform can be expressed as a matrix-vector product using the 4-by-4 DFT matrix $\boldsymbol{\Omega}_4$ as shown below.

$$(12.6) \qquad \begin{bmatrix} y_0 \\ y_3 \\ y_6 \\ y_9 \end{bmatrix} = \begin{bmatrix} 1 & 1 & 1 & 1 \\ 1 & \omega_4 & \omega_4^2 & \omega_4^3 \\ 1 & \omega_4^2 & \omega_4^4 & \omega_4^6 \\ 1 & \omega_4^3 & \omega_4^6 & \omega_4^9 \end{bmatrix} \begin{bmatrix} x_0 \\ x_3 \\ x_6 \\ x_9 \end{bmatrix},$$

$$(12.7) \qquad \begin{bmatrix} y_1 \\ y_4 \\ y_7 \\ y_{10} \end{bmatrix} = \begin{bmatrix} 1 & 1 & 1 & 1 \\ 1 & \omega_4 & \omega_4^2 & \omega_4^3 \\ 1 & \omega_4^2 & \omega_4^4 & \omega_4^6 \\ 1 & \omega_4^3 & \omega_4^6 & \omega_4^9 \end{bmatrix} \begin{bmatrix} x_1 \\ x_4 \\ x_7 \\ x_{10} \end{bmatrix},$$

$$(12.8) \qquad \begin{bmatrix} y_2 \\ y_5 \\ y_8 \\ y_{11} \end{bmatrix} = \begin{bmatrix} 1 & 1 & 1 & 1 \\ 1 & \omega_4 & \omega_4^2 & \omega_4^3 \\ 1 & \omega_4^2 & \omega_4^4 & \omega_4^6 \\ 1 & \omega_4^3 & \omega_4^6 & \omega_4^9 \end{bmatrix} \begin{bmatrix} x_2 \\ x_5 \\ x_8 \\ x_{11} \end{bmatrix}.$$

The same results can be obtained if we directly multiply the vector $\boldsymbol{x}$ by a 12-by-12 sparse matrix $\boldsymbol{F}_{12}$ which contains the elements of the 4-by-4 DFT matrix $\boldsymbol{\Omega}_4$ at appropriate locations:

$$(12.9) \quad
\begin{bmatrix} y_0 \\ y_1 \\ y_2 \\ y_3 \\ y_4 \\ y_5 \\ y_6 \\ y_7 \\ y_8 \\ y_9 \\ y_{10} \\ y_{11} \end{bmatrix}
=
\begin{bmatrix}
1 & 0 & 0 & 1 & 0 & 0 & 1 & 0 & 0 & 1 & 0 & 0 \\
0 & 1 & 0 & 0 & 1 & 0 & 0 & 1 & 0 & 0 & 1 & 0 \\
0 & 0 & 1 & 0 & 0 & 1 & 0 & 0 & 1 & 0 & 0 & 1 \\
1 & 0 & 0 & \omega_4 & 0 & 0 & \omega_4^2 & 0 & 0 & \omega_4^3 & 0 & 0 \\
0 & 1 & 0 & 0 & \omega_4 & 0 & 0 & \omega_4^2 & 0 & 0 & \omega_4^3 & 0 \\
0 & 0 & 1 & 0 & 0 & \omega_4 & 0 & 0 & \omega_4^2 & 0 & 0 & \omega_4^3 \\
1 & 0 & 0 & \omega_4^2 & 0 & 0 & \omega_4^4 & 0 & 0 & \omega_4^6 & 0 & 0 \\
0 & 1 & 0 & 0 & \omega_4^2 & 0 & 0 & \omega_4^4 & 0 & 0 & \omega_4^6 & 0 \\
0 & 0 & 1 & 0 & 0 & \omega_4^2 & 0 & 0 & \omega_4^4 & 0 & 0 & \omega_4^6 \\
1 & 0 & 0 & \omega_4^3 & 0 & 0 & \omega_4^6 & 0 & 0 & \omega_4^9 & 0 & 0 \\
0 & 1 & 0 & 0 & \omega_4^3 & 0 & 0 & \omega_4^6 & 0 & 0 & \omega_4^9 & 0 \\
0 & 0 & 1 & 0 & 0 & \omega_4^3 & 0 & 0 & \omega_4^6 & 0 & 0 & \omega_4^9
\end{bmatrix}
\begin{bmatrix} x_0 \\ x_1 \\ x_2 \\ x_3 \\ x_4 \\ x_5 \\ x_6 \\ x_7 \\ x_8 \\ x_9 \\ x_{10} \\ x_{11} \end{bmatrix}.
$$

The properly constructed sparse matrix $\boldsymbol{F}_{12}$ can now be expressed as the Kronecker product of the 4-by-4 DFT matrix and a 3-by-3 identity matrix:

$$(12.10) \quad
\boldsymbol{F}_{12} = \boldsymbol{\Omega}_4 \otimes \boldsymbol{I}_3 =
\begin{bmatrix}
1 \cdot \boldsymbol{I}_3 & 1 \cdot \boldsymbol{I}_3 & 1 \cdot \boldsymbol{I}_3 & 1 \cdot \boldsymbol{I}_3 \\
1 \cdot \boldsymbol{I}_3 & \omega_4 \cdot \boldsymbol{I}_3 & \omega_4^2 \cdot \boldsymbol{I}_3 & \omega_4^3 \cdot \boldsymbol{I}_3 \\
1 \cdot \boldsymbol{I}_3 & \omega_4^2 \cdot \boldsymbol{I}_3 & \omega_4^4 \cdot \boldsymbol{I}_3 & \omega_4^6 \cdot \boldsymbol{I}_3 \\
1 \cdot \boldsymbol{I}_3 & \omega_4^3 \cdot \boldsymbol{I}_3 & \omega_4^6 \cdot \boldsymbol{I}_3 & \omega_4^9 \cdot \boldsymbol{I}_3
\end{bmatrix}.
$$

Accordingly, the computation performed in Step 1 of the two-factor mixed-radix FFT can be compactly represented by a single matrix equation:

$$(12.11) \quad \boldsymbol{y} = (\boldsymbol{\Omega}_{N_1} \otimes \boldsymbol{I}_{N_0}) \cdot \boldsymbol{x}.$$

Observe that because $\boldsymbol{x} = \mathbf{vec}\,\boldsymbol{A}[N_0, N_1]$, and $\boldsymbol{y} = \mathbf{vec}\,\boldsymbol{A}_1[N_0, N_1]$ after each row of matrix $\boldsymbol{A}$ is multiplied by the DFT matrix $\boldsymbol{\Omega}_{N_1}$ (which is symmetric), the Kronecker product can also be understood in terms of $\boldsymbol{A}$ and $\boldsymbol{A}_1$:

$$(12.12) \quad \boldsymbol{y} = (\boldsymbol{\Omega}_{N_1} \otimes \boldsymbol{I}_{N_0})\,\mathbf{vec}\,\boldsymbol{A} = \mathbf{vec}\left\{\left(\boldsymbol{\Omega}_{N_1} \cdot \boldsymbol{A}^T\right)^T\right\} = \mathbf{vec}\left\{\boldsymbol{A} \cdot \boldsymbol{\Omega}_{N_1}\right\} = \mathbf{vec}\,\boldsymbol{A}_1.$$

To rework Step 2, observe that the computed length-N vector $\boldsymbol{y}$ is contained in the N_0-by-N_1 matrix $\boldsymbol{A}_1$ at the end of Step 1, which is shown below for $N = 3 \times 4$:

$$(12.13) \quad
\boldsymbol{A}_1 =
\begin{bmatrix}
a_{0,0} & a_{0,1} & a_{0,2} & a_{0,3} \\
a_{1,0} & a_{1,1} & a_{1,2} & a_{1,3} \\
a_{2,0} & a_{2,1} & a_{2,2} & a_{2,3}
\end{bmatrix}
=
\begin{bmatrix}
y_0 & y_3 & y_6 & y_9 \\
y_1 & y_4 & y_7 & y_{10} \\
y_2 & y_5 & y_8 & y_{11}
\end{bmatrix}.
$$

In Step 2, each element $a_{n_0, \hat{n}_1}$ in $\boldsymbol{A}_1$ must be multiplied by a twiddle factor $\omega_N^{\hat{n}_1 n_0}$ at rs t. Our task is to construct an N-by-N sparse matrix $\boldsymbol{D}_N$ so that the same results can be obtained by the matrix-vector product $\boldsymbol{z} = \boldsymbol{D}_N \cdot \boldsymbol{y}$. Since the elements in vector $\boldsymbol{y}$ are scaled by the diagonal elements of a diagonal matrix, we know that we must have

$$z_\ell = D_N[\ell, \ell] \cdot y_\ell = \omega_N^{n_0 \hat{n}_1}\, y_\ell \quad \text{if } \ell = \hat{n}_1 N_0 + n_0, \qquad (\because\, y_\ell = a_{n_0, \hat{n}_1})$$

and we can generate the diagonal matrix $\boldsymbol{D}_N$ by advancing ℓ from 0 to $N-1$ according to the column-major index mapping scheme:

$$
\begin{array}{ll}
\textbf{for } k := 0 \textbf{ to } N - 1 \textbf{ do} & \text{initialize } D_N \text{ to be an} \\
\quad \textbf{for } i := 0 \textbf{ to } N - 1 \textbf{ do} & N\text{-by-}N \text{ zero matrix} \\
\quad\quad D_N[i, k] := 0 & \text{column by column} \\
\quad \textbf{end for} & \\
\textbf{end for} & \\
\ell := 0 & \\
\textbf{for } \hat{n}_1 := 0 \textbf{ to } N_1 - 1 \textbf{ do} & \text{assign twiddle factors} \\
\quad \textbf{for } n_0 := 0 \textbf{ to } N_0 - 1 \textbf{ do} & \text{to diagonal elements} \\
\quad\quad D_N[\ell, \ell] := \omega_N^{n_0 \hat{n}_1} & \because y_\ell = a_{n_0, \hat{n}_1} = A_1[n_0, \hat{n}_1], \\
\quad\quad \ell := \ell + 1 & \text{and } \ell = \hat{n}_1 N_0 + n_0 \\
\quad \textbf{end for} & \\
\textbf{end for} &
\end{array}
$$

For our example with $N = 12$, $N_0 = 3$, and $N_1 = 4$, the matrix-vector product $z = D_{12} \cdot y$ is de n ed by

$$
(12.14) \quad
\begin{bmatrix} z_0 \\ z_1 \\ z_2 \\ z_3 \\ z_4 \\ z_5 \\ z_6 \\ z_7 \\ z_8 \\ z_9 \\ z_{10} \\ z_{11} \end{bmatrix}
=
\begin{bmatrix}
1 & 0 & 0 & 0 & 0 & 0 & 0 & 0 & 0 & 0 & 0 & 0 \\
0 & 1 & 0 & 0 & 0 & 0 & 0 & 0 & 0 & 0 & 0 & 0 \\
0 & 0 & 1 & 0 & 0 & 0 & 0 & 0 & 0 & 0 & 0 & 0 \\
0 & 0 & 0 & 1 & 0 & 0 & 0 & 0 & 0 & 0 & 0 & 0 \\
0 & 0 & 0 & 0 & \omega_N & 0 & 0 & 0 & 0 & 0 & 0 & 0 \\
0 & 0 & 0 & 0 & 0 & \omega_N^2 & 0 & 0 & 0 & 0 & 0 & 0 \\
0 & 0 & 0 & 0 & 0 & 0 & 1 & 0 & 0 & 0 & 0 & 0 \\
0 & 0 & 0 & 0 & 0 & 0 & 0 & \omega_N^2 & 0 & 0 & 0 & 0 \\
0 & 0 & 0 & 0 & 0 & 0 & 0 & 0 & \omega_N^4 & 0 & 0 & 0 \\
0 & 0 & 0 & 0 & 0 & 0 & 0 & 0 & 0 & 1 & 0 & 0 \\
0 & 0 & 0 & 0 & 0 & 0 & 0 & 0 & 0 & 0 & \omega_N^3 & 0 \\
0 & 0 & 0 & 0 & 0 & 0 & 0 & 0 & 0 & 0 & 0 & \omega_N^6
\end{bmatrix}
\begin{bmatrix} y_0 \\ y_1 \\ y_2 \\ y_3 \\ y_4 \\ y_5 \\ y_6 \\ y_7 \\ y_8 \\ y_9 \\ y_{10} \\ y_{11} \end{bmatrix} .
$$

Since the vector z can overwrite y, we assume that the modi ed vector z is similarly mapped to the 3-by-4 matrix A_1; i.e., we have

$$
(12.15) \quad
A_1 = \begin{bmatrix} a_{0,0} & a_{0,1} & a_{0,2} & a_{0,3} \\ a_{1,0} & a_{1,1} & a_{1,2} & a_{1,3} \\ a_{2,0} & a_{2,1} & a_{2,2} & a_{2,3} \end{bmatrix}
= \begin{bmatrix} z_0 & z_3 & z_6 & z_9 \\ z_1 & z_4 & z_7 & z_{10} \\ z_2 & z_5 & z_8 & z_{11} \end{bmatrix} ,
$$

and each short DFT-like computation in Step 2 transforms a column of $N_0 = 3$ elements from the matrix A_1 (which now contains the N elements of vector z.) Each short transform can again be expressed as a matrix-vector product using a 3-by-3 DFT matrix Ω_3 as shown below.

$$
(12.16) \quad
\begin{aligned}
&\begin{bmatrix} c_0 \\ c_1 \\ c_2 \end{bmatrix} = \begin{bmatrix} 1 & 1 & 1 \\ 1 & \omega_3 & \omega_3^2 \\ 1 & \omega_3^2 & \omega_3^4 \end{bmatrix} \begin{bmatrix} z_0 \\ z_1 \\ z_2 \end{bmatrix} ,
&&\begin{bmatrix} c_6 \\ c_7 \\ c_8 \end{bmatrix} = \begin{bmatrix} 1 & 1 & 1 \\ 1 & \omega_3 & \omega_3^2 \\ 1 & \omega_3^2 & \omega_3^4 \end{bmatrix} \begin{bmatrix} z_6 \\ z_7 \\ z_8 \end{bmatrix} , \\[2em]
&\begin{bmatrix} c_3 \\ c_4 \\ c_5 \end{bmatrix} = \begin{bmatrix} 1 & 1 & 1 \\ 1 & \omega_3 & \omega_3^2 \\ 1 & \omega_3^2 & \omega_3^4 \end{bmatrix} \begin{bmatrix} z_3 \\ z_4 \\ z_5 \end{bmatrix} ,
&&\begin{bmatrix} c_9 \\ c_{10} \\ c_{11} \end{bmatrix} = \begin{bmatrix} 1 & 1 & 1 \\ 1 & \omega_3 & \omega_3^2 \\ 1 & \omega_3^2 & \omega_3^4 \end{bmatrix} \begin{bmatrix} z_9 \\ z_{10} \\ z_{11} \end{bmatrix} .
\end{aligned}
$$

The same results can be obtained if we directly multiply the vector z by a block-diagonal matrix G_{12} formed by repeating the 3-by-3 DFT matrix Ω_3 on its diagonal:

$$(12.17) \quad \begin{bmatrix} a_{0,0} \\ a_{1,0} \\ a_{2,0} \\ a_{0,1} \\ a_{1,1} \\ a_{2,1} \\ a_{0,2} \\ a_{1,2} \\ a_{2,2} \\ a_{0,3} \\ a_{1,3} \\ a_{2,3} \end{bmatrix} \begin{bmatrix} c_0 \\ c_1 \\ c_2 \\ c_3 \\ c_4 \\ c_5 \\ c_6 \\ c_7 \\ c_8 \\ c_9 \\ c_{10} \\ c_{11} \end{bmatrix} = \begin{bmatrix} 1 & 1 & 1 & 0 & 0 & 0 & 0 & 0 & 0 & 0 & 0 & 0 \\ 1 & \omega_3 & \omega_3^2 & 0 & 0 & 0 & 0 & 0 & 0 & 0 & 0 & 0 \\ 1 & \omega_3^2 & \omega_3^4 & 0 & 0 & 0 & 0 & 0 & 0 & 0 & 0 & 0 \\ 0 & 0 & 0 & 1 & 1 & 1 & 0 & 0 & 0 & 0 & 0 & 0 \\ 0 & 0 & 0 & 1 & \omega_3 & \omega_3^2 & 0 & 0 & 0 & 0 & 0 & 0 \\ 0 & 0 & 0 & 1 & \omega_3^2 & \omega_3^4 & 0 & 0 & 0 & 0 & 0 & 0 \\ 0 & 0 & 0 & 0 & 0 & 0 & 1 & 1 & 1 & 0 & 0 & 0 \\ 0 & 0 & 0 & 0 & 0 & 0 & 1 & \omega_3 & \omega_3^2 & 0 & 0 & 0 \\ 0 & 0 & 0 & 0 & 0 & 0 & 1 & \omega_3^2 & \omega_3^4 & 0 & 0 & 0 \\ 0 & 0 & 0 & 0 & 0 & 0 & 0 & 0 & 0 & 1 & 1 & 1 \\ 0 & 0 & 0 & 0 & 0 & 0 & 0 & 0 & 0 & 1 & \omega_3 & \omega_3^2 \\ 0 & 0 & 0 & 0 & 0 & 0 & 0 & 0 & 0 & 1 & \omega_3^2 & \omega_3^4 \end{bmatrix} \begin{bmatrix} z_0 \\ z_1 \\ z_2 \\ z_3 \\ z_4 \\ z_5 \\ z_6 \\ z_7 \\ z_8 \\ z_9 \\ z_{10} \\ z_{11} \end{bmatrix}.$$

The left-hand-side vector c in the equation above contains the elements of matrix $A_2[N_0, N_1]$ (stacked column by column) as prescribed by Step 2 of the two-factor mixed-radix FFT algorithm. The properly constructed sparse matrix G_{12} can now be expressed as the Kronecker product of a 4-by-4 identity matrix and a 3-by-3 DFT matrix:

$$(12.18) \quad G_{12} = I_4 \otimes \Omega_3 = \begin{bmatrix} 1 \cdot \Omega_3 & 0 \cdot \Omega_3 & 0 \cdot \Omega_3 & 0 \cdot \Omega_3 \\ 0 \cdot \Omega_3 & 1 \cdot \Omega_3 & 0 \cdot \Omega_3 & 0 \cdot \Omega_3 \\ 0 \cdot \Omega_3 & 0 \cdot \Omega_3 & 1 \cdot \Omega_3 & 0 \cdot \Omega_3 \\ 0 \cdot \Omega_3 & 0 \cdot \Omega_3 & 0 \cdot \Omega_3 & 1 \cdot \Omega_3 \end{bmatrix} = \begin{bmatrix} \Omega_3 & & & \\ & \Omega_3 & & \\ & & \Omega_3 & \\ & & & \Omega_3 \end{bmatrix}.$$

Accordingly, the computation performed in Step 2 of the two-factor mixed-radix FFT can be compactly represented by a single matrix equation:

$$(12.19) \quad c = (I_{N_1} \otimes \Omega_{N_0}) \cdot z = (I_{N_1} \otimes \Omega_{N_0}) \cdot D_N \cdot y.$$

Observe that because $z = \text{vec }A_1[N_0, N_1]$, and $c = \text{vec }A_2[N_0, N_1]$ after the columns of A_1 have been multiplied by the DFT matrix Ω_{N_0}, the Kronecker product can also be understood in terms of A_1 and A_2:

$$(12.20) \quad c = (I_{N_1} \otimes \Omega_{N_0}) \text{ vec }A_1 = \text{vec }\{\Omega_{N_0} \cdot A_1\} = \text{vec }A_2.$$

Combining Equations (12.11) and (12.19), the computation performed by Steps 1 and 2 together can still be expressed by a single matrix equation:

$$(12.21) \quad c = (I_{N_1} \otimes \Omega_{N_0}) \cdot \underbrace{D_N \cdot \overbrace{(\Omega_{N_1} \otimes I_{N_0}) \cdot x}^{\text{vetor } y}}_{\text{vector } z}.$$

The last step, Step 3, of the two-factor mixed-radix FFT involves mapping $A_2[\hat{n}_0, \hat{n}_1]$ to X_r for $r = \hat{n}_0 N_1 + \hat{n}_1$. To rework Step 3, it is again useful to display the required mapping in matrix form:

$$(12.22) \quad A_2 = \begin{bmatrix} a_{0,0} & a_{0,1} & a_{0,2} & a_{0,3} \\ a_{1,0} & a_{1,1} & a_{1,2} & a_{1,3} \\ a_{2,0} & a_{2,1} & a_{2,2} & a_{2,3} \end{bmatrix} = \begin{bmatrix} X_0 & X_1 & X_2 & X_3 \\ X_4 & X_5 & X_6 & X_7 \\ X_8 & X_9 & X_{10} & X_{11} \end{bmatrix}.$$

Using the known relationship between $a_{\hat{n}_0,\hat{n}_1}$ and c_ℓ shown on the left-hand side of Equation (12.17), together with the known relationship between $a_{\hat{n}_0,\hat{n}_1}$ and X_r as identified by (12.22), we can now determine the relationship between vector c and the naturally ordered output vector X. For our example with $N=12$, $N_0=3$, and $N_1=4$, the matrix equation

$$(12.23) \qquad\qquad c = P_N \cdot X,$$

expresses vector c as the product of a sparse 12-by-12 permutation matrix P_{12} and the naturally ordered output vector X:

$$(12.24) \quad
\begin{bmatrix} a_{0,0} \\ a_{1,0} \\ a_{2,0} \\ a_{0,1} \\ a_{1,1} \\ a_{2,1} \\ a_{0,2} \\ a_{1,2} \\ a_{2,2} \\ a_{0,3} \\ a_{1,3} \\ a_{2,3} \end{bmatrix}
\begin{bmatrix} c_0 \\ c_1 \\ c_2 \\ c_3 \\ c_4 \\ c_5 \\ c_6 \\ c_7 \\ c_8 \\ c_9 \\ c_{10} \\ c_{11} \end{bmatrix}
=
\begin{bmatrix} X_0 \\ X_4 \\ X_8 \\ X_1 \\ X_5 \\ X_9 \\ X_2 \\ X_6 \\ X_{10} \\ X_3 \\ X_7 \\ X_{11} \end{bmatrix}
=
\begin{bmatrix}
1 & 0 & 0 & 0 & 0 & 0 & 0 & 0 & 0 & 0 & 0 & 0 \\
0 & 0 & 0 & 0 & 1 & 0 & 0 & 0 & 0 & 0 & 0 & 0 \\
0 & 0 & 0 & 0 & 0 & 0 & 0 & 0 & 1 & 0 & 0 & 0 \\
0 & 1 & 0 & 0 & 0 & 0 & 0 & 0 & 0 & 0 & 0 & 0 \\
0 & 0 & 0 & 0 & 0 & 1 & 0 & 0 & 0 & 0 & 0 & 0 \\
0 & 0 & 0 & 0 & 0 & 0 & 0 & 0 & 0 & 1 & 0 & 0 \\
0 & 0 & 1 & 0 & 0 & 0 & 0 & 0 & 0 & 0 & 0 & 0 \\
0 & 0 & 0 & 0 & 0 & 0 & 1 & 0 & 0 & 0 & 0 & 0 \\
0 & 0 & 0 & 0 & 0 & 0 & 0 & 0 & 0 & 0 & 1 & 0 \\
0 & 0 & 0 & 1 & 0 & 0 & 0 & 0 & 0 & 0 & 0 & 0 \\
0 & 0 & 0 & 0 & 0 & 0 & 0 & 1 & 0 & 0 & 0 & 0 \\
0 & 0 & 0 & 0 & 0 & 0 & 0 & 0 & 0 & 0 & 0 & 1
\end{bmatrix}
\begin{bmatrix} X_0 \\ X_1 \\ X_2 \\ X_3 \\ X_4 \\ X_5 \\ X_6 \\ X_7 \\ X_8 \\ X_9 \\ X_{10} \\ X_{11} \end{bmatrix}.$$

$$\underbrace{\qquad}_{\text{vector } c} \underbrace{\qquad}_{\text{vector } c} \qquad \underbrace{\qquad\qquad\qquad}_{\text{Permutation Matrix } P_N} \underbrace{\quad}_{\text{vector } X}$$

Observe that $c = \text{vec } A_2$, and $X = \text{vec } A_2^T$; hence, the same permutation matrix can also be used to relate $\text{vec } A_2$ and $\text{vec } A_2^T$:

$$(12.25) \qquad\qquad \text{vec } A_2 = P_N \cdot \text{vec } A_2^T.$$

To generate the N-by-N permutation matrix P_N, observe that in order to satisfy

$$(12.26) \qquad c_\ell = a_{\hat{n}_0,\hat{n}_1} = A_2[\hat{n}_0,\hat{n}_1] = X_r, \text{ where } \ell = \hat{n}_1 N_0 + \hat{n}_0, \quad r = \hat{n}_0 N_1 + n_1,$$

we must permute X_r, the $(r+1)^{st}$ element in the naturally ordered vector X to the $(\ell+1)^{st}$ position in vector c, which dictates that $P_N[\ell,r] = 1$ for $\ell = \hat{n}_1 N_0 + \hat{n}_0$, $r = \hat{n}_0 N_1 + \hat{n}_1$. The permutation matrix P_N can thus be generated by the pseudo-code:

```
for k := 0 to N − 1 do                    initialize P_N to be an
    for i := 0 to N − 1 do                N-by-N zero matrix
        P_N[i, k] := 0                    column by column
    end for
end for
ℓ := 0
for n̂_1 := 0 to N_1 − 1 do
    for n̂_0 := 0 to N_0 − 1 do
        r := n̂_0 N_1 + n̂_1            construct P_N to permute
        P_N[ℓ, r] := 1                    X_r to the position of c_ℓ
        ℓ := ℓ + 1                        ∵ ℓ = n̂_1 N_0 + n̂_0
    end for
end for
```

Because $c = P_N \cdot X$, we can replace c on the left-hand side of Equation (12.21) by $P_N \cdot X$, and we can now express the entire two-factor mixed-radix FFT by one matrix equation which relates input vector x directly to output vector X:

$$(12.27) \qquad \underbrace{P_N \cdot X}_{\text{vector } c} = (I_{N_1} \otimes \Omega_{N_0}) \cdot D_N \cdot \overbrace{(\Omega_{N_1} \otimes I_{N_0}) \cdot x}^{\text{vector } y} .$$

Note that the naturally ordered output vector X can be recovered by multiplying both sides by the inverse of the permutation matrix $P_N^{-1} = P_N^T$, i.e.,

$$(12.28) \qquad X = \underbrace{P_N^T \cdot (I_{N_1} \otimes \Omega_{N_0}) \cdot D_N \cdot \overbrace{(\Omega_{N_1} \otimes I_{N_0}) \cdot x}^{\text{vector } y}}_{\text{vector } c} .$$

Recall that $z = \textbf{vec } A_1[N_0, N_1]$, and $c = \textbf{vec } A_2[N_0, N_1]$ after each column of matrix A_1 is multiplied by the DFT matrix Ω_{N_0}. Using the result from Equation (12.25) on $c = \textbf{vec } A_2$, we obtain

$$(12.29) \qquad X = P_N^T \cdot c = P_N^T \cdot \textbf{vec } A_2 = P_N^T \cdot \left(P_N \cdot \textbf{vec } A_2^T \right) = \textbf{vec } A_2^T ;$$

the same holds for $z = \textbf{vec } A_1$:

$$(12.30) \qquad P_N^T \cdot z = P_N^T \cdot \textbf{vec } A_1 = P_N^T \cdot \left(P_N \cdot \textbf{vec } A_1^T \right) = \textbf{vec } A_1^T .$$

Therefore, the result $X = \textbf{vec } A_2^T$ can be obtained by transforming $P_N^T \cdot z = \textbf{vec } A_1^T$ directly, which requires that we transpose A_1 in advance and multiply the rows of A_1^T (instead of the columns of A_1) by the DFT matrix Ω_{N_0}, i.e.,

$$(12.31) \qquad X = P_N^T \cdot c = \textbf{vec } A_2^T = (\Omega_{N_0} \otimes I_{N_1}) \cdot \textbf{vec } A_1^T = (\Omega_{N_0} \otimes I_{N_1}) \cdot P_N^T z,$$

which leads to the nal expression:

$$(12.32) \qquad X = \underbrace{(\Omega_{N_0} \otimes I_{N_1}) \cdot P_N^T \cdot D_N \cdot (\Omega_{N_1} \otimes I_{N_0})}_{\text{Factors of the DFT Matrix } \Omega_N} \cdot x.$$

Comparing the right-hand side of this equation with that of the de ning equation $X = \Omega_N \cdot x$, we have produced the Kronecker product factorization of the N-by-N DFT matrix Ω_N, namely,

$$(12.33) \qquad \Omega_N = (\Omega_{N_0} \otimes I_{N_1}) \cdot P_N^T \cdot D_N \cdot (\Omega_{N_1} \otimes I_{N_0}) .$$

Note that Ω_N is shown to be the product of four very *sparse* N-by-N matrices tw o of them are de ned by the Kronecker products, one is a permutation matrix, and one is a diagonal matrix.

12.2   From Two-Factor to Multi-Factor Mixed-Radix FFT

For multi-factor composite N, we may always begin with two factors assuming that

$$(12.34) \qquad N = \overbrace{F_0 \times \cdots \times F_{\nu-1}}^{N_0} \times \overbrace{F_\nu}^{N_1} = N_0 \times N_1.$$

With $N_0 = N/F_\nu$ and $N_1 = F_\nu$, we apply the results from the last section to factor the N-by-N DFT matrix $\boldsymbol{\Omega}_N$, and we express

$$(12.35) \qquad \boldsymbol{X} = \underbrace{(\boldsymbol{\Omega}_{N_0} \otimes \boldsymbol{I}_{F_\nu}) \cdot \boldsymbol{P}_N^T \cdot \boldsymbol{D}_N \cdot (\boldsymbol{\Omega}_{F_\nu} \otimes \boldsymbol{I}_{N_0})}_{\boldsymbol{\Omega}_N} \cdot \boldsymbol{x}, \text{ where } N_0 = N/F_\nu.$$

To further decompose the N_0-by-N_0 DFT matrix $\boldsymbol{\Omega}_{N_0}$, we again apply the two-factor results, because we may express the composite N_0 to be the product of two factors:

$$(12.36) \qquad N_0 = \overbrace{F_0 \times \cdots \times F_{\nu-2}}^{M_{\nu-1}} \times \overbrace{F_{\nu-1}}^{N_1} = M_{\nu-1} \times F_{\nu-1}.$$

With $N_0 = M_{\nu-1} \times F_{\nu-1}$, we may express

$$(12.37) \qquad \boldsymbol{\Omega}_{N_0} = \left(\boldsymbol{\Omega}_{M_{\nu-1}} \otimes \boldsymbol{I}_{F_{\nu-1}}\right) \cdot \boldsymbol{P}_{N_0}^T \cdot \boldsymbol{D}_{N_0} \cdot \left(\boldsymbol{\Omega}_{F_{\nu-1}} \otimes \boldsymbol{I}_{M_{\nu-1}}\right).$$

To show how the Kronecker product $\left(\boldsymbol{\Omega}_{N_0} \otimes \boldsymbol{I}_{F_\nu}\right)$ can be expanded and simplified factor by factor in a systematic manner, we shall use $N = F_0 \times F_1 \times F_2 \times F_3 \times F_4$, and we define $M_4 = N/F_4$, $M_3 = M_4/F_3 = N/(F_3 F_4)$, $M_2 = M_3/F_2 = N/(F_2 F_3 F_4)$, $M_1 = M_2/F_1 = N/(F_1 F_2 F_3 F_4)$, and $M_0 = M_1/F_0 = 1$. We begin with $N = M_4 \times F_4$:

$$(12.38) \qquad \boldsymbol{\Omega}_N = \underbrace{(\boldsymbol{\Omega}_{M_4} \otimes \boldsymbol{I}_{F_4})}_{\text{expanded below}} \cdot \boldsymbol{P}_N^T \cdot \boldsymbol{D}_N \cdot (\boldsymbol{\Omega}_{F_4} \otimes \boldsymbol{I}_{M_4}),$$

and we continue to factor $\boldsymbol{\Omega}_{M_4}$ (because $M_4 = M_3 \times F_3$), so that we can expand and simplify the product

$$(12.39) \qquad \boldsymbol{\Omega}_{M_4} \otimes \boldsymbol{I}_{F_4} = \left\{ (\boldsymbol{\Omega}_{M_3} \otimes \boldsymbol{I}_{F_3}) \cdot \boldsymbol{P}_{M_4}^T \cdot \boldsymbol{D}_{M_4} \cdot (\boldsymbol{\Omega}_{F_3} \otimes \boldsymbol{I}_{M_3}) \right\} \otimes \boldsymbol{I}_{F_4}$$

using some Kronecker product properties which we are now motivated to learn about in the next section. After the detour we shall return to Equation (12.38) and derive the matrix equation for multi-factor mixed-radix FFT.

12.2.1 Selected properties and rules for Kronecker products

Recall that a Kronecker product is defined for matrices of arbitrary dimensions, but standard product is defined only for conformable matrices. Therefore, when standard product $\boldsymbol{A} \cdot \boldsymbol{B}$ occurs in any formula that follows, it is assumed that the number of columns in $\boldsymbol{A}$ is equal to the number of rows in $\boldsymbol{B}$ so that the expression is valid of course, $\boldsymbol{A}$ itself may be the result of Kronecker products and/or standard products, and so is $\boldsymbol{B}$.

The list of Kronecker product properties given below is not exhaustive, and the selection is based on our needs to decompose the DFT matrix analytically. While many of these results can be stated as stand alone theorems to be proved directly by showing that the (i, j)th element of the matrix in the left-hand side is equal to the (i, j)th element of the matrix in the right-hand side, such a formal proof could be laborious and may not shed light on the meaning of the equation or how it could be used. For our purpose it is more important to understand what the equations (representing the properties) mean and how they can be used in the context of factoring the DFT matrix, and we shall use well-understood examples from the last section to provide such context in our discussion. Readers interested in the proofs of these properties may consult [26].

1. An identity matrix of order $N = N_0 \times N_1$ can be expressed as

    (12.40) $$I_N = I_{N_0} \otimes I_{N_1} = I_{N_1} \otimes I_{N_0}.$$

    The same holds for zero matrices.

2. If α is a scalar, then

    (12.41) $$A \otimes (\alpha \cdot B) = \alpha \cdot (A \otimes B).$$

3. The Kronecker product is distributive with respect to addition, that is,

    (12.42) $$(A + B) \otimes C = A \otimes C + B \otimes C; \quad A \otimes (B + C) = A \otimes B + A \otimes C.$$

4. The Kronecker product is associative, that is,

    (12.43) $$A \otimes (B \otimes C) = (A \otimes B) \otimes C.$$

5. While the rule $(A \cdot B)^T = B^T \cdot A^T$ holds for standard matrix products, a different rule is required by the Kronecker product:

    (12.44) $$(A \otimes B)^T = A^T \otimes B^T.$$

6. The following rule applies to mixed products:

    (12.45) $$(A \otimes B) \cdot (C \otimes D) = (A \cdot C) \otimes (B \cdot D).$$

    Observe that one standard product occurs on the left-hand side, and two standard products occur on the right-hand side. As stated at the beginning of this section, we shall assume that the matrices involved make each standard product valid.

    The cases involving identity matrixes warrant special attention:

    (12.46) $$(A \otimes I) \cdot (I \otimes D) = A \otimes D = (I \otimes D) \cdot (A \otimes I).$$

    (12.47) $$(A \cdot C) \otimes I = (A \cdot C) \otimes (I \cdot I) = (A \otimes I) \cdot (C \otimes I).$$

    The result can be easily extended to the product of more than two matrices:

    (12.48) $$(A \cdot C \cdot D) \otimes I = ((A \cdot C) \otimes I) \cdot (D \otimes I) = (A \otimes I) \cdot (C \otimes I) \cdot (D \otimes I).$$

7. While the rule $(A \cdot B)^{-1} = B^{-1} \cdot A^{-1}$ holds for the standard product of nonsingular matrices A and B, a different rule is required by the Kronecker product:

    (12.49) $$(A \otimes B)^{-1} = A^{-1} \otimes B^{-1}.$$

8. If each column of A is multiplied by matrix Z, then

    (12.50) $$\mathbf{vec}\{Z \cdot A\} = (I \otimes Z) \cdot \mathbf{vec}\, A.$$

    Recall that $(I \otimes Z)$ represents a block diagonal matrix, and that we have demonstrated this result for I_4 and $Z = \Omega_3$ when deriving formula (12.20) in Section 12.1. If each row of A is multiplied by matrix Y, then

    (12.51) $$\mathbf{vec}\left\{A \cdot Y^T\right\} = (Y \otimes I) \cdot \mathbf{vec}\, A.$$

    Recall that we have demonstrated this result for I_3 and $Y = \Omega_4$ (which is symmetric) when deriving formula (12.12) in Section 12.1.

9. If each row of A is multiplied by matrix Y, and each column of A is multiplied by matrix Z, then the result is the standard product $Z \cdot A \cdot Y^T$, which can also be expressed using Kronecker products and the vector-valued operator **vec**:

$$\begin{aligned}
\textbf{vec}\left\{Z \cdot A \cdot Y^T\right\} &= (I \otimes Z) \cdot (Y \otimes I) \cdot \textbf{vec}\, A \\
(12.52) &\\
&= (Y \otimes Z) \cdot \textbf{vec}\, A. \qquad \text{(by (12.46), property 6)}
\end{aligned}$$

10. There exist permutation matrices P and Q such that

$$(12.53) \qquad\qquad A \otimes B = P \cdot (B \otimes A) \cdot Q.$$

12.2.2 Complete factorization of the DFT matrix

After taking a detour in the Kronecker product algebra in the last subsection, we now return to complete the factorization of the DFT matrix Ω_N of order $N = F_0 \times F_1 \times F_2 \times F_3 \times F_4$. We have obtained the first factor $P_N^T \cdot D_N \cdot (\Omega_{F_4} \otimes I_{M_4})$ by Equation (12.38). After relabeling P_N^T as $Q_N^{F_4}$ and D_N as $D_N^{F_4}$ to reflect the dimension N as well as the two factors being $N_0 = N/F_4$ and $N_1 = F_4$, we have

$$(12.54) \qquad \Omega_N = (\Omega_{M_4} \otimes I_{F_4}) \cdot \underbrace{Q_N^{F_4} \cdot D_N^{F_4} \cdot (\Omega_{F_4} \otimes I_{N/F_4})}_{\text{The first factor } R_1} \cdot$$

Next, using $M_4 = M_3 \times F_3$, we proceed to *simplify* the expansion of $(\Omega_{M_4} \otimes I_{F_4})$ given by Equation (12.39):

$$\begin{aligned}
&\Omega_{M_4} \otimes I_{F_4} \\
&= \left\{ (\Omega_{M_3} \otimes I_{F_3}) \cdot Q_{M_4}^{F_3} \cdot D_{M_4}^{F_3} \cdot (\Omega_{F_3} \otimes I_{M_3}) \right\} \otimes I_{F_4} && \because (12.39) \\
(12.55) \quad &= \left\{ (\Omega_{M_3} \otimes I_{F_3}) \otimes I_{F_4} \right\} \cdot \left\{ (Q_{M_4}^{F_3} \cdot D_{M_4}^{F_3}) \otimes I_{F_4} \right\} \cdot \left\{ (\Omega_{F_3} \otimes I_{M_3}) \otimes I_{F_4} \right\} && \because (12.48) \\
&= (\Omega_{M_3} \otimes I_{F_3 \cdot F_4}) \cdot \left((Q_{M_4}^{F_3} \cdot D_{M_4}^{F_3}) \otimes I_{F_4} \right) \cdot (\Omega_{F_3} \otimes I_{M_3 \cdot F_4}) && \because (12.43),(12.40) \\
&= (\Omega_{M_3} \otimes I_{F_3 \cdot F_4}) \cdot \underbrace{\left((Q_{M_4}^{F_3} \cdot D_{M_4}^{F_3}) \otimes I_{F_4} \right) \cdot (\Omega_{F_3} \otimes I_{N/F_3})}_{\text{The second factor } R_2} \cdot
\end{aligned}$$

By repeating the same expansion and reduction process on $(\Omega_{M_3} \otimes I_{F_3 \cdot F_4})$ using $M_3 = M_2 \times F_2$, we obtain

$$\begin{aligned}
&\Omega_{M_3} \otimes I_{F_3 \cdot F_4} \\
&= \left\{ (\Omega_{M_2} \otimes I_{F_2}) \cdot Q_{M_3}^{F_2} \cdot D_{M_3}^{F_2} \cdot (\Omega_{F_2} \otimes I_{M_2}) \right\} \otimes I_{F_3 \cdot F_4} \\
(12.56) \quad &= \left\{ (\Omega_{M_2} \otimes I_{F_2}) \otimes I_{F_3 \cdot F_4} \right\} \cdot \left\{ (Q_{M_3}^{F_2} \cdot D_{M_3}^{F_2}) \otimes I_{F_3 \cdot F_4} \right\} \cdot \left\{ (\Omega_{F_2} \otimes I_{M_2}) \otimes I_{F_3 \cdot F_4} \right\} \\
&= (\Omega_{M_2} \otimes I_{F_2 \cdot F_3 \cdot F_4}) \cdot \left((Q_{M_3}^{F_2} \cdot D_{M_3}^{F_2}) \otimes I_{F_3 \cdot F_4} \right) \cdot (\Omega_{F_2} \otimes I_{M_2 \cdot F_3 \cdot F_4}) \\
&= (\Omega_{M_2} \otimes I_{F_2 \cdot F_3 \cdot F_4}) \cdot \underbrace{\left((Q_{M_3}^{F_2} \cdot D_{M_3}^{F_2}) \otimes I_{F_3 \cdot F_4} \right) \cdot (\Omega_{F_2} \otimes I_{N/F_2})}_{\text{The third factor } R_3} \cdot
\end{aligned}$$

Following the pattern established by the last two factors, we can express

$$(12.57) \quad \mathbf{\Omega}_{M_2} \otimes \mathbf{I}_{F_2 \cdot F_3 \cdot F_4} = \left(\mathbf{\Omega}_{M_1} \otimes \mathbf{I}_{F_1 \cdot F_2 \cdot F_3 \cdot F_4} \right) \cdot \underbrace{ \left(\left(\mathbf{Q}_{M_2}^{F_1} \cdot \mathbf{D}_{M_2}^{F_1} \right) \otimes \mathbf{I}_{F_2 \cdot F_3 \cdot F_4} \right) \cdot \left(\mathbf{\Omega}_{F_1} \otimes \mathbf{I}_{N/F_1} \right) }_{\text{The fourth factor } \mathbf{R}_4}.$$

Recall that $M_1 = M_2/F_1 = N/(F_1 F_2 F_3 F_4) = F_0$; hence, the fth factor is given by

$$(12.58) \quad \mathbf{\Omega}_{M_1} \otimes \mathbf{I}_{F_1 \cdot F_2 \cdot F_3 \cdot F_4} = \underbrace{ \mathbf{\Omega}_{F_0} \otimes \mathbf{I}_{N/F_0} }_{\text{The fth factor } \mathbf{R}_5}.$$

In summary, we have shown that for $N = F_0 \times F_1 \times F_2 \times F_3 \times F_4$, the DFT matrix can be expressed as the product of ve N-by-N sparse matrices:

$$(12.59) \quad \mathbf{\Omega}_N = \mathbf{R}_5 \cdot \mathbf{R}_4 \cdot \mathbf{R}_3 \cdot \mathbf{R}_2 \cdot \mathbf{R}_1,$$

where each sparse-matrix factor is de ned by mixed products:

$$
\begin{aligned}
\mathbf{R}_1 &= \mathbf{Q}_N^{F_4} \cdot \mathbf{D}_N^{F_4} \cdot \left(\mathbf{\Omega}_{F_4} \otimes \mathbf{I}_{N/F_4} \right), \quad N = \prod_{k=0}^{4} F_k, \\[2mm]
\mathbf{R}_2 &= \left(\left(\mathbf{Q}_{M_4}^{F_3} \cdot \mathbf{D}_{M_4}^{F_3} \right) \otimes \mathbf{I}_{F_4} \right) \cdot \left(\mathbf{\Omega}_{F_3} \otimes \mathbf{I}_{N/F_3} \right), \quad M_4 = \prod_{k=0}^{3} F_k, \\[2mm]
\mathbf{R}_3 &= \left(\left(\mathbf{Q}_{M_3}^{F_2} \cdot \mathbf{D}_{M_3}^{F_2} \right) \otimes \mathbf{I}_{F_3 \cdot F_4} \right) \cdot \left(\mathbf{\Omega}_{F_2} \otimes \mathbf{I}_{N/F_2} \right), \quad M_3 = \prod_{k=0}^{2} F_k, \\[2mm]
\mathbf{R}_4 &= \left(\left(\mathbf{Q}_{M_2}^{F_1} \cdot \mathbf{D}_{M_2}^{F_1} \right) \otimes \mathbf{I}_{F_2 \cdot F_3 \cdot F_4} \right) \cdot \left(\mathbf{\Omega}_{F_1} \otimes \mathbf{I}_{N/F_1} \right), \quad M_2 = \prod_{k=0}^{1} F_k, \\[2mm]
\mathbf{R}_5 &= \mathbf{\Omega}_{F_0} \otimes \mathbf{I}_{N/F_0}.
\end{aligned}
$$

(12.60)

Recall that the permutation matrix $\mathbf{Q}_N^{N_1}$ and the diagonal matrix $\mathbf{D}_N^{N_1}$ are both fully speci ed given the values for the dimension N and the (single) factor N_1, because the two-factor $N = (N/N_1) \times N_1$. For easy reference, we restate the pseudo-code de nitions for $\mathbf{Q}_N^{N_1}$ and $\mathbf{D}_N^{N_1}$. The following code segment generates $\mathbf{Q}_N^{N_1}$, which is the *transpose* of the previously de ned $\mathbf{P}_N$:

```
for k := 0 to N − 1 do                    initialize Q_N^{N_1} to be an
    for i := 0 to N − 1 do                    N-by-N zero matrix
        Q_N^{N_1}[i, k] := 0                     column by column
    end for
end for
N_0 := N/N_1                            ∵ two-factor N = (N/N_1) × N_1
r := 0
for n̂_0 := 0 to N_0 − 1 do
    for n̂_1 := 0 to N_1 − 1 do
        ℓ := n̂_1 N_0 + n̂_0
        Q_N^{N_1}[r, ℓ] := 1                   ∵ P_N[ℓ, r] = 1 and Q_N^{N_1} = P_N^T
        r := r + 1                            ∵ r = n̂_0 N_1 + n̂_1
    end for
end for
```

We also repeat here the pseudo-code that generates the diagonal matrix $D_N^{N_1}$:

```
for k := 0 to N − 1 do                    initialize D_N^{N1} to be an
    for i := 0 to N − 1 do                      N-by-N zero matrix
        D_N^{N1}[i, k] := 0                        column by column
    end for
end for
N_0 := N/N_1                          ∵ two-factor N = (N/N_1) × N_1
ℓ := 0
for n̂_1 := 0 to N_1 − 1 do                    assign twiddle factors
    for n_0 := 0 to N_0 − 1 do                 to diagonal elements
        D_N^{N1}[ℓ, ℓ] := ω_N^{n_0 n̂_1}      ∵ y_ℓ = a_{n_0, n̂_1} = A_1[n_0, n̂_1],
        ℓ := ℓ + 1                             and ℓ = n̂_1 N_0 + n_0
    end for
end for
```

With the DFT matrix $\mathbf{\Omega}_N$ completely factored, the five-factor mixed-radix FFT can be cast as five sparse-matrix-vector products, that is,

(12.61)
$$X = \overbrace{\mathbf{\Omega}_N \cdot x}^{\text{DFT}} = \underbrace{R_5 \cdot \Big(R_4 \cdot (R_3 \cdot (R_2 \cdot (R_1 \cdot x))) \Big)}_{\text{mixed-radix FFT}}.$$

The five steps of the mixed-radix FFT can now be represented compactly by five matrix equations:

(12.62)
$$y_1 = R_1 \cdot x = \overbrace{Q_N^{F_4} \cdot D_N^{F_4} \cdot (\mathbf{\Omega}_{F_4} \otimes I_{N/F_4}) \cdot x}^{\text{Step 1}},$$

$$y_2 = R_2 \cdot y_1 = \overbrace{\big((Q_{M_4}^{F_3} \cdot D_{M_4}^{F_3}) \otimes I_{F_4} \big) \cdot (\mathbf{\Omega}_{F_3} \otimes I_{N/F_3}) \cdot y_1}^{\text{Step 2}},$$

$$y_3 = R_3 \cdot y_2 = \overbrace{\big((Q_{M_3}^{F_2} \cdot D_{M_3}^{F_2}) \otimes I_{F_3 \cdot F_4} \big) \cdot (\mathbf{\Omega}_{F_2} \otimes I_{N/F_2}) \cdot y_2}^{\text{Step 3}},$$

$$y_4 = R_4 \cdot y_3 = \overbrace{\big((Q_{M_2}^{F_1} \cdot D_{M_2}^{F_1}) \otimes I_{F_2 \cdot F_3 \cdot F_4} \big) \cdot (\mathbf{\Omega}_{F_1} \otimes I_{N/F_1}) \cdot y_3}^{\text{Step 4}},$$

$$X = R_5 \cdot y_4 = \overbrace{(\mathbf{\Omega}_{F_0} \otimes I_{N/F_0}) \cdot y_4}^{\text{Step 5}}.$$

Observe that the algorithm described by these five steps actually represents a self-sorting variant of the mixed-radix algorithm described previously by recursive equations in Section 11.3.2, because the former produces naturally ordered X, whereas the latter produces scrambled output $c = P \cdot X$ at the end of Step 5 and all of the re-ordering work to recover $X = P^T \cdot c$ is done in Step 6 regardless of the number of factors. We discuss how to obtain the particular factorization which leads to $P \cdot X$ in Section 12.4.

12.3   Other Forms by Alternate Index Splitting

Recall that the alternate index splitting strategy presented in Section 11.4 in Chapter 11 leads to the DIF FFT. In this section we shall work out the corresponding Kronecker product factorization of the DFT matrix. We again begin with the two-factor FFT in multidimensional formulation based on the recursive equation established for arbitrary ν-factor composite N in Section 11.4 of Chapter 11.

Step 0. Map x_ℓ to $A[n_0, n_1]$ for $\ell = n_1 N_0 + n_0$.

Step 1. Compute $N/N_1 = N_0$ short DFT-like transforms of length N_1:

$$
\begin{aligned}
&\underline{\textbf{for }} n_0 := 0 \textbf{ to } N_0 - 1 \underline{\textbf{ do}} \\
&\quad \underline{\textbf{for }} \hat{n}_1 := 0 \textbf{ to } N_1 - 1 \underline{\textbf{ do}} \\
&\qquad A_1[n_0, \hat{n}_1] = \sum_{n_1=0}^{N_1-1} \left\{ A[n_0, n_1]\, \omega_N^{\hat{n}_1 n_0} \right\} \omega_{N_1}^{\hat{n}_1 n_1} \\
&\quad \underline{\textbf{end for}} \\
&\underline{\textbf{end for}}
\end{aligned}
$$

Step 2. Compute $N/N_0 = N_1$ short DFT transforms of length N_0:

$$
\begin{aligned}
&\underline{\textbf{for }} \hat{n}_1 := 0 \textbf{ to } N_1 - 1 \underline{\textbf{ do}} \\
&\quad \underline{\textbf{for }} \hat{n}_0 := 0 \textbf{ to } N_0 - 1 \underline{\textbf{ do}} \\
&\qquad A_2[\hat{n}_0, \hat{n}_1] = \sum_{n_0=0}^{N_0-1} A_1[n_0, \hat{n}_1]\, \omega_{N_0}^{\hat{n}_0 n_0} \\
&\quad \underline{\textbf{end for}} \\
&\underline{\textbf{end for}}
\end{aligned}
$$

Step 3. Map $A_2[\hat{n}_0, \hat{n}_1]$ to X_r for $r = \hat{n}_0 N_1 + \hat{n}_1$.

Comparing this algorithm with the two-factor algorithm given in the last section, we see that the twiddle factors contained in the diagonal matrix $\boldsymbol{D}_N$ are now applied to $\textbf{vec}\,\boldsymbol{A} = \boldsymbol{x}$ in Step 1 rather than $\textbf{vec}\,\boldsymbol{A}_1$ in Step 2. Since this is the only difference, we can adapt the matrix equation from last section without repeating the reformulating process. That is, corresponding to equation (12.32), we have

$$(12.63) \qquad \boldsymbol{X} = (\boldsymbol{\Omega}_{N_0} \otimes \boldsymbol{I}_{N_1}) \cdot \boldsymbol{P}_N^T \cdot (\boldsymbol{\Omega}_{N_1} \otimes \boldsymbol{I}_{N_0}) \cdot \boldsymbol{D}_N \cdot \boldsymbol{x}.$$

To obtain the complete factorization of the DFT matrix $\boldsymbol{\Omega}_N$ of order $N = \prod_{k=0}^{4} F_k$, we have, corresponding to Equation (12.54),

$$(12.64) \qquad \boldsymbol{\Omega}_N = (\boldsymbol{\Omega}_{M_4} \otimes \boldsymbol{I}_{F_4}) \cdot \underbrace{\boldsymbol{Q}_N^{F_4} \cdot (\boldsymbol{\Omega}_{F_4} \otimes \boldsymbol{I}_{N/F_4}) \cdot \boldsymbol{D}_N^{F_4}}_{\text{The rst factor } \boldsymbol{R}_1}.$$

The expansion and reduction of $(\boldsymbol{\Omega}_{M_4} \otimes \boldsymbol{I}_{F_4})$ follow the same steps as our derivation of Equation (12.55), and we obtain the second matrix factor:

$$
(12.65)
\begin{aligned}
&\boldsymbol{\Omega}_{M_4} \otimes \boldsymbol{I}_{F_4} \\
&= \left\{ (\boldsymbol{\Omega}_{M_3} \otimes \boldsymbol{I}_{F_3}) \cdot \boldsymbol{Q}_{M_4}^{F_3} \cdot (\boldsymbol{\Omega}_{F_3} \otimes \boldsymbol{I}_{M_3}) \cdot \boldsymbol{D}_{M_4}^{F_3} \right\} \otimes \boldsymbol{I}_{F_4} \qquad\qquad \because (12.64) \\
&= (\boldsymbol{\Omega}_{M_3} \otimes \boldsymbol{I}_{F_3 \cdot F_4}) \cdot \underbrace{(\boldsymbol{Q}_{M_4}^{F_3} \otimes \boldsymbol{I}_{F_4}) \cdot (\boldsymbol{\Omega}_{F_3} \otimes \boldsymbol{I}_{N/F_3}) \cdot (\boldsymbol{D}_{M_4}^{F_3} \otimes \boldsymbol{I}_{F_4})}_{\text{The second factor } \boldsymbol{R}_2}.
\end{aligned}
$$

The remaining factors can now be easily adapted from the results given by Equations (12.56), (12.57), and (12.58), and the complete factorization of the DFT matrix $\boldsymbol{\Omega}_N$ is given by

$$(12.66) \qquad \boldsymbol{\Omega}_N = \boldsymbol{R}_5 \cdot \boldsymbol{R}_4 \cdot \boldsymbol{R}_3 \cdot \boldsymbol{R}_2 \cdot \boldsymbol{R}_1,$$

where each sparse-matrix factor is de ned by mixed products:

$$R_1 = Q_N^{F_4} \cdot (\Omega_{F_4} \otimes I_{N/F_4}) \cdot D_N^{F_4}, \quad N = \prod_{k=0}^{4} F_k,$$

$$R_2 = (Q_{M_4}^{F_3} \otimes I_{F_4}) \cdot (\Omega_{F_3} \otimes I_{N/F_3}) \cdot (D_{M_4}^{F_3} \otimes I_{F_4}), \quad M_4 = \prod_{k=0}^{3} F_k,$$

(12.67)
$$R_3 = (Q_{M_3}^{F_2} \otimes I_{F_3 \cdot F_4}) \cdot (\Omega_{F_2} \otimes I_{N/F_2}) \cdot (D_{M_3}^{F_2} \otimes I_{F_3 \cdot F_4}), \quad M_3 = \prod_{k=0}^{2} F_k,$$

$$R_4 = (Q_{M_2}^{F_1} \otimes I_{F_2 \cdot F_3 \cdot F_4}) \cdot (\Omega_{F_1} \otimes I_{N/F_1}) \cdot (D_{M_2}^{F_1} \otimes I_{F_2 \cdot F_3 \cdot F_4}), \quad M_2 = \prod_{k=0}^{1} F_k,$$

$$R_5 = \Omega_{F_0} \otimes I_{N/F_0}.$$

12.4 Factorization Results by Alternate Expansion

We commented at the end of Section 12.2.2 that there is more than one way to factor the DFT matrix even when the indices are mapped and split the same way. In this section we derive the factorization results which lead to the unordered mixed-radix FFT algorithms given by recursive equations in Sections 11.3.2 and 11.4.1 in Chapter 11.

12.4.1 Unordered mixed-radix DIT FFT

We rst show how to obtain a different factorization result directly from Equation (12.28) in Section 12.1, which is repeated here:

(12.68)
$$X = P_N^T \cdot (I_{N_1} \otimes \Omega_{N_0}) \cdot D_N \cdot (\Omega_{N_1} \otimes I_{N_0}) \cdot x.$$

To obtain the complete factorization of the DFT matrix Ω_N of order $N = \prod_{k=0}^{4} F_k$, we begin with $N = M_4 \times F_4$ and we express Equation (12.68) using $N_0 = M_4$ and $N_1 = F_4$ as we have done many times before. The result is

(12.69)
$$\Omega_N = \underbrace{Q_N^{F_4}}_{P_1} \cdot (I_{F_4} \otimes \Omega_{M_4}) \cdot \underbrace{D_N^{F_4} \cdot (\Omega_{F_4} \otimes I_{M_4})}_{\text{factor } R_1}, \quad N = M_4 \times F_4.$$

The expansion and reduction of $(I_{F_4} \otimes \Omega_{M_4})$, using $M_4 = M_3 \times F_3$, follow the same steps as our derivation of Equation (12.55), and the result is

$$I_{F_4} \otimes \Omega_{M_4}$$
$$= I_{F_4} \otimes \left\{ Q_{M_4}^{F_3} \cdot (I_{F_3} \otimes \Omega_{M_3}) \cdot D_{M_4}^{F_3} \cdot (\Omega_{F_3} \otimes I_{M_3}) \right\} \qquad \because (12.69)$$

(12.70)
$$= \left\{ I_{F_4} \otimes \left(Q_{M_4}^{F_3} \cdot (I_{F_3} \otimes \Omega_{M_3}) \right) \right\} \cdot \left\{ I_{F_4} \otimes \left(D_{M_4}^{F_3} \cdot (\Omega_{F_3} \otimes I_{M_3}) \right) \right\} \quad \because (12.48)$$

$$= \underbrace{(I_{F_4} \otimes Q_{M_4}^{F_3})}_{\text{factor } P_2} \cdot (I_{F_4 \cdot F_3} \otimes \Omega_{M_3}) \cdot \underbrace{\left\{ I_{F_4} \otimes \left(D_{M_4}^{F_3} \cdot (\Omega_{F_3} \otimes I_{M_3}) \right) \right\}}_{\text{factor } R_2} \quad \because (12.43), (12.40)$$

After expanding and reducing $I_{F_4 \cdot F_3} \otimes \Omega_{M_3}$ as well as the resulting $I_{F_4 \cdot F_3 \cdot F_2} \otimes \Omega_{M_2}$, we obtain the complete factorization of the DFT matrix Ω_N:

$$(12.71) \quad \Omega_N = \underbrace{(P_1 \cdot P_2 \cdot P_3 \cdot P_4)}_{\text{matrix } P^T} \cdot (R_5 \cdot R_4 \cdot R_3 \cdot R_2 \cdot R_1) = P^T \cdot (R_5 \cdot R_4 \cdot R_3 \cdot R_2 \cdot R_1),$$

where P^T is the permutation matrix defined by the product of

$$(12.72) \quad \begin{aligned} P_1 &= Q_N^{F_4}, \quad N = \prod_{k=0}^{4} F_k, \\ P_2 &= I_{F_4} \otimes Q_{M_4}^{F_3}, \quad M_4 = \prod_{k=0}^{3} F_k, \\ P_3 &= I_{F_4 \cdot F_3} \otimes Q_{M_3}^{F_2}, \quad M_3 = \prod_{k=0}^{2} F_k, \\ P_4 &= I_{F_4 \cdot F_3 \cdot F_2} \otimes Q_{M_2}^{F_1}, \quad M_2 = \prod_{k=0}^{1} F_k; \end{aligned}$$

and each sparse-matrix factor R_k is defined by mixed products:

$$(12.73) \quad \begin{aligned} R_1 &= D_N^{F_4} \cdot (\Omega_{F_4} \otimes I_{M_4}), \\ R_2 &= I_{F_4} \otimes \left(D_{M_4}^{F_3} \cdot (\Omega_{F_3} \otimes I_{M_3}) \right), \\ R_3 &= I_{F_4 \cdot F_3} \otimes \left(D_{M_3}^{F_2} \cdot (\Omega_{F_2} \otimes I_{M_2}) \right), \\ R_4 &= I_{F_4 \cdot F_3 \cdot F_2} \otimes \left(D_{M_2}^{F_1} \cdot (\Omega_{F_1} \otimes I_{M_1}) \right), \\ R_5 &= I_{F_4 \cdot F_3 \cdot F_2 \cdot F_1} \otimes \Omega_{F_0}. \end{aligned}$$

Observe that with the DFT matrix now factored as $\Omega_N = P^T \cdot R_5 \cdot R_4 \cdot R_3 \cdot R_2 \cdot R_1$, we can rewrite the DFT as

$$(12.74) \quad X = \overbrace{\Omega_N \cdot x}^{\text{DFT}} = P^T \cdot \left\{ R_5 \cdot \left(R_4 \cdot (R_3 \cdot (R_2 \cdot (R_1 \cdot x))) \right) \right\}.$$

Since P^T denotes a permutation matrix, we have $P^T = P^{-1}$, and we obtain the mixed-radix FFT which computes $P \cdot X$:

$$(12.75) \quad P \cdot X = \underbrace{R_5 \cdot \left(R_4 \cdot (R_3 \cdot (R_2 \cdot (R_1 \cdot x))) \right)}_{\text{unordered mixed-radix DIT FFT}}.$$

The five steps represented by $y_k = R_k \cdot y_{k-1}$, for $k = 1, \ldots, 5$, with $y_0 = x$, now corresponds to the steps of the mixed-radix algorithm which produces the scrambled output $P \cdot X$ and was previously described by the recursive equations in Sections 11.3.2 in Chapter 11.

12.4.2 Unordered mixed-radix DIF FFT

Recall that the alternate index splitting strategy presented in Section 11.4 in Chapter 11 leads to the DIF FFT. Following our explanation in Section 12.3, we obtain the matrix equation representing the two-factor DIF FFT directly from Equation (12.68), that is,

$$(12.76) \qquad \boldsymbol{X} = \boldsymbol{P}_N^T \cdot (\boldsymbol{I}_{N_1} \otimes \boldsymbol{\Omega}_{N_0}) \cdot (\boldsymbol{\Omega}_{N_1} \otimes \boldsymbol{I}_{N_0}) \cdot \boldsymbol{D}_N \cdot \boldsymbol{x}.$$

Note that because the change from $\boldsymbol{D}_N \cdot (\boldsymbol{\Omega}_{N_1} \otimes \boldsymbol{I}_{N_0})$ in Equation (12.68) to $(\boldsymbol{\Omega}_{N_1} \otimes \boldsymbol{I}_{N_0}) \cdot \boldsymbol{D}_N$ in Equation (12.76) is con ned within the factor $\boldsymbol{R}_1$, it has no effect on the expansion process. Therefore, for $N = F_0 \times F_1 \times \boldsymbol{F}_2 \times F_3 \times \boldsymbol{F}_4$, we can obtain the factorization results matching the unordered DIF FFT directly from Equation (12.73) by making the same change within each factor:

$$(12.77) \qquad \begin{aligned} \boldsymbol{R}_1 &= \left(\boldsymbol{\Omega}_{F_4} \otimes \boldsymbol{I}_{M_4} \right) \cdot \boldsymbol{D}_N^{F_4}, \\ \boldsymbol{R}_2 &= \boldsymbol{I}_{F_4} \otimes \left(\left(\boldsymbol{\Omega}_{F_3} \otimes \boldsymbol{I}_{M_3} \right) \cdot \boldsymbol{D}_{M_4}^{F_3} \right), \\ \boldsymbol{R}_3 &= \boldsymbol{I}_{F_4 \cdot F_3} \otimes \left(\left(\boldsymbol{\Omega}_{F_2} \otimes \boldsymbol{I}_{M_2} \right) \cdot \boldsymbol{D}_{M_3}^{F_2} \right), \\ \boldsymbol{R}_4 &= \boldsymbol{I}_{F_4 \cdot F_3 \cdot F_2} \otimes \left(\left(\boldsymbol{\Omega}_{F_1} \otimes \boldsymbol{I}_{M_1} \right) \cdot \boldsymbol{D}_{M_2}^{F_1} \right), \\ \boldsymbol{R}_5 &= \boldsymbol{I}_{F_4 \cdot F_3 \cdot F_2 \cdot F_1} \otimes \boldsymbol{\Omega}_{F_0}. \end{aligned}$$

Since the expansion process is not affected by the changes we made, the permutation matrix $\boldsymbol{P}^T$ is de n ed by Equation (12.72) as before, and we have obtained the desired factorization

$$(12.78) \qquad \boldsymbol{X} = \overbrace{\boldsymbol{\Omega}_N \cdot \boldsymbol{x}}^{\text{DFT}} = \boldsymbol{P}^T \cdot \underbrace{\boldsymbol{R}_5 \cdot \left(\boldsymbol{R}_4 \cdot \left(\boldsymbol{R}_3 \cdot \left(\boldsymbol{R}_2 \cdot (\boldsymbol{R}_1 \cdot \boldsymbol{x}) \right) \right) \right)}_{\text{unordered mixed-radix DIF FFT}},$$

which leads to the unordered mixed-radix DIF FFT that computes $\boldsymbol{P} \cdot \boldsymbol{X}$ b y

$$(12.79) \qquad \boldsymbol{P} \cdot \boldsymbol{X} = \underbrace{\boldsymbol{R}_5 \cdot \left(\boldsymbol{R}_4 \cdot \left(\boldsymbol{R}_3 \cdot \left(\boldsymbol{R}_2 \cdot (\boldsymbol{R}_1 \cdot \boldsymbol{x}) \right) \right) \right)}_{\text{unordered mixed-radix DIF FFT}}.$$

12.5 Unordered FFT for Scrambled Input

In the last section we derived two unordered mixed-radix FFT algorithms which transform naturally ordered input $\boldsymbol{x}$ to scrambled output $\boldsymbol{c} = \boldsymbol{P} \cdot \boldsymbol{X}$, and separate re-ordering steps are required to unscramble $\boldsymbol{c}$ to recover $\boldsymbol{X} = \boldsymbol{P}^T \cdot \boldsymbol{c}$. The re-ordering steps are speci ed by the Kronecker product factorization of the permutation matrix $\boldsymbol{P}^T$, that is, $\boldsymbol{X} = \boldsymbol{P}_1 \cdot (\boldsymbol{P}_2 \cdot (\boldsymbol{P}_3 \cdot (\boldsymbol{P}_4 \cdot \boldsymbol{c})))$.

Because of the symmetry of the DFT matrix, i.e., $\boldsymbol{\Omega} = \boldsymbol{\Omega}^T$, a different FFT algorithm can be produced from the Kronecker product factorization of $\boldsymbol{\Omega}^T$ to get that, we simply *transpose* the factors of $\boldsymbol{\Omega}$. That is, corresponding to

$$(12.80) \quad \boldsymbol{\Omega}_N = \underbrace{(\boldsymbol{P}_1 \cdot \boldsymbol{P}_2 \cdot \boldsymbol{P}_3 \cdot \boldsymbol{P}_4)}_{\text{matrix } \boldsymbol{P}^T} (\boldsymbol{R}_5 \cdot \boldsymbol{R}_4 \cdot \boldsymbol{R}_3 \cdot \boldsymbol{R}_2 \cdot \boldsymbol{R}_1) = \boldsymbol{P}^T \cdot (\boldsymbol{R}_5 \cdot \boldsymbol{R}_4 \cdot \boldsymbol{R}_3 \cdot \boldsymbol{R}_2 \cdot \boldsymbol{R}_1),$$

we now have

$$(12.81)\ \Omega_N^T = \left(R_1^T \cdot R_2^T \cdot R_3^T \cdot R_4^T \cdot R_5^T\right) \underbrace{\left(P_4^T \cdot P_3^T \cdot P_2^T \cdot P_1^T\right)}_{\text{matrix } P} = \left(R_1^T \cdot R_2^T \cdot R_3^T \cdot R_4^T \cdot R_5^T\right) \cdot P.$$

Using the factors of Ω_N^T, we can rewrite the DFT as

$$(12.82)\qquad X = \overbrace{\Omega_N \cdot x}^{\text{DFT}} = \Omega_N^T \cdot x = \underbrace{R_1^T \cdot \left(R_2^T \cdot \left(R_3^T \cdot \left(R_4^T \cdot (R_5^T \cdot (P \cdot x))\right)\right)\right)}_{\text{unordered mixed-radix FFT}}.$$

Observe that the resulting mixed-radix FFT algorithm transforms scrambled input $P \cdot x$ to naturally ordered output X.

For the DIT FFT, we obtain the factors R_k^T from R_k in Equation (12.73). In deriving each R_k^T given below, we apply the standard product rule $(A \cdot B)^T = B^T \cdot A^T$, the Kronecker product rule $(A \otimes B)^T = A^T \otimes B^T$, and we make use of the fact that the DFT matrix Ω, diagonal matrix D, and identity matrix I are all symmetric matrices regardless of their order.

$$R_1^T = \left(D_N^{F_4} \cdot (\Omega_{F_4} \otimes I_{M_4})\right)^T = (\Omega_{F_4} \otimes I_{M_4}) \cdot D_N^{F_4},$$

$$R_2^T = I_{F_4}^T \otimes \left(D_{M_4}^{F_3} \cdot (\Omega_{F_3} \otimes I_{M_3})\right)^T = I_{F_4} \otimes \left((\Omega_{F_3} \otimes I_{M_3}) \cdot D_{M_4}^{F_3}\right),$$

$$(12.83)\ R_3^T = I_{F_4 \cdot F_3}^T \otimes \left(D_{M_3}^{F_2} \cdot (\Omega_{F_2} \otimes I_{M_2})^T\right) = I_{F_4 \cdot F_3} \otimes \left((\Omega_{F_2} \otimes I_{M_2}) \cdot D_{M_3}^{F_2}\right),$$

$$R_4^T = I_{F_4 \cdot F_3 \cdot F_2}^T \otimes \left(D_{M_2}^{F_1} \cdot (\Omega_{F_1} \otimes I_{M_1})\right)^T = I_{F_4 \cdot F_3 \cdot F_2} \otimes \left((\Omega_{F_1} \otimes I_{M_1}) \cdot D_{M_2}^{F_1}\right),$$

$$R_5^T = I_{F_4 \cdot F_3 \cdot F_2 \cdot F_1} \otimes \Omega_{F_0}.$$

For the DIF FFT, we obtain the factors R_k^T directly from the factors R_k in Equation (12.77) by similar steps:

$$R_1^T = D_N^{F_4} \cdot (\Omega_{F_4} \otimes I_{M_4}),$$

$$R_2^T = I_{F_4} \otimes \left(D_{M_4}^{F_3} \cdot (\Omega_{F_3} \otimes I_{M_3})\right),$$

$$(12.84)\qquad R_3^T = I_{F_4 \cdot F_3} \otimes \left(D_{M_3}^{F_2} \cdot (\Omega_{F_2} \otimes I_{M_2})\right),$$

$$R_4^T = I_{F_4 \cdot F_3 \cdot F_2} \otimes \left(D_{M_2}^{F_1} \cdot (\Omega_{F_1} \otimes I_{M_1})\right),$$

$$R_5^T = I_{F_4 \cdot F_3 \cdot F_2 \cdot F_1} \otimes \Omega_{F_0}.$$

For both DIT and DIF algorithms, the permutation matrix $P = P_4^T \cdot P_3^T \cdot P_2^T \cdot P_1^T$, where each factor P_k^T can be obtained from the factor P_k in Equation (12.72):

$$\begin{aligned}
&P_1^T = \left(Q_N^{F_4}\right)^T = Q_N^{M_4}, && \because N = M_4 \times F_4, \\
&P_2^T = \left(I_{F_4} \otimes Q_{M_4}^{F_3}\right)^T = I_{F_4} \otimes Q_{M_4}^{M_3}, && \because M_4 = M_3 \times F_3, \\
(12.85)\ &P_3^T = \left(I_{F_4 \cdot F_3} \otimes Q_{M_3}^{F_2}\right)^T = I_{F_4 \cdot F_3} \otimes Q_{M_3}^{M_2}, && \because M_3 = M_2 \times F_2, \\
&P_4^T = \left(I_{F_4 \cdot F_3 \cdot F_2} \otimes Q_{M_2}^{F_1}\right)^T = I_{F_4 \cdot F_3 \cdot F_2} \otimes Q_{M_2}^{F_0}, && \because M_2 = M_1 \times F_1 = F_0 \times F_1.
\end{aligned}$$

12.6   Utilities of the Kronecker Product Factorization

In the preceding sections we have derived six different factorization results of the DFT matrix, which lead to the *compact* expressions of six different mixed-radix FFTs, and they correspond to the six canonical forms of the radix-2 FFTs derived in Chapters 7, 8, and 9 in [13]. How do we make use of these compact, but somewhat abstract, expressions in our study, development, and implementation of the various mixed-radix FFTs? We address this question below.

First of all, it allows us to characterize each algorithm by the matrix equation which represents a typical step. For the six algorithms we have derived for $N = \prod_{k=0}^{k=4} F_k$, a typical step is represented by $y_3 = R_3 \cdot y_2$, which is sufﬁcient for many purposes when all ﬁve steps have been laid out to provide context and allow immediate generalization to arbitrary number of factors if needed.

1. **Self-sorting mixed-radix DIT$_{NN}$ FFT:** The naturally ordered input x and naturally ordered output X are indicated by the subscript NN.

    (12.86) $$ y_3 = \left(\left(Q_{M_3}^{F_2} \cdot D_{M_3}^{F_2} \right) \otimes I_{F_3 \cdot F_4} \right) \cdot \left(\Omega_{F_2} \otimes I_{N/F_2} \right) \cdot y_2. $$

2. **Self-sorting mixed-radix DIF$_{NN}$ FFT:**

    (12.87) $$ y_3 = \left(Q_{M_3}^{F_2} \otimes I_{F_3 \cdot F_4} \right) \cdot \left(\Omega_{F_2} \otimes I_{N/F_2} \right) \cdot \left(D_{M_3}^{F_2} \otimes I_{F_3 \cdot F_4} \right) \cdot y_2. $$

3. **Unordered mixed-radix DIT$_{NR}$ FFT:** The naturally ordered input x and scrambled output $P \cdot X$ are indicated by the subscript NR.

    (12.88) $$ y_3 = I_{F_4 \cdot F_3} \otimes \left(D_{M_3}^{F_2} \cdot \left(\Omega_{F_2} \otimes I_{M_2} \right) \right) \cdot y_2. $$

4. **Unordered mixed-radix DIF$_{NR}$ FFT:**

    (12.89) $$ y_3 = I_{F_4 \cdot F_3} \otimes \left(\left(\Omega_{F_2} \otimes I_{M_2} \right) \cdot D_{M_3}^{F_3} \right) \cdot y_2. $$

5. **Unordered mixed-radix DIT$_{RN}$ FFT:** The scrambled input $P \cdot x$ and naturally ordered output X are indicated by the subscript RN.

    (12.90) $$ y_3 = I_{F_4 \cdot F_3} \otimes \left(\left(\Omega_{F_2} \otimes I_{M_2} \right) \cdot D_{M_3}^{F_3} \right) \cdot y_2. $$

6. **Unordered mixed-radix DIF$_{RN}$ FFT:**

    (12.91) $$ y_3 = I_{F_4 \cdot F_3} \otimes \left(D_{M_3}^{F_2} \cdot \left(\Omega_{F_2} \otimes I_{M_2} \right) \right) \cdot y_2. $$

When each algorithm is characterized by a single matrix equation, we have a mathematical basis to study and compare different algorithms in a systematic manner.

For the four unordered FFT, the re-ordering step can be studied separately by analyzing the Kronecker product factorization of the permutation matrix P^T or P. For example, a careful study of how to re-distribute the permutations imposed by the factors of P^T throughout all steps results in the *in-place* self-sorting mixed-radix FFT proposed by Temperton in [49].

Second, each matrix equation tells us, in precise (but abstract) mathematical terms, exactly what needs to be computed in a typical step, which helps us focus our thoughts and efforts when developing (or trying to understand or evaluate) the computer program implementing

each algorithm. Note that the focus would be on deciphering the matrix equation and performing equivalent operations directly on the 1-D array y_2, because *no* sparse matrix used in the
mathematical equation would actually be formed in an ef cien t implementation.

Third, the extension from the two-factor case to the multi-factor case was made easy using
the rules of matrix algebra for Kronecker products.

Fourth, recall that the DIT_{RN} FFT and DIF_{RN} FFT were generated directly from rearranging
the factors which previously form the matrix equation for the DIT_{NR} and DIF_{NR} FFT. Therefore, it is not surprising that many other FFT variants can be derived using the matrix algebra
of Kronecker products. For example, researchers have tailored the mixed-radix FFT to composite transform length N formed by certain factors   e.g., power of a speci c factor, product
of speci c factors, product combining mixed factors and power of speci c factors, product
of pairwise-prime factors, . . . , etc.   by direct manipulation of the matrix equation and/or by
incorporating specially designed index mapping schemes into the matrix equation.

Chapter 13

The Family of Prime Factor FFT Algorithms

The prime factor algorithms (PFAs) are specialized mixed-radix algorithms which are based on factoring the transform length into pairwise prime factors. For example, if we factor $N = 12 = 3 \times 4$, then the two factors $N_0 = 3$ and $N_1 = 4$ are relatively prime because their greatest common divisor $\mathbf{gcd}(3, 4) = 1$; if we factor $N = 60 = 3 \times 4 \times 5$, then the three factors $N_0 = 3$, $N_1 = 4$, and $N_2 = 5$ are said to be pairwise prime, because we have $\mathbf{gcd}(3, 4) = 1$, $\mathbf{gcd}(4, 5) = 1$, and $\mathbf{gcd}(3, 5) = 1$. Note that in the multi-factor case, meeting the pairwise prime condition guarantees that any two products of arbitrary factors are also relatively prime for example, $\mathbf{gcd}(3 \times 4, 5) = 1$, $\mathbf{gcd}(3, 4 \times 5) = 1$, and $\mathbf{gcd}(3 \times 5, 4) = 1$. Although the rst prime factor algorithm was published by Good [24] prior to the introduction of the decimation-in-time mixed-radix FFT by Cooley and Tukey [16], we shall derive in Section 13.2 the two-factor PFA by adding bells and whistles to the simpler and more easily understood mixed-radix FFT, using also the various algorithmic and/or mathematical tools we have established and thoroughly explained in deriving the family of mixed-radix FFTs in Chapters 11 and 12.

Although a DFT of composite length N can be computed by a mixed-radix FFT whether the factors of N are pairwise prime or not, it was recognized by Cooley et. al. in [15] that the prime factor algorithm can be extremely useful when used in combination with the mixed-radix algorithm. They further clari ed in [15] that the prime factor algorithm described by Good [24] had been mistakenly said to be equivalent to Cooley and Tukey s arbitrary-factor (mixed-radix) FFT algorithm, and they stressed the importance of distinguishing between these two algorithms since each has its particular advantages which can be exploited in appropriate circumstances. This is indeed the case, and the family of prime factor algorithms has continued to grow with further development by Kolba and Parks [31], Winograd [55], Burrus and Eschenbacher [9], Nussbaumer [35], Rothweiler [41], Otto [36], and Temperton [45, 46, 47, 48, 50]. After we discuss the design and implementation of the prime factor algorithm, it will be clear that the PFA has the following advantages:

- When the factors are pairwise prime, the PFA incurs fewer arithmetic operations than the mixed-radix FFT, because the twiddle factors are eliminated and the associated scaling operations are not required in the PFA.

- It can be useful to combine the PFA with a radix-2 FFT. For example, if $N = N_0 \times N_1 =$

512×77, then the two factors $N_0 = 512$ and $N_1 = 77$ are relatively prime, we can use prime factor algorithm to set up DFTs of lengths N_0 and N_1. Then each DFT of length $N_0 = 512 = 2^9$ can be computed by a radix-2 FFT, and each DFT of length $N_1 = 77 = 7 \times 11$ can be computed by a prime factor algorithm.

- It can also be useful to combine the PFA with an arbitrary factor (mixed-radix) FFT. For example, if $N = 4 \times 9 \times 25 \times 49$, we can factor $N = N_0 \times N_1$ with $N_0 = N_1 = 2 \times 3 \times 5 \times 7$, and use the mixed-radix FFT to set up DFTs of lengths N_0 and N_1. The PFA can then be used to compute each DFT of length $N_0 = N_1 = 2 \times 3 \times 5 \times 7$, because the factors $F_0 = 2$, $F_1 = 3$, $F_2 = 5$, and $F_3 = 7$ are pairwise prime.

- The indexing schemes used by the PFA can be made simpler than that required for the mixed-radix FFT.

- The PFA is easy to program because of the elimination of twiddle factors and the simplicity of the indexing scheme.

- A self-sorting in-place PFA is available and it is equally easy to program.

As to the theory behind the PFA, in Section 13.5 we shall formally introduce a few relevant concepts from elementary number theory concerning the properties of integers, and we proceed to prove the Chinese Remainder Theorem (CRT), because CRT and CRT-related index maps are responsible for the number-theoretic splitting of the DFT matrix, which gives rise to the PFA. Although the matrix equation representing the two-factor PFA can be expanded (using the rules of matrix algebra for Kronecker products) to represent the multi-factor PFA as previously done for the mixed-radix FFT, it remains important to acquire the background in number theory to understand the theoretical aspects of the multi-factor index map, because an ef cien t implementation of the multi-factor PFA depends on a number of theoretical results which are signi cant in their own right.

In the remainder of this chapter we shall consider some practical issues related to the performance of the prime factor algorithms, including the ef cient implementation of the multi-factor PFA and the computation of short DFTs or short rotated DFTs in the PFA.

13.1   Connecting the Relevant Ideas

We shall   rst review and connect the relevant ideas from the preceding two chapters in the context of deriving and extending a two-factor mixed-radix FFT expressed by Equation (12.27):

$$(13.1) \quad \boldsymbol{P}_N \cdot \boldsymbol{X} = \mathbf{vec}\boldsymbol{A}_2 = (\boldsymbol{I}_{N_1} \otimes \boldsymbol{\Omega}_{N_0}) \cdot \boldsymbol{D}_N \cdot (\boldsymbol{\Omega}_{N_1} \otimes \boldsymbol{I}_{N_0}) \cdot \underbrace{\mathbf{vec}\boldsymbol{A}}_{\text{input}\mathbf{x}}, \quad \text{where } N = N_0 \times N_1.$$

Recall that we went through the following stages before we arrived at Equation (13.1):

Stage I We assumed that the input sequence $\boldsymbol{x}$ was stored in matrix $\boldsymbol{A}[N_0, N_1]$ by standard column-major index mapping scheme, and the output sequence $\boldsymbol{X}$ was stored in matrix $\boldsymbol{B}[N_1, N_0]$ (which is the transpose of the intermediate matrix $\boldsymbol{A}_2[N_0, N_1]$) also by column-major index mapping scheme.

Stage II We combine the two predetermined index mapping schemes with the chosen index splitting scheme to *decouple* the DFT computation into multiple short DFT or DFT-like transforms along the columns and rows of the input matrix $A[N_0, N_1]$ and the intermediate matrix $A_1[N_0, N_1]$.

Stage III We then derive the Kronecker product factorization of the DFT matrix Ω_N by *translating* the operations performed on columns and rows of matrices A and A_1 to equivalent matrix operations on the two vectors **vec** A and **vec** A_1, which were *extended* from the two matrices column by column.

Stage IV Based on the predetermined index mapping schemes for input x and output X, we were able to identify $x = $ **vec** A and $X = $ **vec** $B = $ **vec** A_2^T. The latter leads us to define a permutation matrix P_N such that $P_N \cdot X = $ **vec** A_2.

Stage V We obtain the factorization of the DFT matrix Ω_N by writing (13.1) as

$$(13.2) \qquad X = \underbrace{P_N^T \cdot (I_{N_1} \otimes \Omega_{N_0}) \cdot D_N \cdot (\Omega_{N_1} \otimes I_{N_0})}_{\text{Kronecker product factorization of } \Omega_N} \cdot x = \Omega_N \cdot x.$$

Once we have factored Ω_N, we may use the rules of matrix algebra for Kronecker products to incorporate the similarly factored Ω_{N_0} into (13.2) if N_0 remains composite, and this process can be repeated until we obtain the formula for the desired mixed-radix FFT.

We show next that the prime factor algorithms may be derived via stages in parallel to those we have just reviewed in this section.

13.2 Deriving the Two-Factor PFA

In this section our objective is to derive the two-factor prime factor FFT given by

$$(13.3) \qquad \underbrace{P_N \cdot X}_{\text{vec} B} = (I_{N_1} \otimes \Omega_{N_0}) \cdot (\Omega_{N_1} \otimes I_{N_0}) \cdot \underbrace{Q_N \cdot x}_{\text{vec} A},$$

where $N = N_0 \times N_1$ and $\mathbf{gcd}(N_0, N_1) = 1$; P_N and Q_N are specially designed permutation matrices to be discussed in detail below. Comparing Equation (13.3) with Equation (13.1), we see that all scaling operations involving twiddle factors on the diagonal of D_N have been eliminated this is the key feature of the prime factor FFT.

With the five-stage road map set up in the last section, we now derive (13.3) stage by stage. Whenever a concrete example is needed, we shall use $N_0 = 3$ and $N_1 = 4$ as before for easy comparison this is possible because the two factors are relatively prime.

13.2.1 Stage I: Nonstandard index mapping schemes

As described by Burrus and Eschenbacher in [9], the Ruritanian map is used to store input sequence x in matrix $A[N_0, N_1]$, and the Chinese Remainder Theorem (CRT) map is used to store output sequence X in matrix $B[N_0, N_1]$.

The Ruritanian index mapping scheme defines

$$(13.4) \quad x_\ell = A[n_0, n_1] \text{ if } \ell = \left\langle N_1 n_0 + N_0 n_1 \right\rangle_N \overset{\text{def}}{=} \text{ residue of } \left(N_1 n_0 + N_0 n_1 \right) \text{ modulo } N.$$

It is useful to display the mapping in matrix form for $N = N_0 \times N_1 = 3 \times 4$. According to (13.4), $A[n_0, n_1]$ (denoted also by a_{n_0, n_1}) stores x_ℓ for

$$\ell = \left\langle N_1 n_0 + N_0 n_1 \right\rangle_N = \left\langle 4n_0 + 3n_1 \right\rangle_N;$$

we thus have

$$(13.5) \qquad \boldsymbol{A} = \begin{bmatrix} a_{0,0} & a_{0,1} & a_{0,2} & a_{0,3} \\ a_{1,0} & a_{1,1} & a_{1,2} & a_{1,3} \\ a_{2,0} & a_{2,1} & a_{2,2} & a_{2,3} \end{bmatrix} = \begin{bmatrix} x_0 & x_3 & x_6 & x_9 \\ x_4 & x_7 & x_{10} & x_1 \\ x_8 & x_{11} & x_2 & x_5 \end{bmatrix}.$$

Note that $\text{vec}\boldsymbol{A} \neq \boldsymbol{x}$. (This is where $\boldsymbol{Q}_N$, the permutation matrix associated with $\text{vec}\boldsymbol{A}$, comes in.)

The CRT index mapping scheme defines
$$(13.6)$$
$$X_r = B[\hat{n}_0, \hat{n}_1] \text{ if } r = \left\langle \rho N_1 \hat{n}_0 + q N_0 \hat{n}_1 \right\rangle_N \overset{\text{def}}{=} \text{residue of } \left(\rho N_1 \hat{n}_0 + q N_0 \hat{n}_1 \right) \text{ modulo } N,$$

where $0 < \rho < N_0$ and $0 < q < N_1$ are integers satisfying

$$(13.7) \qquad \rho N_1 = s N_0 + 1, \quad q N_0 = t N_1 + 1, \text{ for integers } 0 < s < N_1, \ 0 < t < N_0.$$

This is called the CRT map because the existence of integers ρ, s, q, and t are guaranteed by the Chinese remainder theorem (to be discussed in Section 13.5.3) when N_0 and N_1 are relatively prime. It is also useful to display the mapping in matrix form for $N = N_0 \times N_1 = 3 \times 4$. For this simple example, we can determine the integers $\rho = 1$, $s = 1$, $q = 3$, and $t = 2$ by trial and error, and we know $B[\hat{n}_0, \hat{n}_1]$ (denoted also by $b_{\hat{n}_0, \hat{n}_1}$) stores X_r for

$$r = \left\langle \rho N_1 \hat{n}_0 + q N_0 \hat{n}_1 \right\rangle_N = \left\langle 4 \hat{n}_0 + 9 \hat{n}_1 \right\rangle_N.$$

We thus have

$$(13.8) \qquad \boldsymbol{B} = \begin{bmatrix} b_{0,0} & b_{0,1} & b_{0,2} & b_{0,3} \\ b_{1,0} & b_{1,1} & b_{1,2} & b_{1,3} \\ b_{2,0} & b_{2,1} & b_{2,2} & b_{2,3} \end{bmatrix} = \begin{bmatrix} X_0 & X_9 & X_6 & X_3 \\ X_4 & X_1 & X_{10} & X_7 \\ X_8 & X_5 & X_2 & X_{11} \end{bmatrix}.$$

As pointed out by Temperton in [45], the inverse CRT map determines $\hat{n}_0$ and $\hat{n}_1$ from $r \bmod N_0$ and $r \bmod N_1$ (which are also derived in Section 13.5.3), i.e.,

$$(13.9) \qquad X_r = B[\hat{n}_0, \hat{n}_1] \text{ if } \hat{n}_0 = \left\langle r \right\rangle_{N_0}, \text{ and } \hat{n}_1 = \left\langle r \right\rangle_{N_1}.$$

Note that $\text{vec}\boldsymbol{B} \neq \boldsymbol{X}$, either. (This is where $\boldsymbol{P}_N$, the permutation matrix associated with $\text{vec}\boldsymbol{B}$, comes in.)

With the existence of ρ and q satisfying (13.7) guaranteed by the Chinese remainder theorem, the inverse Ruritanian map for $x_\ell = A[n_0, n_1]$ is given by

$$(13.10) \qquad n_0 = \left\langle \rho \ell \right\rangle_{N_0}, \quad n_1 = \left\langle q \ell \right\rangle_{N_1},$$

where $0 < \rho < N_0$ and $0 < q < N_1$ are integers satisfying (13.7) above, which can also be expressed as

$$(13.11) \qquad \left\langle \rho N_1 \right\rangle_{N_0} = 1, \quad \left\langle q N_0 \right\rangle_{N_1} = 1.$$

The Ruritanian index map is formally presented in Section 13.5.5.

13.2.2 Stage II: Decoupling the DFT computation

Since $\omega_N^N = 1$, it is a fact that $\omega_N^M = \omega_N^{\langle M \rangle_N}$. Accordingly, we may use $\ell = N_1 n_0 + N_0 n_1$ instead of $\ell = \langle N_1 n_0 + N_0 n_1 \rangle_N$ when ℓ appears in the exponent of ω_N. After we substitute $x_\ell = A[n_0, n_1]$ and $\ell = N_1 n_0 + N_0 n_1$ in the DFT formula, the X_r s may be computed by the for-loop:

> **for** $r := 0$ **to** $N - 1$ **do**
> $\qquad X_r := \sum_{n_0=0}^{N_0-1} \sum_{n_1=0}^{N_1-1} A[n_0, n_1]\, \omega_N^{r \cdot (N_1 n_0 + N_0 n_1)}.$
> **end for**

To split the exponent of ω_N, we substitute $r = \rho N_1 \hat{n}_0 + q N_0 \hat{n}_1$, instead of using $r = \langle \rho N_1 \hat{n}_0 + q N_0 \hat{n}_1 \rangle_N$, and we obtain

$$
\begin{aligned}
(13.12) \qquad r \cdot (N_1 n_0 + N_0 n_1) &= (\rho N_1 \hat{n}_0 + q N_0 \hat{n}_1) \cdot (N_1 n_0 + N_0 n_1) \\
&= \rho N_1^2 \hat{n}_0 n_0 + \rho N_1 N_0 \hat{n}_0 n_1 + q N_1 N_0 \hat{n}_1 n_0 + q N_0^2 \hat{n}_1 n_1 \\
&= \rho N_1^2 \hat{n}_0 n_0 + \rho N \hat{n}_0 n_1 + q N \hat{n}_1 n_0 + q N_0^2 \hat{n}_1 n_1.
\end{aligned}
$$

Using $\omega_N^N = 1$, together with $\rho N_1 = s N_0 + 1$ and $q N_0 = t N_1 + 1$ from Equation (13.7), we simplify

$$
\begin{aligned}
(13.13) \qquad \omega_N^{r \cdot (N_1 n_0 + N_0 n_1)} &= \omega_N^{\rho N_1^2 \hat{n}_0 n_0} \cdot \omega_N^{\rho N \hat{n}_0 n_1} \cdot \omega_N^{q N \hat{n}_1 n_0} \cdot \omega_N^{q N_0^2 \hat{n}_1 n_1} \\
&= \omega_N^{\rho N_1^2 \hat{n}_0 n_0} \cdot (\omega_N^N)^{\rho \hat{n}_0 n_1} \cdot (\omega_N^N)^{q \hat{n}_1 n_0} \cdot \omega_N^{q N_0^2 \hat{n}_1 n_1} \\
&= \omega_N^{(\rho N_1) N_1 \hat{n}_0 n_0} \cdot \omega_N^{(q N_0) N_0 \hat{n}_1 n_1} && (\because \omega_N^N = 1) \\
&= \omega_N^{(s N_0 + 1) N_1 \hat{n}_0 n_0} \cdot \omega_N^{(t N_1 + 1) N_0 \hat{n}_1 n_1} \\
&= \omega_N^{N_1 \hat{n}_0 n_0} \cdot \omega_N^{N_0 \hat{n}_1 n_1} && (\because \omega_N^{N_0 N_1} = \omega_N^N = 1) \\
&= \omega_{N_0}^{\hat{n}_0 n_0} \cdot \omega_{N_1}^{\hat{n}_1 n_1}. && (\because \omega_N^{N_0} = \omega_{N_1},\ \omega_N^{N_1} = \omega_{N_0})
\end{aligned}
$$

Substituting $X_r = B[\hat{n}_0, \hat{n}_1]$ and the now simplified $\omega_N^{r \cdot (N_1 n_0 + N_0 n_1)} = \omega_{N_0}^{\hat{n}_0 n_0} \cdot \omega_{N_1}^{\hat{n}_1 n_1}$ in the for-loop indexed by r, we are ready to decouple the DFT which is now described by the double for-loop indexed by $\hat{n}_0$ and $\hat{n}_1$:

> **for** $\hat{n}_1 := 0$ **to** $N_1 - 1$ **do**
> $\quad$ **for** $\hat{n}_0 := 0$ **to** $N_0 - 1$ **do**
> $\qquad B[\hat{n}_0, \hat{n}_1] := \sum_{n_0=0}^{N_0-1} \left(\sum_{n_1=0}^{N_1-1} A[n_0, n_1]\, \omega_{N_1}^{\hat{n}_1 n_1} \right) \omega_{N_0}^{\hat{n}_0 n_0}$
> $\quad$ **end for**
> **end for**

Applying the decoupling technique we have learned in Chapter 11, we can compute the bracketed short DFT in an independent double for-loop indexed by n_0 and $\hat{n}_1$, and we have successfully decoupled the DFT into N_0 short DFTs of length N_1 plus N_1 short DFTs of length N_0:

$$
\begin{aligned}
&\textbf{for } n_0 := 0 \textbf{ to } N_0 - 1 \textbf{ do}\\
&\quad \textbf{for } \hat{n}_1 := 0 \textbf{ to } N_1 - 1 \textbf{ do}\\
&\qquad A_1[n_0, \hat{n}_1] := \sum_{n_1=0}^{N_1-1} A[n_0, n_1]\, \omega_{N_1}^{\hat{n}_1 n_1}\\
&\quad \textbf{end for}\\
&\textbf{end for}\\
&\textbf{for } \hat{n}_1 := 0 \textbf{ to } N_1 - 1 \textbf{ do}\\
&\quad \textbf{for } \hat{n}_0 := 0 \textbf{ to } N_0 - 1 \textbf{ do}\\
&\qquad B[\hat{n}_0, \hat{n}_1] := \sum_{n_0=0}^{N_0-1} A_1[n_0, \hat{n}_1]\, \omega_{N_0}^{\hat{n}_0 n_0}\\
&\quad \textbf{end for}\\
&\textbf{end for}
\end{aligned}
$$

Since a genuine two-dimensional DFT is de ned on input matrix $x[N_0, N_1]$ (which represents, among other possibilities, actual data from a 2-D image of dimensions N_0-by-N_1) and output matrix $X[N_0, N_1]$ by the formula

$$
(13.14) \qquad X[r_0, r_1] = \sum_{\ell_0=0}^{N_0-1} \sum_{\ell_1=0}^{N_1-1} x[\ell_0, \ell_1]\, \omega_{N_1}^{r_1 \ell_1} \omega_{N_0}^{r_0 \ell_0},
$$

for $0 \le r_0 \le N_0 - 1$ and $0 \le r_1 \le N_1 - 1$, we have now successfully mapped the given length-N one-dimensional DFT $X = \Omega_N x$ by permut ations (of both input and output) into a *true* two-dimensional DFT with respect to the input matrix $A[N_0, N_1]$ and output matrix $B[N_0, N_1]$.

13.2.3   Organizing the PFA computation–Part 1

Recall that for our example with $N_0 = 3$ and $N_1 = 4$, the $N = N_0 \times N_2 = 12$ input and output elements are stored in matrices A and B according to the Ruritanian and CRT maps respectively:

$$
(13.15) \qquad A = \begin{bmatrix} x_0 & x_3 & x_6 & x_9 \\ x_4 & x_7 & x_{10} & x_1 \\ x_8 & x_{11} & x_2 & x_5 \end{bmatrix}; \qquad B = \begin{bmatrix} X_0 & X_9 & X_6 & X_3 \\ X_4 & X_1 & X_{10} & X_7 \\ X_8 & X_5 & X_2 & X_{11} \end{bmatrix}.
$$

While it is convenient to describe the two-factor PFA algorithm as performing the short DFT on each column and each row of matrix A, in actual implementation we do not need to physically store the input data in a 2-D matrix, provided that we can access the right group of elements directly from the input array $\{x_\ell\}$ in an equally convenient manner. We shall now use the same example to describe the direct access methods proposed by Temperton [45].

At rst, to help focus our attention on the indices, we replace the input and output maps by two integer maps, namely,

$$
(13.16) \qquad A = \begin{bmatrix} 0 & 3 & 6 & 9 \\ 4 & 7 & 10 & 1 \\ 8 & 11 & 2 & 5 \end{bmatrix}; \qquad B = \begin{bmatrix} 0 & 9 & 6 & 3 \\ 4 & 1 & 10 & 7 \\ 8 & 5 & 2 & 11 \end{bmatrix}.
$$

Our objective is to generate the indices contained in the matrices row by row or column by column without storing the entire matrices. For each mapping scheme, we show how to achieve this objective below.

1. For the Ruritanian map contained in matrix A, we only need to use the Ruritanian formula to obtain the rst row $\{0, 3, 6, 9\}$. After taking a close look of the indices in the second row of matrix A, it appears that $\{4, 7, 10, 1\}$ can be generated by computing

   (13.17) $$\langle 0 + N_1 \rangle_N, \quad \langle 3 + N_1 \rangle_N, \quad \langle 6 + N_1 \rangle_N, \quad \langle 9 + N_1 \rangle_N,$$

   where $N_1 = 4$ and $N = N_0 \times N_1 = 12$. This is indeed the case. Similarly, the third row of indices can be generated from the second row and so on.

   If one deals with columns, the indices in the second column can be generated from the r st by computing

   (13.18) $$\langle 0 + N_0 \rangle_N, \quad \langle 4 + N_0 \rangle_N, \quad \langle 8 + N_0 \rangle_N,$$

   where $N_0 = 3$ and $N = N_0 \times N_1 = 12$.

   The same results are obtained if we use the Ruritanian mapping formula to compute the indices one by one. Suppose

   $$\ell = A[n_0, n_1] = \langle n_0 N_1 + n_1 N_0 \rangle_N,$$

   then

   $$A[n_0 + 1, n_1] = \langle (n_0 + 1)N_1 + n_1 N_0 \rangle_N = \langle \ell + N_1 \rangle_N;$$
   $$A[n_0, n_1 + 1] = \langle n_0 N_1 + (n_1 + 1)N_0 \rangle_N = \langle \ell + N_0 \rangle_N.$$

   For arbitrary $\nu \geq 2$, the ν-dimensional Ruritanian map is formally introduced in Section 13.5.5.

2. For the CRT map contained in matrix B, we only need to use the CRT formula to generate the rst row, and the second row is obtained by

   (a) cyclic shifting the r st row to get $\{3, 0, 9, 6\}$;

   (b) increasing the cyclic shifted indices by one, i.e., obtain

   $$\{3 + 1, 0 + 1, 9 + 1, 6 + 1\} = \{4, 1, 10, 7\}$$

   as the second row.

   Similarly, by adding one to $\{7, 4, 1, 10\}$ (which is the result from cyclic shifting the second row $\{4, 1, 10, 7\}$), we obtain the third row in B. The generation of indices column by column follows the same pattern.

   The same results are obtained if we use the inverse CRT map (see Section 13.5.4) to relate $\hat{n}_0$ and $\hat{n}_1$ (row and column indices of $B[\hat{n}_0, \hat{n}_1]$) to r, the index of the output element X_r mapped to $B[\hat{n}_0, \hat{n}_1]$. Suppose

   $$r = B[\hat{n}_0, \hat{n}_1], \text{ then } \hat{n}_0 = \langle r \rangle_{N_0}, \quad \hat{n}_1 = \langle r \rangle_{N_1},$$

   which implies that

   $$\langle \hat{n}_0 + 1 \rangle_{N_0} = \langle r + 1 \rangle_{N_0}, \text{ and } \langle \hat{n}_1 + 1 \rangle_{N_1} = \langle r + 1 \rangle_{N_1}.$$

   By Chinese remainder theorem (to be covered in Section 13.5.3), the mapping from r to $\hat{n}_0$ and $\hat{n}_1$ is unique, and we have

   $$r + 1 = B[\langle \hat{n}_0 + 1 \rangle_{N_0}, \langle \hat{n}_1 + 1 \rangle_{N_1}].$$

   For arbitrary $\nu \geq 2$, the ν-dimensional CRT map is formally introduced in Section 13.5.4.

13.3   Matrix Formulation of the Two-Factor PFA

13.3.1   Stage III: The Kronecker product factorization

With the twiddle factors (or phase factors) totally absent, the now decoupled length-N DFT can be computed by performing N_0 short DFT of length N_1 along the rows of matrix $\boldsymbol{A}[N_0, N_1]$ (containing $\boldsymbol{Q}_N \cdot x$), followed by N_1 short DFT of length N_0 along the columns of the intermediate matrix $\boldsymbol{A}_1[N_0, N_1]$. These operations are translated to equivalent operations on $\mathbf{vec}\,\boldsymbol{A}$ and $\mathbf{vec}\,\boldsymbol{A}_2$ as before, and we immediately obtain the desired matrix equation:

$$(13.19) \qquad \underbrace{\mathbf{vec}\,\boldsymbol{B}}_{\boldsymbol{P}_N \cdot \boldsymbol{X}} = (\boldsymbol{I}_{N_1} \otimes \boldsymbol{\Omega}_{N_0}) \cdot \overbrace{(\boldsymbol{\Omega}_{N_1} \otimes \boldsymbol{I}_{N_0}) \cdot \underbrace{\mathbf{vec}\,\boldsymbol{A}}_{\boldsymbol{Q}_N \mathbf{x}}}^{\mathbf{vec}\,\boldsymbol{A}_1}.$$

Note that this matrix equation expresses the de n ition of a true two-dimensional DFT on matrices $\boldsymbol{A}[N_0, N_1]$ and $\boldsymbol{B}[N_0, N_1]$ as explained at the end of Stage II.

13.3.2   Stage IV: Defining permutation matrices

To demonstrate the relationship between $\mathbf{vec}\,\boldsymbol{A}$ and the input sequence x and that between $\mathbf{vec}\,\boldsymbol{B}$ and the output sequence $\boldsymbol{X}$ via permutation matrices, we resort to our example for $N = N_0 \times N_1 = 3 \times 4$. Based on the Ruritanian map explicitly given for this example by Equation (13.5), we show that $\mathbf{vec}\,\boldsymbol{A}$ is the product of a sparse 12-by-12 permutation matrix $\boldsymbol{Q}_{12}$ and vector x.

$$(13.20) \qquad
\begin{matrix}
a_{0,0} \\ a_{1,0} \\ a_{2,0} \\ a_{0,1} \\ a_{1,1} \\ a_{2,1} \\ a_{0,2} \\ a_{1,2} \\ a_{2,2} \\ a_{0,3} \\ a_{1,3} \\ a_{2,3}
\end{matrix}
\underbrace{\begin{bmatrix}
x_0 \\ x_4 \\ x_8 \\ x_3 \\ x_7 \\ x_{11} \\ x_6 \\ x_{10} \\ x_2 \\ x_9 \\ x_1 \\ x_5
\end{bmatrix}}_{\mathbf{vec}\,\boldsymbol{A}}
=
\underbrace{\begin{bmatrix}
1 & 0 & 0 & 0 & 0 & 0 & 0 & 0 & 0 & 0 & 0 & 0 \\
0 & 0 & 0 & 0 & 1 & 0 & 0 & 0 & 0 & 0 & 0 & 0 \\
0 & 0 & 0 & 0 & 0 & 0 & 0 & 0 & 1 & 0 & 0 & 0 \\
0 & 0 & 0 & 1 & 0 & 0 & 0 & 0 & 0 & 0 & 0 & 0 \\
0 & 0 & 0 & 0 & 0 & 0 & 0 & 1 & 0 & 0 & 0 & 0 \\
0 & 0 & 0 & 0 & 0 & 0 & 0 & 0 & 0 & 0 & 0 & 1 \\
0 & 0 & 0 & 0 & 0 & 0 & 1 & 0 & 0 & 0 & 0 & 0 \\
0 & 0 & 0 & 0 & 0 & 0 & 0 & 0 & 0 & 0 & 1 & 0 \\
0 & 0 & 1 & 0 & 0 & 0 & 0 & 0 & 0 & 0 & 0 & 0 \\
0 & 0 & 0 & 0 & 0 & 0 & 0 & 0 & 0 & 1 & 0 & 0 \\
0 & 1 & 0 & 0 & 0 & 0 & 0 & 0 & 0 & 0 & 0 & 0 \\
0 & 0 & 0 & 0 & 0 & 1 & 0 & 0 & 0 & 0 & 0 & 0
\end{bmatrix}}_{\text{Permutation Matrix } \boldsymbol{Q}_{12}}
\underbrace{\begin{bmatrix}
x_0 \\ x_1 \\ x_2 \\ x_3 \\ x_4 \\ x_5 \\ x_6 \\ x_7 \\ x_8 \\ x_9 \\ x_{10} \\ x_{11}
\end{bmatrix}}_{\text{vector } \mathbf{x}}$$

Based on the CRT map explicitly given for this example by Equation (13.8), we show that **vec** B is the product of a sparse 12-by-12 permutation matrix P_{12} and vector X.

(13.21)

$$
\underbrace{\begin{matrix} b_{0,0} \\ b_{1,0} \\ b_{2,0} \\ b_{0,1} \\ b_{1,1} \\ b_{2,1} \\ b_{0,2} \\ b_{1,2} \\ b_{2,2} \\ b_{0,3} \\ b_{1,3} \\ b_{2,3} \end{matrix} \begin{bmatrix} X_0 \\ X_4 \\ X_8 \\ X_9 \\ X_1 \\ X_5 \\ X_6 \\ X_{10} \\ X_2 \\ X_3 \\ X_7 \\ X_{11} \end{bmatrix}}_{\text{vec }B}
=
\underbrace{\begin{bmatrix}
1 & 0 & 0 & 0 & 0 & 0 & 0 & 0 & 0 & 0 & 0 & 0 \\
0 & 0 & 0 & 0 & 1 & 0 & 0 & 0 & 0 & 0 & 0 & 0 \\
0 & 0 & 0 & 0 & 0 & 0 & 0 & 0 & 1 & 0 & 0 & 0 \\
0 & 0 & 0 & 0 & 0 & 0 & 0 & 0 & 0 & 1 & 0 & 0 \\
0 & 1 & 0 & 0 & 0 & 0 & 0 & 0 & 0 & 0 & 0 & 0 \\
0 & 0 & 0 & 0 & 0 & 1 & 0 & 0 & 0 & 0 & 0 & 0 \\
0 & 0 & 0 & 0 & 0 & 0 & 1 & 0 & 0 & 0 & 0 & 0 \\
0 & 0 & 0 & 0 & 0 & 0 & 0 & 0 & 0 & 0 & 1 & 0 \\
0 & 0 & 1 & 0 & 0 & 0 & 0 & 0 & 0 & 0 & 0 & 0 \\
0 & 0 & 0 & 1 & 0 & 0 & 0 & 0 & 0 & 0 & 0 & 0 \\
0 & 0 & 0 & 0 & 0 & 0 & 0 & 1 & 0 & 0 & 0 & 0 \\
0 & 0 & 0 & 0 & 0 & 0 & 0 & 0 & 0 & 0 & 0 & 1
\end{bmatrix}}_{\text{Permutation Matrix }P_{12}}
\underbrace{\begin{bmatrix} X_0 \\ X_1 \\ X_2 \\ X_3 \\ X_4 \\ X_5 \\ X_6 \\ X_7 \\ X_8 \\ X_9 \\ X_{10} \\ X_{11} \end{bmatrix}}_{\text{vector }X}
$$

For arbitrarily given $N = N_0 \times N_1$ subject to the condition $\mathbf{gcd}(N_0, N_1) = 1$, the N-by-N permutation matrix Q_N (associated with **vec** A) is constructed by the pseudo-code which implements the Ruritanian index map $\ell = \langle N_1 n_0 + N_0 n_1 \rangle_N$ so that, through the multiplication of Q_N, x_ℓ can be permuted to the position of a_{n_0,n_1} in **vec** A.

```
for i := 0 to N − 1 do                    initialize Q_N to be an
    for k := 0 to N − 1 do                N-by-N zero matrix
        Q_N[k, i] := 0                    column by column
    end for
end for
k := 0
for n_1 := 0 to N_1 − 1 do
    for n_0 := 0 to N_0 − 1 do
        ℓ := ⟨N_1 n_0 + N_0 n_1⟩_N        construct Q_N to
        Q_N[k, ℓ] := 1                    permute x_ℓ to a_{n_0,n_1}
        k := k + 1                        location for next a_{n_0,n_1} in vec A
    end for
end for
```

Accordingly, a short and precise mathematical definition for $Q_N[k, \ell]$, $0 \le k, \ell \le N - 1$, is given by

(13.22)
$$
Q_N[k, \ell] = \begin{cases} 1 & \text{if } k = n_1 N_0 + n_0 \text{ and } \ell = \langle N_1 n_0 + N_0 n_1 \rangle_N; \\ 0 & \text{otherwise.} \end{cases}
$$

Similarly, the permutation matrix P_N (associated with **vec** B) is constructed by the pseudo-code which implements the CRT index map $r = \langle \rho N_1 \hat{n}_0 + q N_0 \hat{n}_1 \rangle_N$ so that, through the multiplication of P_N, X_r can be permuted to the position of $b_{\hat{n}_0,\hat{n}_1}$ in **vec** B.

$$
\begin{aligned}
&\textbf{for } i := 0 \textbf{ to } N-1 \textbf{ do} && \text{initialize } \boldsymbol{P}_N \text{ to be an}\\
&\quad\textbf{for } k := 0 \textbf{ to } N-1 \textbf{ do} && N\text{-by-}N \text{ zero matrix}\\
&\qquad P_N[k,\,i] := 0 && \text{column by column}\\
&\quad\textbf{end for}\\
&\textbf{end for}\\
&k := 0\\
&\textbf{for } \hat{n}_1 := 0 \textbf{ to } N_1 - 1 \textbf{ do}\\
&\quad\textbf{for } \hat{n}_0 := 0 \textbf{ to } N_0 - 1 \textbf{ do}\\
&\qquad r := \langle \rho N_1 \hat{n}_0 + q N_0 \hat{n}_1 \rangle_N && \text{construct } \boldsymbol{P}_N \text{ to}\\
&\qquad P_N[k,\,r] := 1 && \text{permute } X_r \text{ to } b_{\hat{n}_0,\hat{n}_1}\\
&\qquad k := k + 1 && \text{location for next } b_{\hat{n}_0,\hat{n}_1} \text{ in } \textbf{vec } \boldsymbol{B}\\
&\quad\textbf{end for}\\
&\textbf{end for}
\end{aligned}
$$

The definition for $P_N[k,\,r]$, $0 \le k, \ell \le N-1$, can thus be compactly expressed as

(13.23)
$$
P_N[k,\,r] = \begin{cases} 1 & \text{if } k = \hat{n}_1 N_0 + \hat{n}_0 \text{ and } r = \langle \rho N_1 \hat{n}_0 + q N_0 \hat{n}_1 \rangle_N;\\ 0 & \text{otherwise.} \end{cases}
$$

13.3.3 Stage V: Completing the matrix factorization

Since $\boldsymbol{P}_N$ is a permutation matrix, its inverse is simply its transpose, i.e., $\boldsymbol{P}_N^{-1} = \boldsymbol{P}_N^T$, and we may rewrite Equation (13.3) as

(13.24)
$$
\boldsymbol{X} = \underbrace{\boldsymbol{P}_N^T \cdot (\boldsymbol{I}_{N_1} \otimes \boldsymbol{\Omega}_{N_0}) \cdot (\boldsymbol{\Omega}_{N_1} \otimes \boldsymbol{I}_{N_0}) \cdot \boldsymbol{Q}_N}_{\text{Kronecker product factorization of } \boldsymbol{\Omega}_N} \cdot \boldsymbol{x}.
$$

The Kronecker product factorization of the DFT matrix $\boldsymbol{\Omega}_N$ can thus be expressed as

(13.25)
$$
\boldsymbol{\Omega}_N = \boldsymbol{P}_N^T \cdot (\boldsymbol{I}_{N_1} \otimes \boldsymbol{\Omega}_{N_0}) \cdot (\boldsymbol{\Omega}_{N_1} \otimes \boldsymbol{I}_{N_0}) \cdot \boldsymbol{Q}_N = \boldsymbol{P}_N^T \cdot (\boldsymbol{\Omega}_{N_1} \otimes \boldsymbol{\Omega}_{N_0}) \cdot \boldsymbol{Q}_N.
$$

Note that if we multiply both sides of Equation (13.25) by the inverses of the permutation matrices, the same result can be expressed as the Kronecker product factorization of a permuted DFT matrix:

(13.26)
$$
\boldsymbol{P}_N \cdot \boldsymbol{\Omega}_N \cdot \boldsymbol{Q}_N^T = \boldsymbol{\Omega}_{N_1} \otimes \boldsymbol{\Omega}_{N_0}.
$$

This result is referred to as the number-theoretic splitting of the DFT matrix in the FFT literature, because the number theoretic properties of the indices (or addresses) of the data are exploited by the index maps, which are expressed through the permutation matrices $\boldsymbol{Q}_N$ and $\boldsymbol{P}_N$.

13.4 Matrix Formulation of the Multi-Factor PFA

In parallel to our derivation of the five-factor mixed-radix FFT in Section 12.2.2, we now derive the PFA for $N = F_0 \times F_1 \times F_2 \times F_3 \times F_4$, where F_ks are pairwise prime. We define $M_4 = N/F_4$, $M_3 = M_4/F_3 = N/(F_3 F_4)$, $M_2 = M_3/F_2 = N/(F_2 F_3 F_4)$, $M_1 = M_2/F_1 = N/(F_1 F_2 F_3 F_4)$, and $M_0 = M_1/F_0 = 1$. We begin with $N = M_4 \times F_4$. Because M_4 and F_4 are relatively prime, we factor $\boldsymbol{\Omega}_N$ according to (13.25), i.e.,

$$(13.27) \qquad \boldsymbol{\Omega}_N = \boldsymbol{P}_N^T \cdot (\boldsymbol{\Omega}_{F_4} \otimes \boldsymbol{\Omega}_{M_4}) \cdot \boldsymbol{Q}_N,$$

where $\boldsymbol{Q}_N$ is obtained from (13.22) with the two factors N_0 and N_1 replaced by M_4 and F_4, and we use ρ_4 and q_4, instead of ρ and q, to label the constants associated with the CRT map based on $N = M_4 \times F_4$:

$$(13.28) \qquad Q_N[k, \ell] = \begin{cases} 1 & \text{if } k = n_1 M_4 + n_0 \text{ and } \ell = \left\langle F_4 n_0 + M_4 n_1 \right\rangle_N; \\ 0 & \text{otherwise.} \end{cases}$$

$$(13.29) \qquad P_N[k, r] = \begin{cases} 1 & \text{if } k = \hat{n}_1 M_4 + \hat{n}_0 \text{ and } r = \left\langle \rho_4 F_4 \hat{n}_0 + q_4 M_4 \hat{n}_1 \right\rangle_N; \\ 0 & \text{otherwise.} \end{cases}$$

Since $M_4 = M_3 \times F_3$, where M_3 and F_3 are relatively prime, the M_4-by-M_4 DFT matrix $\boldsymbol{\Omega}_{M_4}$ can be similarly factored:

$$(13.30) \qquad \boldsymbol{\Omega}_{M_4} = \boldsymbol{P}_{M_4}^T \cdot (\boldsymbol{\Omega}_{F_3} \otimes \boldsymbol{\Omega}_{M_3}) \cdot \boldsymbol{Q}_{M_4}.$$

Using the right-hand side of (13.30) and the rules of matrix algebra for Kronecker products, we may now *expand* the term $\boldsymbol{\Omega}_{F_4} \otimes \boldsymbol{\Omega}_{M_4}$ in (13.27):

$$\begin{aligned} \boldsymbol{\Omega}_{F_4} \otimes \boldsymbol{\Omega}_{M_4} &= \boldsymbol{\Omega}_{F_4} \otimes \left[\boldsymbol{P}_{M_4}^T \cdot (\boldsymbol{\Omega}_{F_3} \otimes \boldsymbol{\Omega}_{M_3}) \cdot \boldsymbol{Q}_{M_4} \right] \\ &= (\boldsymbol{I}_{F_4} \cdot \boldsymbol{\Omega}_{F_4}) \otimes \left[\boldsymbol{P}_{M_4}^T \cdot (\boldsymbol{\Omega}_{F_3} \otimes \boldsymbol{\Omega}_{M_3}) \cdot \boldsymbol{Q}_{M_4} \right] \\ (13.31) \qquad &= \left(\boldsymbol{I}_{F_4} \otimes \boldsymbol{P}_{M_4}^T \right) \cdot \left\{ \boldsymbol{\Omega}_{F_4} \otimes \left[(\boldsymbol{\Omega}_{F_3} \otimes \boldsymbol{\Omega}_{M_3}) \cdot \boldsymbol{Q}_{M_4} \right] \right\} \\ &= \left(\boldsymbol{I}_{F_4} \otimes \boldsymbol{P}_{M_4}^T \right) \cdot \left\{ (\boldsymbol{\Omega}_{F_4} \cdot \boldsymbol{I}_{F_4}) \otimes \left[(\boldsymbol{\Omega}_{F_3} \otimes \boldsymbol{\Omega}_{M_3}) \cdot \boldsymbol{Q}_{M_4} \right] \right\} \\ &= \left(\boldsymbol{I}_{F_4} \otimes \boldsymbol{P}_{M_4}^T \right) \cdot (\boldsymbol{\Omega}_{F_4} \otimes \boldsymbol{\Omega}_{F_3} \otimes \boldsymbol{\Omega}_{M_3}) \cdot (\boldsymbol{I}_{F_4} \otimes \boldsymbol{Q}_{M_4}). \end{aligned}$$

When the expanded result is incorporated into (13.27), we obtain

$$(13.32) \qquad \boldsymbol{\Omega}_N = \boldsymbol{P}_N^T \cdot \left(\boldsymbol{I}_{F_4} \otimes \boldsymbol{P}_{M_4}^T \right) \cdot \underbrace{(\boldsymbol{\Omega}_{F_4} \otimes \boldsymbol{\Omega}_{F_3} \otimes \boldsymbol{\Omega}_{M_3})}_{\text{to be expanded further}} \cdot (\boldsymbol{I}_{F_4} \otimes \boldsymbol{Q}_{M_4}) \cdot \boldsymbol{Q}_N.$$

Since $M_3 = M_2 \times F_2$, where M_2 and F_2 are relatively prime, the DFT matrix $\boldsymbol{\Omega}_{M_3}$ can be factored in exactly the same manner as $\boldsymbol{\Omega}_{M_4}$, and we proceed to expand the term $\boldsymbol{\Omega}_{F_4} \otimes \boldsymbol{\Omega}_{F_3} \otimes \boldsymbol{\Omega}_{M_3}$. Observe that $\boldsymbol{\Omega}_{F_4} \otimes \boldsymbol{\Omega}_{F_3} = \boldsymbol{G}_{F_4 \times F_3}$ is a matrix of dimension $F_4 \times F_3 = N/M_3$; hence, the expansion result of the Kronecker product $\boldsymbol{G}_{N/M_3} \otimes \boldsymbol{\Omega}_{M_3}$ is readily available from (13.31) if we substitute $\boldsymbol{\Omega}_{F_4}$ and $\boldsymbol{\Omega}_{M_4}$ by $\boldsymbol{G}_{N/M_3}$, $\boldsymbol{\Omega}_{M_3}$ and define the factors and permutation matrices accordingly:

$$\begin{aligned} \boldsymbol{\Omega}_{F_4} &\otimes \boldsymbol{\Omega}_{F_3} \otimes \boldsymbol{\Omega}_{M_3} \\ &= \left(\boldsymbol{\Omega}_{F_4} \otimes \boldsymbol{\Omega}_{F_3} \right) \otimes \boldsymbol{\Omega}_{M_3} \\ &= \boldsymbol{G}_{N/M_3} \otimes \boldsymbol{\Omega}_{M_3} \\ (13.33) \qquad &= \left(\boldsymbol{I}_{N/M_3} \otimes \boldsymbol{P}_{M_3}^T \right) \cdot (\boldsymbol{G}_{N/M_3} \otimes \boldsymbol{\Omega}_{F_2} \otimes \boldsymbol{\Omega}_{M_2}) \cdot (\boldsymbol{I}_{N/M_3} \otimes \boldsymbol{Q}_{M_3}) \qquad \text{(from (13.31))} \\ &= \left(\boldsymbol{I}_{N/M_3} \otimes \boldsymbol{P}_{M_3}^T \right) \cdot \underbrace{(\boldsymbol{\Omega}_{F_4} \otimes \boldsymbol{\Omega}_{F_3} \otimes \boldsymbol{\Omega}_{F_2} \otimes \boldsymbol{\Omega}_{M_2})}_{\text{to be expanded further}} \cdot (\boldsymbol{I}_{N/M_3} \otimes \boldsymbol{Q}_{M_3}). \end{aligned}$$

Since $M_2 = F_0 \times F_1$, where F_0 and F_1 are relatively prime, we apply (13.31) again to complete the expansion:

$$
\begin{aligned}
& \boldsymbol{\Omega}_{F_4} \otimes \boldsymbol{\Omega}_{F_3} \otimes \boldsymbol{\Omega}_{F_2} \otimes \boldsymbol{\Omega}_{M_2} \\
& = \left(\boldsymbol{\Omega}_{F_4} \otimes \boldsymbol{\Omega}_{F_3} \otimes \boldsymbol{\Omega}_{F_2} \right) \otimes \boldsymbol{\Omega}_{M_2} \\
(13.34) \quad & = \left(\boldsymbol{I}_{N/M_2} \otimes \boldsymbol{P}_{M_2}^T \right) \cdot \left[\left(\boldsymbol{\Omega}_{F_4} \otimes \boldsymbol{\Omega}_{F_3} \otimes \boldsymbol{\Omega}_{F_2} \right) \otimes \boldsymbol{\Omega}_{F_1} \otimes \boldsymbol{\Omega}_{F_0} \right] \cdot \left(\boldsymbol{I}_{N/M_2} \otimes \boldsymbol{Q}_{M_2} \right) \\
& = \left(\boldsymbol{I}_{N/M_2} \otimes \boldsymbol{P}_{M_2}^T \right) \cdot \left(\boldsymbol{\Omega}_{F_4} \otimes \boldsymbol{\Omega}_{F_3} \otimes \boldsymbol{\Omega}_{F_2} \otimes \boldsymbol{\Omega}_{F_1} \otimes \boldsymbol{\Omega}_{F_0} \right) \cdot \left(\boldsymbol{I}_{N/M_2} \otimes \boldsymbol{Q}_{M_2} \right).
\end{aligned}
$$

For $N = F_0 \times F_1 \times F_2 \times F_3 \times F_4$, where F_k s are pairwise prime, the complete factorization of the DFT matrix can thus be expressed as

$$
(13.35) \qquad \boldsymbol{\Omega}_N = \boldsymbol{U}^T \cdot \left(\boldsymbol{\Omega}_{F_4} \otimes \boldsymbol{\Omega}_{F_3} \otimes \boldsymbol{\Omega}_{F_2} \otimes \boldsymbol{\Omega}_{F_1} \otimes \boldsymbol{\Omega}_{F_0} \right) \cdot \boldsymbol{V},
$$

where $\boldsymbol{U}^T$ and $\boldsymbol{V}$ are N-by-N permutation matrices defined by the products of sparse permutation matrices used during each step of the expansion process:

$$
\begin{aligned}
(13.36) \quad \boldsymbol{U}^T & = \left[\left(\boldsymbol{I}_{N/M_2} \otimes \boldsymbol{P}_{M_2} \right) \cdot \left(\boldsymbol{I}_{N/M_3} \otimes \boldsymbol{P}_{M_3} \right) \cdot \left(\boldsymbol{I}_{N/M_4} \otimes \boldsymbol{P}_{M_4} \right) \cdot \boldsymbol{P}_N \right]^T \\
& = \boldsymbol{P}_N^T \cdot \left(\boldsymbol{I}_{N/M_4} \otimes \boldsymbol{P}_{M_4}^T \right) \cdot \left(\boldsymbol{I}_{N/M_3} \otimes \boldsymbol{P}_{M_3}^T \right) \cdot \left(\boldsymbol{I}_{N/M_2} \otimes \boldsymbol{P}_{M_2}^T \right);
\end{aligned}
$$

$$
(13.37) \qquad \boldsymbol{V} = \left(\boldsymbol{I}_{N/M_2} \otimes \boldsymbol{Q}_{M_2} \right) \cdot \left(\boldsymbol{I}_{N/M_3} \otimes \boldsymbol{Q}_{M_3} \right) \cdot \left(\boldsymbol{I}_{N/M_4} \otimes \boldsymbol{Q}_{M_4} \right) \cdot \boldsymbol{Q}_N.
$$

The matrix equation representing the five-factor PFA can be obtained directly from (13.35), with permutation matrices $\boldsymbol{U}$ and $\boldsymbol{V}$ defined by (13.36) and (13.37):

$$
(13.38) \qquad \boldsymbol{X} = \overbrace{\boldsymbol{\Omega}_N}^{\text{DFT}} \cdot \boldsymbol{x} = \boldsymbol{U}^T \cdot \left(\boldsymbol{\Omega}_{F_4} \otimes \boldsymbol{\Omega}_{F_3} \otimes \boldsymbol{\Omega}_{F_2} \otimes \boldsymbol{\Omega}_{F_1} \otimes \boldsymbol{\Omega}_{F_0} \right) \cdot \boldsymbol{V} \cdot \boldsymbol{x}.
$$

13.4.1   Organizing the PFA computation—Part 2

The PFA based on (13.38) can be written as

$$
(13.39) \qquad \hat{\boldsymbol{X}} = \left(\boldsymbol{\Omega}_{F_4} \otimes \boldsymbol{\Omega}_{F_3} \otimes \boldsymbol{\Omega}_{F_2} \otimes \boldsymbol{\Omega}_{F_1} \otimes \boldsymbol{\Omega}_{F_0} \right) \hat{\boldsymbol{x}},
$$

where $\hat{\boldsymbol{X}} = \boldsymbol{U}\boldsymbol{X}$ and $\hat{\boldsymbol{x}} = \boldsymbol{V}\boldsymbol{x}$ denote the permuted output and input vectors. In this section we shall discuss how to organize the computation of $\hat{\boldsymbol{X}}$ according to (13.39), assuming $\hat{\boldsymbol{x}}$ is already available. (The generation of scrambled $\hat{\boldsymbol{x}}$ from naturally ordered input $\boldsymbol{x}$, as well as the recovery of naturally ordered $\boldsymbol{X}$ from scrambled output $\hat{\boldsymbol{X}}$, will be examined *after* we study the mathematical theory behind the specially designed index mapping schemes in the next section.)

Observe that the computation of

$$
(13.40) \qquad \hat{\boldsymbol{X}} = (\boldsymbol{A} \otimes \boldsymbol{B}) \hat{\boldsymbol{x}} = \underbrace{(\boldsymbol{I} \otimes \boldsymbol{B})}_{\boldsymbol{R}_2} \cdot \underbrace{(\boldsymbol{A} \otimes \boldsymbol{I})}_{\boldsymbol{R}_1} \hat{\boldsymbol{x}}
$$

can be easily handled by extracting matrix factors $\boldsymbol{R}_1$ and $\boldsymbol{R}_2$ from the Kronecker product $\boldsymbol{A} \otimes \boldsymbol{B}$:

Step 1. Compute $\hat{x}_2 = R_1\,\hat{x}$, where $R_1 = A \otimes I$;

Step 2. Compute $\hat{X} = R_2\,\hat{x}_2$, where $R_2 = I \otimes B$.

Accordingly, we can organize the computation of (13.39) by splitting the Kronecker product

$$(13.41) \qquad \hat{X} = \left(\Omega_{F_4} \otimes \Omega_{F_3} \otimes \Omega_{F_2} \otimes \Omega_{F_1} \otimes \Omega_{F_0}\right)\hat{x} = (A \otimes B)\,\hat{x}$$

in various ways. For example, we may proceed as follows: let $A_{F_4} = \Omega_{F_4}$, let $B_{M_4} = \Omega_{F_3} \otimes \Omega_{F_2} \otimes \Omega_{F_1} \otimes \Omega_{F_0}$, and we express

$$(13.42) \qquad \Omega_{F_4} \otimes B_{M_4} = \underbrace{\left(I_{F_4} \otimes B_{M_4}\right)}_{I \otimes B} \cdot \underbrace{\left(\Omega_{F_4} \otimes I_{M_4}\right)}_{A \otimes I}$$

to extract sparse-matrix factor $R_1 = \Omega_{F_4} \otimes I_{M_4}$.

To extract the second matrix factor R_2, this process can be repeated by splitting

$$(13.43) \qquad I_{F_4} \otimes B_{M_4} = \underbrace{I_{F_4} \otimes \Omega_{F_3}}_{A} \otimes \underbrace{\Omega_{F_2} \otimes \Omega_{F_1} \otimes \Omega_{F_0}}_{B} = A_{N/M_3} \otimes B_{M_3}.$$

Again, because $A \otimes B = (I \otimes B)\cdot(A \otimes I)$, we obtain

$$(13.44) \qquad R_2 = A_{N/M_3} \otimes I_{M_3} = I_{F_4} \otimes \Omega_{F_3} \otimes I_{M_3}.$$

It follows that R_3 can be extracted from splitting

$$(13.45) \qquad I_{N/M_3} \otimes B_{M_3} = \underbrace{I_{N/M_3} \otimes \Omega_{F_2}}_{A} \otimes \underbrace{\Omega_{F_1} \otimes \Omega_{F_0}}_{B}$$

in the same manner, and we obtain

$$(13.46) \qquad R_3 = A_{N/M_2} \otimes I_{M_2} = I_{N/M_3} \otimes \Omega_{F_2} \otimes I_{M_2}.$$

The remaining two factors R_4 and R_5 can be extracted from splitting $I_{N/M_2} \otimes B_{M_2} = I_{N/M_2} \otimes \Omega_{F_1} \otimes \Omega_{F_0} = A_{N/F_0} \otimes \Omega_{F_0}$, resulting in

$$(13.47) \qquad R_4 = A_{N/F_0} \otimes I_{F_0} = I_{N/M_2} \otimes \Omega_{F_1} \otimes I_{F_0}, \quad R_5 = I_{N/F_0} \otimes \Omega_{F_0}.$$

With all ve sparse matrix factors available, we can now express

$$(13.48) \qquad \hat{X} = \left(\Omega_{F_4} \otimes \Omega_{F_3} \otimes \Omega_{F_2} \otimes \Omega_{F_1} \otimes \Omega_{F_0}\right)\hat{x} = \left(R_5\cdot R_4\cdot R_3\cdot R_2\cdot R_1\right)\hat{x},$$

and the computation of $\hat{X}$ can be organized as

$$(13.49) \qquad \hat{X} = \left\{ R_5\cdot\left(R_4\cdot\left(R_3\cdot\left(R_2\cdot\left(R_1\cdot\hat{x}\right)\right)\right)\right)\right\}.$$

13.5 Number Theory and Index Mapping by Permutations

In the last two sections we made use of specially designed index mappings to derive the prime-factor algorithms in matrix form. In this section we study the theory behind these index mappings, and we will see how all ν-factor PFAs ($\nu \geq 2$) can be obtained directly from ν-dimensional index mappings with proven mathematical properties.

13.5.1   Some fundamental properties of integers

The division of a by b: Let a and b be two integers with b positive. We can nd integers q and r to satisfy the equation

$$(13.50) \qquad a = b \times q + r, \quad 0 \le r < b,$$

where b is called the *modulus*, q is called the *quotient*, and r is called the *remainder*. When $r = 0$, b and q are *factors* or *divisors* of a, and b is said to divide a, commonly denoted by $b \,|\, a$. When a has no other divisors than 1 and a, a is prime. In all other cases, a is composite.

The greatest common divisor: The largest number which is a divisor of both a and c is called the greatest common divisor of a and c, and we denote it by **gcd**(a, c).

Relative primality: If **gcd**$(a, c) = 1$, then a and c have no other common divisor than 1; they are called relatively (or mutually) prime.

Congruent modulo b: If $a = b \times q + r$ and $c = b \times s + r$, then a and c are said to be congruent modulo b, denoted by

$$(13.51) \qquad c \equiv a \bmod b \qquad \text{or} \qquad a \equiv c \bmod b.$$

For example, using $b = 7$ as the modulus, we have $19 \equiv 40 \bmod 7$ because $40 = 7 \times 5 + 5$ and $19 = 7 \times 2 + 5$. Note that for each xed modulus b, $a = b \times q + r$ is treated as an equivalence relation, and we could obtain the same remainder r for in nitely many choices of a a ll these choices of a are congruent with respect to modulus b.

For every $c \equiv a \bmod b$, we immediately have $a - c = b \times (q - s) + 0$. Hence, the positive b divides $(a - c)$, and we have

$$(13.52) \qquad b \,|\, (a - c) \text{ iff } c \equiv a \bmod b.$$

That is, two integers with the same residue modulo b must differ by a multiple of b. In the example above we have $19 \equiv 40 \bmod 7$, which implies $7 \,|\, (40 - 19)$, and vice versa.

Observe that if $a = b \times q + r$, then $b \,|\, (a - r)$, and we have $r \equiv a \bmod b$ as expected. For the example $19 \equiv 40 \bmod 7$, we thus have $5 \equiv 40 \bmod 7$ and $5 \equiv 19 \bmod 7$.

Residue modulo b: If $a = b \times q + r$, the arithmetic operation to produce only the residue of a modulo b is denoted by

$$(13.53) \qquad \langle a \rangle_b = r.$$

For example, $\langle 40 \rangle_7 = 5$, $\langle 19 \rangle_7 = 5$, and $\langle 5 \rangle_7 = 5$. The equality of the residues can be directly expressed as $\langle 40 \rangle_7 = \langle 19 \rangle_7 = \langle 5 \rangle_7$

Rules of residue (or modular) arithmetic: Since the residue of a modulo b is restricted to the range $0 \le r \le b - 1$ determined by the modulus, the following rules may be used to simplify $\langle a \rangle_b$ when a is given in a computationally dif cult form.

$$(13.54) \qquad \langle a_1 \pm a_2 \rangle_b = \langle \langle a_1 \rangle_b \pm \langle a_2 \rangle_b \rangle_b$$

(13.55)
$$\langle a_1 \times a_2 \rangle_b = \langle \langle a_1 \rangle_b \times \langle a_2 \rangle_b \rangle_b$$

It was pointed out by McClellan and Rader [33] that in actually performing computations, one can always replace a congruence $c \equiv a \bmod b$, which describes a relation among whole classes of numbers with the same residue, by the equality of the residues $\langle c \rangle_b = \langle a \rangle_b$. Now the rules of computation may be applied to both sides as needed.

Euclid's algorithm: This algorithm uses division method to find $\gcd(A, C)$, where A and C are two positive integers. We shall describe the algorithm, demonstrate how it works, and prove that the result produced by Euclid's algorithm is indeed the greatest common divisor. To determine $\gcd(A, C)$, Euclid's algorithm computes the following sequence of remainders iteratively:

$$\textbf{Compute } R_0 = \langle A \rangle_C$$
$$R_1 = \langle C \rangle_{R_0}$$
$$R_2 = \langle R_0 \rangle_{R_1}$$

(13.56)
$$\vdots$$

$$\textbf{until } R_\mu = \langle R_{\mu-2} \rangle_{R_{\mu-1}} = 0$$
$$\textbf{If } \quad R_0 = 0 \textbf{ then } \gcd(A, C) = C$$
$$\textbf{else } \gcd(A, C) = R_{\mu-1}$$

As shown in the following examples, Euclid's algorithm is easy to apply.

Example 13.1 Let $A = 165$ and $C = 99$; we simply compute the remainders as required by the algorithm:

(13.57)
$$R_0 = \langle 165 \rangle_{99} = 66$$
$$R_1 = \langle 99 \rangle_{66} = 33$$
$$R_2 = \langle 66 \rangle_{33} = 0$$
$$\gcd(165, 99) = R_1 = 33.$$

To find out whether A and C are relatively prime, we may use Euclid's algorithm to determine whether $\gcd(A, C) = 1$.

Example 13.2 Let $A = 195$, $B = 124$, Euclid's algorithm finds $\gcd(195, 124) = 1$ as expected.

(13.58)
$$R_0 = \langle 195 \rangle_{124} = 71$$
$$R_1 = \langle 124 \rangle_{71} = 53$$
$$R_2 = \langle 71 \rangle_{53} = 18$$
$$R_3 = \langle 53 \rangle_{18} = 17$$
$$R_4 = \langle 18 \rangle_{17} = 1$$
$$R_5 = \langle 17 \rangle_{1} = 0$$
$$\gcd(195, 124) = R_4 = 1.$$

Note that Euclid s algorithm assumes that A and C are positive integers without loss of generality, because $\mathbf{gcd}(\pm A, \pm C) = \mathbf{gcd}(|\pm A|, |\pm C|) = \mathbf{gcd}(A, C)$, one only needs to nd the greatest common divisor of two positive integers.

We show next that $\mathbf{gcd}(A, C)$ is indeed computed by Euclid s algorithm. If $R_0 = 0$, then the algorithm terminates, and C is the greatest common divisor as expected. Otherwise, we note the following:

1. The remainders generated by Euclid s algorithm are decreasing in value because
   $0 < R_0 < C, 0 < R_1 < R_0, \cdots, 0 < R_k < R_{k-1}$ implies $0 < R_k < R_{k-1} < \cdots < R_1 < R_0$; hence, $R_\mu = 0$ is expected after nite number of steps.

2. The residues R_k s computed by Euclid s algorithm satisfy the following equations:

$$A = C \times q_0 + R_0$$
$$C = R_0 \times q_1 + R_1$$
$$R_0 = R_1 \times q_2 + R_2$$
(13.59)
$$R_1 = R_2 \times q_3 + R_3$$
$$\vdots$$
$$R_{\mu-3} = R_{\mu-2} \times q_{\mu-1} + R_{\mu-1}$$
$$R_{\mu-2} = R_{\mu-1} \times q_\mu \ + \ 0. \qquad\qquad (\because R_\mu = 0)$$

Since $R_\mu = 0$ in the last equation, we establish that $R_{\mu-1}$ divides $R_{\mu-2}$; i.e., $R_{\mu-1}$ is a factor of $R_{\mu-2}$. Observe that $R_{\mu-3}$ is a linear combination of $R_{\mu-2}$ and $R_{\mu-1}$, where $R_{\mu-2}$ contains $R_{\mu-1}$ as a factor; we thus establish that $R_{\mu-1}$ is also a factor of $R_{\mu-3}$. Since this argument applies to each preceding equation in the system, we conclude that $R_{\mu-1}$ is a factor of C (from the second equation) and A (from the r st equation.)

3. Now that $R_{\mu-1}$ is a common divisor of A and C, we show next that *no* other common divisor of A and C is greater than $R_{\mu-1}$. Letting positive integer D denote an arbitrary common divisor of A and C, we substitute $A = D \times M$ and $C = D \times F$ into the division equation $A = C \times q_0 + R_0$, and we rewrite the system (13.59) as

$$R_0 = A - C \times q_0 = D \times M - D \times F \times q_0$$
$$R_1 = C - R_0 \times q_1 = D \times F - R_0 \times q_1$$
$$R_2 = R_0 - R_1 \times q_2$$
(13.60)
$$R_3 = R_1 - R_2 \times q_3$$
$$\vdots$$
$$R_{\mu-1} = R_{\mu-3} - R_{\mu-2} \times q_{\mu-1}$$
$$0 = R_{\mu-2} - R_{\mu-1} \times q_\mu. \qquad\qquad (\because R_\mu = 0)$$

From the rst equation in the system (13.60) we see that D divides R_0; hence, $D \le R_0$. From the next equation we establish that D divides R_1 because D is a factor in the right-hand side; hence, $D \le R_1$. By continuing this argument with each subsequent equation, we establish that $D \mid R_{\mu-1}$ and $D \le R_{\mu-1}$. This proves that $R_{\mu-1}$ is the greatest common divisor of A and C.

As pointed out by Nussbaumer in [35], an important consequence of Euclid s algorithm is that $\gcd(A, C)$ can be expressed as a linear combination of A and C. That is, there exist integers ρ and q such that

$$(13.61) \qquad \rho A + qC = \gcd(A, C).$$

This result can be easily established from system (13.60): The rst equation expresses R_0 as a linear combination of A and C; the second equation shows that R_1 is a linear combination of A and C using R_0 from the preceding equation. Given R_0 and R_1, the fact that R_{k+2} is a linear combination of R_{k+1} and R_k dictates that $R_{\mu-1} = \gcd(A, C)$ is a linear combination of A and C.

Bezout's relation When A and C are relatively prime, the relation given by (13.61) is known as Bezout s relation: There exist integers ρ and q such that

$$(13.62) \qquad \rho A + qC = 1 \quad \text{if } \gcd(A, C) = 1.$$

Diophantine equation The linear equation with integer coef cients A, C, and K given by

$$(13.63) \qquad Ax + Cy - K = 0 \quad \text{or} \quad Ax + Cy = K$$

is called the Diophantine equation. From (13.61) we know that Diophantine equation has integer solutions ρ and q if $K = D = \gcd(A, C)$. Observe that with the right-hand side $K = D$, if we express $A = D \times M$ and $C = D \times F$, solving (13.63) is equivalent to solving

$$(13.64) \qquad Mx + Fy = 1, \quad \text{where } \gcd(M, F) = 1.$$

We remark that there are an in nite number of integer solutions to the Diophantine equation given by (13.63). To see this, assume that integers ρ and q form a particular solution; by subtracting $A\rho + Cq = K$ from $Ax + Cy = K$ we obtain

$$(13.65) \qquad A(x - \rho) + C(y - q) = 0 \quad \text{or} \quad A(x - \rho) = C(q - y).$$

We may now factor out $D = \gcd(A, C)$ from A and C in (13.65) to obtain

$$(13.66) \qquad M(x - \rho) = F(q - y), \quad \text{where } \gcd(M, F) = 1,$$

and it follows that

$$(13.67) \qquad F \mid (x - \rho) \implies x = \rho + m \times F, \quad \text{where } m \text{ is any integer.}$$

Substituting $x = \rho + m \times F$ into (13.66), we obtain

$$(13.68) \qquad M \times m = q - y \implies y = q - m \times M.$$

Consequently, an in nite number of linearly related solutions may be generated from a particular solution according to (13.67) and (13.68), one for each choice of integer m.

Example 13.3 Suppose that a particular solution to the Diophantine equation $165x + 99y = 33$ is known to be $x = \rho = -1$ and $y = q = 2$. Using $D = \gcd(165, 99) = 33$

found by Euclid s algorithm, we determine $M = 165/D = 5$ and $F = 99/D = 3$. Hence a general solution can be described as

(13.69) $x = -1 + 3m, \quad y = 2 - 5m,$ where m is any integer.

For each nonzero m, a different solution is generated: e.g., for $m = -2$, we obtain $x = -7$ and $y = 12$; for $m = 7$, we obtain $x = 20$ and $y = -33$.

In general, a Diophantine equation of the form $Ax + Cy = K$ has integer solutions *if and only if* $D = \mathbf{gcd}(A, C)$ divides K. To see this, observe that the existence of integers x and y to satisfy $Ax + Cy = D(Mx + Fy) = K$ implies that D divides K; hence, the latter is easily established as a necessary condition. To establish the same as a suf cient condition, we observe that if D divides K, then we may express $K = D \times d$, where d is an integer, and we can rewrite $Ax + Cy = K$ as $Mx + Fy = d$, where $\mathbf{gcd}(M, F) = 1$. The solution x and y now exist because we have already established that integer solution for $\tilde{x} = x/d$ and $\tilde{y} = y/d$ exists for the equivalent problem of solving

(13.70) $M\tilde{x} + F\tilde{y} = 1, \quad \mathbf{gcd}(M, F) = 1.$

From $\tilde{x}$ and $\tilde{y}$ we recover $x = d \times \tilde{x}$ and $y = d \times \tilde{y}$.

Solving Diophantine equation by Euclid's algorithm An iterative process for nding the solutions to

(13.71) $Ax + Cy = \mathbf{gcd}(A, C)$

can be derived from, and explicitly built into, Euclid s algorithm described by system (13.60). We begin with expressing R_0 as a linear combination of A and C:

$$
\begin{aligned}
R_0 &= A - C \times q_0 \\
&= Aa_0 + Cc_0 \\
\implies \quad a_0 &= 1, \quad c_0 = -q_0.
\end{aligned}
$$
(13.72)

Using $R_0 = Aa_0 + Cc_0$ in the second equation, we obtain

$$
\begin{aligned}
R_1 &= C - R_0 \times q_1 \\
&= -Aa_0q_1 + C(1 - c_0q_1) \\
&= Aa_1 + Cc_1 \\
\implies \quad a_1 &= -a_0q_1, \quad c_1 = 1 - c_0q_1.
\end{aligned}
$$
(13.73)

Using $R_k = Aa_k + Cc_k$ for $k = 0$ and $k = 1$ in the third equation, we obtain

$$
\begin{aligned}
R_2 &= R_0 - R_1 \times q_2 \\
&= A(a_0 - a_1q_2) + C(c_0 - c_1q_2) \\
&= Aa_2 + Cc_2 \\
\implies \quad a_2 &= a_0 - a_1q_2, \quad c_2 = c_0 - c_1q_2.
\end{aligned}
$$
(13.74)

Because the rest of Euclid s algorithm computes the same equation $R_k = R_{k-2} - R_{k-1} \times q_k$ for $k = 2, 3, \ldots$, the pattern for computing a_k and c_k from a_{k-1}, c_{k-1} and q_k has been revealed as

(13.75) $$a_k = a_{k-2} - a_{k-1} \times q_k, \quad c_k = c_{k-2} - c_{k-1} \times q_k, \quad k \geq 2.$$

These two equations can now be built into Euclid s algorithm as shown below.

(13.76)
$$R_0 = A - C \times q_0$$
$$a_0 = 1, \quad c_0 = -q_0$$
$$R_1 = C - R_0 \times q_1$$
$$a_1 = -a_0 \times q_1, \quad c_1 = 1 - c_0 \times q_1$$
$$R_2 = R_0 - R_1 \times q_2$$
$$a_2 = a_0 - a_1 \times q_2, \quad c_2 = c_0 - c_1 \times q_2$$
$$R_3 = R_1 - R_2 \times q_3$$
$$\vdots$$
$$R_{\mu-1} = R_{\mu-3} - R_{\mu-2} \times q_{\mu-1}$$
$$a_{\mu-1} = a_{\mu-3} - a_{\mu-2} \times q_{\mu-1}, \quad c_{\mu-1} = c_{\mu-3} - c_{\mu-2} \times q_{\mu-1}$$
$$0 = R_{\mu-2} - R_{\mu-1} \times q_{\mu}. \quad (\because R_{\mu} = 0)$$

Hence, this revised Euclid s algorithm computes $R_{\mu-1} = \mathbf{gcd}(A, C)$ and solves, at the same time, $Ax + Cy = R_{\mu-1}$ with solutions $x = a_{\mu-1}$ and $y = c_{\mu-1}$, where $\mu \geq 1$ assuming that $R_0 \neq 0$. (If $R_0 = 0$, then $C = \mathbf{gcd}(A, C)$, the equation $Ax + Cy = C$ with $A = C \times q_0$ can be solved by inspection. This case can be easily taken care of in the computer program when implementing Euclid s algorithm.)

Example 13.4 The revised Euclid algorithm determines $\mathbf{gcd}(195, 124) = 1$ and solves $195x + 124y = 1$ as shown below.

$$R_0 = A - C \times q_0 = 195 - 124 \times 1 = 71$$
$$a_0 = 1, \quad c_0 = -q_0 = -1$$

$$R_1 = C - R_0 \times q_1 = 124 - 71 \times 1 = 53$$
$$a_1 = -a_0 \times q_1 = -1 \times 1 = -1$$
$$c_1 = 1 - c_0 \times q_1 = 1 - (-1) \times 1 = 2$$

$$R_2 = R_0 - R_1 \times q_2 = 71 - 53 \times 1 = 18$$
$$a_2 = a_0 - a_1 \times q_2 = 1 - (-1) \times 1 = 2$$
$$c_2 = c_0 - c_1 \times q_2 = -1 - 2 \times 1 = -3$$

$$R_3 = R_1 - R_2 \times q_3 = 53 - 18 \times 2 = 17$$
$$a_3 = a_1 - a_2 \times q_3 = -1 - 2 \times 2 = -5$$
$$c_3 = c_1 - c_2 \times q_3 = 2 - (-3) \times 2 = 8$$

$$R_4 = R_2 - R_3 \times q_4 = 18 - 17 \times 1 = 1$$
$$a_4 = a_2 - a_3 \times q_4 = 2 - (-5) \times 1 = 7$$
$$c_4 = c_2 - c_3 \times q_4 = -3 - 8 \times 1 = -11$$

$$R_5 = R_3 - R_4 \times q_5 = 17 - 1 \times 17 = 0$$

The solutions found are $x = a_4 = 7$ and $y = c_4 = -11$. It can be easily verified that $195 \times 7 + 124 \times (-11) = 1365 - 1364 = 1$. Using $x_0 = 7$ and $y_0 = -11$ as a particular solution, we can generate an infinite number of linearly related solutions according to Equations (13.67) and (13.68), and they are

$$x = 7 + 124m, \quad y = -11 - 195m,$$

where m is any integer.

Theorem 13.1 If $Ax + Cy = D = \mathbf{gcd}(A, C)$, then x and y are relatively prime.

Proof: Recall that solving the given equation is equivalent to solving

$$(13.77) \qquad\qquad Mx + Fy = 1, \quad \mathbf{gcd}(M, F) = 1,$$

where $M = A/D$, $F = C/D$. Suppose $\mathbf{gcd}(x, y) = d > 0$. If we substitute $x = d \times \tilde{x}$ and $y = d \times \tilde{y}$ into (13.77), we obtain

$$(13.78) \qquad\qquad d \times (M\tilde{x} + F\tilde{y}) = 1,$$

which dictates that d is a factor of 1; hence, $d = 1$, and $\mathbf{gcd}(x, y) = d = 1$.   ■

Theorem 13.2 The Diophantine equation written in the following form

(13.79) $$Mx = Fz + 1, \quad \mathbf{gcd}(M, F) = 1, \quad M > 0, \; F > 0,$$

has general solutions in the form

(13.80) $$x = x_0 + m \times F, \quad z = z_0 + m \times M,$$

where x_0 and z_0 denote a particular solution, and m is any integer. The given equation has a unique solution $x = \rho$ and $z = s$ if the range is restricted to $0 < \rho \leq F - 1$ and $0 < s \leq M - 1$.

Proof: Observe that we can rewrite the given equation in the form $Mx + Fy = 1$ if we de n e $y = -z$. Hence all of the previous results apply, and we can solve for x and y using Euclid s algorithm as before. The properties of the solutions x and z are then the properties of x and $-y$. Note that for $M > 0$ and $F > 0$, the nonzero integer solutions x and y of the equation $Mx + Fy = 1$ must have opposite signs: i.e., either $x > 0$ and $y < 0$ or $x < 0$ and $y > 0$. We can thus conclude that x and $z = -y$ must be both positive or both negative.

Following (13.67), the general solution $x = x_0 + m \times F$; following (13.68), the general solution $y = y_0 - m \times M$ leads to $z = -y = -(y_0 - m \times M) = z_0 + m \times M$.

To obtain solutions in the range $0 < \rho \leq F - 1$ and $0 < s \leq M - 1$, observe that we can always choose $m > 0$ so that $\tilde{x}_0 = x_0 + m \times F > 0$ and $\tilde{z}_0 = z_0 + m \times M > 0$, and we de ne

(13.81) $$\rho = \langle \tilde{x}_0 \rangle_F, \quad s = \langle \tilde{z}_0 \rangle_M,$$

so that $0 < \rho = \tilde{x}_0 - \alpha F \leq F - 1$ and $0 < s = \tilde{z}_0 - \beta M \leq M - 1$ form unique solutions in the speci ed ranges. ∎

Example 13.5 Recall that $x_0 = 7$ and $y_0 = -11$ satisfying $195x + 124y = 1$ were found by Euclid s algorithm; hence, positive $x_0 = 7$ and positive $z_0 = -y_0 = 11$ satisfy $195x = 124y + 1$. Note that $\rho = 7$ and $s = 11$ form unique solution in the range $0 < \rho < 124$ and $0 < s < 195$.

Since the general solution is given by $x = 7 + 124m$ and $z = 11 + 195m$, for $m = -1$ we have negative $x_1 = -117$ and negative $z_1 = -184$. Since $x_1 < 0$ and $y_1 = -z_1 > 0$ solve $195x + 124y = 1$, the equation $124y = 195t + 1$ for $y > 0$ and $t = -x > 0$ can be solved by $y = y_1 = 184$ and $t = -x_1 = 117$. Note that the unique solution of $195\rho = 124s + 1$ is different from that of $124q = 195t + 1$, although $0 < \rho, t < 124$ and $0 < q, s < 195$. The latter form and its solutions are formalized in the next theorem.

Theorem 13.3 The Diophantine equation written in the following form

(13.82) $$Fy = Mw + 1, \quad \mathbf{gcd}(M, F) = 1, \quad M > 0, \; F > 0,$$

has general solutions in the form

(13.83) $$y = y_0 + m \times M, \quad w = w_0 + m \times F,$$

where y_0 and w_0 denote a particular solution, and m is any integer. The given equation has a unique solution $y = q$ and $w = t$ if the range is restricted to $0 < q \leq M - 1$ and $0 < t \leq F - 1$.

Proof: (Similar to the proof of Theorem 13.2.)

Definition 13.4 The unique solution $0 < \rho \le F - 1$ satisfying

(13.84) $\langle \rho M \rangle_F = 1$ or $\rho M = sF + 1$,   where $\gcd(M, F) = 1$,

is de ned as the reciprocal of M modulo F:

(13.85) $\rho = \langle M \rangle_F^{-1}$.

Similarly, the unique solution $0 \le q \le M - 1$ satisfying

(13.86) $\langle qF \rangle_M = 1$ or $qF = tM + 1$,   where $\gcd(M, F) = 1$,

is de n ed as the reciprocal of F modulo M:

(13.87) $q = \langle F \rangle_M^{-1}$.

Theorem 13.5 If $\gcd(M, F) = 1$, $\rho = \langle M \rangle_F^{-1}$, and $q = \langle F \rangle_M^{-1}$, then

(13.88) $\gcd(\rho, F) = 1$, $\gcd(q, M) = 1$;
(13.89) $\rho M + qF = N + 1$,   where $N = M \times F$.

Proof: Let $\gcd(\rho, F) = D$. If we substitute $\rho = \alpha D$ and $F = \beta D$ into $\rho M = sF + 1$, we obtain

(13.90) $D \cdot (\alpha M - s\beta) = 1$,

which dictates that $D = 1$; hence, we have proved that $\gcd(\rho, F) = 1$. The same can be done to prove $\gcd(q, M) = 1$, and we see no need to repeat it.

We assume next that $r = \langle \rho M + qF \rangle_N$; hence, $\rho M + qF = \alpha N + r$, where $0 \le r < N$. Because $\rho = \langle M \rangle_F^{-1}$, and $q = \langle F \rangle_M^{-1}$, we may express

(13.91) $\rho M = sF + 1$, $qF = tM + 1$,

and it follows that

(13.92)
$$\rho M + qF = sF + 1 + qF = \alpha N + r,$$
$$\rho M + qF = \rho M + tM + 1 = \alpha N + r,$$

from which we obtain

(13.93)
$$(s + q) \cdot F = \alpha N + (r - 1) \Longrightarrow F \,|\, (r - 1)$$
$$(\rho + t) \cdot M = \alpha N + (r - 1) \Longrightarrow M \,|\, (r - 1)$$

Because M and F do not have common factor, $(M \times F) \,|\, (r - 1)$. Since $N = M \times F$, we must have $N \,|\, (r - 1)$, which is only possible when $r - 1 = 0$ because $0 \le r < N$. Hence we have proved that $r = 1$ and $\rho M + qF = \alpha N + 1$.

To show that $\alpha = 1$, note that $0 < \rho M < N$ and $0 < qF < N$ because $0 < \rho < F$ and $0 < q < M$; hence, $\rho M + qF = \alpha N + 1 < 2N$, and we must have $\alpha = 1$.   ■

13.5.2 A simple case of index mapping by permutation

The simple index map given in the next theorem provides the rst link between the DFT and number theory.

Theorem 13.6 [33] For $\ell = 0, 1, 2, \cdots, N-1$, the mapping $f(\ell) = \langle \rho\ell \rangle_N$ is *one-to-one* if ρ and N are relatively prime.

Proof: To show that f is one-to-one, we need to show $f(k) \neq f(j)$ if $k \neq j$ for $0 \leq k, j \leq N-1$. This is to be proved by contradiction: we assume that there exist $k \neq j$ such that $f(k) = f(j)$. This assumption results in

$$\langle \rho k \rangle_N = \langle \rho j \rangle_N, \quad k \neq j, \quad 0 \leq k, j \leq N-1.$$

The equality of the residues dictates that the modulus N divides $(\rho k - \rho j)$; i.e., we must have

$$N \mid \rho(k-j),$$

which contradicts the given conditions $\mathbf{gcd}(\rho, N) = 1$ and $0 < |k-j| < N$, because they forbid N to become a divisor of either term. Hence our assumption is incorrect, and this proves that all N values of $f(\ell)$ for $0 \leq \ell \leq N-1$ are distinct as desired. ∎

Example 13.6 For relatively prime integers $\rho = 3$ and $N = 4$, we may use $f(\ell) = \langle 3\ell \rangle_4$ to map the sequence $\{0, 1, 2, 3\}$ to $\{f(0), f(1), f(2), f(3)\} = \{0, 3, 2, 1\}$, which is a permutation of the original sequence. When the original sequence represents the indices of data samples contained in vector $\boldsymbol{x} = \{x_0, x_1, x_2, x_3\}$, the index mapping results in the permuted $\boldsymbol{y} = \{x_0, x_3, x_2, x_1\}$.

In general, for every ρ relatively prime to N, we can express the permutation $y_\ell = x_{f(\ell)}$ by matrix-vector product $\boldsymbol{y} = \boldsymbol{Q}_N \boldsymbol{x}$, where $\boldsymbol{Q}_N$ is the permutation matrix de n ed by

$$(13.94) \qquad Q_N[\ell, k] = \begin{cases} 1 & \text{if } k = f(\ell) = \langle \rho\ell \rangle_N; \\ 0 & \text{otherwise.} \end{cases}$$

To connect the DFT of the permuted $\boldsymbol{y}$ to the DFT of the original $\boldsymbol{x}$, we denote

$$\boldsymbol{X} = \boldsymbol{\Omega}_N \cdot \boldsymbol{x}, \quad \text{and } \boldsymbol{Y} = \boldsymbol{\Omega}_N \cdot \boldsymbol{y}, \quad \text{where } \boldsymbol{y} = \boldsymbol{Q}_N \cdot \boldsymbol{x}.$$

For $f(r) = \langle \rho r \rangle_N$, we express

$$(13.95) \qquad Y_{f(r)} = \sum_{\ell=0}^{N-1} y_\ell \, \omega_N^{f(r)\cdot\ell} = \sum_{\ell=0}^{N-1} x_{f(\ell)} \, \omega_N^{f(r)\cdot\ell}.$$

Recall that $\omega_N^N = 1$; hence, the exponent of ω_N is evaluated modulo N, that is,

$$(13.96) \qquad \omega_N^M = \omega_N^{\langle M \rangle_N}.$$

Because M and $\langle M \rangle_N$ are interchangeable in the exponent of ω_N, we may use ρr f or $f(r)$ and $\rho\ell$ f or $f(\ell)$, and we obtain

$$(13.97) \qquad \omega_N^{f(r)\cdot\ell} = \omega_N^{\rho\cdot r\cdot\ell} = \omega_N^{r\cdot\rho\cdot\ell} = \omega_N^{r\cdot f(\ell)}.$$

This result connects the number theory to the DFT, and we can now rewrite (13.95) as

$$(13.98) \qquad Y_{f(r)} = \sum_{\ell=0}^{N-1} x_{f(\ell)}\, \omega_N^{f(r)\cdot\ell} = \sum_{\ell=0}^{N-1} x_{f(\ell)}\, \omega_N^{r\cdot f(\ell)} = \sum_{m=0}^{N-1} x_m\, \omega_N^{r\cdot m} = X_r.$$

Therefore, for two sequences related by $y_\ell = x_{f(\ell)}$, the DFTs are related by $Y_{f(r)} = X_r$. Recall that for input data we express $y = Q_N x$; for the output we may relate X and Y by

$$(13.99) \qquad X = Q_N \cdot Y \quad \text{or} \quad Y = Q_N^T \cdot X.$$

13.5.3 The Chinese remainder theorem

To provide application context before this theorem is stated, we shall consider the problem of solving simultaneous linear congruences with respect to different moduli. The problem takes the following form: Solve for unknown integer S to satisfy the ν congruences

$$(13.100) \qquad \begin{aligned} \langle S \rangle_{F_0} &= r_0, \\ \langle S \rangle_{F_1} &= r_1, \\ \langle S \rangle_{F_2} &= r_2, \\ &\ \vdots \\ \langle S \rangle_{F_{\nu-1}} &= r_{\nu-1}. \end{aligned}$$

In other words, the solution S we are seeking must be evaluated to each specied residue modulo each given modulus.

Example 13.7 For $F_0=3$, $F_1=4$, $r_0=2$, and $r_1=3$, the congruences $\langle S \rangle_3 = 2$ and $\langle S \rangle_4 = 3$ may be solved by trial and error. It is obvious that $S = 23$ is the solution.

The Chinese remainder theorem provides a unique solution which simultaneously satises the ν congruences given by (13.100), *provided that the moduli are pairwise prime*. This theorem is formally stated and proved below.

Theorem 13.7 (Chinese remainder theorem) Let $F_0, F_1, \ldots, F_{\nu-1}$ be positive integers that are pairwise prime; i.e,

$$(13.101) \qquad \gcd(F_\ell, F_k) = 1 \quad \text{when } \ell \neq k.$$

Let $N = F_0 \times F_1 \times \cdots \times F_{\nu-1}$, and let $r_0, r_1, \ldots, r_{\nu-1}$ be integers, $0 \le r_k \le F_k - 1$ for $0 \le k \le \nu-1$. Then there is exactly one integer $0 \le S \le N-1$ that satises the congruences $\langle S \rangle_{F_k} = r_k$ for $0 \le k \le \nu-1$.

Proof: As we have pointed out at the beginning of this chapter, the pairwise prime condition (13.101) implies that F_k and $L_k = N/F_k$ are relatively prime, i.e.,

$$(13.102) \qquad \gcd(L_k, F_k) = 1, \quad k = 0, 1, \ldots, \nu-1.$$

Under the condition $\gcd(L_k, F_k) = 1$, we can now nd ρ_k to satisfy the congruence

$$(13.103) \qquad \langle \rho_k L_k \rangle_{F_k} = 1$$

by solving the Diophantine equation

(13.104) $$\rho_k L_k = s_k F_k + 1, \quad \text{where } \mathbf{gcd}(L_k, F_k) = 1$$

for $0 < \rho_k < F_k$, $0 < s_k < L_k$. (The Diophantine equation and its solutions were discussed in detail in the last section.)

Observe that $L_\ell = N/F_\ell$ contains each F_k, $k \neq \ell$, as a factor; hence, F_k divides L_ℓ and $\langle L_\ell \rangle_{F_k} = 0$. Using the prescribed remainders r_ℓ, together with the now available ρ_ℓ and L_ℓ, for $0 \leq \ell \leq \nu - 1$, we let

(13.105) $$\hat{S} = \sum_{\ell=0}^{\nu-1} r_\ell \rho_\ell L_\ell = r_0 \rho_0 L_0 + r_1 \rho_1 L_1 + \cdots + r_{\nu-1} \rho_{\nu-1} L_{\nu-1}.$$

Using the rules of residue arithmetic, we show next that $\langle \hat{S} \rangle_{F_k} = r_k$ for every k:

(13.106)
$$
\begin{aligned}
\langle \hat{S} \rangle_{F_k} &= \left\langle \sum_{\ell=0}^{\nu-1} r_\ell \rho_\ell L_\ell \right\rangle_{F_k} \\
&= \left\langle \sum_{\ell=0}^{\nu-1} \langle r_\ell \rho_\ell L_\ell \rangle_{F_k} \right\rangle_{F_k} \\
&= \left\langle \langle r_k \rho_k L_k \rangle_{F_k} + \sum_{\substack{\ell=0 \\ \ell \neq k}}^{\nu-1} \langle r_\ell \rho_\ell \rangle_{F_k} \cdot \overbrace{\langle L_\ell \rangle_{F_k}}^{0} \right\rangle_{F_k} \\
&= \left\langle r_k \rho_k L_k \right\rangle_{F_k} + 0 \\
&= \left\langle \langle r_k \rangle_{F_k} \cdot \langle \rho_k L_k \rangle_{F_k} \right\rangle_{F_k} \\
&= \langle r_k \rangle_{F_k} \cdot 1 \\
&= r_k.
\end{aligned}
$$

Although $\hat{S}$ is a solution, it will not be the only solution. Because $\langle \hat{S} - \alpha N \rangle_{F_k} = \langle \hat{S} \rangle_{F_k}$ for any integer α, we can define $S = \langle \hat{S} \rangle_N$ so that $0 \leq S \leq N - 1$ is the solution in the desired range.

To prove that $S = \langle \hat{S} \rangle_N$ is the unique solution between 0 and $N - 1$, let us assume that S_2 is another solution. Since $r_k = \langle S_2 \rangle_{F_k} = \langle S \rangle_{F_k}$ implies that F_k divides $(S_2 - S)$, $k = 0, 1, \ldots, \nu - 1$, we establish that $(S_2 - S)$ is a multiple of $N = F_0 \times F_1 \times \cdots \times F_{\nu-1}$; i.e., $S_2 = S + \beta N$, where β is an integer. With $0 \leq S \leq N - 1$ and $0 \leq S_2 \leq N - 1$, we must have $\beta = 0$; hence, $S_2 = S$, and S is the unique solution in the range between 0 and $N - 1$. ∎

13.5.4 The ν-dimensional CRT index map

As we proved in Theorem 13.7 Chinese Remainder Theorem (CRT): under the condition that

$$N = \prod_{k=0}^{\nu-1} F_k, \quad \text{where } \nu \geq 2, \quad \mathbf{gcd}(F_j, F_k) = 1 \text{ if } j \neq k,$$

there is exactly one integer $0 \leq r \leq N - 1$ that satis es the congruences $\langle r \rangle_{F_k} = \hat{n}_k$ for $0 \leq \hat{n}_k \leq F_k - 1, 0 \leq k \leq \nu - 1$, and we have shown that $r = \langle \hat{S} \rangle_N$, where

$$(13.107) \qquad \hat{S} = \sum_{k=0}^{\nu-1} \rho_k L_k \hat{n}_k, \quad \rho_k = \langle L_k \rangle_{F_k}^{-1}, \quad M_k = N/F_k.$$

Observe that if we de ne

$$(13.108) \qquad X_r = B[\hat{n}_0, \hat{n}_1, \ldots, \hat{n}_{\nu-1}], \quad \text{for } r = \langle \hat{S} \rangle_N,$$

then there is exactly one index $0 \leq r \leq N - 1$ corresponding to one location, denoted by $[\hat{n}_0, \hat{n}_1, \ldots, \hat{n}_{\nu-1}]$, in the ν-dimensional array $B[F_0, F_1, \ldots, F_{\nu-1}]$, and we may obtain r using the so-called CRT map:

$$(13.109) \qquad r = \left\langle \sum_{k=0}^{\nu-1} \left\langle \frac{N}{F_k} \right\rangle_{F_k}^{-1} \frac{N}{F_k} \, \hat{n}_k \right\rangle_N, \quad \text{where } \nu \geq 2.$$

The *inverse* CRT map, which determines the corresponding element $B[\hat{n}_0, \hat{n}_1, \ldots, \hat{n}_{\nu-1}]$ for every X_r, is de ned by the ν given linear congruences:

$$(13.110) \qquad \hat{n}_k = \langle r \rangle_{F_k}, \quad k = 0, 1, \ldots, \nu - 1.$$

Example 13.8 For $N = F_0 \times F_1 \times F_2 = 3 \times 4 \times 5$, if the DFT output X of length N has been mapped to the 3-D array $B[F_0, F_1, F_2]$ by the CRT map, we can determine the frequency index of X_r from its location in B according to the formula

$$(13.111) \qquad \begin{aligned} r &= \left\langle \langle 20 \rangle_3^{-1} \cdot 20 \, \hat{n}_0 + \langle 15 \rangle_4^{-1} \cdot 15 \, \hat{n}_1 + \langle 12 \rangle_5^{-1} \cdot 12 \, \hat{n}_2 \right\rangle_{60} \\ &= \langle 40 \, \hat{n}_0 + 45 \, \hat{n}_1 + 36 \, \hat{n}_2 \rangle_{60}. \end{aligned}$$

For example, the element mapped to $B[1, 3, 4]$ is X_{19}, because $r = \langle 40 + 135 + 144 \rangle_{60} = 19$.

For each X_r, we can determine its location in B using the inverse CRT map. For example, the element in B corresponding to X_{47} is $B[2, 3, 2]$, because $\hat{n}_0 = \langle 47 \rangle_3 = 2$, $\hat{n}_1 = \langle 47 \rangle_4 = 3$, and $\hat{n}_2 = \langle 47 \rangle_5 = 2$.

13.5.5   The ν-dimensional Ruritanian index map

The Ruritanian correspondence proposed by Good [24] was also established under the condition that

$$N = \prod_{k=0}^{\nu-1} F_k, \quad \text{where } \nu \geq 2, \quad \mathbf{gcd}(F_j, F_k) = 1 \text{ if } j \neq k.$$

For $0 \leq \ell \leq N - 1$, the Ruritanian correspondence maps x_ℓ to $A[n_0, n_1, \ldots, n_{\nu-1}]$ if

$$(13.112) \qquad \ell = \left\langle \sum_{k=0}^{\nu-1} L_k \, n_k \right\rangle_N, \quad \text{where } L_k = \frac{N}{F_k}, \quad \nu \geq 2,$$

and ℓ is the only solution (in the range from 0 to $N - 1$) that satis es all ν congruences $\langle \rho_k \ell \rangle_{F_k} = n_k$ for $0 \leq n_k \leq F_k - 1, 0 \leq k \leq \nu - 1, \rho_k = \langle L_k \rangle_{F_k}^{-1}$. The proof of this result consists of three parts:

1. Prove that $\hat{T} = \sum_{k=0}^{\nu-1} L_k\, n_k$ is a solution by showing that $\left\langle \rho_k \hat{T} \right\rangle_{F_k} = n_k$ for every k:

$$
\begin{aligned}
\left\langle \rho_k \hat{T} \right\rangle_{F_k} &= \left\langle \rho_k \sum_{\tau=0}^{\nu-1} L_\tau n_\tau \right\rangle_{F_k} \\
&= \left\langle \sum_{\tau=0}^{\nu-1} \left\langle \rho_k L_\tau n_\tau \right\rangle_{F_k} \right\rangle_{F_k} \\
&= \left\langle \left\langle \rho_k L_k n_k \right\rangle_{F_k} + \sum_{\substack{\tau=0 \\ \tau \neq k}}^{\nu-1} \overbrace{\left\langle L_\tau \right\rangle_{F_k}}^{0} \cdot \left\langle \rho_k n_\tau \right\rangle_{F_k} \right\rangle_{F_k} \\
&= \left\langle \rho_k L_k n_k \right\rangle_{F_k} + 0 \qquad (\because F_k \text{ divides } L_\tau) \\
&= \left\langle \left\langle \rho_k L_k \right\rangle_{F_k} \cdot \left\langle n_k \right\rangle_{F_k} \right\rangle_{F_k} \\
&= 1 \cdot \left\langle n_k \right\rangle_{F_k} \qquad \left(\because \rho_k = \left\langle L_k \right\rangle_{F_k}^{-1} \right) \\
&= n_k.
\end{aligned}
$$

(13.113)

2. Prove that $\ell = \left\langle \hat{T} \right\rangle_N$ is a solution: because $\left\langle \hat{T} - \alpha N \right\rangle_{F_k} = \left\langle \hat{T} \right\rangle_{F_k}$ for any integer α, we can define $\ell = \left\langle \hat{T} \right\rangle_N$ so that $0 \leq \ell \leq N-1$ is a solution.

3. Prove $\ell = \left\langle \hat{T} \right\rangle_N$ is the unique solution in the range from 0 to $N-1$.

   To show this, we assume $\hat{\ell}$ is another solution. Since $n_k = \left\langle \rho_k \hat{\ell} \right\rangle_{F_k} = \left\langle \rho_k \ell \right\rangle_{F_k}$ implies that F_k divides $\rho_k(\hat{\ell} - \ell)$, and we know $\mathbf{gcd}(\rho_k, F_k) = 1$ (from Theorem 13.5), we conclude that F_k divides $(\hat{\ell} - \ell)$ for $k = 0, 1, \ldots, \nu - 1$. Hence $(\hat{\ell} - \ell)$ is a multiple of $N = F_0 \times F_1 \times \cdots \times F_{\nu-1}$; i.e., $\hat{\ell} = \ell + \beta N$, where β is an integer. With $0 \leq \ell \leq N-1$ and $0 \leq \hat{\ell} \leq N-1$, we must have $\beta = 0$; hence, $\hat{\ell} = \ell$, and ℓ is the unique solution in the range from 0 to $N-1$.

Again the *inverse* Ruritanian map is defined by the ν given congruences $n_k = \left\langle \rho_k \ell \right\rangle_{F_k}$, $k = 0, 1, \ldots, \nu - 1$.

Example 13.9 For $N = F_0 \times F_1 \times F_2 = 3 \times 4 \times 5$, if the DFT input data sequence x of length N has been mapped to the 3-D array $A[F_0, F_1, F_2]$ by the Ruritanian map, we can determine the time index of x_ℓ from its location in A according to the formula

(13.114) $$\ell = \left\langle 20\, n_0 + 15\, n_1 + 12\, n_2 \right\rangle_{60}.$$

For example, the element mapped to $A[1, 3, 4]$ is x_{53} because $\ell = \left\langle 20 + 45 + 48 \right\rangle_{60} = 53$.

For each x_ℓ, we can determine its location in A using the inverse CRT map $n_k = \left\langle \rho_k \ell \right\rangle_{F_k}$, with $\rho_0 = \left\langle 20 \right\rangle_3^{-1} = 2$, $\rho_1 = \left\langle 15 \right\rangle_4^{-1} = 3$, $\rho_2 = \left\langle 12 \right\rangle_5^{-1} = 3$. For example, the element in A corresponding to x_{47} is $A[1, 1, 1]$, because $n_0 = \left\langle 94 \right\rangle_3 = 1$, $n_1 = \left\langle 141 \right\rangle_4 = 1$, and $n_2 = \left\langle 141 \right\rangle_5 = 1$.

13.5.6   Organizing the ν-factor PFA computation—Part 3

For arbitrary $\nu \geq 2$, we now have the ν-dimensional CRT and Ruritanian index mapping formulas available from Sections 13.5.4 and 13.5.5; hence, the recursive equations representing a multi-factor PFA can be derived in the same manner as the arbitrary factor mixed-radix FFT. For $\nu = 3$, we have $N = F_0 \times F_1 \times F_2$. Assuming that the three factors are pairwise prime, we map the input x_ℓ to $A[n_0, n_1, n_2]$ using the 3-D Ruritanian map from Section 13.5.5, and we map the output X_r to $B[\hat{n}_0, \hat{n}_1, \hat{n}_2]$ using the 3-D CRT map from Section 13.5.4. By repeating the systematic decoupling processes performed in Sections 11.2.1, 11.2.2, and 11.2.3 with the two new mapping formulas, we obtain the recursive equations describing the three-factor PFA:

Step 0. Map x_ℓ to $A[n_0, n_1, n_2]$ using the 3-D Ruritanian mapping formula.

Step 1. Compute $A_1[n_0, n_1, \hat{n}_2] = \sum_{n_2=0}^{N_2-1} A[n_0, n_1, n_2] \, \omega_{F_2}^{\hat{n}_2 n_2}$.

Step 2. Compute $A_2[n_0, \hat{n}_1, \hat{n}_2] = \sum_{n_1=0}^{N_1-1} A_1[n_0, n_1, \hat{n}_2] \, \omega_{F_1}^{\hat{n}_1 n_1}$.

Step 3. Compute $A_3[\hat{n}_0, \hat{n}_1, \hat{n}_2] = \sum_{n_0=0}^{N_0-1} A_2[n_0, \hat{n}_1, \hat{n}_2] \, \omega_{F_0}^{\hat{n}_0 n_0}$.

Step 4. Map $B[\hat{n}_0, \hat{n}_1, \hat{n}_2] = A_3[\hat{n}_0, \hat{n}_1, \hat{n}_2]$ to X_r using the 3-D CRT mapping formula.

The Kronecker matrix equation representing the three-factor PFA is available from Equation (13.48) in Section 13.4.1, namely,

$$(13.115) \qquad\qquad \mathbf{vec}\, B = \left(\mathbf{\Omega}_{F_2} \otimes \mathbf{\Omega}_{F_1} \otimes \mathbf{\Omega}_{F_0} \right) \mathbf{vec}\, A,$$

where matrix A contains input sequence $\{x_\ell\}$ according to the 3-D Ruritanian map, and matrix B contains output $\{X_r\}$ according to the CRT map.

In either form the generalization to arbitrary ν-factor PFA is immediate as we have done in obtaining the mixed-radix FFT for arbitrary composite N.

13.6   The In-Place and In-Order PFA

13.6.1   The implementation-related concepts

In Section 13.2 we provided full details in the derivation of a two-factor PFA, and we illustrated the crucial index mapping steps using an example for $N = N_0 \times N_1$ with $N_0 = 3$ and $N_1 = 4$. In this section we shall use the same example to introduce the concepts of in-place and in-order implementation.

Recall the first version of the two-factor PFA given in Section 13.2:

```
for n₀ := 0 to N₀ − 1 do
    for n̂₁ := 0 to N₁ − 1 do
        A₁[n₀, n̂₁] := ∑ₙ₁₌₀^{N₁−1} A[n₀, n₁] ωₙ₁^{n̂₁n₁}
    end for
end for
for n̂₁ := 0 to N₁ − 1 do
    for n̂₀ := 0 to N₀ − 1 do
        B[n̂₀, n̂₁] := ∑ₙ₀₌₀^{N₀−1} A₁[n₀, n̂₁] ωₙ₀^{n̂₀n₀}
    end for
end for
```

Since each inner for-loop computes a DFT, we can explicitly express it as a matrix-vector product. By updating each vector to contain the computed matrix-vector product, we immediately obtain the in-place implementation of the two-factor PFA. For $N_0 = 3$ and $N_1 = 4$, the in-place PFA is shown below.

$$
\begin{aligned}
&\textbf{for } n_0 := 0 \textbf{ to } 2 \textbf{ do}\\[4pt]
&\qquad
\begin{bmatrix} a_{n_0,0} \\ a_{n_0,1} \\ a_{n_0,2} \\ a_{n_0,3} \end{bmatrix}
:=
\begin{bmatrix}
1 & 1 & 1 & 1 \\
1 & \omega_4^1 & \omega_4^2 & \omega_4^3 \\
1 & \omega_4^2 & \omega_4^4 & \omega_4^6 \\
1 & \omega_4^3 & \omega_4^6 & \omega_4^9
\end{bmatrix}
\begin{bmatrix} a_{n_0,0} \\ a_{n_0,1} \\ a_{n_0,2} \\ a_{n_0,3} \end{bmatrix}\\[6pt]
&\textbf{end for}\\[6pt]
&\textbf{for } n_1 := 0 \textbf{ to } 3 \textbf{ do}\\[4pt]
&\qquad
\begin{bmatrix} a_{0,n_1} \\ a_{1,n_1} \\ a_{2,n_1} \end{bmatrix}
:=
\begin{bmatrix}
1 & 1 & 1 \\
1 & \omega_3^1 & \omega_3^2 \\
1 & \omega_3^2 & \omega_3^4
\end{bmatrix}
\begin{bmatrix} a_{0,n_1} \\ a_{1,n_1} \\ a_{2,n_1} \end{bmatrix}\\[6pt]
&\textbf{end for}
\end{aligned}
$$

As explicitly shown inside each for-loop, the PFA is now in-place because each short DFT computed as a matrix-vector product *overwrites* the data vector *either* a row *or* a column of the matrix $\boldsymbol{A}$, which initially represents the input data sequence according to the Ruritanian map, and at the end of the PFA computation, the updated matrix $\boldsymbol{A}$ represents the DFT output according to the CRT map. That is,

$$(13.116) \qquad \textbf{On input:} \quad \boldsymbol{A} =
\begin{bmatrix}
a_{0,0} & a_{0,1} & a_{0,2} & a_{0,3} \\
a_{1,0} & a_{1,1} & a_{1,2} & a_{1,3} \\
a_{2,0} & a_{2,1} & a_{2,2} & a_{2,3}
\end{bmatrix}
=
\begin{bmatrix}
x_0 & x_3 & x_6 & x_9 \\
x_4 & x_7 & x_{10} & x_1 \\
x_8 & x_{11} & x_2 & x_5
\end{bmatrix}.
$$

$$(13.117) \qquad \textbf{On output:} \quad \boldsymbol{A} =
\begin{bmatrix}
a_{0,0} & a_{0,1} & a_{0,2} & a_{0,3} \\
a_{1,0} & a_{1,1} & a_{1,2} & a_{1,3} \\
a_{2,0} & a_{2,1} & a_{2,2} & a_{2,3}
\end{bmatrix}
=
\begin{bmatrix}
X_0 & X_9 & X_6 & X_3 \\
X_4 & X_1 & X_{10} & X_7 \\
X_8 & X_5 & X_2 & X_{11}
\end{bmatrix}.
$$

As we explained in Section 13.2.3, while it is convenient to describe the two-factor PFA as performing the short DFT on each column and each row of matrix $\boldsymbol{A}$, in actual implementation we do not need to physically store the input data in a 2-D matrix, provided that we can access the right group of elements directly from the input array $\{x_\ell\}$ in an equally convenient manner, and we have used the same example with $N_0 = 3$ and $N_1 = 4$ to develop the direct access methods in Section 13.2.3. Using the direct access method described there we would be overwriting the input elements as shown here:

$$
\begin{bmatrix} x_0 \\ x_3 \\ x_6 \\ x_9 \end{bmatrix}
:=
\begin{bmatrix}
1 & 1 & 1 & 1 \\
1 & \omega_4^1 & \omega_4^2 & \omega_4^3 \\
1 & \omega_4^2 & \omega_4^4 & \omega_4^6 \\
1 & \omega_4^3 & \omega_4^6 & \omega_4^9
\end{bmatrix}
\begin{bmatrix} x_0 \\ x_3 \\ x_6 \\ x_9 \end{bmatrix},
\qquad
\begin{bmatrix} x_4 \\ x_7 \\ x_{10} \\ x_1 \end{bmatrix}
:=
\begin{bmatrix}
1 & 1 & 1 & 1 \\
1 & \omega_4^1 & \omega_4^2 & \omega_4^3 \\
1 & \omega_4^2 & \omega_4^4 & \omega_4^6 \\
1 & \omega_4^3 & \omega_4^6 & \omega_4^9
\end{bmatrix}
\begin{bmatrix} x_4 \\ x_7 \\ x_{10} \\ x_1 \end{bmatrix},
$$

$$(13.118)$$

$$
\begin{bmatrix} x_8 \\ x_{11} \\ x_2 \\ x_5 \end{bmatrix}
:=
\begin{bmatrix}
1 & 1 & 1 & 1 \\
1 & \omega_4^1 & \omega_4^2 & \omega_4^3 \\
1 & \omega_4^2 & \omega_4^4 & \omega_4^6 \\
1 & \omega_4^3 & \omega_4^6 & \omega_4^9
\end{bmatrix}
\begin{bmatrix} x_8 \\ x_{11} \\ x_2 \\ x_5 \end{bmatrix}.
$$

$$(13.119) \quad \begin{bmatrix} x_0 \\ x_4 \\ x_8 \end{bmatrix} := \begin{bmatrix} 1 & 1 & 1 \\ 1 & \omega_3^1 & \omega_3^2 \\ 1 & \omega_3^2 & \omega_3^4 \end{bmatrix} \begin{bmatrix} x_0 \\ x_4 \\ x_8 \end{bmatrix}, \quad \begin{bmatrix} x_3 \\ x_7 \\ x_{11} \end{bmatrix} := \begin{bmatrix} 1 & 1 & 1 \\ 1 & \omega_3^1 & \omega_3^2 \\ 1 & \omega_3^2 & \omega_3^4 \end{bmatrix} \begin{bmatrix} x_3 \\ x_7 \\ x_{11} \end{bmatrix},$$

$$\begin{bmatrix} x_6 \\ x_{10} \\ x_2 \end{bmatrix} := \begin{bmatrix} 1 & 1 & 1 \\ 1 & \omega_3^1 & \omega_3^2 \\ 1 & \omega_3^2 & \omega_3^4 \end{bmatrix} \begin{bmatrix} x_6 \\ x_{10} \\ x_2 \end{bmatrix}, \quad \begin{bmatrix} x_9 \\ x_1 \\ x_5 \end{bmatrix} := \begin{bmatrix} 1 & 1 & 1 \\ 1 & \omega_3^1 & \omega_3^2 \\ 1 & \omega_3^2 & \omega_3^4 \end{bmatrix} \begin{bmatrix} x_9 \\ x_1 \\ x_5 \end{bmatrix}.$$

Consequently, the actual in-place computations are performed directly on the input data array:

(13.120)   **On input:** $x = \{x_0, x_1, x_2, x_3, x_4, x_5, x_6, x_7, x_8, x_9, x_{10}, x_{11}\}$.

Since the input and output are still linked by the now-absent matrix A as explicitly shown in Equation (13.116), we know that the updated array x now contains scrambled output, i.e.,

(13.121)   **On output:** $x = \{X_0, X_7, X_2, X_9, X_4, X_{11}, X_6, X_1, X_8, X_3, X_{10}, X_5\}$.

Therefore, the in-place PFA we obtain is *not* in-order   e.g., the updated x_1 equals X_7, the updated x_3 equals X_9, the updated x_5 equals X_{11}, ..., etc. If we want to have an in-place and in-order PFA, we must have the updated input array x contain the DFT output $X_0, X_1, \ldots,$ X_{11} in consecutive order so that each updated x_k equals X_k for every $k = 0, 1, \ldots, 11$.

Since the ordering of the input elements is explicitly linked to the ordering of the output elements by Equation (13.116), even matrix A is not used in the actual implementation; we can see that in order to have an in-place and in-order PFA, we must use the same index map for input and output. For example, if the Ruritanian map is used for input, then it must also be used for output:

$$(13.122) \quad \textbf{On input:} \quad A = \begin{bmatrix} a_{0,0} & a_{0,1} & a_{0,2} & a_{0,3} \\ a_{1,0} & a_{1,1} & a_{1,2} & a_{1,3} \\ a_{2,0} & a_{2,1} & a_{2,2} & a_{2,3} \end{bmatrix} = \begin{bmatrix} x_0 & x_3 & x_6 & x_9 \\ x_4 & x_7 & x_{10} & x_1 \\ x_8 & x_{11} & x_2 & x_5 \end{bmatrix}.$$

$$(13.123) \quad \textbf{On output:} \quad A = \begin{bmatrix} a_{0,0} & a_{0,1} & a_{0,2} & a_{0,3} \\ a_{1,0} & a_{1,1} & a_{1,2} & a_{1,3} \\ a_{2,0} & a_{2,1} & a_{2,2} & a_{2,3} \end{bmatrix} = \begin{bmatrix} X_0 & X_3 & X_6 & X_9 \\ X_4 & X_7 & X_{10} & X_1 \\ X_8 & X_{11} & X_2 & X_5 \end{bmatrix}.$$

Since we change the output map from CRT to Ruritanian, we need to re-derive the in-order PFA in Section 13.6.2.

If the CRT map is used for output, then it must also be used for input:

$$(13.124) \quad \textbf{On input:} \quad A = \begin{bmatrix} a_{0,0} & a_{0,1} & a_{0,2} & a_{0,3} \\ a_{1,0} & a_{1,1} & a_{1,2} & a_{1,3} \\ a_{2,0} & a_{2,1} & a_{2,2} & a_{2,3} \end{bmatrix} = \begin{bmatrix} x_0 & x_9 & x_6 & x_3 \\ x_4 & x_1 & x_{10} & x_7 \\ x_8 & x_5 & x_2 & x_{11} \end{bmatrix}.$$

$$(13.125) \quad \textbf{On output:} \quad A = \begin{bmatrix} a_{0,0} & a_{0,1} & a_{0,2} & a_{0,3} \\ a_{1,0} & a_{1,1} & a_{1,2} & a_{1,3} \\ a_{2,0} & a_{2,1} & a_{2,2} & a_{2,3} \end{bmatrix} = \begin{bmatrix} X_0 & X_9 & X_6 & X_3 \\ X_4 & X_1 & X_{10} & X_7 \\ X_8 & X_5 & X_2 & X_{11} \end{bmatrix}.$$

Since we change the input map from Ruritanian to CRT, we need to re-derive this version of the in-order PFA in Section 13.6.3.

13.6.2   The in-order algorithm based on Ruritanian map

For $N = N_0 \times N_1$, where N_0 and N_1 are relatively prime, the decoupling process in Section 13.2.2 must now be redone using the Ruritanian mapping formula on both index ℓ and index r; hence, we modify (13.12) as below.

$$
\begin{aligned}
(13.126) \qquad r \cdot \overbrace{(N_1 n_0 + N_0 n_1)}^{\ell} &= (N_1 \hat{n}_0 + N_0 \hat{n}_1) \cdot (N_1 n_0 + N_0 n_1) \\
&= N_1^2 \hat{n}_0 n_0 + N_1 N_0 \hat{n}_0 n_1 + N_1 N_0 \hat{n}_1 n_0 + N_0^2 \hat{n}_1 n_1 \\
&= N_1^2 \hat{n}_0 n_0 + N \hat{n}_0 n_1 + N \hat{n}_1 n_0 + N_0^2 \hat{n}_1 n_1.
\end{aligned}
$$

Using $\omega_N^N = 1$, together with $\omega_N^{N_1} = \omega_{N_0}$ and $\omega_N^{N_0} = \omega_{N_1}$, we obtain

$$
(13.127) \qquad \omega_N^{r \cdot (N_1 n_0 + N_0 n_1)} = \omega_{N_0}^{N_1 \hat{n}_0 n_0} \cdot \omega_{N_1}^{N_0 \hat{n}_1 n_1}.
$$

The modified two-factor PFA can now be easily described:

Step 0. Map x_ℓ to $A[n_0, n_1]$ using the Ruritanian index map.

Step 1. Compute $A_1[n_0, \hat{n}_1] = \sum_{n_1=0}^{N_1-1} A[n_0, n_1] \omega_{N_1}^{N_0 \hat{n}_1 n_1}$.

Step 2. Compute $B[\hat{n}_0, \hat{n}_1] = \sum_{n_0=0}^{N_0-1} A_1[n_0, \hat{n}_1] \omega_{N_0}^{N_1 \hat{n}_0 n_0}$.

Step 3. Map $B[\hat{n}_0, \hat{n}_1]$ to X_r using the Ruritanian index map.

When the direct indexing method described in Section 13.2.3 is used in actual implementation as explained in Section 13.6.1, the in-place implementation of this version of the PFA overwrites every input x_k by output X_k for $0 \le k \le N-1$, and we have obtained an in-order PFA. Note that the decoupled transforms in Step 1 and Step 2 are not exactly DFTs, and we will address how to compute such DFT-like short transforms in Section 13.7.

13.6.3   The in-order algorithm based on CRT map

Alternatively, we may use the CRT index map on both input and output to obtain another in-place and in-order PFA. For $N = N_0 \times N_1$, where N_0 and N_1 are relatively prime, we modify Equation (13.126) as below.

$$
\begin{aligned}
(13.128) \qquad r \cdot \overbrace{(\rho N_1 n_0 + q N_0 n_1)}^{\ell} &\\
= (\rho N_1 \hat{n}_0 + q N_0 \hat{n}_1) &\cdot (\rho N_1 n_0 + q N_0 n_1) \\
= \rho^2 N_1^2 \hat{n}_0 n_0 + \rho q N_0 N_1 \hat{n}_0 n_1 &+ \rho q N_0 N_1 \hat{n}_1 n_0 + q^2 N_0^2 \hat{n}_1 n_1 \\
= \rho N_1 (s N_0 + 1) \hat{n}_0 n_0 + \rho N \hat{n}_0 n_1 &+ q N \hat{n}_1 n_0 + q N_0 (t N_1 + 1) \hat{n}_1 n_1.
\end{aligned}
$$

For $N = N_0 \times N_1$, using $\omega_N^N = 1$, together with $\omega_N^{N_1} = \omega_{N_0}$ and $\omega_N^{N_0} = \omega_{N_1}$, we obtain

$$
(13.129) \qquad \omega_N^{r \cdot (\rho N_1 n_0 + q N_0 n_1)} = \omega_{N_0}^{\rho \hat{n}_0 n_0} \cdot \omega_{N_1}^{q \hat{n}_1 n_1}.
$$

The modified two-factor PFA can now be easily described:

Step 0. Map x_ℓ to $A[n_0, n_1]$ using the CRT index map.

Step 1. Compute $A_1[n_0, \hat{n}_1] = \sum_{n_1=0}^{N_1-1} A[n_0, n_1]\, \omega_{N_1}^{q\hat{n}_1 n_1}$.

Step 2. Compute $B[\hat{n}_0, \hat{n}_1] = \sum_{n_0=0}^{N_0-1} A_1[n_0, \hat{n}_1]\, \omega_{N_0}^{\rho\hat{n}_0 n_0}$.

Step 3. Map $B[\hat{n}_0, \hat{n}_1]$ to X_r using the CRT index map.

Again, when the direct indexing method described in Section 13.2.3 is used in actual implementation as explained in Section 13.6.1, the in-place implementation of this version of the PFA overwrites every input x_k by output X_k for $0 \le k \le N - 1$, and we have obtained another in-order PFA. Note that the decoupled transforms in Step 1 and Step 2 of this version of the in-order PFA are not exactly DFTs, either, and we will address how to compute them in Section 13.7.

13.7   Efficient Implementation of the PFA

When the DFT length N is composite with three or more factors, we know that a mixed-radix FFT can be implemented through a sequence of index mappings to 3-D arrays with the DFT-like transforms computed by nested multiplication (as proposed by de Boor [18] and described by Chu and George with details in [13]). When the three or more factors are pairwise prime, a multi-factor prime factor FFT can be implemented as a sequence of two-factor PFAs as noted by Burrus and Eschenbacher [9] and Temperton [45].

Note that for the arbitrary factor mixed-radix FFT, three-dimensional arrays are used in the actual implementation [13, 18]. However, for the multi-factor prime factor FFT, the data are not physically stored in a two-dimensional matrix, because we can use the direct indexing methods described in Section 13.2.3 to access the 1-D input array x, and we have demonstrated how the direct indexing methods are used in the actual in-place implementation of two-factor PFAs by an example in Section 13.6.1. Recall that both Ruritanian and CRT maps are unique, and they identify each unique group of elements for the DFT or DFT-like computation.

Suppose $N = F_0 \times F_1 \times F_2$, where the three factors are pairwise prime; hence, each factor and the product of the other two factors are also relatively prime, and we can de ne three two-factor PFAs by expressing $N = N_0 \times N_1$ with $N_1 = F_k$ and $N_0 = N/F_k$ for $k = 0, 1, 2$.

The in-place and in-oder implementation of the three-factor PFA consists of the following three stages:

Stage A. Let $N_1 = F_0$, and $N_0 = F_1 \times F_2$; compute the N_0 length N_1 DFT-like transforms in-place. The arithmetic operations required are proportional to $N_0 \times N_1^2 = N \times F_0$.

   Remarks: The groups of elements identi ed by the direct indexing scheme for the length N_1 transforms in this stage and the next two stages are the groups uniquely determined by the chosen three-factor mapping scheme.

Stage B. Let $N_1 = F_1$, and $N_0 = F_0 \times F_2$; compute the N_0 length N_1 DFT-like transforms in-place. The arithmetic operations required are proportional to $N_0 \times N_1^2 = N \times F_1$.

Stage C. Let $N_1 = F_2$, and $N_0 = F_0 \times F_1$; compute the N_0 length N_1 DFT-like transforms in-place. The arithmetic operations required are proportional to $N_0 \times N_1^2 = N \times F_2$.

Accordingly, the total arithmetic operations required by the three-factor PFA are proportional to $N(F_0 + F_1 + F_2)$. If the DFT length N is composite with arbitrary $\nu > 3$ pairwise-prime factors, a sequence of ν two-factor PFA can be de ned with $N_0 = F_k$ and $N_1 = N/F_k$

for $k = 0, 1, \ldots, \nu - 1$, and the ν-stage PFA algorithm would require arithmetic operations proportional to $N(F_0 + F_1 + \cdots + F_{\nu-1})$.

The ef ciency of the PFA depends on the size of each factor, because in every stage the N/F_k DFT or DFT-like transforms are of length F_k; hence, the factors are desired to be relatively small. For example, the collection of the so-called small-n DFTs provided by Nussbaumer in [35] are for factors from the set $\{2, 3, 4, 5, 7, 8, 9, 16\}$. Note that we can choose at most four pairwise-prime factors from this set, and the largest $N = 5 \times 7 \times 9 \times 16 = 5040$. (As mentioned at the beginning of this chapter, the PFA based on these small-n DFT modules can be combined with radix-2 FFT or mixed-radix FFT to perform much larger transforms.)

As to the DFT-like transforms in the two versions of the in-order PFA, they are referred to in the literature as r otated DFT for reasons given below.

Version (i) Ruritanian map based in-order PFA For $N = 3 \times 4$ with $N_0 = 3$ and $N_1 = 4$, the three DFT-like transforms involving $\omega_{N_1}^{N_0 r \ell}$ are

$$(13.130) \qquad \begin{bmatrix} x_0 \\ x_3 \\ x_6 \\ x_9 \end{bmatrix} := \begin{bmatrix} 1 & 1 & 1 & 1 \\ 1 & \omega_4^3 & \omega_4^6 & \omega_4^9 \\ 1 & \omega_4^6 & \omega_4^{12} & \omega_4^{18} \\ 1 & \omega_4^9 & \omega_4^{18} & \omega_4^{27} \end{bmatrix} \begin{bmatrix} x_0 \\ x_3 \\ x_6 \\ x_9 \end{bmatrix}, \quad \cdots, \quad \cdots \quad \text{etc.}$$

The four DFT-like transforms involving $\omega_{N_0}^{N_1 r \ell}$ are

$$(13.131) \qquad \begin{bmatrix} x_0 \\ x_4 \\ x_8 \end{bmatrix} := \begin{bmatrix} 1 & 1 & 1 \\ 1 & \omega_3^4 & \omega_3^8 \\ 1 & \omega_3^8 & \omega_3^{16} \end{bmatrix} \begin{bmatrix} x_0 \\ x_4 \\ x_8 \end{bmatrix}, \quad \cdots, \quad \cdots, \quad \cdots, \quad \text{etc.}$$

Version (ii) CRT map based in-order PFA For $N = N_0 \times N1 = 3 \times 4$, recall $\rho = 1$ and $q = 3$; hence, we have three DFT transforms involving $\omega_{N_1}^{q r \ell} = \omega_{N_1}^{3 r \ell}$

$$(13.132) \qquad \begin{bmatrix} x_0 \\ x_3 \\ x_6 \\ x_9 \end{bmatrix} := \begin{bmatrix} 1 & 1 & 1 & 1 \\ 1 & \omega_4^3 & \omega_4^6 & \omega_4^9 \\ 1 & \omega_4^6 & \omega_4^{12} & \omega_4^{18} \\ 1 & \omega_4^9 & \omega_4^{18} & \omega_4^{27} \end{bmatrix} \begin{bmatrix} x_0 \\ x_3 \\ x_6 \\ x_9 \end{bmatrix}, \quad \cdots, \quad \cdots, \quad \text{etc.}$$

The four DFT-like transform involving $\omega_{N_0}^{\rho r \ell}$ with $\rho = 1$ represents the DFT itself:

$$(13.133) \qquad \begin{bmatrix} x_0 \\ x_4 \\ x_8 \end{bmatrix} := \begin{bmatrix} 1 & 1 & 1 \\ 1 & \omega_3^1 & \omega_3^2 \\ 1 & \omega_3^2 & \omega_3^4 \end{bmatrix} \begin{bmatrix} x_0 \\ x_4 \\ x_8 \end{bmatrix}, \quad \cdots, \quad \cdots, \quad \cdots, \quad \text{etc.}$$

Observe that each of the DFT-like matrices in Equations (13.130), (13.131),(13.132), and (13.133) can be expressed as the product of a DFT matrix and a permutation matrix. For example, by making use of $\omega_4^4 = 1$, we obtain

$$\begin{bmatrix} 1 & 1 & 1 & 1 \\ 1 & \omega_4^3 & \omega_4^6 & \omega_4^9 \\ 1 & \omega_4^6 & \omega_4^{12} & \omega_4^{18} \\ 1 & \omega_4^9 & \omega_4^{18} & \omega_4^{27} \end{bmatrix} = \begin{bmatrix} 1 & 1 & 1 & 1 \\ 1 & \omega_4^3 & \omega_4^6 & \omega_4^9 \\ 1 & \omega_4^2 & \omega_4^4 & \omega_4^6 \\ 1 & \omega_4^1 & \omega_4^2 & \omega_4^3 \end{bmatrix}$$

$$(13.134)$$

$$= \begin{bmatrix} 1 & 0 & 0 & 0 \\ 0 & 0 & 0 & 1 \\ 0 & 0 & 1 & 0 \\ 0 & 1 & 0 & 0 \end{bmatrix} \begin{bmatrix} 1 & 1 & 1 & 1 \\ 1 & \omega_4^1 & \omega_4^2 & \omega_4^3 \\ 1 & \omega_4^2 & \omega_4^4 & \omega_4^6 \\ 1 & \omega_4^3 & \omega_4^6 & \omega_4^9 \end{bmatrix}$$

Because this can be done for every DFT-like matrix in the two in-order PFAs, the DFT-like matrices are referred to as the rotated DFT. If we let P denote the permutation matrix and let Ω denote the DFT matrix, the result of a rotated DFT given by

$$(13.135) \qquad\qquad z = (P\Omega)y$$

can be obtained by permuting the DFT result, that is,

$$(13.136) \qquad\qquad z = P(\Omega y).$$

Therefore, by adding a permutation step, all specially designed small-n DFT modules can be used to compute the rotated DFTs needed in the ef cient implementation of the prime factor FFT algorithms. Among the two in-order PFAs, the one based on CRT map has simpler direct indexing scheme, and the Fortran code implementing the algorithm was described in [45], where the small-n rotated DFTs for factors of 2, 3, and 4 are included and they are explicitly coded to minimize the arithmetic operations.

Chapter 14

On Computing the DFT of Large Prime Length

The conventional FFT usually refers to the family of mixed-radix algorithms for rapidly computing the DFT de ned by the formula (excluding division by N):

$$(14.1) \qquad X_r = \sum_{\ell=0}^{N-1} x_\ell e^{-j2\pi r\ell/N}, \quad \text{for } r = 0, 1, \cdots, N-1,$$

and the IDFT de ned by the formula (including division by N):

$$(14.2) \qquad x_\ell = \frac{1}{N} \sum_{r=0}^{N-1} X_r e^{j2\pi r\ell/N}, \quad \text{for } \ell = 0, 1, \cdots, N-1,$$

where $j \equiv \sqrt{-1}$, and the sequences x_ℓ and X_r each consists of N complex data samples.

The various FFTs are tailored to the DFT/IDFT of different lengths. As discussed in Chapter 11, when the length N is a power of two, the familiar radix-2 FFT achieves a complexity of $O(N\log_2 N)$; when N is not a power of two, the DFT/IDFT can be computed by the FFT generalized for composite $N = \prod_{k=1}^{m} F_k$, and the complexity becomes $O(N(F_1 + F_2 + \cdots + F_m))$. When some or all of the factors are pairwise prime, the prime factor FFT algorithms presented in Chapter 13 can be used, possibly in combination with the radix-2 FFT or mixed-radix FFT. For general composite N, various forms of the mixed-radix FFT and the prime factor FFT were proposed during the four decades from the late 1950s to the early 1990s [4, 9, 16, 18, 24, 31, 41, 44, 45, 50, 55]. These algorithms are particularly ef cient when N is the product of small factors. For example, the mutually prime factors of N must be selected from the set $\{2, 3, 4, 5, 7, 8, 9, 16\}$ in most published prime factor FFT algorithms.

Accordingly, the performance of the FFT for composite N depends on the sizes of the factors: at one extreme, when all factors are identical and equal to two, we have the highly ef cient $O(N\log_2 N)$ radix-2 FFT; at the other extreme, when $N/2$ is a large prime number, the complexity becomes $O(N(2 + N/2))$, and the execution time grows with N^2, which is at the same rate as computing the DFT/IDFT directly according to Equations (14.1) and (14.2). Therefore, for transforms with length N being a large prime or containing large prime factors, little improvement can be expected from the various mixed-radix FFTs. In this chapter we shall present two alternatives which can improve the performance of FFT when its length N

is a large prime or contains large prime factors, and we provide MATLAB implementation to demonstrate the merits of these algorithms implemented in a high-level language.

14.1 Performance of FFT for Prime N

Numerical experiments in MATLAB 5.3 and MATLAB 7.4: While many available FFT programs can handle non-power-of-two lengths, users are usually warned that such data sets may be processed by much slower algorithms [28, 29, 39]. For FFT of large prime length, we could face the situation that *neither* the speed *nor* the accuracy is acceptable. In this section we experiment with built-in FFT codes from two versions of MATLAB®[1]. Our results demonstrate the great improvement made in MATLAB 7.4 in the computation of prime-length FFT. Because the FFT codes provided by both versions of MATLAB are built-in functions, a fair comparison of length-2^n FFT and prime-length FFT can be made using the same executable code in all cases.

In MATLAB 5.3, the FFT code runs at drastically different speeds depending on whether the length is a power of two or a large prime. For example, given in Tables 14.1 and 14.2 are some tests we ran using MATLAB 5.3 built-in FFT and inverse FFT functions, namely, **fft** and **ifft**. In Table 14.1 we compare the cpu times and the total number of floating-point operations (flops) required to compute ifft(fft(x)) for complex series x of length $N_1 = 2^s$ and prime length N_2. For each power-of-two N_1, we select the largest prime $N_2 < N_1$, which can be obtained by $N_2 = \max(\text{primes}(N_1))$ in MATLAB.

Table 14.1 Performance of MATLAB 5.3 built-in FFT.

	Computing ifft(fft(x)) for complex series of length $N_1 = 2^s$ and prime length N_2					
$N_1 = 2^s$	Built-in Code Timings (CPU 1.3 GHz)	Arithmetic Cost Total Flops	Prime N_2	Built-in Code Timings (CPU 1.3 GHz)	Arithmetic Cost Total Flops
2048	0.0010 sec	0.25 million	2039	0.87 sec	67 million
4096	0.0025 sec	0.54 million	4093	3.44 sec	268 million
8192	0.0058 sec	1.16 million	8191	13.73 sec	1 billion
16384	0.0130 sec	2.49 million	16381	54.78 sec	4 billion
32768	0.0290 sec	5.31 million	32749	218.09 sec	16 billion

From the timing results in Table 14.1, we see that the computing time for prime N_2 grows with N_2^2 (instead of $N_1 \log_2 N_1$ for $N_1 = 2^s$), and that for $N_2 = 4093$, the time required is already more than a thousand times longer than the radix-2 FFT time for $N_1 = 2^{12} = 4096$. Since the $O\left(N_2^2\right)$ time quadrupled when N_2 is doubled, it quickly grows to 218 seconds for prime $N_2 = 32749$, which is more than 7,000 times longer than the 0.029 seconds needed to complete both forward and inverse FFT for $N_1 = 2^{15} = 32768$.

Since the number of floating-point operations grows with N_2^2 for prime N_2, we are also concerned with the loss of accuracy in the computed results when N_2 is large. To measure the error, we compare the result $y = \text{ifft}(\text{fft}(x))$ with the input series x, and we report the relative

---

[1]MATLAB is a registered trademark of The MathWorks, Inc.

error

$$E_\infty \equiv \frac{\|x - y\|_\infty}{\|x\|_\infty}$$

in Table 14.2. Because ifft(fft(x)) should reproduce x, the difference between y and x reflects the loss of accuracy in the computing process. From the results in Table 14.2, we see that for $N_1 = 2^s$, the full double-precision accuracy is maintained for all N_1 values ranging from 2048 to 32768; however, for prime N_2 values in the same range, the loss of accuracy is quite significant — only single-precision accuracy remains for $N_2 \geq 16381$. (Note that since MATLAB supports IEEE 16-digit precision, all results are computed in standard double precision floating point arithmetic. If the fft and ifft functions were implemented for a single-precision environment, we risk losing all significant digits in y when $N_2 \geq 16381$.)

Table 14.2 Measuring error in computing ifft(fft(x)) in MATLAB 5.3.

Measuring error in computing y = ifft(fft(x)) for complex series of length $N_1 = 2^s$ and prime length N_2			
$N_1 = 2^s$	Relative Error E_∞	Prime N_2	Relative Error E_∞
2048	2.6407e—15	2039	1.8057e—11
4096	2.3728e—15	4093	7.1580e—10
8192	4.3739e—15	8191	1.1085e—9
16384	4.0839e—15	16381	1.4637e—8
32768	6.0783e—15	32749	5.7189e—8

This is no longer the case with MATLAB 7.4, which includes executable code (based on the FFTW library [20, 30]) for computing the DFT of large prime length N efficiently and accurately as demonstrated by the results in Tables 14.3 and 14.4. Note that the function flops is no longer available in MATLAB 7.4; hence, the total floating-point operations are not reported in Table 14.3.

Table 14.3 Performance of MATLAB 7.4 built-in FFT.

Computing ifft(fft(x)) for complex series of length $N_1 = 2^s$ and prime length N_2			
$N_1 = 2^s$	Built-in Code Timings (CPU 3.2 GHz)	Prime N_2	Built-in Code Timings (CPU 3.2 GHz)
2048	0.0005 sec	2039	0.0019 sec
4096	0.0011 sec	4093	0.0039 sec
8192	0.0025 sec	8191	0.0050 sec
16384	0.0059 sec	16381	0.0128 sec
32768	0.0173 sec	32749	0.0625 sec

To gain the knowledge concerning the design and implementation of fast algorithms for computing the prime-length DFT, we shall explore two approaches in the remainder of this chapter, and we provide MATLAB implementation to demonstrate the merits of these algorithms.

Table 14.4 Measuring error in computing ifft(fft(x)) in MATLAB 7.4.

Measuring error in computing $y=$ifft(fft(x)) for complex series of length $N_1=2^s$ and prime length N_2			
$N_1=2^s$	Relative Error E_∞	Prime N_2	Relative Error E_∞
2048	4.2611e—16	2039	1.0699e—15
4096	5.0700e—16	4093	1.3466e—15
8192	4.8105e—16	8191	1.8412e—15
16384	5.9934e—16	16381	5.4653e—15
32768	6.0071e—16	32749	1.5449e—15

14.2 Fast Algorithm I: Approximating the FFT

To describe this approach, we interpret the DFT results from Equation (14.1) as the function values of

$$(14.3) \qquad F(\theta) = \sum_{\ell=0}^{N-1} x_\ell e^{-j\ell\theta}$$

at equispaced $\theta_r = r(2\pi/N)$ between 0 and 2π, yielding $X_r = F(\theta_r)$ for $r = 0, 1, \cdots, N-1$.

The approximate FFT algorithm proposed by Anderson and Dahleh [2] combines the radix-2 FFT and local Taylor series expansion to approximate $X_r = F(\theta_r)$ in the following manner:

Step 1. Express the kth derivative of $F(\theta)$ as

$$(14.4) \qquad F^{(k)}(\theta) = \sum_{\ell=0}^{N-1}(-j\ell)^k x_\ell e^{-j\ell\theta} = \sum_{\ell=0}^{N-1} \tilde{x}_\ell e^{-j\ell\theta}, \quad \text{where } \tilde{x}_\ell = (-j\ell)^k x_\ell.$$

Note that $F^{(0)}(\theta) = F(\theta)$.

Step 2. Evaluate $F(\theta)$, $F'(\theta)$, $F''(\theta)$, $\cdots$, and $F^{(k)}(\theta)$ at a set of $M = 2^s$ ($M > N$) equispaced $\hat\theta$ values between 0 and 2π; i.e., use $\{\hat\theta_0, \hat\theta_1, \cdots, \hat\theta_{M-1}\}$ with $\hat\theta_r \equiv r(2\pi/M)$, and compute

$$(14.5) \qquad F^{(k)}(\hat\theta_r) = \sum_{\ell=0}^{N-1} \tilde{x}_\ell e^{-j\ell\hat\theta_r} = \sum_{\ell=0}^{N-1} \tilde{x}_\ell e^{-j2\pi r\ell/M}, \quad \text{for } r = 0, 1, \cdots, M - 1.$$

To convert (14.5) to an M-point (and M-term) DFT, we need to add more terms with zero coefficients. That is, we define $\tilde\alpha_\ell = \tilde{x}_\ell$ for $0 \le \ell \le N - 1$, and add $\tilde\alpha_\ell = 0$ for $N \le \ell \le M - 1$ to obtain the properly defined DFT, namely,

$$(14.6) \qquad F^{(k)}(\hat\theta_r) = \sum_{\ell=0}^{M-1} \tilde\alpha_\ell e^{-j2\pi r\ell/M}, \quad \text{for } r = 0, 1, \cdots, M - 1.$$

Since M is a power of two, the function $F(\theta)$ and each of its derivatives can now be evaluated on the M equispaced $\hat\theta_r$ values by a radix-2 FFT algorithm at the cost of $O(M \log_2 M)$ arithmetic operations.

Step 3. For every $\theta_r = r(2\pi/N), 0 \leq r \leq N - 1$, we determine the nearest $\hat{\theta}_n = n(2\pi/M)$, and we approximate $X_r = F(\theta_r)$ by computing $X_r \approx T(\hat{\theta}_n + \delta)$, where $\delta = \theta_r - \hat{\theta}_n$, and

$$(14.7) \quad T(\hat{\theta}_n + \delta) = F(\hat{\theta}_n) + \delta F'(\hat{\theta}_n) + \frac{\delta^2}{2!}F''(\hat{\theta}_n) + \cdots + \frac{\delta^k}{k!}F^{(k)}(\hat{\theta}_n)$$

is the degree-k Taylor polynomial expanded at each chosen $\hat{\theta}_n$.

The complexity of this algorithm is $O(kM\log_2 M)$, where k is the degree of Taylor s polynomial in (14.7), and $M = 2^s > N$ is the length of the extended DFT in (14.6). To approximate the IDFT results from Equation (14.2), the corresponding steps can be similarly developed.

14.2.1 Array-smart implementation in MATLAB

In this section we give the approximate FFT/IFFT algorithms in the form of MATLAB® functions Tfft and iTfft. To obtain an array-smart implementation, we have made use of MATLAB vectorized operations and built-in functions (including the fft/ifft) in processing all data arrays. To make the vectorized algorithm easy to understand, we connect the pseudo-code to our mathematical derivation by using the same Greek letters as array names, and we identify the elements of each array in a comment immediately after the array is named. For example, we have used θ t o name the array containing $[\theta_0, \theta_1, \cdots, \theta_{M-1}]$ in the pseudo-code for function Tfft. Note that θ can be simply replaced by theta in the actual code.

Algorithm 14.1 The approximate FFT algorithm Tfft in MATLAB-style pseudo-code

---

$function\ FX\ =\ Tfft(x, degree)$

begin

$\quad N\ =\ length(x);$

$\quad M\ =\ 2\ nextpow2(N);$ compute smallest $M = 2^p > N$

$\quad \tilde{\alpha}\ =\ [x,\ zeros\ (1, M—N)];$ array $\tilde{\alpha} = [\tilde{\alpha}_0, \tilde{\alpha}_1, \cdots, \tilde{\alpha}_{M-1}]$

$\quad \theta\ =\ linspace(0, 2*\pi, N{+}1);$

$\quad \theta\ =\ \theta(1{:}N);$ array $\theta = [\theta_0, \theta_1, \cdots, \theta_{N-1}]$

$\quad \hat{\theta}\ =\ linspace(0, 2*\pi, M{+}1);$ array $\hat{\theta} = [\hat{\theta}_0, \cdots, \hat{\theta}_{M-1}, \hat{\theta}_M]$

$\quad h\ =\ 2*\pi/M;$

$\quad IDX\ =\ round(\theta/h) + 1;$ compute n for all nearest $\hat{\theta}_n$

$\quad \delta\ =\ (\theta - \hat{\theta}(IDX));$ compute all $\delta = \theta_r - \hat{\theta}_n$

$\quad f\ =\ fft(\tilde{\alpha});$ call built-in fft to compute M function values

$\quad f\ =\ [f,\ f(1)];$ include function value at boundary $\hat{\theta}_M = 2\pi$

$\quad FX\ =\ f(IDX);$ extract all $F(\hat{\theta}_n)$ values

$\quad S\ =\ \delta;$ initialize S by array δ

$\quad$ **for** $k = 1$ **to** $degree$

$\quad\quad \tilde{\alpha}\ =\ \tilde{\alpha}.*(-j*[0{:}M—1]);$ $j = \sqrt{-1}$ is a built-in constant in MATLAB

$\quad\quad fprime\ =\ fft(\tilde{\alpha});$ call built-in fft to compute M derivative values

$\quad\quad fprime\ =\ [fprime,\ fprime(1)];$ include derivative value at $\hat{\theta}_M = 2\pi$

$\quad\quad FX\ =\ FX + S.*fprime(IDX);$ compute $T(\hat{\theta}_n + \delta)$ term by term

$\quad\quad S\ =\ S.*(\delta/(k+1));$

$\quad$ **end for**

end

Algorithm 14.2 The approximate IFFT algorithm iTfft in MATLAB pseudo-code

function FXINV = iTfft(X, degree)

begin

$N = length(X);$

$M = 2\ nextpow2(N);$ compute smallest $M = 2^p > N$

$\tilde{\alpha} = [X,\ zeros\ (1,\ M{-}N)];$ array $\tilde{\alpha} = [\tilde{\alpha}_0, \tilde{\alpha}_1, \cdots, \tilde{\alpha}_{M-1}]$

$\theta = linspace(0, 2{*}\pi, N{+}1);$

$\theta = \theta(1{:}N);$ array $\theta = [\theta_0, \theta_1, \cdots, \theta_{N-1}]$

$\hat{\theta} = linspace(0, 2{*}\pi, M{+}1);$ array $\hat{\theta} = [\hat{\theta}_0, \cdots, \hat{\theta}_{M-1}, \hat{\theta}_M]$

$h = 2{*}\pi/M;$

$IDX = round(\theta/h) + 1;$ compute n for all nearest $\hat{\theta}_n$

$\delta = (\theta - \hat{\theta}(IDX));$ compute all $\delta = \theta_r - \hat{\theta}_n$

$f = ifft(\tilde{\alpha});$ call built-in ifft to compute M function values

$f = [f,\ f(1)];$ include function value at boundary $\hat{\theta}_M = 2\pi$

$FXINV = f(IDX);$ extract all $F(\hat{\theta}_n)$ values

$S = \delta;$ initialize S by array δ

for $k = 1$ **to** _degree_

     $\tilde{\alpha} = \tilde{\alpha}.{*}(j{*}[0{:}M{-}1]);$ $j = \sqrt{-1}$ is a built-in constant in MATLAB

     $fprime = ifft(\tilde{\alpha});$ call built-in ifft to compute M derivative values

     $fprime = [fprime,\ fprime(1)];$ include derivative value at boundary $\hat{\theta}_M = 2\pi$

     $FXINV = FXINV + S.{*}fprime(IDX);$ compute $T(\hat{\theta}_n + \delta)$ term by term

     $S = S.{*}(\delta/(k+1));$

end for

$FXINV = FXINV * M/N;$ including division by N

end

14.2.2 Numerical results

We evaluate the two function M- les Tfft.m and iTfft.m on computing the DFT/IDFT of data sets with prime length N_2, and we present the results in Table 14.5. In coding the algorithm, we have chosen the smallest $M = 2^s > N$ to be the length of the extended DFT/IDFT, and we leave the degree k of Taylor s polynomial as an input parameter. Since the execution time and the accuracy of the algorithm are determined by both M and k, we identify the values used in our experiment in Table 14.5.

In Table 14.5, we choose degree k to gain the maximum accuracy in the results. For computing the DFT/IDFT of the same data (with results more accurate than Table 14.2), the Tfft and iTfft times for prime $N_2 \geq 4093$ are signi cantly faster than those for prime N_2 in Table 14.1, even the user M- le is interpreted and expected to run more slowly than executable code, and they re ect the expected difference between an $O(kM \log_2 M)$ algorithm and an $O(N_2^2)$ algorithm. Compared with the results in Tables 14.3 and 14.4, the approximated results are less accurate than those computed by the built-in FFT in MATLAB 7.4, and they are also less accurate than those computed by the Bluestein s FFT to be presented in the next section.

Table 14.5 Evaluating function M- les Tfft.m and iTfft.m for large prime N.

			MATLAB 5.3	MATLAB 7.4	
Prime	Method Parameters		M-File Times	M-File Times	Relative
N_2	$M = 2^s$	degree k	(CPU 1.3 GHz)	(CPU 3.2 GHz)	Error E_∞
2039	2048	22	0.06 sec	0.03 sec	7.1549e—13
4093	4096	22	0.14 sec	0.06 sec	1.4760e—12
8191	8192	22	0.31 sec	0.13 sec	2.9721e—12
16381	16384	22	0.65 sec	0.28 sec	6.7510e—12
32749	32768	22	1.55 sec	0.67 sec	1.5064e—11

Computing $X = \mathrm{Tfft}(x, k)$ and $y = \mathrm{iTfft}(X, k)$ for complex x of prime length N_2.

14.3   Fast Algorithm II: Using Bluestein's FFT

Shortly after Cooley and Tukey published their original paper [16] on radix-2 FFT and its potential generalization to mixed-radix FFT, Bluestein presented an FFT for arbitrary N including primes [5]. Bluestein s algorithm resurfaced in 1991 through a theoretical study of its performance on the hypercube [43], where it was shown to require fewer communication cycles than Bergland s mixed radix FFT [4] for composite N. Although Bluestein s FFT handles sequences of prime length N with a desirable complexity of $O(M \log_2 M)$, where $M = 2^s \geq 2N - 2$, its implementation and performance results seem to have been absent in the FFT literature.

In this section, we shall derive Bluestein s FFT and provide array-smart implementations for both FFT and its inverse in MATLAB. The MATLAB programs are then used to compute the DFT/IDFT of prime length in the numerical experiments that follow.

14.3.1   Bluestein's FFT and the chirp Fourier transform

We indicated in Chapter 9 that the discrete cyclic convolution is useful in the development of the chirp Fourier transform as well as the fast Fourier transform algorithm for arbitrary (possibly prime) N. The chirp Fourier transform was covered in Section 9.3 in Chapter 9, where we showed it to represent a partial DFT, which can be converted to a partial linear convolution, and the latter can be converted to a partial cyclic convolution computable by two FFTs and one inverse FFT. Bluestein s FFT makes use of the same ideas to turn a DFT of length N into a partial linear convolution, and the latter can be turned into a partial cyclic convolution computable by two FFTs and one inverse FFT of lengths all equal to $M = 2^s \geq 2N - 2$.

14.3.2 The equivalent partial linear convolution

We begin by rewriting the DFT de n ed by (14.1) as

$$X_r = \sum_{\ell=0}^{N-1} x_\ell \, \omega_N^{r\ell}, \quad \omega_N \equiv e^{-j2\pi/N}, \quad \text{for } r = 0, 1, \cdots, N-1,$$

(14.8)
$$= \sum_{\ell=0}^{N-1} x_\ell \, \omega_N^{\frac{1}{2}\left(r^2 + \ell^2 - (r-\ell)^2\right)}$$

$$= \omega_N^{\frac{1}{2}r^2} \sum_{\ell=0}^{N-1} x_\ell \, \omega_N^{\frac{1}{2}\ell^2} \, \omega_N^{-\frac{1}{2}(r-\ell)^2}.$$

To convert the DFT de ned by (14.8) to a partial linear convolution, we de ne

(14.9)
$$Z_r = \omega_N^{-\frac{1}{2}r^2} X_r, \qquad y_\ell = x_\ell \omega_N^{\frac{1}{2}\ell^2}, \qquad h_{r-\ell} = \omega_N^{-\frac{1}{2}(r-\ell)^2},$$

and rewrite (14.8) as

(14.10)
$$Z_r = \sum_{\ell=0}^{N-1} y_\ell \cdot h_{r-\ell}, \quad r = 0, 1, \ldots, N-1.$$

Observe that $\{Z_0, Z_1, \cdots, Z_{N-1}\}$ computed according to Formula (14.10) are the middle N (beginning with the Nth) elements obtained from the linear convolution of the length-N sequence

$$\{y_0, y_1, \cdots, y_{N-1}\}$$

and the length-$(2N-1)$ sequence

$$\{f_0, f_1, \cdots, f_{2N-2}\} = \{h_{-N+1}, h_{-N+2} \cdots, h_{-1}, h_0, h_1, \cdots, h_{N-1}\}.$$

As we did in Section 9.3 before, we have explicitly stored the data

$$\{h_{-N+1}, h_{-N+2}, \cdots, h_{-1}, h_0, h_1, \cdots, h_{N-1}\}$$

in the array f in the speci ed order, so that f_0 refers to the rst element in the sequence, and f_k refers to the $(k+1)$st element in the sequence. Accordingly, for prime $N = 5$, we need the middle ve elements (beginning with the fth) from the linear convolution de ned by the stationary sequence

$$\{y_0, y_1, y_2, y_3, y_4\},$$

which is of length $N=5$, and the moving sequence (to be reversed as shown in Figure 9.11)

$$\{f_0, f_1, f_2, f_3, f_4, f_5, f_6, f_7, f_8\} = \{h_{-4}, h_{-3}, h_{-2}, h_{-1}, h_0, h_1, h_2, h_3, h_4\},$$

which is of length $2N-1 = 9$. (See Figure 9.11 for a very similar example.)

Remark 1: Note that we have $h_n = \omega_N^{-\frac{1}{2}n^2}$ here, which is different from $h_n = \omega_{KN}^{\frac{1}{2}n^2}$ de ned in the chirp Fourier transform presented in Section 9.3, and we now have $h_n = h_{n\pm N}$. Therefore, with $h_n = h_{n+N} = h_{n+5}$, we have

$$\{h_{-4}, h_{-3}, h_{-2}, h_{-1}\} = \{h_1, h_2, h_3, h_4\},$$

and the formula given by (14.10) can be interpreted as a cyclic convolution of $\{y_0, y_2, y_1, y_3, y_4\}$ and $\{h_0, h_1, h_2, h_3, h_4\}$ as de ned in Section 9.2.3. However, since the length $N \neq 2^s$, we still have to extend it to a cyclic convolution of length $M = 2^s$. Note that for $N = 5$, the results $\{Z_0, Z_1, Z_2, Z_3, Z_4\}$ computed by (14.10) are the rs t $N = 5$ elements obtained from the cyclic convolution of the following two sequences of length $M = 2N - 2 = 2^3$: the stationary sequence is given by

$$\{y_0, y_1, y_2, y_3, y_4, 0, 0, 0\},$$

and the moving sequence is given by

$$
\begin{aligned}
&\{f_0, f_1, f_2, f_3, f_4, f_5, f_6, f_7\} \\
&= \{h_0, h_1, h_2, h_3, h_1, h_2, h_3, h_4\} \\
&= \{h_0, h_1, h_2, h_3, h_{-1}, h_{-2}, h_{-3}, h_{-4}\} \quad\quad (\because h_\ell = h_{-\ell}) \\
&= \{h_0, h_1, h_2, h_3, h_4, h_3, h_2, h_1\}, \quad\quad (\because h_{-\ell} = h_{-\ell+N})
\end{aligned}
$$

where $N - 2 = 3$ zeros are appended to $\{y_0, y_1, \ldots, y_4\}$, and $N - 2 = 3$ elements $\{h_3, h_2, h_1\}$ are appended to $\{h_0, h_1, \ldots, h_4\}$.

In general, if $2N - 2$ is a power of two, then $M = 2N - 2 = 2^s$ is the shortest power-of-two length we may use to implement the cyclic convolution of length N.

Remark 2: When $2N - 2 \neq 2^n$, we must choose $M = 2^s > 2N - 1$. It turns out that we obtain the same result (which is to be derived in the next section) whether we treat Formula (14.10) as a partial linear convolution or a cyclic convolution of length N, we have chosen *not* to convert h_{-k} to h_{-k+N} in the development so that the extension (to length $M = 2^s > 2N - 1$) strategy can be easily adapted for implementing the chirp Fourier transform (as de ned in Chapter 9) if it is needed.

14.3.3   The equivalent partial cyclic convolution

We have indicated above that if $2N - 2$ is a power of two, then $M = 2N - 2 = 2^s$ is the shortest power-of-two length we may use to implement the equivalent cyclic convolution of length N. The fact that we shall do exactly that for $N = 5$ does not prevent us from using the same example to *explain* what the algorithm ought to do when we must choose $M = 2^s > 2N - 1$. Indeed, for $N = 5$, if we re-examine the partial linear convolution de ned by (14.10), it is not dif cult to see that $\{Z_0, Z_1, Z_2, Z_3, Z_4\}$ are *also* the rst ve elements resulting from the cyclic convolution of the stationary sequence

$$\{y_0, y_1, y_2, y_3, y_4, 0, 0, 0, 0\}$$

and the moving sequence (to be reversed as shown in Figure 9.12)

$$\{h_0, h_1, h_2, h_3, h_4, h_{-4}, h_{-3}, h_{-2}, h_{-1}\}.$$

(See Figure 9.12 for a very similar example.)

Since both sequences are of length $2N - 1 = 9 \neq 2^n$, we must obtain an equivalent cyclic convolution of length $M = 2^s > 2N - 1$, so that it can be computed by two radix-2 FFTs and one radix-2 IFFT. For $2N - 1 = 9$, we use the next power of two for M; hence, $M = 16$. To obtain the equivalent cyclic convolution, we simply pad the stationary sequence with zeros, and move the last $N-1=4$ elements of the moving sequence to the end; i.e., we perform the cyclic convolution of the sequences

$$\{y_0,\ y_1,\ y_2,\ y_3,\ y_4,\ 0,\ 0,\ 0,\ 0,\ 0,\ 0,\ 0,\ 0,\ 0,\ 0,\ 0,\ 0\}$$

and

$$\{h_0,\ h_1,\ h_2,\ h_3,\ h_4,\ 0,\ 0,\ 0,\ 0,\ 0,\ 0,\ 0,\ h_{-4},\ h_{-3},\ h_{-2},\ h_{-1}\}.$$

Assuming that the moving sequence $\{h_0,\ h_1,\ \ldots,\ h_4,\ 0,\ \ldots,\ 0,\ h_{-4},\ \ldots,\ h_{-1}\}$ is stored as $\{f_0,\ f_1,\ \cdots,\ f_{15}\}$, we can now invoke the time-domain Cyclic Convolution Theorem 9.1, to compute

(14.11) $$\{Z_r\} = \mathbf{IDFT}\left(\{Y_r \cdot F_r\}\right),$$

where

(14.12) $$\{Y_r\} = \mathbf{DFT}\left(\{y_\ell\}\right), \quad \{F_r\} = \mathbf{DFT}\left(\{f_\ell\}\right).$$

Remark: Note that the scaling factor in Theorem 9.1 has been removed because the IDFT de ned by Formula (14.2) includes division by M, while the DFT de n ed by Formula (14.1) excludes division by M.

Therefore, Bluestein s FFT requires the computation of a full cyclic convolution of length $M = 2^s$ via FFT/IFFT at a cost proportional to $M \log_2 M$. We shall take the rst N results from the convolution results $\{Z_0, Z_1, \cdots, Z_{M-1}\}$, and we obtain $X_r = Z_r \omega_N^{\frac{1}{2}r^2}$ for $r = 0, 1, \ldots, N-1$.

14.3.4   The algorithm

Bluestein's Algorithm for computing the discrete Fourier transform

$$X_r = \sum_{\ell=0}^{N-1} x_\ell\, \omega_N^{r\ell}, \quad r = 0, 1, \ldots, N-1,\ N \neq 2^n.$$

Step 1. Compute the elements needed in the moving sequence:

(14.13) $$h_\ell = \omega_N^{-\frac{1}{2}\ell^2}, \quad \ell = 0, 1, \ldots, N-1,$$

where $\omega_N^{-\frac{1}{2}\ell^2} = \omega_{2N}^{-\ell^2}$, and $\omega_{2N} = e^{-j\pi/N}$. Note that because $h_{-\ell} = h_\ell$, only h_ℓ needs to be computed.

Step 2. De n e M as the smallest power of two that is greater than or equal to $2N-2$, and compute the extended moving sequence of length M de ned by

(14.14) $$f_\ell = \begin{cases} h_\ell, & \ell = 0, 1, \ldots, N-1; \\ h_{M-\ell}, & \ell = M-N+1, \ldots, M-1; \\ 0, & \ell = N, \ldots, M-N,\ \text{if}\ M > 2N-2. \end{cases}$$

Step 3. Use the radix-2 FFT to compute the DFT de ned by

(14.15) $$F_r = \sum_{\ell=0}^{M-1} f_\ell\, \omega_M^{r\ell}, \quad r = 0, 1, \ldots, M-1.$$

Step 4. Given $\{x_\ell\}$, compute the zero-padded stationary sequence de ned by

$$(14.16) \qquad y_\ell = \begin{cases} x_\ell \omega_N^{\frac{1}{2}\ell^2}, & \ell = 0, 1, \ldots, N-1, \\ 0, & \ell = N, \ldots, M-1, \end{cases}$$

where $\omega_N^{\frac{1}{2}\ell^2} = \omega_{2N}^{\ell^2}$, and $\omega_{2N} = e^{-j\pi/N}$.

Step 5. Use the radix-2 FFT to compute the DFT de ned by

$$(14.17) \qquad Y_r = \sum_{\ell=0}^{M-1} y_\ell \omega_M^{r\ell}, \quad r = 0, 1, \ldots, M-1,$$

where $\omega_M = e^{-j2\pi/M}$.

Step 6. Compute

$$(14.18) \qquad U_r = Y_r \cdot F_r, \quad r = 0, 1, \ldots, M-1.$$

Step 7. Use the radix-2 IFFT to compute the IDFT de ned by

$$(14.19) \qquad Z_r = \frac{1}{M} \sum_{\ell=0}^{M-1} U_\ell \omega_M^{-r\ell}, \quad r = 0, 1, \ldots, M-1,$$

where $\omega_M = e^{-j2\pi/M}$.

Step 8. Extract the X_r s from the top N elements in $\{Z_r\}$ by

$$(14.20) \qquad X_r = Z_r \omega_N^{\frac{1}{2}r^2}, \quad r = 0, \ldots, N-1,$$

where $\omega_N^{\frac{1}{2}r^2} = \omega_{2N}^{r^2}$, and $\omega_{2N} = e^{-j\pi/N}$.

14.3.5 Array-smart implementation in MATLAB

In this section we give Bluestein s FFT/IFFT algorithms in the form of MATLAB® functions Bfft and iBfft, which implement the steps outlined above. In coding the algorithm, we have made use of MATLAB vectorized operations and built-in functions (including fft/ifft) in processing all data arrays. By examining the steps in the Bfft algorithm, we see that it calls the built-in fft (twice) and ifft (once) on three data sets with length extended to $M = 2^s \geq 2N - 2$. In the iBfft, the roles of fft and ifft are reversed, but the number of calls remains three in total. The complexity of the Bfft and iBfft is thus $O(M\log_2 M)$.

Algorithm 14.3 Bluestein s FFT algorithm Bfft in MATLAB-style pseudo-code

---

function $Z = Bfft(x)$

begin

  $N = length(x)$;

  $\theta = \pi/N$;

  $p = rem((0{:}N{-}1){.}\,2,\ 2*N)$;

  $h = exp(j*\theta*p)$; $j = \sqrt{-1}$ is a built-in constant in MATLAB

  $f = h$;

  $M = 2\ nextpow2(2*N{-}2)$;   compute smallest $M = 2^s \geq 2N - 2$

  $f(M{-}N{+}2{:}M) = h(N{:}{-}12)$;

  if $M > 2 * N - 2$

    $f(N{+}1{:}M{-}N{+}1) = zeros(1,\ M{-}2*N{+}1)$;

  end

  $fout = fft(f)$;   call built-in fft

  $wp = exp(-j*\theta*p)$;

  $y = x.*wp$;

  $y(N{+}1{:}M) = zeros(1, M\ N\ )$;

  $yout = fft(y)$;   call built-in fft

  $u = yout.*fout$;

  $w = ifft(u)$;   call built-in ifft

  $Z = w(1{:}N).*wp$;

end

Algorithm 14.4 Bluestein s IFFT algorithm iBfft in MATLAB-style pseudo-code

---

function $Y = iBfft(z)$

begin

  $N = length(z)$;

  $\theta = -\pi/N$;   (i) change the sign of θ

  $p = rem((0{:}N\ 1){.}\,2,\ 2*N)$;

  $h = exp(j*\theta*p)$; $j = \sqrt{-1}$ in MATLAB

  $f = h$;

  $M = 2\ nextpow2(2*N{-}2)$;   compute smallest $M = 2^s \geq 2N - 2$

  $f(M\ N{+}2{:}M\ ) = h(N{:}\ 1{:}2)$;

  if $M > 2 * N - 2$

    $f(N{+}1{:}M\ N{+}1) = zeros(1,\ M{-}2*N{+}1)$;

  end

  $fout = ifft(f)$;   (ii) change fft to ifft

  $wp = exp(-j*\theta*p)$;

  $g = z.*wp$;

  $g(N{+}1{:}M) = zeros(1, M\ N\ )$;

  $gout = ifft(g)$;   (iii) change fft to ifft

  $u = gout.*fout$;

  $w = fft(u)$;   (iv) change ifft to fft

  $Y = w(1{:}N).*wp$;

  $Y = Y*M/N$;   including division by N

end

14.3.6 Numerical results

We report the performance of function M- les Bfft.m and iBfft.m in this section. For DFT/IDFT of prime lengths N_2, the results for computing iBfft(Bfft(x)) are given in Table 14.6. Note that the values for N_2 in Tables 14.6 are those used to evaluate iTfft/Tfft in Table 14.5, and the same sets of values were used to evaluate ifft/fft in Tables 14.1, 14.2, 14.3, and 14.4. For the same length, an identical complex array x is used in all tables.

Table 14.6 Performance of Bluestein s FFT for large prime N.

Computing $y =$iBfft(Bfft(x)) for complex x of prime length N_2.					
	MATLAB 5.3 (CPU 1.3 GHz)			MATLAB 7.4 (CPU 3.2 GHz)	
Prime N_2	M- le Timings	Relative Error E_∞		M- le Timings	Relative Error E_∞
2039	0.02 sec	6.6436e 15		0.009 sec	1.1239e 15
4093	0.04 sec	1.2546e 14		0.018 sec	1.1887e 15
8191	0.08 sec	1.2104e 14		0.036 sec	1.2467e 15
16381	0.16 sec	1.9542e 14		0.088 sec	1.3126e 15
32749	0.37 sec	3.6524e 14		0.286 sec	1.4063e 15

The performance of the functions Bfft and iBfft in Table 14.6 is consistent with our expectation from an $O(M \log_2 M)$ algorithm, where $M = 2^s \geq 2N_2 - 2$. (For each N_2 given in Table 14.6, the smallest $M = 2^s$ is obtained by setting the exponent $s = $ nextpow$(2 * N_2 - 2)$ in the M- le functions Bfft and iBfft.) To assess the different approaches for prime N_2, we compare Table 14.6 with Tables 14.1 and 14.2, and we see that very signi cant improvement in both execution time and accuracy is gained by functions Bfft and iBfft when using MAT-LAB 5.3. When comparing the MATLAB 7.4 results in Table 14.6 with those in Tables 14.3 and 14.4, note that the interpreted M- le functions Bfft and iBfft are expected to run more slowly than the built-in executable code. Compared with the approximate FFT results in Table 14.5, Bluestein s FFT runs faster and provides more accurate results in all cases.

Bibliography

[1] A. Ambardar. *Analog and Digital Signal Processing*. Brooks/Cole Publishing Company, Paci c Grove, CA, second edition, 1999.

[2] C. Anderson and M. D. Dahleh. Rapid computation of the discrete Fourier transform. *SIAM J. Sci. Comput.*, 17(4):913 919, 1996.

[3] J. Arsac. *Fourier Transforms and the Theory of Distributions*. Prentice-Hall, Inc., Englewood Cliffs, NJ, 1966.

[4] G. D. Bergland. The fast Fourier transform recursive equations for arbitrary length records. *Math. Comp.*, 21(98):236 238, 1967.

[5] L. L. Bluestein. A linear lter ing approach to the computation of discrete Fourier transform. *IEEE Transactions on Audio and Electroacoustics*, AU-18:451 455, 1970. Reprinted in *Digital Signal Processing*, Eds. L. R. Rabimer and C. M. Rader, pp. 317 321, IEEE Press, New York, 1972.

[6] R. N. Bracewell. *The Fourier Transform and Its Applications*. McGraw-Hill Inc., San Francisco, CA, third edition, 2000.

[7] W. L. Briggs and V. E. Hensen. *The DFT: An Owner's Manual for the Discrete Fourier Transform*. The Society for Industrial and Applied Mathematics, Philadelphia, PA, 1995.

[8] E. O. Brigham. *The Fast Fourier Transform and Its Applications*. Prentice-Hall, Inc., Upper Saddle River, NJ, 1988.

[9] C. S. Burrus and P. W. Eschenbacher. An in-place, in-order prime factor FFT algorithm. *IEEE Transactions on Acoustics, Speech, and Signal Processing*, ASSP-29:806 817, 1981.

[10] T. Butz. *Fourier Transformation for Pedestrians*. Springer-Verlag, Berlin, 2006.

[11] D. C. Champeney. *Fourier Transforms and Their Physical Applications*. Academic Press Inc., London, UK, 1973.

[12] D. C. Champeney. *A Handbook of Fourier Theorems*. Cambridge University Press, Cambridge, UK, 1987.

[13] E. Chu and A. George. *Inside the FFT Black Box: Serial and Parallel Fast Fourier Transform Algorithms*. CRC Press, Boca Raton, FL, 2000.

[14] R. V. Churchill. *Fourier Series and Boundary Value Problems*. McGraw-Hill Book Company, Inc., New York, second edition, 1963.

[15] J. W. Cooley, P. A. W. Lewis, and P. D. Welch. Historical notes on the fast Fourier transform. *IEEE Trans. Audio and Electroacoustics*, 15:76 79, 1967.

[16] J. W. Cooley and J. W. Tukey. An algorithm for the machine calculation of complex Fourier series. *Math. Comp.*, 19:297 301, 1965.

[17] H. F. Davis. *Fourier Series and Orthogonal Functions*. Allyn and Bacon, Inc., Boston, MA, 1963.

[18] C. de Boor. FFT as nested multiplications, with a twist. *SIAM J. Sci. Stat. Comput.*, 1:173 178, 1980.

[19] A. Deitmar. *A First Course in Harmonic Analysis*. Springer-Verlag, New York, second edition, 2005.

[20] M. Frigo and S. G. Johnson. FFTW: An adaptive software architecture for the FFT. In *Proceedings of the International Conference on Acoustics, Speech, and Signal Processing*, volume 3, pp. 1381 1384, 1998.

[21] D. Gabor. Theory of communication. *J.I.E.E.*, 93(Part III):492 444, 1946.

[22] R. R. Goldberg. *Fourier Transforms*. Cambridge University Press, Cambridge, UK, 1965.

[23] R. C. Gonzales and R. E. Woods. *Digital Image Processing*. Addison-Wesley Publishing Company, Reading, MA, 1992.

[24] I. J. Good. The interaction algorithm and practical Fourier analysis. *J. Roy. Statist. Soc.,* Ser. B, 20:361 372, 1958. Addendum, 22:372 375, 1960.

[25] J. C. Goswami and A. K. Chan. *Fundamentals of Wavelets*. John Wiley & Sons, Inc., New York, 1999.

[26] A. Graham. *Kronecker Products and Matrix Calculus with Applications*. Ellis Horwood Limited, West Sussex, UK, 1981.

[27] R. W. Hamming. *Digital Filters*. Prentice-Hall, Inc., Englewood Cliffs, NJ, third edition, 1989.

[28] D. J. Higham and N. J. Higham. *The MATLAB Guide*. The Society for Industrial and Applied Mathematics, Philadelphia, PA, 2000.

[29] D. J. Higham and N. J. Higham. *The MATLAB Guide*. 2nd ed., The Society for Industrial and Applied Mathematics, Philadelphia, PA, 2005.

[30] FFTW: The Fastest Fourier Transform in the West. http://www.fftw.org/.

[31] D. P. Kolba and T. W. Parks. A prime factor algorithm using high-speed convolution. *IEEE Trans. Acoust. Speech Signal Process*, ASSP-25:281 294, 1977.

[32] C. F. Van Loan. *Computational Frameworks for the Fast Fourier Transform*. The Society for Industrial and Applied Mathematics, Philadelphia, PA, 1992.

[33] J. H. McClellan and C. M. Rader. *Number Theory in Digital Signal Processing*. Prentice-Hall, Inc., Englewood Cliffs, NJ, 1979.

[34] Y. Nievergelt. *Wavelets Made Easy*. Birkhäuser Boston, Cambridge, MA, 1999.

[35] H. J. Nussbaumer. *Fast Fourier Transform and Convolution Algorithms*. Springer Series in Information Sciences. Springer-Verlag, Berlin, 1981.

[36] J. S. Otto. Symmetric prime factor fast Fourier transform algorithms. *SIAM J. Sci. Stat. Comput.*, 19:419 431, 1989.

[37] A. Papoulis. *The Fourier Integral and Its Applications*. McGraw-Hill Book Company, Inc., New York, 1962.

[38] B. Porat. *A Course in Digital Signal Processing*. John Wiley & Sons, Inc., New York, 1997.

[39] W. H. Press, S. A. Teukolsky, W. T. Vetterling, and B. P. Flannery. *Numerical Recipes in C++: The Art of Scientific Computing*. Cambridge University Press, Cambridge, UK, second edition, 2001.

[40] J. N. Rayner. *An Introduction to Spectral Analysis*. Pion Limited, London, UK, 1971.

[41] J. H. Rothweiler. Implementation of the in-order prime factor transform for variable sizes. *IEEE Transactions on Acoustics, Speech, and Signal Processing*, ASSP-30:105 107, 1982.

[42] E. M. Stein and G. Weiss. *Introduction to Fourier Analysis on Euclidean Spaces*. Princeton University Press, Princeton, NJ, 1971.

[43] P. N. Swarztrauber, R. A. Sweet, W. L. Briggs, V. E. Henson, and J. Otto. Bluestein s FFT for arbirary N on the hypercube. *Parallel Computing*, 17:607 617, 1991.

[44] C. Temperton. Self-sorting mixed-radix fast Fourier transforms. *J. of Computational Physics*, 52:1 23, 1983.

[45] C. Temperton. Implementation of a self-sorting in-place prime factor FFT algorithm. *J. of Computational Physics*, 58:283 299, 1985.

[46] C. Temperton. Implementation of a prime factor FFT algorithm on Cray-1. *Parallel Computing*, 6:99 108, 1988.

[47] C. Temperton. A new set of minimum-add small-n rotated DFT modules. *J. of Computational Physics*, 75:190 198, 1988.

[48] C. Temperton. A self-sorting in-place prime factor real/half-complex FFT algorithm. *J. of Computational Physics*, 75:199 216, 1988.

[49] C. Temperton. Self-sorting in-place fast Fourier transforms. *SIAM J. Sci. Comput.*, 12:808 823, 1991.

[50] C. Temperton. A generalized prime factor FFT algorithm for any $n = 2^p 3^q 5^r$. *SIAM J. Sci. Comput.*, 13:676 686, 1992.

[51] E. C. Titchmarsh. *Introduction to the Theory of Fourier Integrals.* Clarendon Press, Oxford, UK, 1962.

[52] J. S. Walker. *Fourier Analysis.* Oxford University Press, New York, 1988.

[53] H. J. Weaver. *Applications of Discrete and Continuous Fourier Analysis.* John Wiley & Sons, Inc., New York, 1983.

[54] H. J. Weaver. *Theory of Discrete and Continuous Fourier Analysis.* John Wiley & Sons, Inc., New York, 1989.

[55] S. Winograd. On computing the DFT. *Math. Comp.*, 32:175 199, 1978.

[56] C. R. Wylie. *Advanced Engineering Mathematics.* McGraw-Hall Book Company, New York, fourth edition, 1975.

Index

Milton Keynes UK
Ingram Content Group UK Ltd.
UKHW050455071024
449327UK00015B/387